ANTIGENIC VARIATION

ANTIGENIC VARIATION

Edited by

ALISTER CRAIG
Liverpool School of Tropical Medicine, UK

ARTUR SCHERF
Institut Pasteur, Paris, France

ACADEMIC PRESS
An imprint of Elsevier

Amsterdam   Boston   Heidelberg   London   New York   Oxford
Paris   San Diego   San Francisco   Singapore   Sydney   Tokyo

This book is printed on acid-free paper

Copyright © 2003 by ELSEVIER Ltd
except Chapter 16 *Surface antigenic variation in* Giardia Lamblia by Theodore E. Nash

All rights reserved.
No part of this publication may be reproduced or transmitted in any form or by any means, electronic or mechanical, including photocopying, recording, or any information storage and retrieval system, without permission in writing from the publisher.

Academic Press
An Imprint of Elsevier
84 Theobald's Road, London WC1X 8RR, UK
http://www.academicpress.com

Academic Press
An Imprint of Elsevier
525 B Street, Suite 1900 San Diego, California 92101–4495, USA
http://www.academicpress.com

ISBN 0–12–194851–X

Library of Congress Catalog Number: 20031 08095

British Library Cataloguing in Publication Data
Antigenic variation
    1. Antigens   2. Biological variation
    I. Craig, Alister G.   II. Scherf, Arthur
    571.9'645

Composition by Genesis Typesetting Limited, Rochester, Kent
Printed and bound in England by Biddles Ltd, Guildford & Kings Lynn

03 04 05 06 07 B 9 8 7 6 5 4 3 2 1

Contents

Contributors	vii
Foreword	xi
Preface	xv

1. **Mechanisms of antigenic variation: an overview** — 1
   Piet Borst

2. **HIV variation – a question of signal-to-noise** — 16
   Simon Wain-Hobson

3. **Calicivirus** — 33
   Alan D. Radford, Susan Dawson and Rosalind M. Gaskell

4. **Influenza – the chameleon virus** — 52
   John Oxford, Ramani Eswarasaran, Alex Mann and Robert Lambkin

5. **Rotavirus** — 84
   C. Anthony Hart and Nigel A. Cunliffe

6. ***Haemophilus influenzae*** — 102
   Derek W. Hood and E. Richard Moxon

7. **Phase variation in *Helicobacter pylori* lipopolysaccharide** — 122
   Ben J. Appelmelk and Christina M.J.E. Vandenbroucke-Grauls

8. **Genetic variation in the pathogenic *Neisseria* species** — 142
   Thomas F. Meyer and Stuart A. Hill

9. ***Candida albicans*** — 165
   David R. Soll

10. **The MSG gene family and antigenic variation in the fungus *Pneumocystis carinii*** — 202
    James R. Stringer

11 **Trypanosome antigenic variation – a heavy investment in the evasion of immunity** J. David Barry and Richard McCulloch	224
12 **Antigenic variation in *Anaplasma marginale* and *Ehrlichia* (*Cowdria*) *ruminantium*** Suman M. Mahan	243
13 **Antigenic variation and its significance to *Babesia*** David R. Allred, Basima Al-Khedery and Roberta M. O'Connor	273
14 **Antigenic variation in *Plasmodium falciparum* and other *Plasmodium* species** Mallika Kaviratne, Victor Fernandez, William Jarra, Deirdre Cunningham, Mary R. Galinski, Mats Wahlgren and Peter R. Preiser	291
15 **Antigenic variation in *Borrelia*: relapsing fever and Lyme borreliosis** Alan G. Barbour	319
16 **Surface antigenic variation in *Giardia lamblia*** Theodore E. Nash	357
17 **Free-living and parasitic ciliates** Theodore G. Clark and James D. Forney	375
18 **The impact of antigenic variation on pathogen population structure, fitness and dynamics** Neil M. Ferguson and Alison P. Galvani	403
Index	433

Contributors

David R. Allred, Department of Pathobiology, University of Florida, Gainesville, Florida, USA

Basima Al-Khedery, Department of Pathobiology, University of Florida, Gainesville, Florida, USA

Ben J. Appelmelk, Department of Medical Microbiology and Infection Prevention, Vrije Universiteit Medical Center, Amsterdam, The Netherlands

Alan G. Barbour, Departments of Microbiology and Molecular Genetics and Medicine, University of California Irvine College of Medicine, Irvine, California, USA

J. David Barry, Wellcome Centre for Molecular Parasitology, Anderson College, University of Glasgow, Glasgow, UK

Piet Borst, The Netherlands Cancer Institute, Division of Molecular Biology, and Centre of Biomedical Genetics, Amsterdam, The Netherlands

Theodore G. Clark, Department of Microbiology and Immunology, College of Veterinary Medicine, Cornell University, Ithaca, New York, USA

Nigel A. Cunliffe, Department of Medical Microbiology and Genitourinary Medicine, University of Liverpool, Liverpool, UK

Deirdre Cunningham, National Institute for Medical Research, The Ridgeway, Mill Hill, London, UK

Susan Dawson, Department of Veterinary Clinical Sciences, University of Liverpool Veterinary Teaching Hospital, Neston, South Wirral, UK

Ramani Eswarasaran, Academic Virology and Retroscreen Ltd, St Bart's and The London, Queen Mary's School of Medicine and Dentistry, London, UK

Neil M. Ferguson, Department of Infectious Disease Epidemiology, Faculty of Medicine, Imperial College London, London, UK

Victor Fernandez, Microbiology and Tumorbiology Centre (MTC), Karolinska Institutet, Stockholm; Swedish Institute for Infectious Disease Control, Sweden

James D. Forney, Department of Biochemistry, Purdue University, West Lafayette, Indiana, USA

Mary R. Galinski, Vaccine Research Center at Yerkes, Emory University, Atlanta, Georgia, USA

Alison P. Galvani, Department of Integrative Biology, University of California, Berkeley, California, USA

Rosalind M. Gaskell, Department of Veterinary Pathology, University of Liverpool Veterinary Teaching Hospital, Neston, South Wirral, UK

C. Anthony Hart, Department of Medical Microbiology and Genitourinary Medicine, University of Liverpool, Liverpool, UK

Stuart A. Hill, Department of Biological Sciences, Northern Illinois University, DeKalb, Illinois, USA

Derek W. Hood, Molecular Infectious Diseases Group, University of Oxford Department of Paediatrics, Weatherall Institute of Molecular Medicine, John Radcliffe Hospital, Headington, Oxford, UK

William Jarra, National Institute for Medical Research, The Ridgeway, Mill Hill, London, UK

Mallika Kaviratne, National Institute for Medical Research, The Ridgeway, Mill Hill, London, UK

Robert Lambkin, Academic Virology and Retroscreen Ltd, St Bart's and The London, Queen Mary's School of Medicine and Dentistry, London, UK

Suman M. Mahan, Heartwater Control Research Program, University of Florida, College of Veterinary Medicine, Department of Pathobiology, Gainesville, Florida, USA

Alex Mann, Academic Virology and Retroscreen Ltd, St Bart's and The London, Queen Mary's School of Medicine and Dentistry, London, UK

Richard McCulloch, Wellcome Centre for Molecular Parasitology, Anderson College, University of Glasgow, Glasgow, UK

Thomas F. Meyer, Department of Molecular Biology, Max Planck Institute for Infection Biology, Berlin, Germany

E. Richard Moxon, Molecular Infectious Diseases Group, University of Oxford Department of Paediatrics, Weatherall Institute of Molecular Medicine, John Radcliffe Hospital, Headington, Oxford, UK

Theodore E. Nash, Laboratory of Parasitic Diseases, National Institutes of Allergy and Infectious Diseases, National Institutes of Health, Bethesda, Maryland, USA

Roberta M. O'Connor, Division of Geographic Medicine and Infectious Diseases, New England Medical Center, Boston, Massachusetts, USA

John Oxford, Academic Virology and Retroscreen Ltd, St Bart's and The London, Queen Mary's School of Medicine and Dentistry, London, UK

Peter R. Preiser, National Institute for Medical Research, The Ridgeway, Mill Hill, London, UK

Alan D. Radford, Department of Veterinary Clinical Sciences, University of Liverpool Veterinary Teaching Hospital, Neston, South Wirral, UK

David R. Soll, Department of Biological Sciences, University of Iowa, Iowa City, Iowa, USA

James R. Stringer, Department of Molecular Genetics, Biochemistry and Microbiology, University of Cincinnati, Cincinnati, USA

Christina M.J.E. Vandenbroucke-Grauls, Department of Medical Microbiology and Infection Prevention, Vrije Universiteit Medical Center, Amsterdam, The Netherlands

Mats Wahlgren, Microbiology and Tumorbiology Centre (MTC), Karolinska Institutet, Stockholm; Swedish Institute for Infectious Disease Control, Sweden

Simon Wain-Hobson, Unite de Rétrovirologie Moléculaire, Institut Pasteur, Paris, France

Foreword

Conflicts of various sizes and duration have been a hallmark of human history. The outcome of such battles will depend on many factors including the strategies and tactics employed by the different combatants. Historically, deception and camouflage have been used to avoid the detection of troops and equipment and to deploy fighting elements closer to the enemy. In many ways the ongoing battle between the host and the potential pathogen during tissue colonization and infection resembles the picture in the larger arena of human conflict. Pathogens can deploy vast numbers of individual fighting units or weapons such as toxins and adhesions to penetrate, damage and disable the host defences. Perhaps the most powerful defence the mammalian host possesses is the immune system. It is important to be able to identify the enemy in any conflict. The immune system is demarked into the innate (early warning) and adaptive systems. The innate immune system appears early in the evolution of multicellular organisms and is the early warning system through which the host recognizes 'infectious non-self' (the presence of pathogens) as against 'non-infectious non-self' (the presence of environmental antigens). Cells of the innate immune system, including macrophages, dendritic cells and even some epithelial cells, harbour specific receptors that can recognize so-called Pathogen Associated Molecular Patterns (PAMPs). PAMPs are generic macromolecules that are essential for the viability of pathogens but are not present in host cells or tissues. PAMPs identified to date include bacterial DNA (unmodified CpG motifs), double-stranded RNA, flagella and various glycolipids including lipopolysaccharides (LPS). PAMPs are recognized by specific receptors including Toll receptors, which, once bound, become activated and send signals down specific intracellular signalling pathways. Toll-mediated signalling activates killing or protection mechanisms including the production of proinflammatory cytokines, radical formation (O and NO), chemokines and antimicrobial peptides. These signalling events have also been shown to trigger the production of specific cytokines, which interfere with the action of regulatory T cells and facilitate antigen-specific T cell stimulation. Historically we have 'accidentally' activated Toll receptors through the use of crude adjuvants, such as Complete Freud's Adjuvant, which can enhance the immunogenicity of vaccines but are too toxic for use in humans.

The innate immune system is also critically important for triggering adaptive immunity. Adaptive immunity appears much later in evolution and is the mechanism by which the mammalian immune system identifies individual antigenic components of pathogens. Consequently, mammals are able to mount a long-lasting specific, as against a short-duration non-specific, immune response. The cell central to the activities of the acquired immune system is the lymphocyte. The acquired immune response is dependent on the recognition of individual epitopes by B lymphocytes (B cell epitopes recognized by antibodies) and T lymphocytes (T cell epitopes recognized by the T cell receptor). Since most proteins or pathogens harbour multiple T and B cell epitopes the host has an excellent chance of identifying pathogen-associated structures. Unfortunately microorganisms can multiply rapidly to generate enormous numbers of offspring, and through interactions with the host immune system they have evolved countermeasures for their own defence. Pathogens often present epitopes that are 'favoured' in terms of recognition by the host immune system (so-called immunodominant epitopes). This means that the availability of the critical repertoire of so-called 'protective' epitopes is reduced. Further pathogens have also evolved cunning mechanisms for actually varying to different degrees the type of protective epitope they display. This phenomenon is known as antigenic variation and is the subject of this text.

Antigenic variation appears to be a fundamental mechanism by which pathogens avoid immune clearance and lengthen the duration of persistence in the host. Variation can occur in epitopes recognized by B cells and T cells and is one of the many mechanisms by which pathogens avoid antibody binding and the attentions of activated T cells. Hence, any antigenic variation can have a profound effect on the ability of the host immune system to protect the host. The rate and degree of antigenic variation associated with pathogens differs enormously. The degree of antigenic variation a pathogen is able to display will be limited by the biological constraints associated with the tough *in vivo* lifestyle pathogens face. For example, a viral or parasite surface protein has to serve a biological function associated with maintaining the integrity of the surface coat, promote host cell interaction or even facilitate interactions with other pathogen proteins or macromolecules. Even in the face of these constraints we now have evidence for the occurrence of antigenic variation in pathogens on a significant scale. It is also important to note that we are even now just discovering significant potential for antigenic variation in many pathogens. For example, examination of the genome sequence of mycobacteria has identified some previously unknown large gene families, suggesting we may have underestimated the potential for antigenic variation in this group of pathogens. Nevertheless, the ability to vary antigens is far from limitless in different pathogens. Herein lies the enormous intrigue in this area. Some of the mechanisms evolved by pathogens, particularly some bacteria and parasites, are incredibly sophisticated and are amongst the most interesting genetic systems so far investigated. Many examples of such systems are discussed in this text.

Antigenic variation has enormous consequences for disease therapy. The design of diagnostic reagents and new vaccines can be readily compromised by antigenic variation. Many of the infectious agents for which we still do not have effective vaccines (HIV, malaria, etc.) exhibit significant degrees of antigenic variation and we need to learn as much as possible about the mechanisms involved and the rate of variations in populations. Such information can only benefit those involved in designing vaccines against such highly variable targets. In some cases we may even be able to target drugs that interfere with the capacity of a pathogen to display antigenic variation by targeting enzymes such as specific recombinases.

Overall, antigenic variation is one of the most exciting areas of modern molecular biology, covering genetics, immunology and cell biology, and this text provides an excellent summary of the current state of this field.

Professor Gordon Dougan
Centre for Molecular Microbiology and Infection
Department of Biological Sciences
Imperial College London
UK

Preface

The ability to adapt to an environment is one of the most important factors in the survival of an organism. Infection represents an extreme example of this, with the need to grow under highly variable conditions as well as to degrade the host defence mechanisms sufficiently to allow for transmission. Infectious agents have developed a range of mechanisms over many generations, allowing them to interact with the host through surface-expressed ligands while protecting themselves from the resulting immune response. This book examines the ways in which a diverse range of pathogens have evolved to address this problem, from viruses through to protozoan parasites. In calling this book *Antigenic Variation* we were mindful that this term can imply a complex mechanism (such as seen in the human malaria parasite, *Plasmodium falciparum*), but we were keen to be inclusive, and some of the molecular mechanisms that we thought might contribute to this process are described below.

- Allelic variation – variation in a single locus (usually through mutation and selection).
- 'Simple' switching – usually involving simple transcriptional control (e.g. flip-flop, frameshifting).
- Antigenic variation (*sensu stricto*) – switching between members of a multigene family, which involves a complicated mechanism of activation and silencing.

However, we soon learned from our contributors that not only had organisms developed other ways of protecting themselves inside the host, but also that a consideration of how these pathways act to improve the success of the organism (in terms of either growth or transmission) was equally important. Therefore the fact that pathogens have developed these programmed or random DNA rearrangements through a wide range of genetic, epigenetic, post-transcriptional and even inhibitory RNA mechanisms should be no surprise. The unravelling of these techniques to subvert or overcome host defence systems has often been a slow and difficult process and the recent deluge of genomic sequence information will hopefully accelerate the process. However, it will be, as always, through a careful consideration of the biology of the organism that progress will come and with it the development of interventions for the diseases caused by these pathogens, which often exert an unequal burden on the resource-poor nations of the developing world.

We would like to thank all our contributors for their time and effort in providing testaments to the resourcefulness of their 'favourite' pathogens. We would also like to thank the Liverpool School of Tropical Medicine (A.C.) and Institut Pasteur (CNRS) (A.S.) for supporting us in our desire to look at a broader world of variation. To all these people go our thanks and we hope that this volume will encourage other researchers to delve into this challenging but fascinating world of threat and counter-threat. Finally, we would like to thank our publishers for staying calm when deadlines drifted over the horizon.

Alister Craig
Artur Scherf

1

MECHANISMS OF ANTIGENIC VARIATION: AN OVERVIEW

PIET BORST

INTRODUCTION

There are two ways in which antigens undergo variation. First, **randomly**, through DNA alterations, introduced by:

- errors in DNA (or RNA) replication/repair;
- recombination between non-identical genes;
- reassortment of gene segments if the genome is not in one piece.

For want of a better name, I call this random or unprogrammed variation. This does not imply that this type of variation cannot be selected for. Obviously, the replication machinery of small RNA viruses has been selected in evolution to generate a high rate of errors, causing enormous population diversity. The reassortment of genes made possible by genomes in pieces can lead to the antigenic shift that is at the basis of the influenza pandemics (see Chapter 4).

Secondly, through **programmed variation**, also called phase variation, multiphasic antigenic variation, true antigenic variation, or, more modestly, antigenic variation *sensu stricto*, which is characterized by two properties:

- a family of genes encoding proteins with the same or similar functions (also called paralogous genes);
- the ability to express only one of the gene family members at a time and to alter the member expressed from time to time.

In principle, mechanism 1 (random variation) alters the gene templates as it diversifies the gene repertoire of the parasite population, whereas mechanism 2 (programmed variation) may operate without changing the repertoire. However, the repertoire in mechanism 2 is not static either, and does evolve by random change. In essence, however, mechanism 2 provides diversity in a single clone, mechanism 1 only between members of the parasite population. It should be stressed that programmed variation does not imply a developmental program, like the precisely timed switch from fetal to adult hemoglobin. Switching in antigenic variation is stochastic and the program offers only the potential for variation, not a blueprint.

Although this subdivision into two major mechanisms is useful to delineate how antigenic variation works, nature usually does not bend to our simple categories and there are examples of antigenic variation that could be grouped under either mechanism 1 or mechanism 2. For instance, somatic mutation is an important step in diversifying immunoglobulins (Gearhart, 2002). It is highly programmed, but it uses up the pre-existing template. In this subdivision, it belongs in category 1, but I can imagine that some people would put it in category 2.

It should also be pointed out that the differential expression of a family of genes (mechanism 2) may be controlled itself by random mutagenesis (mechanism 1). For instance, expression of some gene families is controlled by random slippage during replication of repeat sequences. The presence of the repeats is programmed, but the alteration in repeat length, which may change transcription initiation or alter the reading frame, occurs randomly. The genes themselves are not varied, only their on/off state.

RANDOM 'UNPROGRAMMED' VARIATION

Darwin called it 'descent with modification': genome transfer from parents to progeny is not perfect. This imperfection creates antigen drift, the substrate for selection and evolution. Descent with modification is the inherent consequence of imperfect DNA transactions, that is, DNA replication, DNA repair, DNA recombination and the invasion of genomes by extraneous DNAs. In fact, replication is often programmed to be imperfect. I still remember the shock when John Drake found that it is possible to isolate T4 bacteriophage mutants that copy their DNA more precisely than wild-type phage (Drake et al., 1969). This was the first indication that sloppy genome replication can be advantageous.

Later, Charles Weissmann and colleagues showed that the replication of the RNA bacteriophage Qβ is so imprecise that on average one copying mistake is made in a genome of only 3000 nucleotides (Domingo et al., 1978). There is no wild-type Qβ genome, only an average genome. Manfred Eigen introduced the term quasi-species to describe the extreme variability caused by sloppy genome replication.

Viruses obviously make the most of 'descent with modification'. Short replication times and large populations make it possible to generate sufficient antigenic variation without additional programmed rearrangements within the lifetime of a single organism. Drake and Holland (1999) have calculated that the mutation rate of RNA viruses is close to the theoretical maximum: even a modest increase would self-eliminate the viral population. Retroviruses, such as HIV, have become most notorious for their variability, but as Simon Wain-Hobson points out in Chapter 2, 'HIV is near the bottom of the first division compared to other RNA viruses, such as polio and influenza A'. His conclusion is that HIV 'does not need variation to cause disease', although variation does help the virus to escape from chemotherapy.

As pointed out by Wain-Hobson, even influenza virus does not need antigenic variation to replicate in a single patient. Antigenic variation allows it to get a foothold in a host population intensely exposed to the virus and carrying neutralizing antibodies that block viral entry. If the host is unable to make fully effective neutralizing antibodies, as is the case with HIV, antigenic variation is not even required for spread of the virus through a previously exposed host population.

Viruses employ random variation to escape from the host immune defence, which may allow the virus to establish chronic infections, or to evade life-long host immunity due to neutralizing antibodies. Organisms with larger genomes than viruses cannot afford sloppy DNA replication, and random variation of antigen genes therefore makes a more modest contribution to antigenic variability. It is conceivable, however, that organisms would evolve a mechanism for 'somatic mutation' that acts locally, only on antigen genes as the host used for antibody genes (Gearhart, 2002).

Such a mechanism has been described in *Borrelia* surface protein genes (see Chapter 15). The rapid drift of the transcribed gene does not appear to be due to a mutagenic polymerase, however, but to small-patch (segmental) gene conversion from nearby donor genes.

To my knowledge no example is known of a true mutagenic polymerase acting locally on antigen genes in any non-viral parasite. Donelson and coworkers (Rice-Ficht *et al.*, 1982; Lu *et al.*, 1993, 1994) have repeatedly proposed that imperfect copying of the VSG genes of *Trypanosoma brucei* by a mutagenic polymerase contributes to the antigenic variation of this parasite (see also McKenzie and Rosenberg, 2001). The evidence for this proposal has always remained indirect and can be explained in a more plausible way, i.e. by strong selection for occasional point mutations introduced by normal DNA transactions, as discussed by Graham and Barry (1996). The possibility that mutator polymerases activated by stress responses could add to antigenic variation has recently been raised by McKenzie and Rosenberg (2001). There is no concrete evidence as yet supporting this possibility.

PROGRAMMED ANTIGENIC VARIATION: GENERAL PRINCIPLES

Programmed antigenic variation is due to differential control of a family of surface antigen genes. As summarized in Table 1.1, a gene can be activated by

Table 1.1 Programmed antigenic variation

A	Recombinational control
	A.1 Inversion
	A.2 Reciprocal recombination
	A.3 Gene conversion (cassette mechanism)
	A.4 Deletion (preceded by gene conversion)
B	*In situ* control
	B.1 By varying the length of simple repeat tracts
	B.1.a Transcriptional control
	B.1.b Translational control
	B.2 By varying DNA modification
	B.3 Without DNA alterations
	B.3.a Telomeric silencing
	B.3.b Expression site body

moving it downstream of an active promoter (group A), or by activating it where it resides, i.e. *in situ* (group B). Group A mechanisms invariably involve DNA recombination; group B, alterations in DNA repeat length (B.1), in DNA modification (B.2), or no DNA alterations whatsoever (B.3).

Obviously, there are other ways to subdivide mechanisms of programmed antigenic variation (Moxon *et al.*, 1994; Borst *et al.*, 1996; Deitsch *et al.*, 1997; Borst, 2002b). The scheme in Table 1.1, however, provides a simple framework to introduce the other chapters in this book.

RECOMBINATIONAL MECHANISMS OF ANTIGENIC VARIATION

The simplest way to move genes relative to a fixed promoter is by gene inversion (A.1), but promoter inversion is another way to obtain a similar result (Figure 1.1). The inversion is catalysed by a site-specific DNA recombinase that recognizes a short DNA segment flanking the invertible segment on both sides (van de Putte and Goosen, 1992; Johnson, 2002). This mechanism looks simple, but can be used in a sophisticated way. As shown by Komano *et al.* (1987), the *Escherichia coli* plasmid R64 contains four contiguous DNA segments that can create seven different open reading frames by multi-inversion. Such multi-inversion systems are called 'shufflons' and many examples of complex shufflons are known (Johnson, 2002).

In theory, one could also envisage a mechanism employing a movable promoter, located on a site-specific transposon, which is able to insert in front of a large number of genes. Such a transposon could differentially control the expression of a large gene family. No example of such a mechanism has been reported, however.

Mechanism A.2 (Table 1.1; Figure 1.1), reciprocal recombination, is more widespread. It is used for genes near telomeres, and provides a versatile

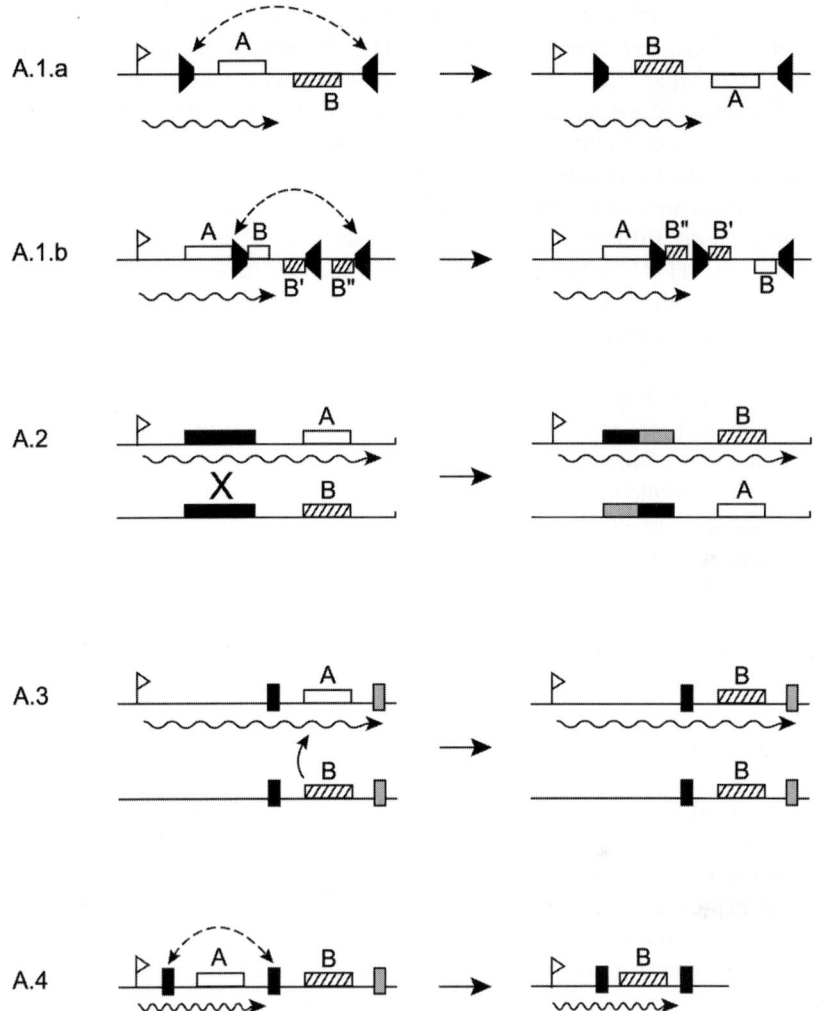

Figure 1.1 Recombinational versions of gene switching (highly schematic).

(A.1.a) is a simple inversion switch, (A.1.b) is a multiple-choice version, also called a shufflon. Here only two choices for an alternative end B for gene A are given; in nature, shufflons are often much more complex (Johnson, 2002). The promoter is indicated by the flag; the filled skewed blocks are short homologous sequences recognized by a site-specific recombinase.

(A.2) shows a reciprocal recombination between two telomeres, the upper one containing an active expression site. The recombination occurs between homologous sequences (black and grey blocks). The vertical line indicates a chromosome end.

(A.3) is a gene conversion, in which a copy of gene B replaces gene A in an active expression site. The recombination occurs between (short) blocks of homologous sequences 5' (black) and 3' (grey) of the genes.

(A.4) shows a deletion, removing the first (A) of two genes, resulting in a switch of A to B. Each gene ends with a strong transcriptional stop, limiting transcription to the promoter-proximal gene. The deletion occurs by crossovers in the black homology tracks.

See text for further explanation.

mechanism in organisms with large numbers of (mini-) chromosomes or linear plasmids, e.g. trypanosomes (Chapter 11), *Pneumocystis* (Chapter 10) and *Borrelia* (Chapter 15). Reciprocal recombination even appears to be the main mechanism used for antigenic variation in *Pneumocystis*, although a contribution of gene conversion cannot yet be excluded. An unusual form of reciprocal recombination controlling chromosome-internal genes has been reported by Blaser and coworkers (Dworkin and Blaser, 1997). They studied the surface layer proteins (SLPs) of *Campylobacter* and found that a single promoter controls a set of genes. Inactive copies can exchange with the actively transcribed gene by a reciprocal recombination, using homology in the promoter area and downstream of the coding region. As the active and silent copies are in inverse orientation relative to each other, the recombination leads to inversion of large segments of DNA. Obviously, if all genes were in the same orientation, reciprocal recombination would result in excision of a circular DNA element and the risk of losing such DNA. The *Campylobacter* SLP system could also be brought under A.1 (Table 1.1), as it involves an inversion. However, the inversion is not carried out by a site-specific recombinase, but by a *recA* generalized recombination system. It is therefore more akin to mechanism A.2 (Table 1.1).

A recombination mechanism akin to reciprocal recombination is used by retroviruses. There are two viral genomes packaged in a virion and reverse transcriptase switches template about three times per round of replication. In cells containing genetically distinct proviruses, this represents an effective mechanism for diversity generation (Chapter 2). The recombination rates in RNA viruses with segmented genomes are much lower.

Gene conversion (A.3; Table 1.1) is the most versatile of all mechanisms of recombinational forms of antigenic variation. It can handle both telomeric and chromosome-internal genes and the repertoire of donor genes that can be moved into an expression site is in principle unlimited. Another formidable advantage of gene conversion is that this mechanism can also be used for patch-wise, segmental gene conversion (Roth *et al.*, 1989), allowing the use of pseudo-gene donors and the construction of highly chimeric expressed genes (see Chapters 11 and 15; Figure 1.1, mechanism A.3.b.). No other form of programmed antigenic variation has this option. Whether successful chimeric genes generated in an active expression site can be added to the repertoire by a reverse gene conversion is not known.

'Cassette mechanism' is another name for the gene conversion mechanism, as the gene copies inserted behind a single promoter in an expression site are like the cassettes inserted in a tape recorder. The first cassette mechanism described, the mating type switch in yeast (Haber, 2002), involves a site-specific endonuclease, but whether a dedicated endonuclease is also involved in antigenic variation in African trypanosomes, the organism using this mechanism most exuberantly, is under debate (see Chapter 11; Borst *et al.*, 1996; Borst, 2002b). The high rate of switching and the limited sequence homology available for gene conversion favour (but do not prove) the involvement of a dedicated endonuclease (Borst *et al.*, 1996), but this enzyme remains to be found.

A peculiar problem with the cassette mechanism is that a diversity of genes has to be slotted into an expression site by recombination. As recombination usually prefers substantial stretches of perfect homology, this would favour donor genes most similar to the acceptor gene already present in the expression site, defeating the purpose of the gene replacement. In practice, trypanosomes manage to avoid this pitfall and they readily activate genes with only short stretches of flanking homology with the expression site, as schematically indicated in Figure 1.1, mechanism A.3.a. How they do this is still in the realm of speculation (Borst *et al.*, 1996).

Finally, *Borrelia* species have been shown to switch the antigen gene expressed by DNA deletion (Chapter 15) (Barbour, 2002), as outlined in Figure 1.1, mechanism A.4. Arrays of antigen genes can be introduced in an active expression site by recombination. If each gene is followed by a strong transcriptional stop, removing the first gene in the array by deletion activates the second one. Such a mechanism can only work if antigen genes are introduced into the expression site in an asymmetric fashion, i.e. an array of genes replacing a single gene. If not, the gene will be entirely lost from the genome and the gene repertoire would rapidly shrink.

All mechanisms in class A use a simple device to restrict expression to a single member of the gene family: there is only a single promoter and only the gene directly downstream of that promoter is active. The available evidence indicates that this elegant system for mutually exclusive expression of the members of a large gene family is used in its austere form by *Pneumocystis* for controlling about 100 copies of the MSG family of surface antigens (Chapter 10). *Borrelia* species have more than one expression site for the expression of their *vsp* and *vlp* gene families and only one is active at a time (Chapter 15). The most extensive multiplicity of expression sites is found in African trypanosomes: there are about 20 bloodstream expression sites and only one can be stably activated at a time (Chapter 11). Two possible explanations, not mutually exclusive, have been advanced for this multiplicity of expression sites:

1 Silent expression sites allow the stepwise assembly of intact *VSGs* from pseudogenes (Chapter 11).
2 We have argued that the *raison d'être* of the multiple expression sites of *T. brucei* does not reside in the additional opportunity they provide to assemble or switch *VSGs*, but in the expression site associated genes, which are co-transcribed with the *VSG* gene in the active expression site (Borst and Ulbert, 2001). Expression sites encode *inter alia* genes for the heterodimeric transferrin receptor of *T. brucei*. Each site encodes a slightly different receptor and we have shown that this variability in receptor allows the trypanosome to cope with the diversity in transferrins found in its mammalian hosts. A receptor with high affinity for bovine transferrin, for instance, may not bind

dog transferrin at all. By switching to another expression site the trypanosome can switch on the synthesis of a receptor that does bind dog transferrin (Bitter *et al.*, 1998; Gerrits *et al.*, 2002).

IN SITU ACTIVATION OF SURFACE ANTIGEN GENES BY ALTERING THE SIZE OF SIMPLE DNA REPEATS OR BY DNA MODIFICATION

As summarized in Table 1.1, the mechanisms for *in situ* activation/inactivation are diverse. Variation of the size of a simple repeat can be used either to activate transcription, or to allow translation to avoid an early stop (Figure 1.2, mechanism B.2). Repeat control of transcription was discovered in *Bordetella pertussis* (Willems *et al.*, 1990), in which a poly (dC) tract controls the effect of an upstream activator. This activator must be precisely positioned relative to the transcriptional start site to get transcription. One bp more or less in this tract can inactivate the gene. Similar results were obtained for the surface variable lipoproteins of *Mycoplasma hyorhinis*, which requires a poly (dA) tract of exactly 17 bp for optimal transcription initiation (Borst, 1992).

Control of gene expression by length variation of simple repeats was discovered in *Neisseria* in a gene family, the *opa* genes, encoding the opacity

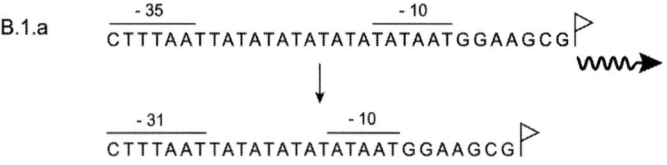

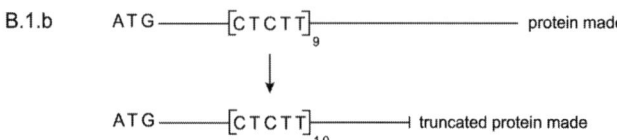

Figure 1.2 Reversible control of gene expression by variation of the size of simple repeat sequences.

In (B.1.a) the activity of the promoter (flag) is determined by sequences positioned at minus 35 and minus 10 relative to the start of transcription. A change in the 5'–TA repeat (only one strand of the DNA is shown) from 8 to 6 TA units shifts the minus 35 sequence to minus 31, resulting in a block of transcription initiation.

In (B.1.b) the coding sequence contains a pentanucleotide repeat. Alterations in this repeat may throw translation out of the frame, resulting in a truncated protein.

See text and Chapters 6 and 8 for details.

protein, a minor outer membrane protein. Part of the signal sequence of the opacity protein is encoded by an mRNA segment consisting of a series of CTCTT pentamers. Addition or subtraction of one or two pentamers by DNA replication errors will throw translation out of frame and will result in the synthesis of a truncated protein (Chapter 8).

The bacterial genome sequences have shown how intensively the length variation of simple sequence repeats is used by bacteria for phase variation, i.e. the high-frequency, stochastic, reversible on–off switching of gene expression. Hood and Moxon discuss many examples in Chapter 6 on *Haemophilus influenzae*, of both transcriptional control (pili, outer membrane protein) and translational control (virulence genes, genes involved in the synthesis of the lipopolysaccharide component of the cell wall). Another example is reported in Chapter 7 by Appelmelk, who discusses variation of the surface lipopolysaccharide (LPS) of *Helicobacter pylori*. The sugar side chains of LPS are highly variable in their composition and antigenic property and this is due to phase variation of the enzymes involved in their biosynthesis. The genes encoding these enzymes contain poly (dC) tracts, and variation of the length of these tracts controls the complete translation of these genes.

The repeat length variations that are central to mechanism B.1 (Table 1.1) are thought to be mediated by slipped strand mispairing during DNA replication. The DNA replication machinery has a problem with the precise duplication of simple repeat sequences, resulting in frequent insertions and deletions of repeat monomers. The mistakes are random, and mistakes made in the replication of one gene of a family do not affect the possibility of making mistakes in other genes. Hence, there is no cross-talk between the genes and no mechanisms for allelic exclusion, unlike in mechanism A. In fact, genes controlled at the translation level (mechanism B.1.b) will be active one-third of the time, because of the triplet code. In larger gene families this means that several members will always be active at the same time. This is not elegant. Moreover, regulation at the translational level is wasteful, because a lot of genes are being transcribed and partially translated without yielding a functional protein product. This may be the reason why mechanism B.1 (Table 1.1) has only been seen in bacteria and not in organisms with more complex genomes.

I realize that the categorization of mechanism B.1 under *in situ* control in Table 1.1 is debatable, because *in situ* control usually implies epigenetic control without DNA sequence changes, whereas B.1 does involve minor changes. As the genes do not move and as switching is due to DNA replication errors rather than a recombination event, I have put repeat control under the *in situ* mechanisms, but a separate category would also be reasonable.

In contrast to the widespread use of mechanism B.1 for creating surface diversity in bacteria, reversible control of gene expression by variations in DNA modification seems to be rare. To my knowledge this mechanism (B.2) has only been associated with the control of bacterial pili-adhesin complexes. It was discovered by David Low and colleagues as the mechanism underlying the phase variation of pyelonephritis-associated pili (the *pap* system) of *E. coli* (Braaten *et al.*, 1991). This on–off system depends on the alternative methylation of two nearby DNA adenine methylase (Dam) sites. The system is self-perpetuating, but has a finite chance of switching during DNA replication. The system integrates input from several global transcriptional regulators and is even sensitive to DNA packaging by the histone-like structuring protein of *E. coli*. The *pap* system also regulates synthesis of type I pili and flagella and therefore appears to have a profound effect on the surface structures of *E. coli*. The advantage of this on–off system is that it allows formation of surface structures only when needed. Making these structures is energetically expensive and they tend to be highly immunogenic. A recent discussion of this sophisticated control system is provided by Hernday *et al.* (2002) and another system using Dam methylation in a different way to control gene expression is analysed by Wallecha *et al.* (2002).

GENE FAMILIES CONTROLLED WITHOUT DETECTABLE DNA ALTERATIONS

Table 1.2 recapitulates how parasites using programmed antigenic variation in a grand style with large gene families accomplish this. It is remarkable that a substantial number of gene families are controlled by mechanisms that do not involve any detectable alteration in DNA. In these cases the activity of genes must be controlled either by chromatin structure or by intranuclear location, or by a combination of both.

In 1990, Dan Gottschling and coworkers (Gottschling *et al.*, 1990) suggested that the expression sites of trypanosomes are controlled by telomeric silencing.

Table 1.2 Antigenic variation of eukaryotic parasites: mechanisms used to control expression of surface antigen genes

Organism	Surface antigen	Number of genes (approx.)	Location of expressed genes	Mechanism to change gene expressed
Trypanosoma brucei	VSG	1000	Telomeric	Recombinational *In situ* (minor)
Plasmodium falciparum	PfEMP1 (var)[a]	50	Anywhere[b]	*In situ*
Pneumocystis carinii	MSG	100	Telomeric	Recombinational
Giardia lamblia	VSP	150	Anywhere	*In situ*

Abbreviations: VSG, variant surface glycoprotein; PfEMP1, *Plasmodium falciparum* erythrocyte membrane protein 1; MSG, major surface glycoprotein; VSP, variant specific surface protein. See text and Chapters 11 (*T. brucei*), 14 (*P. falciparum*), 10 (*P. carinii*) and 16 (*G. lamblia*) for details.
[a]The PfEMP1 is not on the surface of the parasite, but on the surface of the erythrocyte in which it resides.
[b]Predominantly (sub)-telomeric, but with some well-documented chromosome-internal transcription units.

This form of stochastic control of gene expression was discovered in yeast, and it affects genes close to chromosome ends. The mechanism is complex and involves two elements:

1 Competition between an active and a silenced chromatin state. The silenced chromatin state is associated with (sub)telomeric repeats and may spread inward to adjacent genes. The silencing is counteracted by transcription, strong promoters being more effective than weak ones. Silencing is an all or none effect: either silencing wins and the gene is completely off; or transcription wins and silencing is completely relieved. The system may be reset during DNA replication. Resetting is stochastic, and the period that the promoter is active is determined by the balance between promoter strength and silencing strength. With high concentrations of silencing proteins and a weak promoter, the gene is usually off (Aparicio and Gottschling, 1994).
2 Telomeres are usually in the nuclear periphery and peripheral location is required for silencing (Maillet et al., 1996).

Following Gottschling's suggestion that differential control of trypanosomal telomeric *VSG* expression sites is accomplished by telomeric silencing, this model enjoyed a period of popularity. The enthusiasm waned after it was shown that trypanosomes cannot maintain full activity of two expression sites over time. In Gottschling's model for silencing, the transcriptional activity of each telomere is regulated independently of all other telomeres and one would therefore expect to find trypanosomes with two or more expression sites fully active. Such trypanosomes do not exist. Although it is possible to devise more complex versions of telomeric silencing that include cross-talk between expression sites, resulting in efficient allelic exclusion, I find such models far-fetched and unattractive (Borst and Ulbert, 2001). Moreover, the overview in Table 1.2 shows that most gene families that are controlled without DNA alterations do not exclusively reside at telomeres. Hence, telomeric silencing cannot explain the control of these gene families. This does not mean, however, that silenced gene copies might not be in a 'silencing compartment' in the nucleus, from which they have to be retrieved for activation. Transcription of trypanosome *VSG* gene promoters and of *Plasmodium var* gene promoters can be activated by putting the promoters on a plasmid without their flanking or internal repressing sequences. These sequences might mediate segregation into a nuclear silencing compartment even if the promoters are not near telomeres, as is the case for some *var* gene promoters (see Chapters 11 and 14).

The demise of the telomeric silencing model has focused attention on other mechanisms that involve competition for limiting amounts of positively acting factors. An interesting example is the expression site body (ESB) discovered by Navarro and Gull (2001) in bloodstream trypanosomes. They found that the expressed VSG gene present in an active expression site is located in a transcriptional body, which survives removal of DNA with DNase I. They conclude that the ESB is a unique subnuclear structure, required for complete high-level expression of VSG genes, and not an ad hoc assembly of all components required for transcribing an expression site. The ESB provides a simple explanation for limiting expression to one site at a time (mono-allelic expression). If there is space for only one expression site in each ESB, mono-allelic expression is guaranteed.

This 'privileged location' model is a specific example of a general model in which all genes in a family are repressed and are competing for one or more limiting factors to be derepressed. These factors could be in a privileged location, but it is conceivable that this is not required. The limiting factors might act at transcription initiation, or relieve blocks beyond initiation, as appears to be the case with trypanosome bloodstream expression sites. An interesting silencing mechanism was recently found by Dr H.J. Lujan (personal communication). They found that all VSP genes in *Giardia lamblia* are transcribed, but that the silent ones are kept silent by RNA interference. Although this does not answer the question how one of the 150 VSP genes escapes silencing, the use of RNA_i for silencing represents a new and unexpected twist to *in situ* control of antigenic variation.

As I have pointed out elsewhere (Borst, 2002a), the ESB model is a useful hypothesis, but leaves many questions unanswered. The major challenge is to identify proteins that are specific for the active expression site and to determine how they work. Since *in situ* control of complex gene families is used by several major parasites (Table 1.2), the mechanism of this control is a major unsolved problem in molecular parasitology.

AS A SURVIVAL STRATEGY, PROGRAMMED ANTIGENIC VARIATION IS NOT AS SIMPLE AS IT LOOKS

In textbooks, antigenic variation is usually depicted as a simple and elegant strategy for a parasite to evade the host immune system. In practice, it is not that simple and at least four requirements must be met:

1. A large repertoire of surface antigens to be able to survive for extended times in a host with an aggressive immune system.
2. Mechanisms for switching the surface antigen expressed in a subfraction of the parasite population, before antibodies hit it.
3. The ability to express surface antigen genes in a semi-defined sequence to avoid gross population heterogeneity, which would rapidly squander the antigen repertoire. This is a plausible, but admittedly theoretical, argument, which has never been tested by infecting a host with a mixture of 100 different parasite coat variants. It therefore remains possible that the immune system would not be able to make a sufficient dose of antibody against all 100

variants. Nevertheless, I find it hard to believe that sequential expression would not provide an advantage to the parasite.
4 The ability to combine antigenic variation with substrate uptake. Translocators for small hydrophilic molecules, which can diffuse through the surface coat, can be hidden in the plasma membrane of the parasite below the surface coat. However, parasites that take up large host macromolecules with relatively invariant surface receptors, such as African trypanosomes, need special tricks to avoid the targeting of these receptors by the host immune system. The trypanosome accomplishes this feat by hiding the receptors in an invagination of the plasma membrane, the flagellar pocket; by producing high-affinity receptors shielded from antibodies by loading with substrate; and by high rates of internalization. Given this complexity, it is understandable that many protozoa prefer to hide in host cells rather than brave the immunological storm outside by antigenic variation.

UNSOLVED PROBLEMS AND CHALLENGES

The list is long, and I lift out only some of the most pressing issues.

1 The biological advantages of surface antigen variation are often unclear. There are a few examples, such as the VSG coat of trypanosomes, where the variation only serves to allow the parasite to escape immune destruction. In most cases, however, surface variation serves other purposes – attachment to host cells, altering motility, etc. Examples discussed in this book are the variable surface antigens of *Haemophilus, Helicobacter, Pneumocystis, Giardia*, and even HIV. Examples of variable surface antigens that clearly do not predominantly serve to evade the host immune response are some of the surface appendages of bacteria and the transferrin receptor repertoire of *T. brucei*. In an elegant review, Moxon *et al.* (1994) have discussed how the highly mutable loci of pathogenic bacteria, called contingency genes, 'facilitate the efficient exploration of phenotypic solutions to unpredictable aspects of the host environment'. The nature of these solutions requires further analysis of host–parasite interactions.
2 For parasites that stick around for long periods of time in the host bloodstream, large repertoires of surface coats are essential. If every parasite in the population decides to put on a different coat, the host will rapidly make antibodies against the entire repertoire. The order in which different coats are used by the population should therefore be controlled somehow. There are some indications on how this might be accomplished by African trypanosomes (see Chapter 11), but this central problem in antigenic variation deserves more systematic attention.
3 How organisms manage to control large families of surface antigens *in situ* (mechanism B.3 in Table 1.1) is still far from clear. This is a fundamental

biological problem far exceeding the boundaries of antigenic variation (Borst, 2002a) and it would be nice if the solution came from the field of molecular microbiology.

ACKNOWLEDGEMENTS

I thank the authors of several other chapters in this book, students and postdocs in my laboratory, and Dr Gloria Rudenko (The Peter Medawar Building for Pathogen Research, Oxford, UK) for helpful comments. The experimental work on trypanosomes in my laboratory is supported in part by grants from the Netherlands Foundation for Chemical Research (CW), with financial support from The Netherlands Organization for Scientific Research (NWO).

REFERENCES

Aparicio, O. M. and Gottschling, D. E. (1994) *Genes Devel.* 8, 1133–1146.
Barbour, A. (2002) In *Mobile DNA II* (eds N. L. Craig *et al.*), ASM Press, Washington, DC, pp. 972–994.
Bitter, W., Gerrits, H., Kieft, R. and Borst, P. (1998) *Nature* 391, 499–502.
Borst, P. (1992) *Curr. Biol.* 2, 304–306.
Borst, P. (2002a) *Cell* 109, 5–8.
Borst, P. (2002b) In *Antigenic Variation to Mobile DNA II* (eds L. C. Craig *et al.*), ASM Press, Washington, DC, pp. 953–971.
Borst, P. and Ulbert, S. (2001) *Mol. Biochem. Parasitol.* 114, 17–27.
Borst, P., Rudenko, G., Taylor, M. C., Blundell, P. A., Van Leeuwen, F., Bitter, W., Cross, M. and McCulloch, R. (1996) *Arch. Med. Res.* 27, 379–388.
Braaten, B. A., Blyn, L. B., Skinner, B. S. and Low, D. A. (1991) *J. Bacteriol.* 173, 1789–1800.
Deitsch, K. W., Moxon, E. R. and Wellems, T. E. (1997) *Microbiol. Mol. Biol. Rev.* 61, 281–293.
Domingo, E., Sabo, D., Taniguchi, T. and Weissmann, C. (1978) *Cell* 13, 735–744.
Drake, J. W. and Holland, J. J. (1999) *Proc. Natl Acad. Sci. USA* 96, 13910–13913.
Drake, J. W., Allen, E. F., Forsberg, S. A., Preparata, R.-M. and Greening, E. O. (1969) *Nature* 1128–1131.
Dworkin, J. and Blaser, M. J. (1997) *Mol. Microbiol.* 26, 433–440.
Gearhart, P. J. (2002) *Nature* 419, 29–31.
Gerrits, H., Mussmann, R., Bitter, W., Kieft, R. and Borst, P. (2002) *Mol. Biochem. Parasitol.* 119, 237–247.
Gottschling, D. E., Aparicio, O. M., Billington, B. L. and Zakian, V. A. (1990) *Cell* 63, 751–762.
Graham, V. S. and Barry, J. D. (1996) *Mol. Biochem. Parasitol.* 79, 35–45.
Haber, J. E. (2002) In *Mobile DNA II* (eds N. L. Craig *et al.*), ASM Press, Washington, DC, pp. 927–952.
Hernday, A., Krabbe, M., Braaten, B. and Low, D. (2002) *Proc. Natl Acad. Sci. USA* 99, 16470–16476.

Johnson, R. C. (2002) In *Mobile DNA II* (eds N. L. Craig *et al.*), ASM Press, Washington, DC, pp. 230–271.

Komano, T., Kubo, A. and Nisioka, T. (1987) *Nucleic Acids Res.* 15, 1165–1172.

Lu, Y., Alarcon, C. M., Hall, T., Reddy, L. V. and Donelson, J. E. (1994) *Mol. Cell Biol.* 14, 3971–3980.

Lu, Y., Hall, T., Gay, L. S. and Donelson, J. E. (1993) *Cell* 72, 397–406.

Maillet, L., Boscheron, C., Gotta, M., Marcand, S., Gilson, E. and Gasser, S. M. (1996) *Genes Devel.* 10, 1796–1811.

McKenzie, G. J. and Rosenberg, S. M. (2001) *Curr. Opin. Microbiol.* 4, 586–594.

Moxon, E. R., Rainey, P. B., Nowak, M. A. and Lenski, R. E. (1994) *Curr. Biol.* 4, 24–33.

Navarro, M. and Gull, K. (2001) *Nature* 414, 759–763.

Rice-Ficht, A. C., Chen, K. K. and Donelson, J. E. (1982) *Nature* 298, 676–679.

Roth, C., Bringaud, F., Layden, R. E., Baltz, T. and Eisen, H. (1989) *Proc. Natl Acad. Sci. USA* 86, 9375–9379.

van de Putte, P. and Goosen, N. (1992) *Trends Genet.* 8, 457–462.

Wallecha, A., Munster, V., Correnti, J., Chan, T. and van der Woude, M. (2002) *J. Bacteriol.* 184, 3338–3347.

Willems, R., Paul, A., Van der Heide, H. G. J., Ter Avest, A. R. and Mooi, F. R. (1990) *EMBO J.* 9, 2803–2809.

2

HIV Variation – A Question of Signal-To-Noise

Simon Wain-Hobson

Is the spectacular genetic variability exhibited by the human immunodeficiency virus (HIV), the AIDS virus, important from the point of view of pathology? Although my neighbour Claude, who works at the Assemblée Nationale in Paris, knows little about HIV, he 'knows' that the virus mutates like crazy and outruns the immune system. This he has picked up from the informed press. HIV does intrinsically mutate quickly compared to most microbes, yet it is near the bottom of the first division compared to other RNA viruses, such as polio and influenza A (Drake and Holland, 1999). Nonetheless, as HIV is a life-long infection, variation is accumulated over time such that cross-sectional variation in any population dwarfs that inherent to the annual influenza A epidemic or an outbreak of polio. It is simply stunning; we have never handled something like this before. Yet, the idea that HIV endlessly outruns and outguns the immune system has precious little support. The key question is whether the abundance of variation undergoes strong positive selection. This chapter attempts to square the abundant genetic variation observed for HIV with the biology of the virus. The answer is that HIV is doing very well and doesn't need variation to cause disease.

THE HIV LENTIVIRUSES

The metamorphosis of RNA into DNA is called reverse transcription and is apparent among a variety of plant and mammalian viruses, which, accordingly, can all be viewed as retroviruses. In common parlance, however, 'retroviruses' refer to the group that gave rise to the identification of reverse transcriptase, that singular DNA polymerase that takes both RNA and DNA as template. The virion is enveloped by a lipid bilayer wrought from the host cell plasma membrane. Within the nucleocapsid structure are two copies of plus strand, polyA+ genomic RNA, which are reverse transcribed into a single DNA copy and integrated into the host cell chromosome by the retroviral enzyme, integrase. This is referred to as the provirus.

Genetically retroviruses are an extraordinarily varied group of viruses in terms of biology and genetic organization (Coffin *et al.*, 1997; Renoux-Elbé *et al.*, 2002). Some apparently cause no disease whatsoever; others result in leukaemias and sarcomas, haematological and neurological disorders and immunosuppression, of which AIDS is the most frightening. HIV belongs to the lentiviral group, which has isogenic counterparts in primates and more distant relatives in cats, horses, sheep, goats and cattle.

HIV has come from nowhere to occupy the number one slot in terms of annual mortality of any infectious disease in something like 50 years or so. On a virological Richter scale, HIV would come in at over nine (Figure 2.1). And, of course, the epidemic is expanding. When referring to the pandemic, one is alluding to HIV-1, whose epicentre was around the Great Lakes region of East Africa. There is a second virus, HIV-2, originally confined to West Africa around Guinea Bissau, which has spread since. Although pathogenic, it is less virulent, needing more years to produce AIDS. HIV-1 and HIV-2 differ by mutually exclusive genes *vpu* and *vpx* (Figure 2.2) and a variety of subtler genetic differences.

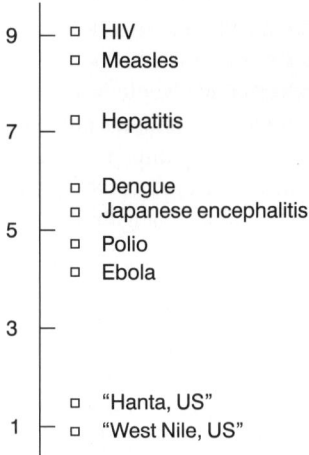

Figure 2.1 A virological Richter scale: 10 = all deaths worldwide from infectious diseases in 1999. HIV/AIDS checks in at 9.2. A projection of 6 million dead from HIV/AIDS in year 2001 would move HIV/AIDS up to around 9.5 on this logarithmic scale. Hanta US and West Nile US refer to outbreaks in the USA. Data for these and for Ebola are estimates and do not figure in the WHO report. For reference, tuberculosis and malaria measure 8.7 and 8.4 on the same scale. (*Source:* The World Health Report 2000 http://www.who.int/whr/2000/en/statistics.htm.)

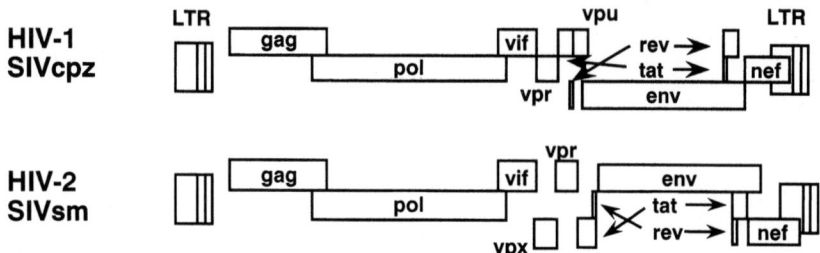

Figure 2.2 Genomic organization of the human immunodeficiency viruses and their isogenic primate counterparts – SIVcpz and SIVsm simian immunodeficiency virus of chimpanzees and sooty mangabeys respectively. Note that these primate viruses have not been shown to be pathogenic in their hosts.

HIV-2 came to light primarily through Western blot analyses indicating sera positive to the core (Gag) proteins yet negative for the envelope proteins (Clavel et al., 1986). Careful serology later helped identify variants of HIV-1, mainly in Cameroon and Gabon, that diverged strongly in the envelope proteins (Huet et al., 1989; Simon et al., 1998). These viruses belong to the HIV-1 O and N groups (Figure 2.3). Albeit pathogenic, their prevalence is far below that of the group M viruses, as the variants from all over the world are now called.

HIV-1 has an isogenic counterpart in the simian immunodeficiency virus of chimpanzees (SIVcpz) (Peeters et al., 1989; Huet et al., 1990) and its origin in the central Africa rain forests – chimpanzees from west Africa have proved to be systematically negative for SIVcpz. For HIV-2 the sooty mangabey harbours an isogenic counterpart (SIVsm). Just when SIVcpz and SIVsm crossed over to humans is probably too hard a call, although it is possibly as recently as the first half of the twentieth century (Korber et al., 2000). Note that if HIV-1 and HIV-2 had been in humans for millennia one would have expected them to have crossed the Atlantic with the slave trade, like another human retrovirus, HTLV-1. Based on the phylogeny of the human, chimp and sooty mangabey viruses, it is generally agreed that there have been at least three crossovers from chimpanzees and even six or more from the sooty mangabey to man (Hahn et al., 2000).

MUTATION, RECOMBINATION AND FIXATION RATES

In terms of point mutations there are two classes of virus depending on whether replication errors are corrected (i.e. DNA viruses) or not (retroviruses and RNA viruses). Even though retroviral reverse transcriptase (RT) uniquely synthesizes DNA, there is no associated exonuclease activity associated with RT. Furthermore, there are no proof-reading enzymes associated with the virion or replication complex. Not surprisingly RT is able to elongate beyond mismatches more readily than non-viral DNA polymerases. RNA viruses too are unable to

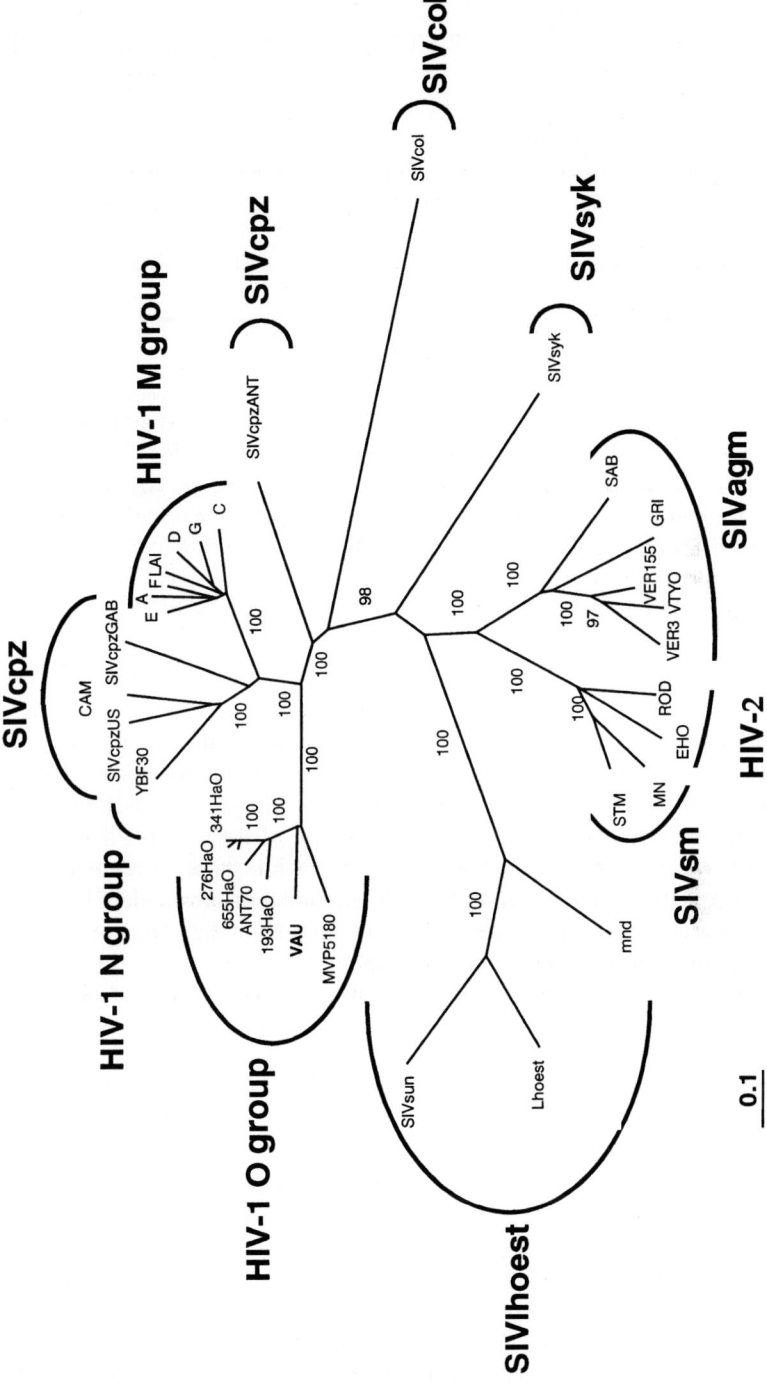

Figure 2.3 Phylogenic tree of the human and simian immunodeficiency viruses based on the *pol* gene sequence.

edit or proof read RNA replication, so it comes as no surprise to learn that the mutation rates of retroviruses and RNA viruses are 10^3 to 10^6 greater than those of DNA viruses and DNA-based genomes (Table 2.1). Mutation rates are invariably expressed as the probability of making a point mutation per base per round of replication (Drake and Holland, 1999). This is intrinsic to the organism and does not take into account the vagaries of selection (see later).

Table 2.1 Point mutation and recombination rates for some RNA viruses and retroviruses

Point mutation	
RNA viruses	
Measles	~1/genome/cycle
Poliovirus	~0.9/genome/cycle
Vesicular stomatitis virus	~1/genome/cycle
Retroviruses	
Human immunodeficiency virus type 1 (HIV-1)	~0.25/genome/cycle
Bovine leukaemia virus (BLV)	~0.01/genome/cycle
Spleen necrosis virus (SNV)	~0.08/genome/cycle
Homologous recombination	
Human immunodeficiency virus type 1 (HIV-1)	~3/genome/cycle
Spleen necrosis virus (SNV)	~0.3/genome/cycle

As is apparent from Table 2.1, the mutation rate of HIV is certainly at the upper end for retroviruses. However, the mutation rates of retroviruses are at least an order of magnitude lower than those of RNA viruses such as influenza A, poliovirus and VSV. When normalized to the size of the genome, the mutation rate for RNA viruses turns out to be ~1 per cycle, while those for retroviruses are between 0.3–0.01 per genome. HIV is not the sultan of substitution; the reason why HIV genetic variation is so noteworthy must lie in the tempo of infection, not in the mechanics of mismatch formation.

The other intrinsic rate to be aware of is the recombination rate. As two copies of genomic RNA are packaged in a virion, recombination can occur during reverse transcription. In fact retroviral recombination was first noted in 1971, just one year after the identification of reverse transcriptase (Vogt, 1971; Kawai and Hanafusa, 1972; Weiss *et al.*, 1973). Recombination occurs by a template switching mechanism called 'copy choice' (Coffin, 1979). The recombination rate, the number of crossovers per round of replication, for HIV turns out to be approximately three crossovers per round, which took most in the field by surprise (Yu *et al.*, 1998; Jetzt *et al.*, 2000). While this is 10× the point mutation rate, a similar situation pertains to spleen necrosis virus, so the finding is not exceptional (Table 2.1). Obviously if the two genomes packaged are identical, that is, derived from the same provirus, then the effects of recombination will be non-existent. If, however, a cell harbours ≥2 genetically distinct proviruses then recombination will greatly contribute to the complexity. This is so important an issue that it will be dealt with in a subsequent section. Suffice to say that recombination is observed at all levels of HIV genetics without exception (Li *et al.*, 1988; Robertson *et al.*, 1995; Wooley *et al.*, 1997; Carr *et al.*, 1998; Simon *et al.*, 1998; Gao *et al.*, 1999; Peeters *et al.*, 1999; Takehisa *et al.*, 1999; McCutchan, 2000; Hoelscher *et al.*, 2001). This is in sharp contrast to influenza viruses and other negatively stranded RNA viruses, where recombination is scarce.

The fixation rate refers to the number of amino acid changes accumulated, or fixed, per year. It is far less precise than the mutation rate, for it does not take into account the number of rounds of replication per year, the number of individuals infected. When making comparisons among viruses with very different lifestyles, fixation rates lose even more meaning, particularly when the comparisons are among non-orthologous proteins. For all their weaknesses, the most rapidly evolving regions of the influenza A and HIV genomes are fixing amino acid substitutions at around 1% per year.

There are clearly considerable qualitative and quantitative differences between HIV and influenza A in mutation and recombination rates. Despite comparable maximal fixation rates this is the softest parameter with little explanatory power. We need to look at influenza A and the lifestyle and dynamics of HIV replication *in vivo* to understand why HIV variation is so stunning.

A TOUCH OF 'FLU

The influenza A virus paradigm is known to all. Every year a new strain invades most countries, oscillating roughly between winters in the northern and southern hemispheres. The strain usually carries a small number of mutations with respect to last year's strain. Importantly, a few of these substitutions are in B cell epitopes in the haemagglutinin and/or neuraminidase surface proteins, which means that the new strain is not neutralized by antibodies as fast as the previous strain. This slight edge is enough to allow the virus to replicate efficiently and spread within the population. This annual iterative process is referred to as antigenic drift.

There are two points of interest to us here. First, viral strains have historically been characterized by antisera. This is true for influenza A, poliovirus with three strains, hepatitis A virus (one strain) and human rhinovirus (>100 strains). Since then monoclonal antibodies have become the reagents of choice, often increasing the resolving power. With serologically defined strains, correlations with virulence could sometimes be established. With serology the phrase antigenic variation made sense. For HIV the virus is far more genetically heterogeneous than for influenza A. Antisera are poorly neutralizing, meaning that the genesis of neutralizing groups and the subtypes is not possible. Furthermore, genetic subtypes of clades do not correlate readily with serological parameters (Moore *et al.*, 1996). This is not without precedent, for antibodies to the caprine arthritis/encephalitis lentivirus are not neutralizing.

The second point emerging from influenza A is that antigenic variation allowed spread within the human population. Antigenic variation is not necessary to escape the innate or adaptive immune responses within an individual. Symptoms result within a few days of infection and within three weeks or so the virus is eliminated completely. The same is true for other acute viral infections. In other words, once the virus infects a host it executes its genetic program, its bauplan. If it has functions that allow it to interfere with the interferon pathways, mannose-binding proteins, complement etc., they were inherent to the strain at entry, having evolved many generations earlier.

AN INSIDE LOOK AT HIV REPLICATION

Although HIV replicates in CD4+ T lymphocytes, dendritic cells and macrophages, cross-sectional quantitative analyses always show that the T lymphocytes carry the greatest viral burden (McIlroy *et al.*, 1995). Furthermore, the virus is more abundant in the memory, as opposed to the naive T cell compartment, hence we will focus principally on these cells (Schnittman *et al.*, 1990). T-cells are schizophrenic in that they spend most of their time in a resting state roaming the body in search of cognate antigen. When they encounter antigen in the context of antigen-presenting cells, they proliferate. Most of these effector lymphoblasts have a short life span of a few days and die. A small fraction will move back to a resting memory cell state. What happens to the virus during these events?

When a virus infects a lymphoblast productive infection may ensue, which is complete in about 18 to 24 hours. When a resting cell harbours a provirus, transcription is initiated but stalls after formation of a 56 nucleotide RNA stem loop called TAR at the 5' of all viral transcripts. As no viral protein can be made the infected resting cell is not detected by HIV-specific cellular immunity. When the cell is activated by antigen and moves to proliferation, basal transcription increases and breaks through the TAR transcription block, leading to the synthesis of the early viral proteins Tat, Rev and Nef. Tat is a transcription enhancer working through TAR. Transcription and elongation are both greatly enhanced allowing a shift to the expression of late-phase proteins, notably the structural polyproteins Gag and Env as well as Pol, which encodes the three crucial retroviral enzymes protease, reverse transcriptase and integrase. Even though the host effector cell is bound to die, enough virus can be produced before the inevitable implosion of its effector host.

In the event of a productively infected cell moving back to a memory cell, it must be assumed that the HIV provirus is silenced. This is technically a very difficult area and has been poorly studied. However, it follows from the observation that the infection of resting cells is highly inefficient. Thus some viruses may be replicating in a series of consecutive rounds of replication while others may be turning over more slowly with T lymphocytes. Presently it is not possible to weight the relative contribution of the two compartments. Yet, if HIV moved only from one lymphoblast to another then it is hard to appreciate why the virus evolved the sophisticated TAR/Tat off/on switch coupled to the activation/proliferation state of the host cell. A number of studies have shown that latently infected cells drawn to inflammatory sites result in local bursts of viral replication (Blancou *et al.*, 2001). Given the large number of different antigens the immune system comes across, many distinct bursts occur in parallel. Certainly vaccination results in an increase in free viral load in the blood (Ho, 1992; Staprans *et al.*, 1995; Brichacek *et al.*, 1996). What fraction of virus escapes from such a burst and how fast is virus turning over?

TEMPO OF HIV REPLICATION

There are two complementary approaches to the study of viral dynamics *in vivo*. The first and most widely used approach is an experimental classic – perturb the system. HIV plasma viraemia is remarkably stable over time, varying little from one day to another. Using multiple anti-HIV drugs, the decay of viraemia could be measured and modelled (Ho *et al.*, 1995; Wei *et al.*, 1995; Perelson *et al.*, 1996). The area has been extraordinarily intense and controversial, for inevitably assumptions had to be made. However, it has been enormously fertile and moved virology on by leaps and bounds. Despite the apparent stability of plasma viraemia, it turns over with a half-life of hours. As much as 90% of virus turns over per day while most of the virus is not infectious, being associated with immunoglobulin. Modelling of T cell turnover proved particularly difficult to ascertain, although it is fair to say now that there is increased turnover of CD4 T lymphocytes over the course of disease. In short, immunosuppression follows from T cell activation. Furthermore, it was estimated that HIV goes through about 200 consecutive cycles of replication per year (Perelson *et al.*, 1996).

The second approach involves the rate of fixation of mutations. Again assumptions have to be made, but in this case they are fewer (Wain-Hobson, 1993a,b; Pelletier *et al.*, 1995). The first two hypervariable regions of the envelope protein are probably the most variable regions of the entire genome. While the rate is variable the upper limit is close to 1% nucleotide changes per year. These regions are under very weak selection pressure (Plikat *et al.*, 1997; Kils-Hütten *et al.*, 2001), meaning that the fixation rate approximates to the product of the point mutation rate and the annual number of consecutive cycles of replication. As the point mutation rate is known (approx. 2.5×10^{-5} per base per round) (Mansky and Temin, 1995), the remaining unknown, the number of cycles per year, turns out to be around 400. The big unknown in this calculation is the effect of recombination on the accumulation of mutation. Since this value is within the range of that derived from a totally different approach, measuring T cell dynamics, it suggests that the correction factor necessary to accommodate recombination is not huge, probably less than a factor of two.

The viral burst size is the number of viruses produced by an infected cell. Electron micrographs show that hundreds, if not thousands, of viruses can be produced by a cell. Nonetheless, assuming that a cell produces but two viruses that give rise to productively infected cells, then with 200–400 consecutive cycles per year the cumulative number of infected cells must be 2^{200}–2^{400}, or 10^{60}–10^{120}. Such numbers are simply impossible – after all there are 'only' about 10^{80} atoms in the universe. The only way out of this situation is to reduce the burst size to a non-integer close to one – ~1.01. Yet viral burst sizes must be integers. Hence the unavoidable corollary is that the vast majority of infected cells cannot produce virus, while the vast majority of virus produced cannot infect cells. In other words, only a small fraction of infected cells produce virus. What is restricting so efficiently viral replication? The only candidates are the innate and adaptive immune responses. Antibody, complement and mannose-binding proteins can clear free virus, while innate (Schwartz *et al.*, 1998) and specific cellular anti-HIV immune responses can clear infected cells. In favour of this, CD8 T cell depletion in rhesus macaques infected by SIV resulted in vastly increased virus production (Schmitz *et al.*, 1999).

Hence control of HIV replication is superb and probably explains why it takes years for HIV to demolish the immune system. The crucial point is the following. This phenomenal degree of control occurs over the entire course of HIV infection, during which intrapatient genetic heterogeneity grows remorselessly and can attain as much as 20% in the hypervariable regions of envelope. This represents a phenomenal degree of sequence complexity and an abundance of epitopes that should overwhelm the immune system – certainly more than T cells in the body. Yet viraemia remains tightly controlled given what is formally possible.

GULLIVER AND HIVLAND

Robin Weiss first invoked Gulliver and HIVland in an ironic yet highly focused critique of our blindness to the unfurling AIDS pandemic (Weiss, 2001). It is another, pathological side to this analogy that interests us here. How did the Lilliputians contain Gulliver? They tied him down with many little ropes. How might a system tie down a rapidly evolving virus? By targeting many sites such that the probability of escape is less than the inverse of a typical population size.

Some numbers. The point mutation rate for HIV is ~0.25×10^{-5} per base per cycle. Assuming (1) that a single mutation is sufficient to efficiently abrogate the effects of a host antiviral strategy, and (2) that all mutations are assumed to arise with equal probabilities, then within a population of 40 000 genomes there will be a very good chance of finding one genome encoding that resistance mutation. The probability of having a variant resistant to three antiviral strategies would require a population of ~64×10^{12} genomes. In an HIV-infected individual there are simply not this number of virions of infected cells at any one time. Hence theory predicts that the virus should succumb. Of course viruses invest enormous resources trying to subvert antiviral responses (Ploegh, 1998) but then in terms of an arms race the mammalian host has evolved a gigantic armamentarium that needs little expanding. In short, acute infections are just that while persistent infections persist, for they have evolved a strategy of living with the wrath of the host. This scenario explains why there is so much destruction of virus and infected somatic cells, which can regenerate.

The reader may recognize this as the principle behind multidrug therapy, not only in the treatment of HIV infection, but also TB. HIV infection is curtailed by a myriad of small ropes, which, together, tie the virus up to the point that the course of disease is long and protracted. We know, however, that HIV wins in the end. However, this is a far cry from the virus outrunning the immune system by mutation. If mutation does occur in an epitope recognizable by host cytotoxic T lymphocytes (CTLs) the infected cell is still expressing numerous other viral epitopes, meaning that the infected cell is still visible to the immune system.

Much has been written in the past decade about escape from CTLs (Meyerhans et al., 1991; Phillips et al., 1991; Nietfield et al., 1995; Price et al., 1997; Borrow and Shaw, 1998; McMichael, 1998; O'Connor et al., 2002). Nearly all the arguments in favour of escape are based solely on correlations. Proof is in short supply, although it must be acknowledged that there are hard experiments to undertake. We still do not know if a virus with a CTL epitope escape sequence is fitter than its parental virus in the same immunological background. Logically, if the virus were endlessly escaping CTLs then with the extraordinary dynamics of HIV replication – about 90% of virus is turning over daily – there should be strong pressure to rapidly select for escape mutants in a matter of days, accompanied by a remorseless increase in plasma viraemia. This is not seen. However, as soon as CD8 T cells are depleted viraemia bounces back. In the same way as soon as multidrug therapy is stopped viraemia generally bounces back to pre-treatment levels. There are many ropes tying the virus down such that there is almost a steady-state in the short term between Gulliver and the Lilliputians. Gulliver eventually gets away not by mutation but by slowly and remorselessly killing off the soldiers, alias CD4 T cells.

Using the drug analogy, can the virus 'escape' from the adaptive immune system, for we know that monotherapy is no longer in vogue? Certainly in the beginning of primary infection, when CTL responses are being induced and amplified, there is a window in which immune responses will be far more oligoclonal. It is precisely in these cases that the few convincing cases of CTL escape have been described (Borrow and Shaw, 1998; O'Connor et al., 2002). Following primary infection, however, the evidence is poor.

RAMPANT RECOMBINATION

There is a further aspect of the viral escape scenario that has surfaced recently. In one's mind's eye, and in textbook figures, one 'sees' a single virion infecting a susceptible cell and executing a single round of replication generating large numbers of progeny. But in fact how many viruses infect a cell *in vivo*? By using fluorescent *in situ* hybridization a recent report showed that HIV-1 infected splenocytes from two patients harboured multiple proviruses (Jung *et al.*, 2002). The range was between 1 and 8 proviruses per cell with a mean of 3–4 per cell. The frequency distributions were remarkably similar despite different clinical presentations and viral load. With some 75–80% of infected cells harbouring ≥2 proviruses the stage is set for rampant recombination.

Be that as it may, if the proviruses were identical or very similar then recombination might not have a massive effect on the dynamics of HIV genome evolution. However, if the proviruses were highly divergent, then the picture would be otherwise. To get at this, single HIV-positive interphase nuclei were laser microdissected and transferred to PCR tubes and a segment encompassing the first two hypervariable regions of the Env protein amplified. Intracellular sequence divergence was massive and a collection of sequences derived from three cells all harbouring three or four proviruses is shown in Figure 2.4. The most striking features are the numbers of distinct sequences per cell and the extent of genetic variation within a single cell – up to 29% amino acid difference for cell R5 (compare sequences R5–2 and R5–3). Such a degree of variation is typical for inter-isolate comparisons. It may be noted that all but two sequences from patient R have an additional pair of cysteine residues, indicated by asterisks over R4–1, which is exceptional. Only sequences R5–3 and R5–4 show the standard pattern of cysteine residues in V1V2. This supports the notion that patient R was infected by two distinct HIV-1 strains and that cell R5 harboured genomes of each. A number of sequences are arguably recombinants. For example, B7–2*, B7–4 and B7–5* all have identical 3' sequences while their 5' halves are all different. For R5–3 and R5–4 there is a segment between residues 55 and 82 that differs by 9 base changes, suggesting that one of the two might well be a recombinant.

With around 200 consecutive rounds of replication per year (Pelletier *et al.*, 1995; Perelson *et al.*, 1996), most of which involve multiply infected cells as shown here, then by fifteen years some descendants of the initial genome should have experienced up to 9000 crossovers (200 rounds × 3 crossovers/round × 15 years). Assuming that there are few recombination hot spots, then a 9200 nucleotide genome should be marked by approximately as many recombination sites as nucleotides in the genome. The effects of such rampant recombination on the analyses of HIV variation needs to be urgently addressed. For example, simulations of sequence evolution allowing recombination have indicated that branch lengths appear to be overestimated (Schierup and Hein, 2000).

It is here that the story crosses that of CTL escape. With the majority of infected cells harbouring ≥2 genetically different proviruses, the number of distinct antigenic epitopes that are presented by a single cell is increased.

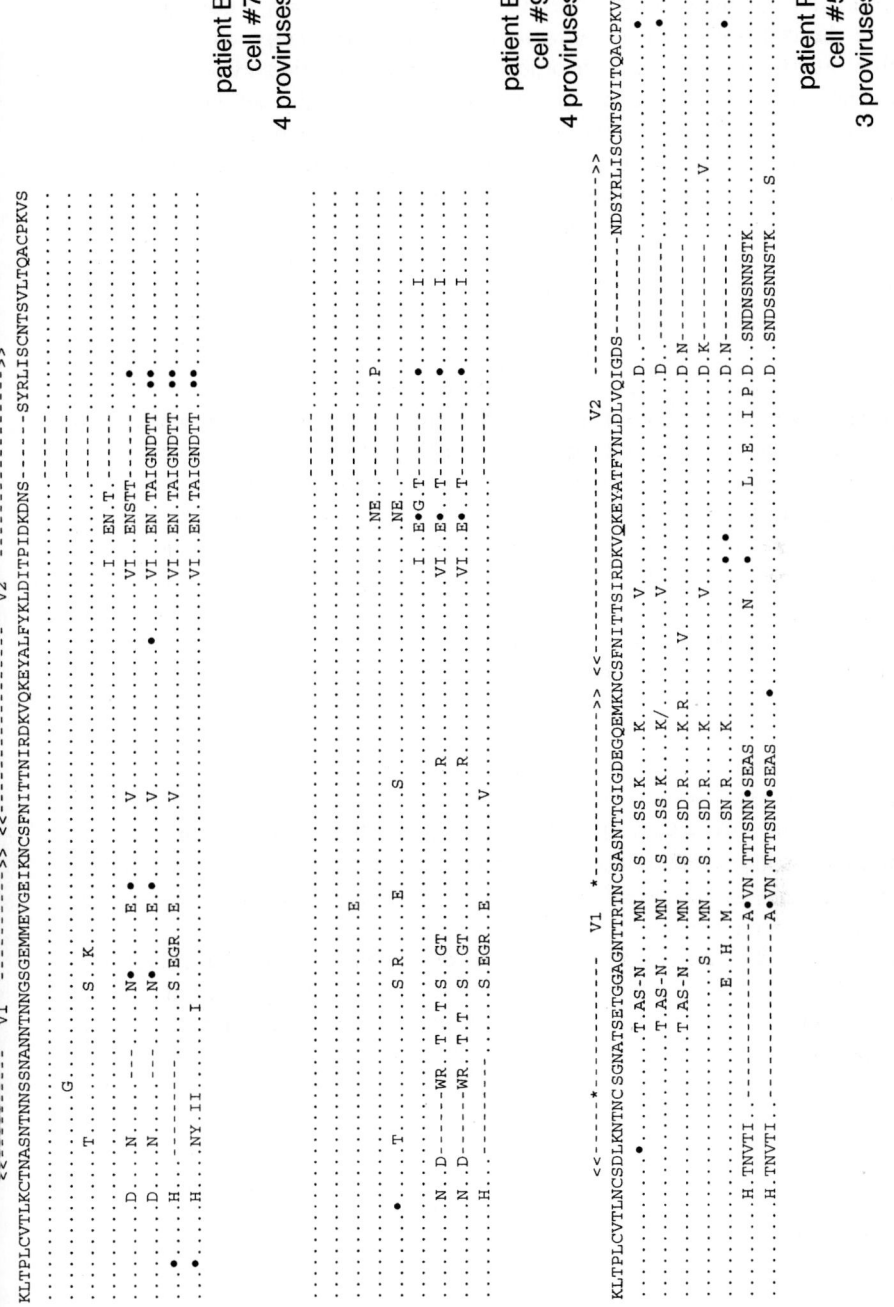

Figure 2.4 Sequence diversity within single infected CD4 T cells. The segment corresponds to the hypervariable V1V2 region of the gp120 surface envelope. Only differences are scored with respect to an arbitrarily defined reference sequence. Dashes indicate gaps, solid circles synonymous substitutions, slash (/) single base frame shifts. On the left are the sequence codes. The prefixes B7, R5 etc. refer to the individual cell; the proviral copy number is given as p/c. The asterisks above two cysteine (C) residues for R5–5 indicate the additional amino acid pair, which is exceptional.

Furthermore, a mutation in an epitope encoded by one provirus would still leave the cell vulnerable to recognition of the same epitope encoded by the other proviruses. Therefore, in order to escape recognition by a cytotoxic T lymphocyte clone, mutations in the epitope encoded by all proviruses are necessary, which is improbable given a mutation rate of only 0.25/base/cycle (Mansky and Temin, 1995). The tremendous intracellular genetic diversity means that there will be more epitopes per cell than if the proviruses were homogeneous, again stacking the cards against CTL escape. While recombination would allow HIV to recover from deleterious mutations and the effects of relentless bottlenecking inherent to the chronic phase of HIV infection (Wain-Hobson, 1993a,b; Ho *et al.*, 1995), the price to pay for multi-infection may be a broadening of immune recognition.

How are multiply infected cells generated? Given that the frequency of HIV-1 infected CD4+ splenocytes is of the order of 1% (McIlroy *et al.*, 1995; Gratton *et al.*, 2000), the simplest explanation is that they result from cell-to-cell spread involving numerous virions (Phillips and Bourinbaiar, 1992; Fais *et al.*, 1995). In the rarer event of an uninfected lymphoblast having two infected neighbours harbouring viruses from different lineages, for example the two distinct strains noted for patient R (Figure 2.3), subtypes or even groups (i.e. M and O), mixed infections with highly divergent HIV variants will arise (Carr *et al.*, 1998; Peeters *et al.*, 1999; Takehisa *et al.*, 1999; Hoelscher *et al.*, 2001).

Only as a provirus in a latently infected CD4 T cell, i.e. in a non-productive form, does an HIV genome have a long half-life. Otherwise, an HIV genome is an ephemeral entity which 'exists' for a few hours from transcription to reverse transcription, for during the latter step it has recombined around 3 times with another genome. Coupled with the rapid dynamics of virus production and destruction the description and analysis of such an ethereal situation will tax our imagination and computer power to the hilt.

IF NOT BY MUTATION, HOW DOES HIV PERSIST?

There is no proven answer to this question, so the following is more in the realm of deduction. HIV infection is an endless guerrilla war between the virus and the immune system. HIV can circulate unimpeded in the form of latently infected T cells. If the cell encounters antigen and is activated, so is the HIV provirus. What is crucial here is the relative rate of virus transcription and assembly to the rate of infiltration of anti-HIV specific immune responses to the infectious centres (Blancou *et al.*, 2001). In short, it is like a highly localized secondary virus infection. Perhaps the immune system may succeed in clearing the majority of infectious centres. If the virus is able to achieve productive infection in just a few centres, then with the massive burst size typical of most viral infections, the losses can be recovered. Persistence would seem to come from the fact that HIV can get into, and out of, resting memory T cells.

IN VIVO MOST VIRAL VARIATION IS BASICALLY NOISE

A virus cannot have emerged without having been positively selected by Darwinian evolution. However, once a success story, all a virus has to do is to maintain fitness. As HIV is indisputably a success story from its point of view, is all of HIV's genetic variation noise? The question is better rephrased as what fraction of HIV genetic variation is exploited by the virus in a natural setting? Certainly HIV can rapidly adapt to a new situation and develop resistance to a drug, but this is an unnatural situation. When taking into account the phylogenic relatedness of sequences it has been very difficult to show that positive selection (Plikat *et al.*, 1997; Kils-Hütten *et al.*, 2001). If phylogeny is not respected, then it is possible to identify positively selected residues. Furthermore, the effect of recombination on all present methods designed to ascertain positive selection pressures, including codon-based analyses, has not been explored.

Increasingly it is being appreciated that a minority of substitutions are the 'stuff of evolution' (Haydon *et al.*, 1998; Sala and Wain-Hobson, 2000). Probably a minority is something of the order of a <1%. Even so, given the enormity of sequence space, viable solutions are not rare. Evidence for this comes from the molecular clock-like behaviour in the accumulation of amino acid substitutions over time (Hayashida *et al.*, 1985; Gojobori *et al.*, 1990; Querat *et al.*, 1990; Villaverde *et al.*, 1991; Easteal, 1992; Elena *et al.*, 1992; Sanchez *et al.*, 1992; McGeoch *et al.*, 1995; Yang *et al.*, 1995; Leitner *et al.*, 1997; Plikat *et al.*, 1997), purifying selection being ever operative. It comes as a surprise that most virus variation is of little adaptive consequence. Inevitably the why question comes up. Perhaps it is not for us to answer this question; suffice it to note that if viruses were sitting at the top of a peak in the adaptive fitness landscape then the only way would be down. We have seen just how hostile the host environment is. Furthermore, population sizes are small compared to sequence space, meaning that razor edge survival cannot depend on access to vast numbers of mutants. Only if the fitness landscape open to a virus is extensive and highly interconnected can a virus survive.

In short, most of the exuberant genetic variation characteristic of HIV is noise. The signal must be there and can be seen as soon as drugs are used sub-optimally. It is just that the signal-to-noise ratio is low. A bit like the Internet.

REFERENCES

Blancou, P., Chenciner, N., Cumont, M. C., Wain-Hobson, S., Hurtrel, B. and Cheynier, R. (2001) *Proc. Natl Acad. Sci USA* 98, 13237–13242.

Borrow, P. and Shaw, G. M. (1998) *Immunol. Rev.* 164, 37–51.

Brichacek, B., Swindells, S., Janoff, E. N., Pirruccello, S. and Stevenson, M. (1996) *J. Infect. Dis.* 174, 1191–1199.

Carr, J. K., Salminen, M. O., Albert, J., Sanders-Buell, E., Gotte, D., Birx, D. L. and McCutchan, F. E. (1998) *Virology* 247, 22–31.

Clavel, F., Guetard, D., Brun-Vezinet, F., Chamaret, S., Rey, M. A., Santos-Ferreira, M. O., Laurent, A. G., Dauguet, C., Katlama, C., Rouzioux, C. *et al.* (1986) *Science* 233, 343–346.

Coffin, J. M. (1979) *J. Gen. Virol.* 42, 1–26.

Coffin, J. M., Hughes, S. H. and Varmus, H. E. (1997) *Retroviruses*. Cold Spring Harbor Laboratory Press.

Drake, J. W. and Holland, J. J. (1999) *Proc. Natl Acad. Sci. USA* 96, 13910–13913.

Easteal, S. (1992) *BioEssays* 14, 415–419.

Elena, S. F., Gonzalez-Candelas, F. and Moya, A. (1992) *J. Mol. Evol.* 35, 223–229.

Fais, S., Capobianchi, M. R., Abbate, I., Castilletti, C., Gentile, M., Cordiali Fei, P., Ameglio, F. and Dianzani, F. (1995) *AIDS* 9, 329–335.

Gao, F., Bailes, E., Robertson, D. L., Chen, Y., Rodenburg, C. M., Michael, S. F., Cummins, L. B., Arthur, L. O., Peeters, M., Shaw, G. M., Sharp, P. M. and Hahn, B. H. (1999) *Nature* 397, 436–441.

Gojobori, T., Moriyama, E. N. and Kimura, M. (1990) *Proc. Natl Acad. Sci. USA* 87, 10015–10018.

Gratton, S., Cheynier, R., Dumaurier, M. J., Oksenhendler, E. and Wain-Hobson, S. (2000) *Proc. Natl Acad. Sci. USA* 97, 14566–14571.

Hahn, B. H., Shaw, G. M., De Cock, K. M. and Sharp, P. M. (2000) *Science* 287, 607–614.

Hayashida, H., Toh, H., Kikuno, R. and Miyata, T. (1985) *Mol. Biol. Evol.* 2, 289–303.

Haydon, D., Lea, S., Fry, L., Knowles, N., Samuel, A. R., Stuart, D. and Woodhouse, M. E. (1998) *J. Mol. Evol.* 46, 465–474.

Ho, D. D. (1992) *Lancet* 339, 1549.

Ho, D. D., Neumann, A. U., Perelson, A. S., Chen, W., Leonard, J. M. and Markowitz, M. (1995) *Nature* 373, 123–126.

Hoelscher, M., Kim, B., Maboko, L., Mhalu, F., von Sonnenburg, F., Birx, D. L. and McCutchan, F. E. (2001) *AIDS* 15, 1461–1470.

Huet, T., Cheynier, R., Meyerhans, A., Roelants, G. and Wain-Hobson, S. (1990) *Nature* 345, 356–359.

Huet, T., Dazza, M. C., Brun-Vezinet, F., Roelants, G. E. and Wain-Hobson, S. (1989) *AIDS* 3, 707–715.

Jetzt, A. E., Yu, H., Klarmann, G. J., Ron, Y., Preston, B. D. and Dougherty, J. P. (2000) *J. Virol.* 74, 1234–1240.

Jung, A., Maier, R., Vartanian, J. P., Bocharov, G., Jung, V., Fischer, U., Meese, E., Wain-Hobson, S. and Meyerhans, A. (2002) *Nature* 418, 144.

Kawai, S. and Hanafusa, H. (1972) *Virology* 49, 37–44.

Kils-Hütten, L., Cheynier, R., Wain-Hobson, S. and Meyerhans, A. (2001) *J. Gen. Virol.* 82, 1621–1627.

Korber, B., Muldoon, M., Theiler, J., Gao, F., Gupta, R., Lapedes, A., Hahn, B. H., Wolinsky, S. and Bhattacharya, T. (2000) *Science* 288, 1789–1796.

Leitner, T., Kumar, S. and Albert, J. (1997) *J. Virol.* 71, 4761–4770.

Li, W. H., Tanimura, M. and Sharp, P. M. (1988) *Mol. Biol. Evol.* 5, 313–330.

Mansky, L. M. and Temin, H. M. (1995) *J. Virol.* 69, 5087–5094.

McCutchan, F. E. (2000) *AIDS* 14, S31–S44.

McGeoch, D. J., Cook, S., Dolan, A., Jamieson, F. E. and Telford, E. A. (1995) *J. Mol. Biol.* 247, 443–458.

McIlroy, D., Autran, B., Cheynier, R., Wain-Hobson, S., Clauvel, J. P., Oksenhendler, E., Debre, P. and Hosmalin, A. (1995) *J. Virol.* 69, 4737–4745.

McMichael, A. (1998) *Cell* 93, 673–676.

Meyerhans, A., Dadaglio, G., Vartanian, J. P., Langlade-Demoyen, P., Frank, R., Asjo, B., Plata, F. and Wain-Hobson, S. (1991) *Eur. J. Immunol.* 21, 2637–2640.

Moore, J. P., Cao, Y., Leu, J., Qin, L., Korber, B. and Ho, D. D. (1996) *J. Virol.* 70, 427–444.

Nietfield, W., Bauer, M., Fevrier, M., Maier, R., Holzwarth, B., Frank, R., Maier, B., Riviere, Y. and Meyerhans, A. (1995) *J. Immunol.* 154, 2189–2197.

O'Connor, D. H., Allen, T. M., Vogel, T. U., Jing, P., DeSouza, I. P., Dodds, E., Dunphy, E. J., Melsaether, C., Mothe, B., Yamamoto, H., Horton, H., Wilson, N., Hughes, A. L. and Watkins, D. I. (2002) *Natl Med.* 8, 493–499.

Peeters, M., Honore, C., Huet, T., Bedjabaga, L., Ossari, S., Bussi, P., Cooper, R. W. and Delaporte, E. (1989) *AIDS* 3, 625–630.

Peeters, M., Liegeois, F., Torimiro, N., Bourgeois, A., Mpoudi, E., Vergne, L., Saman, E., Delaporte, E. and Saragosti, S. (1999) *J. Virol.* 73, 7368–7375.

Pelletier, E., Saurin, W., Cheynier, R., Letvin, N. L. and Wain-Hobson, S. (1995) *Virology* 208, 644–652.

Perelson, A. S., Neumann, A. U., Markowitz, M., Leonard, J. M. and Ho, D. D. (1996) *Science* 271, 1582–1586.

Phillips, D. M. and Bourinbaiar, A. S. (1992) *Virology* 186, 261–273.

Phillips, R. E., Rowland-Jones, S., Nixon, D. F., Gotch, F. M., Edwards, J. P., Ogunlesi, A. O., Elvin, J. G., Rothbard, J. A., Bangham, C. R. M., Rizza, C. R. and McMichael, A. J. (1991) *Nature* 354, 453–459.

Plikat, U., Nieselt-Struwe, K. and Meyerhans, A. (1997) *J. Virol.* 71, 4233–4240.

Ploegh, H. L. (1998) *Science* 280, 248–253.

Price, D. A., Goulder, P. J., Klenerman, P., Sewell, A. K., Easterbrook, P. J., Troop, M., Bangham, C. R. and Phillips, R. E. (1997) *Proc. Natl Acad. Sci. USA* 94, 1890–1895.

Querat, G., Audoly, G., Sonigo, P. and Vigne, R. (1990) *Virology* 175, 434–447.

Renoux-Elbé, C., Cheynier, R. and Wain-Hobson, S. (2002) *J. Mol. Evol.* 54, 376–385.

Robertson, D. L., Sharp, P. M., McCutchan, F. E. and Hahn, B. H. (1995) *Nature* 374, 124–126.

Sala, M. and Wain-Hobson, S. (2000) *J. Mol. Evol.* 51, 12–20.

Sanchez, C. M., Gebauer, F., Sune, C., Mendez, A., Dopazo, J. and Enjuanes, L. (1992) *Virology* 190, 92–105.

Schierup, M. H. and Hein, J. (2000) *Genetics* 156, 879–891.

Schmitz, J. E., Kuroda, M. J., Santra, S., Sasseville, V. G., Simon, M. A., Lifton, M. A., Racz, P., Tenner-Racz, K., Dalesandro, M., Seallon, B. J., Ghrayeb, J., Forman, M. A., Montefiori, D. C., Rieber, E. P., Letvin, N. L. and Reimann, K. A. (1999) *Science* 283, 857–860.

Schnittman, S. M., Lane, H. C., Greenhouse, J., Justement, J. S., Baseler, M. and Fauci, A. S. (1990) *Proc. Natl Acad. Sci. USA* 87, 6058–6062.

Schwartz, O., Marechal, V., Friguet, B., Arenzana-Seisdedos, F. and Heard, J.-M. (1998) *J. Virol.* 72, 3845–3850.

Simon, F., Mauclere, P., Roques, P., Loussert-Ajaka, I., Muller-Trutwin, M. C., Saragosti, S., Georges-Courbot, M. C., Barre-Sinoussi, F. and Brun-Vezinet, F. (1998) *Natl Med.* 4, 1032–1037.

Staprans, S. I., Hamilton, B. L., Follansbee, S. E., Elbeik, T., Barbosa, P., Grant, R. M. and Feinberg, M. B. (1995) *J. Exp. Med.* 182, 1727–1737.

Takehisa, J., Zekeng, L., Ido, E., Yamaguchi-Kabata, Y., Mboudjeka, I., Harada, Y., Miura, T., Kaptu, L. and Hayami, M. (1999) *J. Virol.* 73, 6810–6820.

Villaverde, A., Martinez, M. A., Sobrino, F., Dopazo, J., Moya, A. and Domingo, E. (1991) *Gene* 103, 147–153.

Vogt, P. K. (1971) *Virology* 46, 947–952.

Wain-Hobson, S. (1993a) *Curr. Opin. Genet. Devel.* 3, 878–883.

Wain-Hobson, S. (1993b) *Nature* 366, 22.

Wei, X., Ghosh, S. K., Taylor, M. E., Johnson, V. A., Emini, E. A., Deutsch, P., Lifson, J. D., Bonhoeffer, S., Nowak, M. A., Hahn, B. H., Saag, M. S. and Shaw, G. M. (1995) *Nature* 373, 117–122.

Weiss, R. A. (2001) *Nature* 410, 963–967.

Weiss, R. A., Mason, W. S. and Vogt, P. K. (1973) *Virology* 52, 535–552.

Wooley, D. P., Smith, R. A., Czajak, S. and Desrosiers, R. C. (1997) *J. Virol.* 71, 9650–9653.

Yang, Z., Lauder, I. J. and Lin, H. J. (1995) *J. Mol. Evol.* 41, 587–596.

Yu, H., Jetzt, A. E., Ron, Y., Preston, B. D. and Dougherty, J. P. (1998) *J. Biol. Chem.* 273, 167–174.

3

CALICIVIRUS

ALAN D. RADFORD, SUSAN DAWSON AND ROSALIND M. GASKELL

INTRODUCTION

The *Caliciviridae* is a family of small, non-enveloped, icosahedral viruses, with a positive sense, single-stranded RNA genome of approximately 7.5 kilobases. The family name (*calyx* is Latin for cup or chalice) comes from the appearance of the majority of caliciviruses by electronmicroscopy, where the capsid surface appears to be covered with cup-like depressions.

The *Caliciviridae* contains viruses responsible for a wide range of human and animal diseases. The subclassification of these viruses has proved difficult, both because of their inherent diversity and also because of initial difficulties in characterizing a number of viruses that do not grow in cell culture. However, there is now sequence data available for many caliciviruses and this has enabled the classification of the family into four genera (Figure 3.1) (Green *et al.*, 2000b).

The vesivirus genus contains those viruses for which vesicle formation tends to be a prominent feature of the pathology and includes vesicular exanthema of swine virus, San Miguel Sea Lion virus (SMSV), and feline and canine caliciviruses.

The lagovirus genus currently only contains viruses found in lagomorphs – rabbit haemorrhagic disease virus (RHDV) and European brown hare syndrome virus (EBHSV) – hence its name.

The norovirus (previously 'Norwalk-like' viruses) and sapovirus (previously 'Sapporo-like' viruses) genera contain predominantly gastroenteric pathogens of humans and are named according to the origins of the type species of virus in each genus. These two genera may also be broadly defined based on capsid morphology as observed by electronmicroscopy (EM). The sapoviruses have classic calicivirus morphology and have also been termed classic human caliciviruses. In contrast, the noroviruses tend to appear smooth and rounded by EM, giving rise to the

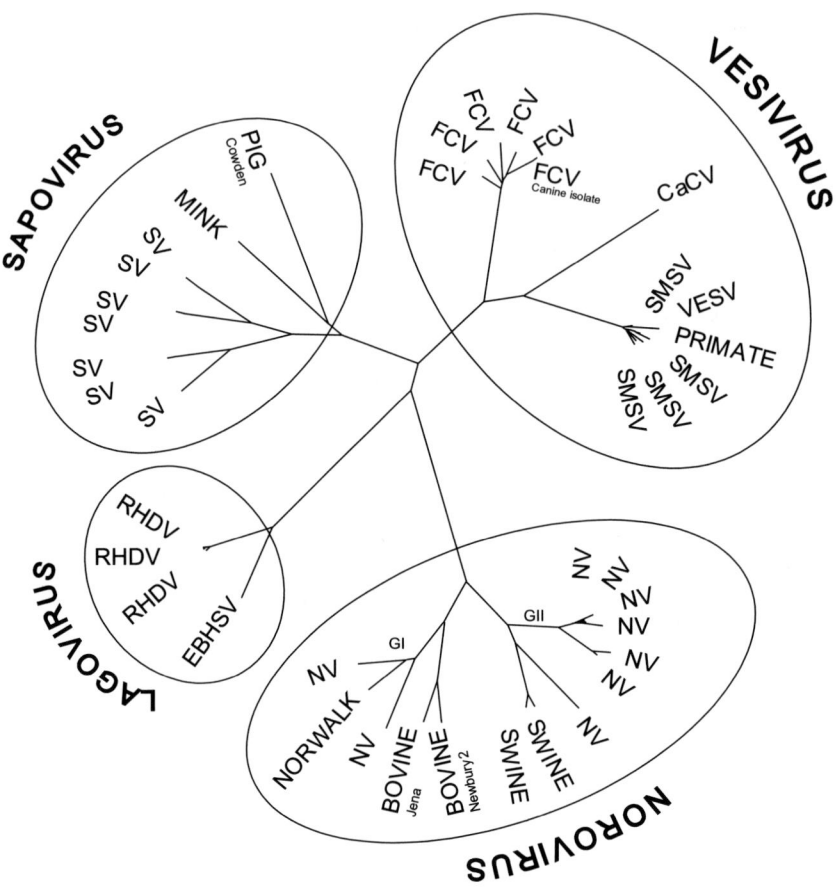

Figure 3.1 The *Caliciviridae*. The *Caliciviridae* can be divided into four genera (Green *et al.*, 2000b). Unrooted and uncorrected distance analysis of viruses in the *Caliciviridae* based on partial sequences from the polymerase region of ORF1. The norovirus genus contains primarily enteric pathogens of man (NV), although more recently related viruses have also been found in cattle and pigs. The human noroviruses have been further subdivided into genogroups I and II. The sapovirus genus contains primarily enteric pathogens of man (SV). As with the noroviruses, related viruses have also been isolated from non-humans, including pigs. The lagovirus genus contains only viruses of rabbits and hares (rabbit haemorrhagic disease virus (RHDV) and European brown hare syndrome virus (EBHSV)). The vesivirus genus contains San Miguel sea lion virus (SMSV) and SMSV-like viruses (includes vesicular exanthema of swine virus, (VESV)), canine calicivirus (CaCV) and the feline calicivirus (FCV) group. The FCVs form a closely related single cluster, particularly when compared to the human noroviruses and sapoviruses.

alternative name of small round structured viruses (SRSVs). Until recently, the only viruses belonging to the norovirus and sapovirus genera were those isolated from humans. However, sequence analysis of enteric viruses isolated from pigs and cattle has suggested that these viruses are more closely related to the human pathogens than to other animal caliciviruses in the vesisvirus and lagovirus genera

(Sugieda *et al.*, 1998; Dastjerdi *et al.*, 1999; Guo *et al.*, 1999; Liu *et al.*, 1999; van Der Poel *et al.*, 2000). To date, no direct zoonotic potential for these viruses has been identified.

All calicivirus genomes encode the non-structural proteins at the 5' end of the genome and the major capsid protein towards the 3' end (Clarke and Lambden, 1997). A separate open reading frame (ORF) encodes a minor structural protein at the 3' end of the genome (Herbert *et al.*, 1996; Glass *et al.*, 2000; Sosnovtsev and Green, 2000). Members of the *Caliciviridae* use one of two different strategies for expressing the non-structural proteins and the major capsid protein from the genome. In the lagoviruses and sapoviruses, the non-structural proteins and major capsid protein are produced from a single large ORF. In contrast, in the vesiviruses and the noroviruses they are encoded by two separate ORFs (Clarke and Lambden, 1997) (Figure 3.2).

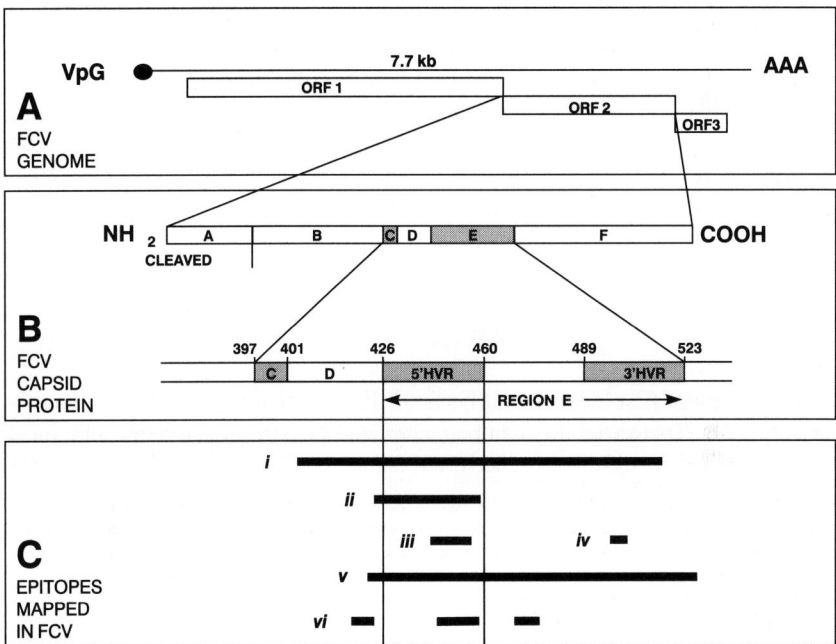

Figure 3.2 Summary of FCV genomic and antigenic structure. (A) The FCV genome contains three ORFs coding for the non-structural proteins, the major capsid protein and a minor capsid protein respectively. (B) The major capsid protein is divided into six conserved regions. Region A is cleaved to produce the mature capsid. Regions B, D and F and the central domain of region E are conserved whilst regions C and the 5' and 3'HVRs of region E are variable. (C) Antigenic regions identified in the capsid protein. (i) amino acids 408–517 (Guiver *et al.*, 1992); (ii) 422–458 (Milton *et al.*, 1992); (iii) 441–455 (Tohya *et al.*, 1997); (iv) 493–494 (Tohya *et al.*, 1997); (v) 420–529 (Neill *et al.*, 2000); (vi) 415–421, 445–451, 451–457 and 475–479 (Radford *et al.*, 1999). All amino acids are numbered according to published FCV F9 capsid sequence (Carter *et al.*, 1992a). (Reproduced with permission of the American Society for Microbiology from Radford *et al.* (1999) *Journal of Virology* 73, 8496–8502.)

In this chapter, we will focus on antigenic variation in feline calicivirus (FCV). Where appropriate, comparisons with other caliciviruses will be made. Feline calicivirus was first identified in domestic cats in 1957. It is a relatively easy virus to study in the laboratory in that, unlike some other caliciviruses, it grows readily in feline cell cultures. The pathogenesis of the disease has also largely been determined since infections can easily be established under defined experimental conditions in specific-pathogen-free cats. The natural history of the disease in its host has also been well studied, the virus being a highly successful pathogen that is widespread in the cat population.

There are a number of features of feline calicivirus that make it an interesting pathogen in which to consider virus variation in general and more specifically antigenic variation. First, FCV is an RNA virus that, as with many other RNA viruses, is likely to be prone to mutation and selection through a number of genetic mechanisms. Secondly, there is a wide spectrum of antigenic and genetic diversity within FCV, although this appears to be restricted to within specific limits. Thirdly, the virus is an exceptionally successful virus in its host, largely because of an asymptomatic carrier state that develops in a significant proportion of clinically recovered animals. The host and viral mechanisms that allow FCV to persist are not yet clear, but antigenic variation may well play a role. Finally, unlike many other RNA viruses, vaccination against FCV has been widespread for many years. This practice may have led to various immunological selection pressures operating on the virus at the population level.

CLINICAL, ANTIGENIC AND GENETIC VARIABILITY

Soon after it was first identified, it became apparent that FCV was an inherently variable virus. This variability can be recognized by clinical, antigenic and, most recently, genomic differences between isolates. Whilst the former two are undoubtedly a reflection of genetic differences, it is convenient to consider each separately.

Clinical disease

Feline calicivirus is a relatively host-specific pathogen, infecting predominantly domestic cats. There is more limited evidence to support infection of some non-domestic members of the *Felidae* (Sabine and Hyne, 1970; Hofmann-Lehmann *et al.*, 1996; Kadoi *et al.*, 1997). Viruses that are antigenically and genetically similar to FCV have also been detected in dogs (Sangabriel *et al.*, 1996; Hashimoto *et al.*, 1999), and there is some epidemiological evidence of an association for FCV between these two species (Binns *et al.*, 2000). However, dogs also have their own genetically distinct calicivirus (Roerink *et al.*, 1999a,b), and the real significance of these FCV-like isolates to either dogs or cats remains uncertain.

Infection with FCV is extremely widespread in the cat population, largely because of the existence of clinically healthy carriers (see below). The prevalence of infection ranges from 8 to 40% depending on the population of cats studied (Wardley *et al.*, 1974; Harbour *et al.*, 1991; Coutts *et al.*, 1994; Binns *et al.*, 2000). The highest infection rates are usually seen where large numbers of cats are grouped together, such as occurs in breeding colonies or rescue catteries. Most primary infections in colonies are in young kittens that become infected as their maternally derived (colostral) antibody declines between 10 and 14 weeks of age (Johnson and Povey, 1983). Infections in older animals may occur, particularly if they enter high-risk situations such as boarding catteries. Apart from the potential, but still unproven, transmission of FCV-like viruses between dogs and cats, there are no known reservoir or alternative hosts for FCV, and *in utero* transmission does not generally seem to occur.

Most isolates of FCV induce a fairly typical syndrome, predominantly characterized by ulceration of the tongue (Gaskell and Dawson, 1994, 1998). Early signs of disease include pyrexia and depression, and mild upper respiratory and conjunctival signs are also generally seen. Oral ulcers begin as vesicles, which subsequently rupture, with necrosis of the overlying epithelium and infiltration of inflammatory cells at the periphery and base. They generally heal over a period of 2–3 weeks. Ulceration on the lips or nose or on other parts of the body may be seen but occurs more rarely.

Whilst most FCV isolates are able to induce some degree of oral ulceration and upper respiratory disease, there appear to be considerable differences between the tissue tropism and pathogenicity of some strains of FCV. Thus, some isolates have a predilection for the lung, causing a primary interstitial pneumonia (Hoover and Kahn, 1975). However, others have been found within macrophages in the synovial membrane of joints, associated with a lameness and pyrexia syndrome (Pedersen *et al.*, 1983; Dawson *et al.*, 1994). This has led to the suggestion that FCV is able to cause a spectrum of disease affecting both the oral and respiratory tract and the joints. In this model of the pathogenesis, individual FCV strains with these tropisms would be capable of affecting both areas but to differing degrees (TerWee *et al.*, 1997).

Feline calicivirus variability is also associated with a range of virulence from non-pathogenic to more severe disease, which may be fatal in young kittens. Recently, an apparently isolated outbreak of severe haemorrhagic fever has been described, caused by a highly virulent strain of FCV (Pedersen *et al.*, 2000). Whilst this is a potentially worrying development in FCV pathogenesis, such severe disease currently appears to be very rare. However, these cases highlight the potential for the development of new patterns of virulence following infection with this virus and the need for continual vigilance amongst veterinarians. A similar situation seems to have existed in rabbits, with the emergence in recent years of the highly virulent RHDV. The virus appeared to originate in China, from where it spread worldwide, but there is evidence that non-pathogenic RHD-like viruses existed in rabbit populations before exposure to RHDV (Capucci *et al.*, 1996; Chasey, 1997; Nagesha *et al.*, 2000).

Antigenic variation

Because FCV can be cultivated *in vitro*, antigenic relationships between isolates have been extensively examined in cell culture using virus neutralization tests. Such studies have shown that although demonstrable antigenic variation exists between viruses, significant cross-reactivity also occurs to a greater or lesser extent. Thus all FCV isolates appear to be serological variants of a single diverse serotype (Povey, 1974; Kalunda *et al.*, 1975; Burki *et al.*, 1976). Varying degrees of cross protection have also been shown *in vivo* between strains of FCV, and this has led to the selection of certain cross-reactive strains (e.g. F9) for use in vaccines (Bittle and Rubic, 1976; Scott, 1977; Povey *et al.*, 1980; Gaskell *et al.*, 1982; Pedersen and Hawkins, 1995).

Although distinct serotypes of FCV cannot be identified, many strains can still be differentiated serologically and several studies have sought to correlate antigenic differences using both polyclonal and monoclonal antibodies, with observed differences between strains (Knowles *et al.*, 1990; Dawson *et al.*, 1993; McArdle *et al.*, 1996; Geissler *et al.*, 1997; Lauritzen *et al.*, 1997; Poulet *et al.*, 2000). In some of these studies, possible weak correlations have been observed with clinical disease or geographic relatedness. However, in general, no clear-cut associations have been found between particular antigenic variants and the geographic, temporal or clinical origins of the strains.

The extent of antigenic variation amongst other caliciviruses, such as the human and rabbit caliciviruses, has been more difficult to determine given the inability to culture these viruses *in vitro*. As a result, antigenic comparison of isolates has required the use of a variety of techniques including immune electron microscopy (EM), solid phase immune EM and enzyme immunoassays. In essence, human noroviruses appear to be antigenically much more diverse than feline caliciviruses, though without access to *in vitro* neutralization tests, actual serotypes are difficult to define. Several systems describing antigenic relationships between human noroviruses have been developed but the most commonly used scheme is that based on UK antigenic types 1–4, which has been extended to contain at least nine types (Cubitt *et al.*, 1987; Carter and Cubitt, 1995; Lewis *et al.*, 1995). Discrepancies between antigenic and genetic typing for noroviruses, and differences in antigenic typing between laboratories may be due in part to the quality of reagents, which initially comprised mainly human convalescent sera. More recently, the specificity of assays has improved with the use of polyclonal and monoclonal sera raised from recombinant capsid antigens (Atmar and Estes, 2001). Increasingly, however, genetic typing methods using PCR are being used to distinguish between isolates for epidemiological and other purposes.

In contrast to the widely divergent human caliciviruses, RHDV isolates appear to be closely related serologically and do not seem to form distinct serotypes, although some antigenic variants have been reported (Berninger and House, 1995; Chasey, 1997; Schirrmeier *et al.*, 1999). In the absence of *in vitro* neutralization tests, however, precise relationships are difficult to determine. Other RHDV strains with differing biological properties, including a non-virulent strain, and one with different haemagglutination characteristics have also been identified (Capucci *et al.*, 1996, 1998).

Genetic variation

Phylogenetic analyses of feline caliciviruses show that their genetic relationships mirror the serological findings. Thus FCVs belong to a single but diverse genogroup, with no evidence for the existence of distinct subgroups or genotypes (Geissler *et al.*, 1997; Glenn *et al.*, 1999; Horimoto *et al.*, 2001) (Figure 3.3).

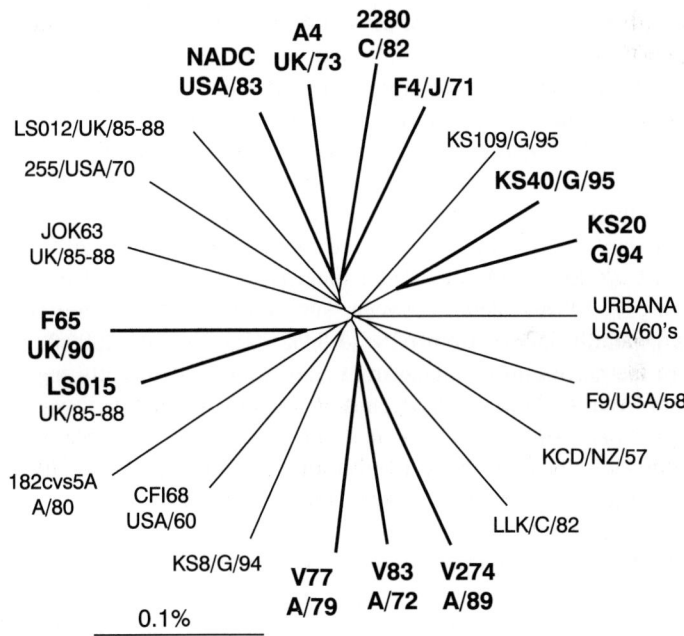

Figure 3.3 Unrooted phylogenetic tree of FCV isolates showing limited evidence of genetic clustering of individual viruses. Instead, each virus is approximately equidistant from the others, producing a marked radial or 'star-like' pattern. The tree was constructed by distance analysis of nucleotide sequences derived from conserved regions of the capsid gene. Each isolate is named according to its local identifier, followed by the country of origin and the year of isolation (A = Australia, C = Canada, G = Germany, J = Japan, UK = United Kingdom, USA = United States of America). For clarity, nodes supported by bootstrap values >70 are marked in bold. Evolutionary distances are to scale. (Reproduced by permission of Elsevier from Glenn *et al.* (1999) *Veterinary Microbiology* 67, 175–193.)

Such findings are largely based on capsid gene analysis, although limited data on ORF 1 and ORF 3 also confirm these observations (Glenn, 1997; Sommerville, 2001). Based on whole capsid sequences, nucleotide and amino acid similarities between epidemiologically unrelated FCV isolates ranged from 75–82% and 81–91% respectively (Glenn, 1997).

The phylogenetic analysis of these sequences shows that FCVs isolated over a number of years group in a 'star-like' cluster, regardless of their geographic or clinical origins (Figure 3.3). All isolates appear to be relatively equidistant from one another, with the genetic diversity restricted to within specific limits. This would suggest that all FCV isolates examined to date, regardless of their time or place of origin, are of equal ecological fitness. Such an observation is consistent with rapid geographical movement of the virus through a globally mobile host population. The rate of this virus translocation would need to be faster than any adaptive evolution to a given geographical location, and may be facilitated by clinically normal carriers, which are responsible for maintaining a high prevalence of infection (see below). However, it is also possible that FCV has not been in the cat population long enough for distinct clades to evolve.

In contrast to feline calicivirus, the considerable antigenic diversity seen in some other caliciviruses is also reflected in their genetic relationships. For example, in the human noroviruses two distinct genogroups have been identified and within these a number of different genomic subgroups (genotypes) (Figure 3.1) (Ando *et al.*, 2000; Green *et al.*, 2000a; Vinje *et al.*, 2000b). Based on complete capsid sequence, amino acid identities between strains of the two distinct human norovirus genogroups ranged from 37% to 44%, whereas within genogroups they ranged from approximately 60% to 100% (Green *et al.*, 2000a). Within genomic subgroups, amino acid identities ranged from 88% to 100%, while between subgroups of the same genogroup, identities generally ranged from 62% to 83% (Green *et al.*, 2000a). Sequence analysis of the noroviruses isolated from cattle and pigs currently suggests they are distinct from the human genogroups (Sugieda *et al.*, 1998; Dastjerdi *et al.*, 1999; Liu *et al.*, 1999; van Der Poel *et al.*, 2000). Considerable genetic and antigenic diversity has also been identified for the human (Matson *et al.*, 1995; Jiang *et al.*, 1997; Vinje *et al.*, 2000a) and animal (Guo *et al.*, 1999, 2001) sapoviruses. Genetic diversity for both RHDV and EBHSV appears to be more limited. However, the non-pathogenic variant of RHDV may be genetically distinguished from the pathogenic forms of the virus (Capucci *et al.*, 1996). There is also some evidence for the existence of genogroups among the pathogenic forms of the virus (Nowotny *et al.*, 1997; Le Gall *et al.*, 1998; Asgari *et al.*, 1999; Schirrmeier *et al.*, 1999).

In summary, many caliciviruses, including feline calicivirus, show considerable antigenic and genetic diversity. However, in FCV, such variability appears to be restricted to within specific limits. Thus, in contrast to human caliciviruses, there is no support for the subdivision of feline caliciviruses into separate serotypes or genotypes.

ANTIGENIC STRUCTURE OF THE CAPSID

Most of the genetic variability between FCV isolates is found in the major capsid gene. Early comparison of capsid gene sequences from different strains of SMSV and FCV allowed the capsid to be divided into six regions, largely on the basis of sequence conservation (Neill, 1992; Seal et al., 1993) (Figure 3.2). In FCV, region A is cleaved by a viral protease in ORF1 to produce the mature capsid protein (Carter et al., 1992b; Sosnovtsev et al., 1998). Regions B, D and F are relatively conserved between isolates whereas regions C and E are variable. As more sequence data has become available, region E has been further divided into a 5' and 3' hypervariable (HVR) region separated by a central, highly conserved domain.

Prior to any epitope-mapping studies, the observed antigenic differences between strains led to the assumption that a number of neutralizing epitopes would be located in variable regions C and E of the capsid. This speculation is now supported by experimental data, as summarized in Figure 3.2, in which the vast majority of identified B cell epitopes map to variable regions of the capsid. Briefly, a recombinant peptide corresponding to amino acids (aa) 408–517 of FCV strain F9 induced the formation of neutralizing polyclonal antisera in rabbits, and cats vaccinated with F9 produced a polyclonal antisera that reacted with this peptide (Guiver et al., 1992). The peptide corresponding to aa 422–458 contained the epitopes for two neutralizing mouse monoclonal antibodies (mAbs) (Milton et al., 1992). Amino acids 441–445 and 493–494 were shown to contain four linear and three conformational epitopes respectively, using FCV escape mutants to identify those sites critical for epitope recognition by mAbs (Tohya et al., 1997). Using similar methods, aa 449 was shown to be crucial to the formation of a further linear neutralizing epitope (Radford et al., 1999). Two linear epitopes identified by antibodies in feline antisera have been mapped using an overlapping peptide library to amino acids 445–451 (antigenic site (ags) 2) and 451–457 (ags3) (Radford et al., 1999). Finally, aa 420–529, when exchanged between two FCV isolates, convey neutralization characteristics of the donor virus to the recipient (Neill et al., 2000).

The location of these epitopes within variable regions of the capsid may explain the spectrum of antigenicities associated with different FCV isolates and goes some way to explaining the lack of serotypes in FCV. However, the evidence of some degree of antigenic cross-reactivity between most virus strains suggests the possibility of epitopes in conserved regions of the capsid. Whilst there is no definitive evidence for the presence of such conserved epitopes, there is now considerable experimental data to support their existence. In one study, a monoclonal antibody that targeted a conformational epitope was characterized. This antibody was also able to neutralize all FCV isolates tested, suggesting that its epitope was conserved. In contrast to all other antibodies in the study, attempts to map this conserved epitope by producing neutralization-resistant escape mutants in cell culture were unsuccessful (Tohya et al., 1991, 1997). This supports the requirement of the sequence that makes up this epitope to be absolutely conserved.

In a separate study, linear epitopes that were recognized by antibodies in feline polyclonal antisera were mapped to conserved region D (ags1) and in the central conserved region of E (ags4) (Radford et al., 1999). Whilst ags1 was only recognized by a minority of antisera, the epitope of ags4 was recognized by antisera to all of the viruses tested (antisera from 11 of 12 cats infected with one of six different strains of virus tested positive for the presence of ags4-specific antibodies). However, it was not possible in this study to determine whether or not these two conserved epitopes induced neutralizing antibodies.

Although immunity to FCV is largely considered to be through virus-neutralizing antibody, protection from clinical disease has been seen with no, or only low levels of antibody. This has led to the suggestion that cell-mediated and local immunity may also play a role in clinical protection (Tham and Studdert, 1987; Knowles et al., 1991). However, the location of T cell epitopes within the virus has not been determined.

In summary, the majority of mapped neutralizing epitopes are located within the HVRs of the FCV capsid. Broadly speaking, linear B cell epitopes have been found in the 5'HVR whilst conformational epitopes are found in the 3'HVR. Whilst there is some evidence to support the existence of conserved neutralizing epitopes, these are yet to be definitively mapped.

VIRUS EVOLUTION

Variable antigenic capsid domains in viruses with RNA genomes are often indicators of an underlying capacity for rapid rates of antigenic evolution. Such a capacity has knock-on effects in all areas of a virus interaction with its host. In the next section, we will review the evidence for FCV antigenic and genetic evolution and discuss the implications of virus evolution for an individual host and for the host population.

The carrier state and evolution in individuals

After recovery from the acute phase of clinical disease, most cats continue to shed FCV from the oropharynx. When shedding is still detectable 30 days following infection, such animals are defined as carriers (Figure 3.4). In some experimental studies, it has been shown that most cats become carriers. Subsequently, there is an exponential decline in the proportion of animals that remain infected, with approximately 50% of the cats still shedding the virus 75 days after infection (Wardley and Povey, 1977). Whilst this most likely represents an over-simplification of the true dynamics of the FCV carrier state, it is a useful guide. It follows that although individual FCV carriers may shed

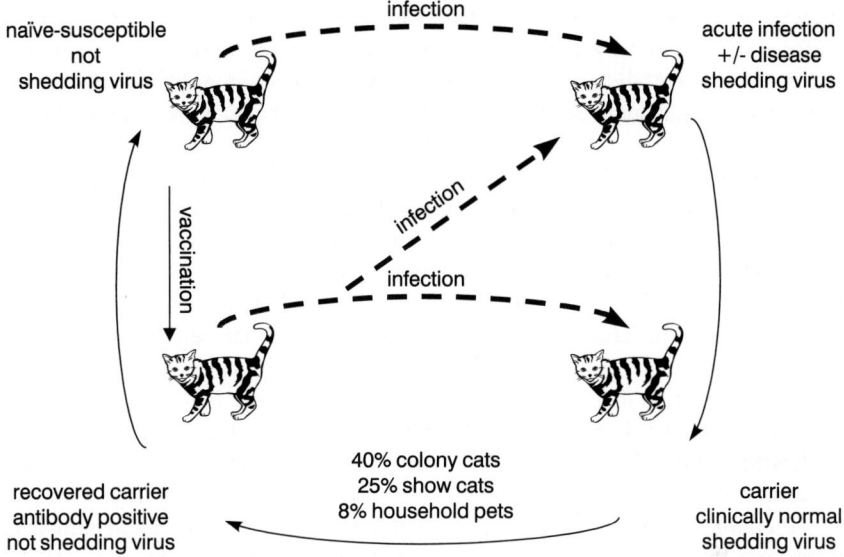

Figure 3.4 Epidemiology of feline calicivirus. Following primary infection, antibody-negative susceptible cats typically develop acute clinical disease. On recovery from disease, most develop an asymptomatic carrier phase. These carriers are largely responsible for the high prevalence of infection in the domestic cat population and are an important source of virus to other cats. Owing to an apparent half-life for carriers of 75 days (see text), most animals eventually cease to shed the virus. Although pre-existing antibody is likely to reduce clinical disease following challenge, it does not prevent infection and antibody-positive animals may still be infected and develop the carrier state. This is true for recovered carriers, vaccinated animals and kittens with maternally derived antibody.

virus for life, most cats appear to eliminate the virus. However, in the field, re-infection is likely to be common because any immunity that develops to FCV infection appears to only reduce clinical disease and does not prevent subsequent infection.

The FCV carrier state is characterized by more-or-less continuous virus shedding and such animals are likely to represent a constant source of infection to other cats. Carriers have been arbitrarily divided into high-, medium- and low-level shedders, each shedding a fairly constant amount of virus, which fluctuates round a mean for that individual cat (Wardley, 1976). High-level excretors are likely to have greater epidemiological significance in that they may more easily, following only brief contact, infect susceptible individuals. Low-level shedders are probably not as infectious and may also be more difficult to detect since the level of virus shed may sometimes fall below the sensitivity of virus isolation.

The site of persistence of FCV in carrier cats is not fully characterized. Although virus has been identified in the epithelium and stratum germinativum of tonsillar fossa mucosa of carriers (Dick *et al.*, 1989), other sites are likely to be involved since tonsilectomy does not terminate the carrier state (Wardley, 1974). Similarly, the mechanism of persistence for FCV has not yet been fully elucidated. However, changes in the neutralization profiles of FCV isolates obtained serially from persistently infected cats have led to the suggestion that antigenic variation and consequent host immune evasion may play a role in the establishment and maintenance of persistent infections (Pedersen et al., 1983; Johnson, 1992; Kreutz et al., 1998; Radford et al., 1998).

Sequence analysis of the epitope-rich variable regions of the capsid has given further insight into the interaction of FCV with the host immune response. Clonal sequence analysis has suggested that, as for other RNA viruses, FCV exists as a mixed population or quasi-species of closely related sequences (Radford et al., 1998). Such mixed viral populations can act as reservoirs from which fitter individuals can be selected in response to any external selection pressures (Domingo et al., 1997). During FCV infection, the quasi-species appears to evolve by a process that involves both the loss of sequence motifs present in the infecting isolate and the gain of both synonymous and non-synonymous nucleotide substitutions. The non-synonymous substitutions (i.e. those which result in a change in amino acid sequence) appear to be located largely in the hypervariable regions of the capsid, which are known to contain B cell epitopes. Such a co-localization of non-synonymous substitutions in antigen-rich HVRs could be explained by immune selection (Johnson, 1992; Kreutz et al., 1998; Radford et al., 1998). There is also limited evidence to suggest that there may be a reduction in quasi-species heterogeneity in the virus over the course of infection, which may lead ultimately to an inability of the virus to further evade the immune response, and its eventual elimination (Radford et al., 1998).

It has been suggested that the rate of FCV evolution in a feline immunodeficiency virus (FIV)-infected cat may be slower that that seen in non-FIV-infected animals (Radford et al., 1998). In addition, it has been shown that FIV infection may potentiate FCV shedding (Dawson et al., 1991; Reubel et al., 1994). It is therefore possible that the immunosuppresssion associated with FIV may lead to a reduction in immune selection pressure on FCV, and this in turn may lead to an inability on the part of the host to eliminate virus.

Other RNA viruses also use inter-genomic recombination as a further method for generating genetic diversity (Romanova et al., 1986; Lemon et al., 1991; Lai, 1992). Whilst individual cats may be infected with more than a single virus at the same time (Radford et al., 2000), there is currently no good evidence for inter-genomic recombination in FCV (Geissler et al., 1997; Glenn et al., 1997). In contrast, there is limited evidence for potential recombination within the noroviruses (Vinje et al., 2000b).

Population evolution

The identification of discrete regions of high variability in the FCV capsid gene has led to the use of these regions for genetic typing of isolates either by restriction fragment length polymorphism of amplicons or by sequence analysis

(Radford *et al.*, 1997, 2000, 2001b; Sykes *et al.*, 1998, 2001). Whilst both methods have been shown to have good utility, sequence analysis has been mostly responsible for the increase in our knowledge of the epidemiological, phylogenetic and evolutionary behaviours of this virus.

Use of an uncorrected (or observed) nucleotide distance value to compare two sequences from the 5'HVR or the whole of region E has allowed the relatedness of two isolates to be mostly divided into one of two categories that correlate well with their epidemiology. Those isolates that share a clear and recent epidemiological link are generally less than 5% different from each other (Radford *et al.*, 1997, 2000). In contrast, those isolates that share no epidemiological link are generally between 20% and 40% different.

Distances of between 5% and 20% have only rarely been identified and are almost entirely limited to the comparison of isolates obtained from large groups of cats, where transmission can readily occur between animals (Radford *et al.*, 2000, 2001b). The groups in which such distance values are found have a high prevalence of infection (up to 60%) and, from results of repeat sampling, are likely to be persistently infected. In ongoing studies by the authors, when isolates within such colonies are compared, occasional distance values of up to 16% have been identified, possibly representing an early stage towards the evolution of a new strain.

Evidence for a potential mechanism for this virus evolution can be sought in the pattern of observed nucleotide substitutions. An analysis of the pattern of nucleotide substitutions observed between isolates within a colony shows the majority of substitutions are located in the 5' and 3'HVRs, as would perhaps be expected from the variable nature of these regions and the known location of B cell epitopes. Interestingly, the vast majority of these substitutions are non-synonymous, whereas the majority of substitutions observed in the conserved regions are synonymous or silent, and therefore do not affect the coding potential of the capsid. This pattern of non-synonymous substitutions outweighing synonymous substitutions is highly suggestive of positive selection driving evolution. Because of the antigenic nature of this region of the genome, it seems likely that any positive selection for evolution will involve the cat's antibody immune response.

These observations have led us to propose a model for the evolution of FCV in groups of cats. In this model, the antibody-mediated immune response in an individual cat positively selects for non-synonymous substitutions in epitope-rich domains of the capsid. These amino acid changes alter the antigenic structure of the capsid and allow the virus to persist within the individual. Opportunities for the virus to escape clearance from the colony are further enhanced by its ability to transmit to other members of the colony in which any immune response is poorly matched to the specific viral variant. In this way, colony persistence is ensured through evolution in carriers and transmission to susceptible contacts. Repetition of this process over time could allow the virus to attain the high levels of genetic variability observed in these colonies. Indeed, studies on the evolution of viruses within such colonies over time should give valuable insight into the evolution of new strains of FCV.

Vaccination in relation to antigenic variation and evolution

Vaccination against FCV has been available for almost 30 years (Gaskell and Dawson, 1998). In the UK, available vaccines are either modified live or inactivated adjuvanted, and are given systemically. Modified live intranasal vaccines are also marketed in some other countries. In addition, the potential use of genetically engineered FCV vaccines is being explored, including the incorporation of all or parts of the FCV capsid gene into recombinant feline herpesvirus, myxoma virus, baculovirus and DNA (plasmid) vaccines (DeSilver et al., 1997; Yokoyama et al., 1998; McCabe and Spibey, 2002; Sommerville et al., 2002).

Conventional modified live or inactivated vaccines induce reasonable protection against disease, but not against infection or the carrier state (Figure 3.4). This failure to protect against infection is important epidemiologically as it maintains seropositive animals as part of the susceptible population. It also allows cats to shed virus without ever having shown signs of clinical disease. It is also clear that this inability of vaccines to protect against infection contributes to the continued high prevalence of infection in the general cat population. Before vaccines were introduced, surveys showed that approximately 8% of household pets, 25% of cats attending cat shows, and 40% of colony cats were shedding the virus (Wardley et al., 1974). However, despite almost 30 years of vaccination, it has been found that approximately 20–25% of cats in a variety of husbandry situations still shed FCV (Harbour et al., 1991; Coutts et al., 1994; Binns et al., 2000). Modelling studies have predicted that, in the absence of a vaccine protecting sufficiently against persistent infection, the pathogen will persist in such groups of cats (Reade et al., 1998). Only when very high vaccine coverage of the population is achieved in situations where carrier rates are low would FCV be eliminated. The continued shedding of field viruses by vaccinated animals may also lead to increased opportunities for virus evolution.

One of the most important considerations for FCV in relation to vaccination is the question of antigenic diversity and strain selection. Although FCV isolates belong to a single diverse serotype, important antigenic differences between FCV isolates mean that no single vaccine strain is likely to protect equally well against all field viruses. Nevertheless, a number of strains of FCV have been selected for use in vaccines on the basis that they are considered broadly cross-reactive. FCV F9 was the first strain shown in the early 1970s and again in the 1990s to neutralize the majority of current FCV isolates (Dawson et al., 1993; Pedersen and Hawkins, 1995; Lauritzen et al., 1997). Subsequently, other strains such as FCV 255 have also been used in vaccines. However, it is probable, at least for the foreseeable future, that strains of FCV that are either poorly protected or not protected against by individual vaccines will always exist in the field.

One might predict that the continued use of a limited repertoire of vaccine strains may lead to the selection of vaccine-resistant isolates over time. There is now some limited evidence to suggest this may be happening for FCV, with recent field isolates appearing to be less cross-reactive with F9 than was originally found (Pedersen and Hawkins, 1995; Lauritzen et al., 1997). Currently, however, there is no evidence to suggest that this is leading to an increase in clinical disease in vaccinated animals. However, it would seem important to develop a surveillance scheme with appropriate sampling strategies to enable the efficacy of commercially available vaccines to be regularly monitored against panels of contemporary FCV field isolates.

Another possible problem with live vaccines is that they may have the potential to infect and cause disease in vaccinated animals. Safety studies required before vaccines are marketed suggest that the modified live viruses in the widely used injectable FCV vaccines are unlikely to spread and persist in the population, and even if they do, they should not cause disease. However, it is clear from sequencing studies that some isolates obtained from recently vaccinated cats appear to have originated from vaccine virus, and in some cases these animals have also been showing signs of disease (Radford et al., 1997, 2000). Until recently, all of these isolations had only been made from recently vaccinated animals, and there was no evidence that vaccine virus could spread or persist in the cat population. However, we have recently identified a single colony of cats that appear to be endemically infected with an FCV-vaccine-like virus, raising the possibility that, in rare cases, vaccine virus may persist in individual cat populations, possibly causing disease (Radford et al., 2001a). Indeed there is some evidence from this study that such persistence may lead to evolution of vaccine virus itself. Whilst persistence and evolution of virus from live vaccines in the cat population is far from desirable, we have only found such isolates in one colony, suggesting that this is likely to be a rare occurrence.

CONCLUSION

FCV is a highly variable pathogen, not only in terms of its clinical presentation but also in terms of its antigenicity and genomic sequence. This variability allows most isolates to be distinguished and this has led to an increase in our understanding of the epidemiology of the disease. However, despite this variability, there is no evidence for any subspecies classification of FCVs, at either the antigenic or the genetic level. There is also sufficient antigenic cross-reactivity between isolates to have allowed the development of reasonably effective vaccines. However, whilst these vaccines reduce clinical disease, they do not appear to protect against infection.

Sequence variability appears to be generated in individual carrier cats and in persistently infected populations and there is good, albeit indirect, evidence to suggest that viral evolution is a consequence of positive immune selection. As such, it has been suggested that the quasi-species nature of this RNA virus and the adaptability that this mixed population conveys may contribute to the ability of the virus to persist in immune recovered cats. More recently there is some evidence to suggest that widespread population immunity in vaccinated animals may be driving the evolution of FCV towards more vaccine-resistant strains.

Study of this highly successful RNA virus of cats provides a valuable insight into how such variable and adaptable pathogens respond to their host both at the individual animal level and more widely in the host population.

REFERENCES

Ando, T., Noel, J. S. and Fankhauser, R. L. (2000) *J. Infect. Dis.* 181 Suppl 2, S336–S348.
Asgari, S., Hardy, J. R. and Cooke, B. D. (1999) *Arch. Virol.* 144, 135–145.
Atmar, R. L. and Estes, M. K. (2001) *Clin. Microbiol. Rev.* 14, 15–37.
Berninger, M. L. and House, C. (1995) *Vet. Microbiol.* 47, 157–165.
Binns, S. H., Dawson, S., Speakman, A. J., Cuevas, L. E., Hart, C. A., Gaskell, C. J., Morgan, K. L. and Gaskell, R. M. (2000) *J. Feline Med. Surg.* 2, 123–133.
Bittle, J. L. and Rubic, W. J. (1976) *Am. J. Vet. Res.* 37, 275–278.
Burki, F., Starustka, B. and Ruttner, O. (1976) *Infect. Immun.* 14, 876–881.
Capucci, L., Fallacara, F., Grazioli, S., Lavazza, A., Pacciarini, M. L. and Brocchi, E. (1998) *Virus Res.* 58, 115–126.
Capucci, L., Fusi, P., Lavazza, A., Pacciarini, M. L. and Rossi, C. (1996) *J. Virol.* 70, 8614–8623.
Carter, M. J. and Cubitt, W. D. (1995) *Curr. Opin. Infect. Dis.* 8, 403–409.
Carter, M. J., Milton, I. D., Meanger, J., Bennett, M., Gaskell, R. M. and Turner, P. C. (1992a) *Virology,* 190, 443–448.
Carter, M. J., Milton, I. D., Turner, P. C., Meanger, J., Bennett, M. and Gaskell, R. M. (1992b) *Arch. Virol.* 122, 223–235.
Chasey, D. (1997) *Laboratory Animals,* 31, 33–44.
Clarke, I. N. and Lambden, P. R. (1997) *J. Gen. Virol.* 78, 291–301.
Coutts, A. J., Dawson, S., Willoughby, K. and Gaskell, R. M. (1994) *Vet. Rec.* 135, 555–556.
Cubitt, W. D., Blacklow, N. R., Herrmann, J. E., Nowak, N. A., Nakata, S. and Chiba, S. (1987) *J. Infect. Dis.* 156, 806–814.
Dastjerdi, A. M., Green, J., Gallimore, C. I., Brown, D. W. and Bridger, J. C. (1999) *Virology* 254, 1–5.
Dawson, S., Bennett, D., Carter, S. D., Bennett, M., Meanger, J., Turner, P. C., Carter, M. J., Milton, I. and Gaskell, R. M. (1994) *Res. Vet. Sci.* 56, 133–143.
Dawson, S., McArdle, F., Bennett, M., Carter, M., Milton, I. P., Turner, P., Meanger, J. and Gaskell, R. M. (1993) *Vet. Rec.* 133, 13–17.
Dawson, S., Smyth, N. R., Bennett, M., Gaskell, R. M., McCracken, C. M., Brown, A. and Gaskell, C. J. (1991) *AIDS* 5, 747–750.
DeSilver, D. A., Guimond, P. M., Gibson, J. K., Thomsen, D. R., Wardley, R. C. and Lowery, D. E. (1997) In *First International Symposium on Caliciviruses. Proceedings of a European Society for Veterinary Virology (ESVV) Symposium, Reading* (eds D.

Chasey, R. M. Gaskell and I. N. Clarke), ESVV and Central Veterinary Laboratory, pp. 131–143.
Dick, C. P., Johnson, R. P. and Yamashiro, S. (1989) *Res. Vet. Sci.* 47, 367–373.
Domingo, E., Menéndez-Arias, L. and Holland, J. J. (1997) *Rev. Med. Virol.* 7, 87–96.
Gaskell, R. M. and Dawson, S. (1994) In *Feline Medicine and Therapeutics* (eds E. A. Chandler, C. J. Gaskell and R. M. Gaskell), Blackwell, Oxford, pp. 453–472.
Gaskell, R. M. and Dawson, S. D. (1998) In *Infectious Diseases of the Dog and Cat* (ed. C. E. Greene), W.B. Saunders, Philadelphia, pp. 97–106.
Gaskell, C. J., Gaskell, R. M., Dennis, P. E. and Woolridge, M. J. A. (1982) *Res. Vet. Sci.* 32, 23–26.
Geissler, K., Schneider, K., Platzer, G., Truyen, B., Kaaden, O.-R. and Truyen, U. (1997) *Virus Res.* 48, 193–206.
Glass, P. J., White, L. J., Ball, J. M., Leparc-Goffart, I., Hardy, M. E. and Estes, M. K. (2000) *J. Virol.* 74, 6581–6591.
Glenn, M. A. (1997) PhD thesis, University of Liverpool.
Glenn, M. A., Gaskell, R. M., Carter, M. J., Lowery, D., Radford, A. D., Bennett, M. and Turner, P. C. (1997) In *First International Symposium on Caliciviruses. Proceedings of a European Society for Veterinary Virology (ESVV) Symposium, Reading* (eds D. Chasey, R. M. Gaskell and I. N. Clarke), ESVV and Central Veterinary Laboratory, pp. 106–110.
Glenn, M., Radford, A. D., Turner, P. C., Carter, M., Lowery, D., DeSilver, D. A., Meanger, J., Baulch-Brown, C., Bennett, M. and Gaskell, R. M. (1999) *Vet. Microbiol.* 67, 175–193.
Green, J., Vinje, J., Gallimore, C. I., Koopmans, M., Hale, A., Brown, D. W., Clegg, J. C. and Chamberlain, J. (2000a) *Virus Genes* 20, 227–236.
Green, K. Y., Ando, T., Balayan, M. S., Berke, T., Clarke, I. N., Estes, M. K., Matson, D. O., Nakata, S., Neill, J. D., Studdert, M. J. and Thiel, H. J. (2000b) *J. Infect. Dis.* 181 Suppl 2, S322–S330.
Guiver, M., Littler, E., Caul, E. O. and Fox, A. J. (1992) *J. Gen. Virol.* 73, 2429–2433.
Guo, M., Chang, K. O., Hardy, M. E., Zhang, Q., Parwani, A. V. and Saif, L. J. (1999) *J. Virol.* 73, 9625–9631.
Guo, M., Evermann, J. F. and Saif, L. J. (2001) *Arch. Virol.* 146, 479–493.
Harbour, D. A., Howard, P. E. and Gaskell, R. M. (1991) *Vet. Rec.* 128, 77–80.
Hashimoto, M., Roerink, F., Tohya, Y. and Mochizuki, M. (1999) *J. Vet. Med. Sci.* 61, 603–608.
Herbert, T. P., Brierley, I. and Brown, T. D. K. (1996) *J. Gen. Virol.* 77, 123–127.
Hofmann-Lehmann, R., Fehr, D., Grob, M., Elgizoli, M., Packer, C., Martenson, J. S., O'Brien, S. J. and Lutz, H. (1996) *Clin. Diagn. Lab. Immunol.* 3, 554–562.
Hoover, E. A. and Kahn, D. E. (1975) *J. Am. Vet. Med. Assoc.* 166, 463–468.
Horimoto, T., Takeda, Y., Iwatsuki-Horimoto, K., Sugii, S. and Tajima, T. (2001) *Virus Genes* 23, 171–174.
Jiang, X., Cubitt, W. D., Berke, T., Zhong, W., Dai, X., Nakata, S., Pickering, L. K. and Matson, D. O. (1997) *Arch. Virol.* 142, 1813–1827.
Johnson, R. P. (1992) *Can. J. Vet. Res.* 56, 326–330.
Johnson, R. P. and Povey, R. C. (1983) *Can. Vet. J.* 24, 6–9.
Kadoi, K., Kiryu, M., Iwabuchi, M., Kamata, H., Yukawa, M. and Inaba, Y. (1997) *New Microbiol.* 20, 141–148.
Kalunda, M., Lee, K. M., Holmes, D. F. and Gillespie, J. H. (1975) *Am. J. Vet. Res.* 36, 353–356.

Knowles, J. O., Dawson, S., Gaskell, R. M., Gaskell, C. J. and Harvey, C. E. (1990) *Vet. Rec.* 127, 125–127.
Knowles, J. O., McArdle, F., Dawson, S., Carter, S. D., Gaskell, C. J. and Gaskell, R. M. (1991) *Vet. Microbiol.* 27, 205–219.
Kreutz, L. C., Johnson, R. P. and Seal, B. S. (1998) *Vet. Microbiol.* 59, 229–236.
Lai, M. M. C. (1992) *Microbiol. Rev.* 56, 61–79.
Lauritzen, A., Jarrett, O. and Sabara, M. (1997) *Vet. Microbiol.* 56, 55–63.
Le Gall, G., Arnauld, C., Boilletot, E., Morisse, J. P. and Rasschaert, D. (1998) *J. Gen. Virol.* 79, 11–16.
Lemon, S. M., Murphy, P. C., Shields, P. A., Ping, L.-H., Feinstone, S. M., Cromeans, T. and Jansen, R. W. (1991) *J. Virol.* 65, 2056–2065.
Lewis, D., Ando, T., Humphrey, C. D., Monroe, S. S. and Glass, R. I. (1995) *J. Clin. Microbiol.* 33, 501–504.
Liu, B. L., Lambden, P. R., Gunther, H., Otto, P., Elschner, M. and Clarke, I. N. (1999) *J. Virol.* 73, 819–825.
Matson, D. O., Zhong, W. M., Nakata, S., Numata, K., Jiang, X., Pickering, L. K., Chiba, S. and Estes, M. K. (1995) *J. Med. Virol.* 45, 215–222.
McArdle, F., Dawson, S., Carter, M. J., Milton, I. D., Turner, P. C., Meanger, J., Bennett, M. and Gaskell, R. M. (1996) *Vet. Microbiol.* 51, 197–206.
McCabe, V. J. and Spibey, N. (2002) *Res. Vet. Sci.* 72, 24.
Milton, I. D., Turner, J., Teelan, A., Gaskell, R., Turner, P. C. and Carter, M. J. (1992) *J. Gen. Virol.* 73, 2435–2439.
Nagesha, H. S., McColl, K. A., Collins, B. J., Morrissy, C. J., Wang, L. F. and Westbury, H. A. (2000) *Arch. Virol.* 145, 749–757.
Neill, J. D. (1992) *Virus Res.* 24, 211–222.
Neill, J. D., Sosnovtsev, S. V. and Green, K. Y. (2000) *J. Virol.* 74, 1079–1084.
Nowotny, N., Bascuñana, C. R., Ballagi-Pordány, A., Gavier-Widén, D., Uhlén, M. and Belák, S. (1997) *Arch. Virol.* 142, 657–673.
Pedersen, N. C. and Hawkins, K. F. (1995) *Vet. Microbiol.* 47, 141–156.
Pedersen, N. C., Elliott, J. B., Glasgow, A., Poland, A. and Keel, K. (2000) *Vet. Microbiol.* 73, 281–300.
Pedersen, N. C., Laliberte, L. and Ekman, S. (1983) *Feline Pract.* 13, 26–35.
Poulet, H., Brunet, S., Soulier, M., Leroy, V., Goutebroze, S. and Chappuis, G. (2000) *Arch. Virol.* 145, 243–261.
Povey, R. C. (1974) *Infect. Immun.* 10, 1307–1314.
Povey, R. C., Koonse, H. and Hays, M. B. (1980) *J. Am. Vet. Med. Assoc.* 177, 347–350.
Radford, A. D., Bennett, M., McArdle, F., Dawson, S., Turner, P. C., Glenn, M. A. and Gaskell, R. M. (1997) *Vaccine* 15, 1451–1458.
Radford, A. D., Dawson, S., Wharmby, C., Ryvar, R. and Gaskell, R. M. (2000) *Vet. Rec.* 146, 117–123.
Radford, A. D., Sommerville, L. M., Dawson, S., Kerins, A. M., Ryvar, R. and Gaskell, R. M. (2001b) *Vet. Rec.* 149, 477–481.
Radford, A. D., Sommerville, L., Ryvar, R., Cox, M. B., Johnson, D. R., Dawson, S. and Gaskell, R. M. (2001a) *Vaccine* 19, 4358–4362.
Radford, A. D., Turner, P. C., Bennett, M., McArdle, F., Dawson, S., Glenn, M. A., Williams, R. A. and Gaskell, R. M. (1998) *J. Gen. Virol.* 79, 1–10.
Radford, A. D., Willoughby, K., Dawson, S., McCracken, C. and Gaskell, R. M. (1999) *J. Virol.* 73, 8496–8502.

Reade, B., Bowers, R. G., Begon, M. and Gaskell, R. M. (1998) *J. Theoret. Biol.* 190, 355–367.
Reubel, G. H., George, J. W., Higgins, J. and Pedersen, N. C. (1994) *Vet. Microbiol.* 39, 335–351.
Roerink, F., Hashimoto, M., Tohya, Y. and Mochizuki, M. (1999a) *Vet. Microbiol.* 69, 69–72.
Roerink, F., Hashimoto, M., Tohya, Y. and Mochizuki, M. (1999b) *J. Gen. Virol.* 80, 929–935.
Romanova, L. I., Blinov, V. M., Tolskaya, E. A., Viktorova, E. G., Kolesnikova, M. S., Guseva, E. A. and Agol, V. I. (1986) *Virology* 155, 202–213.
Sabine, M. and Hyne, R. H. J. (1970) *Vet. Rec.* 87, 794–796.
Sangabriel, M. C. S., Tohya, Y. and Mochizuki, M. (1996) *J. Vet. Med. Sci.* 58, 1041–1043.
Schirrmeier, H., Reimann, I., Kollner, B. and Granzow, H. (1999) *Arch. Virol.* 144, 719–735.
Scott, F. W. (1977) *Am. J. Vet. Res.* 38, 229–234.
Seal, B. S., Ridpath, J. F. and Mengeling, W. L. (1993) *J. Gen. Virol.* 74, 2519–2524.
Sommerville, L. M. (2001) PhD thesis, University of Liverpool.
Sommerville, L. M., Radford, A. D., Glenn, M., Dawson, S., Gaskell, C. J., Kelly, D. F., Cripps, P. J., Porter, C. J. and Gaskell, R. M. (2002) *Vaccine* 20, 1787–1796.
Sosnovtsev, S. V. and Green, K. Y. (2000) *Virology* 277, 193–203.
Sosnovtsev, S. V., Sosnovtseva, S. A. and Green, K. Y. (1998) *J. Virol.* 72, 3051–3059.
Sugieda, M., Nagaoka, H., Kakishima, Y., Ohshita, T., Nakamura, S. and Nakajima, S. (1998) *Arch. Virol.* 143, 1215–1221.
Sykes, J. E., Allen, J. L., Studdert, V. P. and Browning, G. F. (2001) *Vet. Microbiol.* 81, 95–108.
Sykes, J. E., Studdert, V. P. and Browning, G. F. (1998) *Arch. Virol.* 143, 1321–1334.
TerWee, T., Lauritzen, A., Sabara, M., Dreier, K. J. and Kokjohn, K. (1997) *Vet. Microbiol.* 56, 33–45.
Tham, K. M. and Studdert, M. J. (1987) *J. Vet. Med. B* 34, 640–654.
Tohya, Y., Masuoka, K., Takahashi, E. and Mikami, T. (1991) *Arch. Virol.* 117, 173–181.
Tohya, Y., Yokoyama, N., Maeda, K., Kawaguchi, Y. and Mikami, T. (1997) *J. Gen. Virol.* 78, 303–305.
van Der Poel, W. H., Vinje, J., van Der Heide, R., Herrera, M. I., Vivo, A. and Koopmans, M. P. (2000) *Emerg. Infect. Dis.* 6, 36–41.
Vinje, J., Deijl, H., van der Heide, R., Lewis, D., Hedlund, K. O., Svensson, L. and Koopmans, M. P. (2000a) *J. Clin. Microbiol.* 38, 530–536.
Vinje, J., Green, J., Lewis, D. C., Gallimore, C. I., Brown, D. W. and Koopmans, M. P. (2000b) *Arch. Virol.* 145, 223–241.
Wardley, R. C. (1974) PhD thesis, University of Bristol.
Wardley, R. C. (1976) *Arch. Virol.* 52, 243–249.
Wardley, R. C. and Povey, R. C. (1977) *Res. Vet. Sci.* 23, 7–14.
Wardley, R. C., Gaskell, R. M. and Povey, R. C. (1974) *J. Small Anim. Pract.* 15, 579–586.
Yokoyama, N., Fujita, K., Damiani, A., Sato, E., Kurosawa, K., Miyazawa, T., Ishiguro, S., Mochizuki, M., Maeda, K. and Mikami, T. (1998) *J. Vet. Med. Sci.* 60, 717–723.

4

INFLUENZA – THE CHAMELEON VIRUS

JOHN OXFORD, RAMANI ESWARASARAN, ALEX MANN
AND ROBERT LAMBKIN

Meaningful scientific investigations into human influenza virus could not start until the virus was first isolated and identified in the UK in 1933 (Smith *et al.*, 1933). Of course, even the casual observer had noted the annual winter outbreaks of respiratory disease. Two of the scientists involved in this first virological experiment of recovering influenza from throat washes using the ferret had experienced influenza first-hand during training at St Bartholomew's Hospital. Other viruses such as polio or the respiratory virus RSV were also epidemic, the former in the summer and the latter during wintertime. Nevertheless, there was no forewarning of the striking properties of the influenza virus yet to be discovered. A prerequisite to the work on the virus was the discovery of a vital protein of the virus, haemagglutinin (HA), the haemagglutinating ability of the influenza virus for chicken and mammalian erythrocytes and, more importantly, the haemagglutination inhibition (HI) test, which, 60 years later, is still the gold standard in antigenic analysis for influenza (Hirst, 1942). The first serological studies in animal models and post-infection human sera indicated a surprising degree of antigenic change or drift of the HA. At this stage neither the haemagglutinin (HA) nor the neuraminidase (NA) spike had been visualized and nor was it clear that the virus had an internal structure of nucleoprotein (NP), matrix protein (M) and polymerase proteins (PA, PB1, PB2). An epidemic in 1947 which circumvented vaccine-induced immunity demonstrated very clearly that antigenic change or drift had very important practical consequences. But it was the emergence of a completely novel influenza A virus in Asia in 1957 and another in 1968 which confirmed the suspicion harboured since the Great Spanish Pandemic of 1918, namely that the influenza A virus had two epidemiological faces or modes, epidemic and pandemic (Oxford, 2002). In turn both the pandemics and epidemics were related to changing HA or NA proteins: in the former case the proteins were completely different (there are 15 unique HAs and 9 NAs called subtypes) whereas in the latter case there were small antigenic differences on the same proteins.

It was not until the advent of new techniques of protein chemistry and electron microscopy that it became clear that the HA was the most important antigen of the virus and was made up of two polypeptides, HA1 and HA2. These two constituents of HA were separated and studied by peptide mapping, when it was discovered that antigenic change occurred in the HA1 molecule almost entirely. Laver and Webster (1972) also found that the HAs of the different subtypes of influenza A had completely different peptide maps: each was independent and could not transform into other subtypes by mutation. In fact, genetic reassortment or gene shuffling was the technique exploited by influenza A virus to reinvent itself as a new pandemic virus, possibly from the reservoir of 15 HAs and 9 NAs residing in birds, horses, pigs, seals and whales (Webby and Webster, 2001). In contrast, the yearly epidemic or drift viruses differed by only one or two spots on the HA peptide map. These and other studies quickly established that pandemic strains could retain the genes that coded internally situated proteins whilst, at the same time, changing the HA and NA genes. More recent methodologies of nucleotide sequencing alongside the use of monoclonal antibodies and study of escape mutants established that single amino acid changes in exposed regions or epitopes of the HA protein could enable a virus to escape from antibody pressure and these observations provided the scientific explanation of the yearly epidemics, called antigenic drift.

Although these important studies over the seven decades since true virology was able to start have been considerable achievements, a lot remains to be learned about virulence, the genes that regulate pathogenicity and, as regards the host response, the role of the B and T cell immune reactions to the different virus proteins. Important questions are the origin of brand new antigenic variants, namely pandemic strains, and the nature of the biological or immunological forces which drive antigenic change (reviewed by Francis, 1953). Strangely, information on physical structure of the virion may still be incomplete. Figure 4.1 shows the first observation of a natural uncultivated virion with a fringe of HA spikes. This chapter will focus on the HA protein of influenza A virus. The NA protein is important and undergoes the same phenomenon of shift and drift and therefore many of the principles discussed below will apply. All 11 or so viral structural proteins elicit B and T cell response but most is known about anti-HA responses. Antibodies against the NP and M proteins cross-react amongst all influenza viruses of one type but do not extend across types. Similarly, a broad reactivity is detected in T cell responses to M and NP. That new discoveries can still be made, and that our knowledge is incomplete, is illustrated by the report of a new open reading frame of the PB1 protein, coding for a second small peptide some 97 amino acids in length (Chen *et al.*, 2002). This short protein

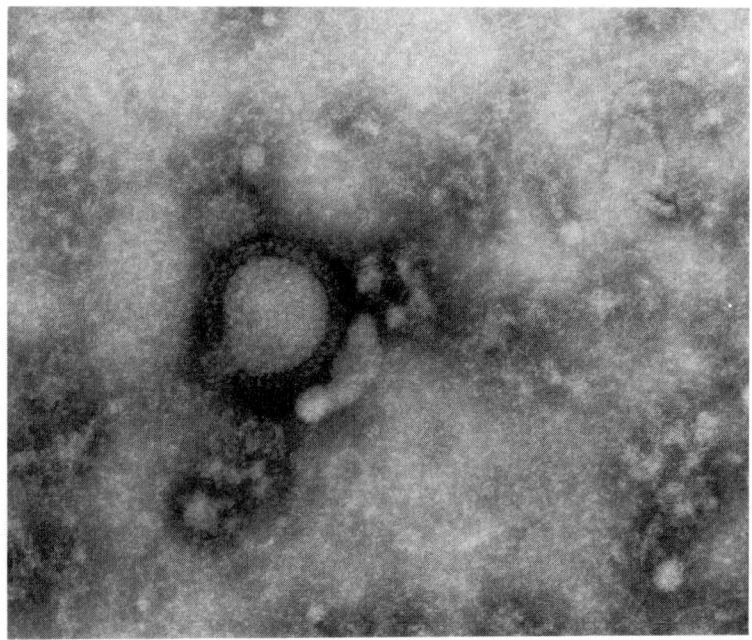

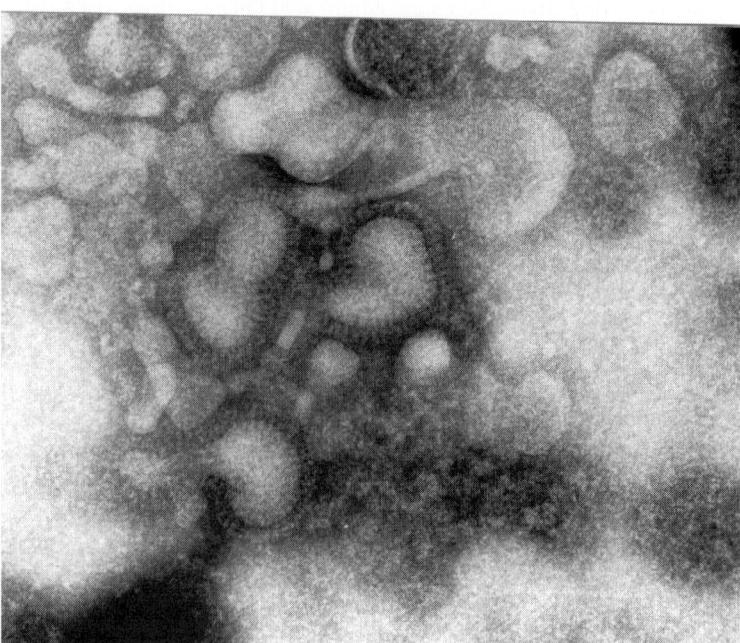

Figure 4.1 Electron micrograph of uncultivated virions in a natural state in a nasal wash from a volunteer (×130 000). (Courtesy of Dr David Hockley, NIBSC.)

induces apoptosis and may therefore be an additional virus virulence factor. Similarly, influenza B will only be discussed sketchily. It is of less clinical impact than the A type and it is not pandemic, but otherwise the same basic rules apply also to these viruses. Equally, influenza C is a well-established respiratory virus pathogen of children but causing mainly mild illness and not pandemics, and will barely be mentioned further here.

The finding of Brand and Skehel (1972) that HA could be removed from virions by a pineapple protease was a major breakthrough, which enabled protein chemists to detach the HA and purify the HA spikes in a sucrose gradient. The NA protein was digested entirely by the bromelain enzyme and so was not a complicating factor in subsequent analysis. The first analyses by X-ray crystallography showed a trimeric (HA) molecule (Wilson *et al.*, 1981), as we shall discuss below. Later, a remarkable structural mobility in the HA molecule was discovered. The HA molecule has been likened to a mouse trap that under certain conditions of low pH during virus infection of a cell could spring and so take a completely different configuration (Bullough *et al.*, 1994). For example, epitopes previously covered up could be exposed and vice versa, whilst functional areas such as the hydrophobic fusion motif would now find itself in a more sensible physical location, where it could mediate fusion between the virus lipid membranes and those of a host cell. Fusion between host cell lipid membranes could be catalysed, which would enable the viral RNA to be released from intracellular vacuoles, to enter the cell nucleus and then to undergo transcription.

THE CLINICAL IMPACT OF INFLUENZA A AND B VIRUSES

Since 1933, when the laboratory study of influenza virus was first established, recurrent epidemics of influenza A and of influenza B have been identified in all countries where surveillance has been carried out. In countries for which epidemiological information is available, years in which no influenza outbreaks are recorded are rare. Most epidemics (more than 350 cases of influenza per 100 000 in the population) and all pandemics are due to the influenza A virus. Influenza A, perhaps most characteristically an infection of children and young adults (reviewed by Stuart-Harris *et al.*, 1983; Fleming, 2001), does affect all age groups, the highest mortality being in those aged 65 years or more, and in people with certain chronic illnesses called the 'at risk' group. In contrast, influenza B outbreaks are four to six times less frequent than those caused by influenza A virus, and are relatively localized, mainly affecting children. There is often an excess of deaths during influenza A outbreaks but this is not a constant feature of influenza B epidemics, although it has certainly been documented on at least two occasions, for example in the USA during 1984. Table 4.1 summarizes the recent impact of influenza, pneumonia or bronchitis in the UK, which can vary

Table 4.1 Excess mortality due to influenza and associated pneumonia or bronchitis in England and Wales

Season	No. excess deaths
1988/89	150
1989/90	25 786
1990/91	6552
1991/92	4807
1992/93	1051
1993/94	9480
1994/95	0
1995/96	13 579
1996/97	28 987
1997/98	790
1998/99	17 783
1999/2000	19 543
2000/01	0
Total	**128 508**

from zero deaths to more than 20 000 per year in the UK alone. Worldwide these figures would increase to hundreds of thousands of deaths annually. Influenza is a much under-estimated viral pathogen and even the Great Pandemic of 1918 has, more or less, been forgotten (Crosby, 1989).

ANTIGENIC MAPPING OF THE HAEMAGGLUTININ

The importance of the HA as a major antigen and virulence factor stimulated a very wide-ranging series of scientific investigations into epitopes, morphology and atomic structure. The basic method of mapping was to produce a panel of monoclonal antibodies to the HA protein. When an individual monoclonal antibody was then used to neutralize virus inevitably some virions in such a mixed population would 'escape' neutralization, presumably because of an amino acid substitution in or around the epitope in question. The HA gene of this 'escape' mutant was sequenced and amino acid changes were identified both on an X-ray crystallographic map and on the linear amino acid sequence of the protein. In this way the precise positions of several epitopes or antigenic areas were located on the HA (Figures 4.2 and 4.3). Finally, the HA of naturally occurring 'drifted' viruses was sequenced and again 'hot spots' or highly changeable areas of the HA were identified. In this manner five important antigenic sites (designated A–E) were identified on the HA protein (Wiley *et al.*, 1981; Wilson *et al.*, 1981). On the NA protein the antigenic sites are much more diffuse and cover a large area of the head of the protein. Much less is known about the atomic structure of M and NP and other polymerase PA, PB1 and PB2 proteins because no X-ray crystallographic

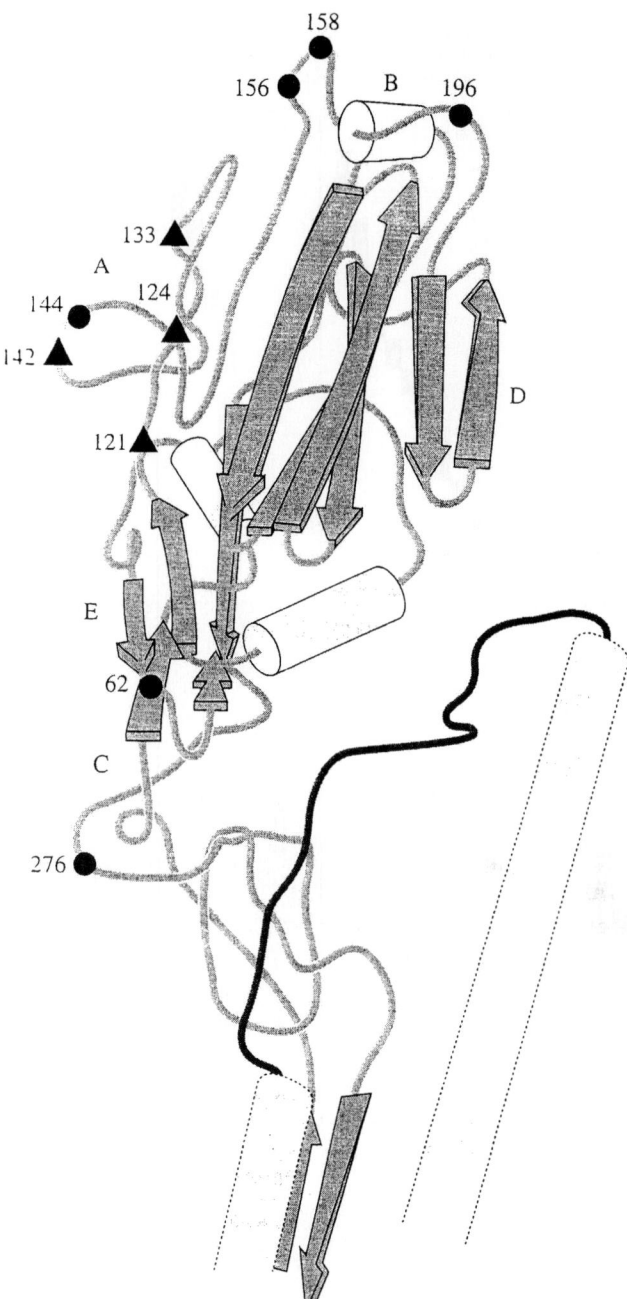

Figure 4.2 Structure of the tip of the influenza HA protein and location of significant amino acid changes at sites A, B, C and D. (Reproduced with permission from Hay *et al.* (2001) *Phil. Trans. R. Soc. Lond.* 356, 1861–1870.)

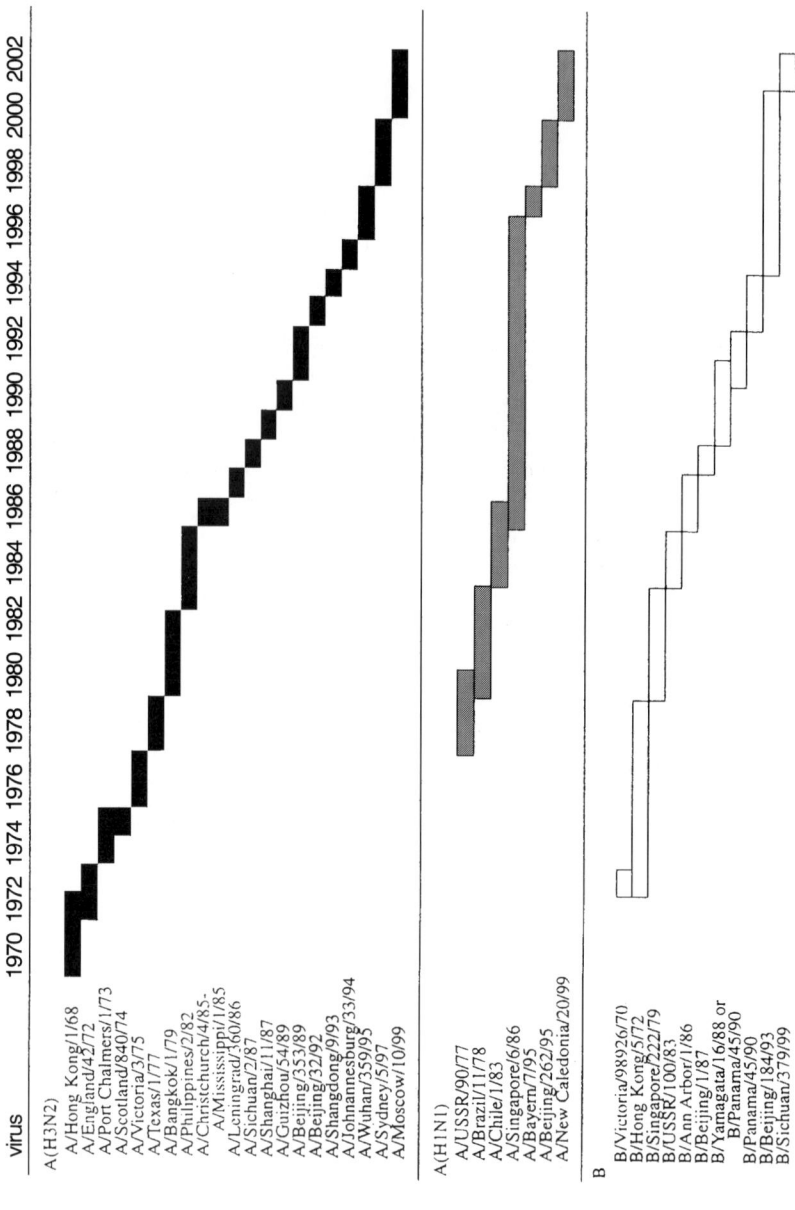

Figure 4.3 Changes in the influenza vaccine compositions recommended by the WHO (1973–2000), which reflects the degree of antigenic change. (Reproduced with permission from Hay et al. (2001) *Phil. Trans. R. Soc. Lond.* 356, 1861–1870.)

data are extant. Humans produce B and T cell responses to all structural and non-structural proteins of the virus, such as NP, but the most powerful immune responses, correlating with protection and recovery from infection, are directed towards the HA and NA proteins.

X-RAY CRYSTALLOGRAPHY OF INFLUENZA HAEMAGGLUTININ

In each virus particle there are some 500 haemagglutinin 'spikes' 14 nm in length and 4–6 nm in diameter, and which appear to be of triangular cross-section. Figure 4.2 shows a representation of the three-dimensional structure of the tip of the haemagglutinin as proposed by Wilson *et al.* (1981). Examination of the structure allows estimation of antigenic change in this area of the molecule. Detailed studies of the HA of the influenza A (H3N2) virus by X-ray crystallography, together with earlier sequencing data, allowed the construction of accurate three-dimensional models of the HA and the precise localization of the cell-binding site, antigenic determinants and fusion sequences. Each haemagglutinin 'spike' has a hydrophobic tail of 25–28 amino acids, by which it is inserted in the virus membrane and by means of which the spikes aggregate when isolated in pure form (Laver and Valentine, 1969). A small hydrophilic domain of approximately 10 amino acids was identified, which presumably makes contact with the underlying M protein.

The trimeric structure of the HA molecule contains two distinct regions: the first is a triple-stranded coiled coil of helices extending 8 nm from the virus membrane; this supports a second portion, which is a globular region of antiparallel β-sheet containing the site reacting with erythrocytes or with receptors on the host cell and also the important antigenic determinant surfaces of the HA.

As regards the important cell-binding site, the distribution of amino acids on the 0.3 nm structure revealed a highly conserved region in a surface pocket on the distal end of the molecule which seemed correctly situated for binding to host cell oligosaccharide. This receptor-binding pocket includes conserved residues Try 98, His 183, Glu 190, Trp 153 and Leu 194, and also residues behind the pocket, including Arg 299. The X-ray crystal structures of haemagglutinins complexed with sialic acid receptors were solved later.

The most highly conserved sequence in the haemagglutinin is the amino terminus of HA2, where only one substitution in the first 11 and 5 in the first 23 amino acids have been observed in comparisons of different viruses. This sequence is associated with the fusion activity by which the virus penetrates a host cell internal lysosomal membrane to initiate infection. Cleavage of HA into HA1 and HA2 is required for infectious penetration of virions and for membrane fusion *in vitro* (Steinhauer and Wharton, 1998). The connecting peptide between HA1 and HA2 is an important determinant of virus virulence, particularly for avian influenza A viruses (Perdue *et al.*, 1997). The HA2 N-terminal sequence is strongly non-polar, and it is unexpectedly rich in glycines, which could indicate flexibility or the presence of an unusual conformation. We will return to this important portion of the HA molecule below. As regards antigenic binding sites, five epitopes (A, B, C, D and E) have been identified by analysing mutants selected in the laboratory under immunological pressure from monoclonal antibodies (Figure 4.2). Site A is an unusual protruding loop from amino acids 140 to 146, which projects 0.8 nm from the local molecular surface and forms the centre of the most obvious antibody-binding site. The haemagglutinin of each antigenically distinct virus of epidemic significance has a mutation in this region. Site B comprises the external residues 187–196 of a χ-helix, and adjacent residues along the upper edge of a pocket implicated in virus receptor binding. Site C has a bulge in the tertiary structure at the disulphide bond between cys 52 and cys 277, 6 nm from the distal tip of the molecule, and comprises another antibody-binding site.

In the first three antigenic sites noted above, the amino acid substitutions causing antigenic variation are external. Site D departs from this arrangement. Unexpectedly, several amino acid substitutions in the haemagglutinins of both natural and laboratory-selected antigenic mutants occur in the interface region between subunits in the haemagglutinin trimer. But these amino acids may be recognized as a result of a relative movement of the globular regions of HA1 to expose the interface regions. Bullough *et al.* (1994) described major refolding of the HA dependent upon the intracellular pH. Under these conditions the fusion peptide (N" terminal of HA2) is transported 15 nm to bring it in close contact with the endosomal membrane.

GENETIC VARIATION OF THE HAEMAGGLUTININ

The nucleotide and hence the amino acid sequences of haemagglutinins of numerous influenza virus strains have been studied (reviewed by Hay *et al.*, 2001). The general conclusions are that certain regions of the haemagglutinin molecules of antigenically different subtypes of influenza A show considerable sequence homologies which are attributed to the need for conservation of function and structural requirements. An overall conservation of 33–36% of amino acid sequences between haemagglutinins of different subtypes has been reported. In contrast, areas in the region of the distal tip of the HA associated with antigenic determinants show considerable variability, whereas the HA2 portion of the molecule and more deeply located regions of the HA1 polypeptide are relatively well conserved.

During major antigenic changes in the haemagglutinin molecule (antigenic shift), nucleotide and amino acid sequences of the polypeptides HA1 and HA2 change radically, whereas in minor antigenic changes (antigenic drift) only small modifications of the amino acid composition of these components are detected, particularly in HA1. Table 4.2 summarizes evolutionary rates for the HA1

Table 4.2 Evolutionary rates for the HA1 portion of the HA gene and protein

Virus	Period	Number	Nucleotide changes per nucleotide site per year	Amino acid changes per amino acid site per year
H1N1	1934–1957	24	3.9×10^{-3}	5.3×10^{-3}
H1N1	1977–1996	146	2.7×10^{-3}	4.2×10^{-3}
H3N2	1968–1996	340	3.7×10^{-3}	4.1×10^{-3}
B	1940–1996	168	1.2×10^{-3}	1.1×10^{-3}

Source: Cox and Kawaoka, 1998

protein. The modern sequence data were reliably predicted by the earlier studies of Laver and Webster (1972), who elegantly demonstrated differences between HA1 polypeptides but not between the HA2 polypeptides of antigenically drifted viruses. It is now known that substitutions of even single amino acids in the molecules may be accompanied by changes in antigenic specificity and in properties such as receptor binding.

One further important series of observations should not be omitted, namely that antigenic change in the HA may be powered not only by immune pressure but also by the selective pressure of cellular receptors (Schild *et al.*, 1983). Thus, because of the physical proximity of epitope B at the tip of the HA and the receptor-binding site, amino acid changes in the HA polypeptide shared by these two sites may affect antigenicity and receptor binding simultaneously. If a heterogeneous virus mixture is allowed to replicate both in mammalian cells and in avian cells, different virus subpopulations are selected on the basis of receptor sites, and concomitantly the viruses may differ antigenically in epitope B at the HA tip. We will return to the selective pressures exerted on HA below. Surprisingly, these non-immunological pressures can result in antigenic change in the HA.

X-RAY CRYSTALLOGRAPHY OF INFLUENZA NEURAMINIDASE

X-ray crystallography has revealed both enzyme active sites and antigenic epitopes (Varghese *et al.*, 1983) and has delineated its precise interaction with antibody molecules. Interest has been focused on the NA more recently as a target for chemical inhibitors of the virus replication cycle (Monto *et al.*, 1999; Gubareva *et al.*, 2000; Hayden *et al.*, 2000; Nicholson *et al.*, 2000; Treanor *et al.*, 2000; Welliver *et al.*, 2001). The enzymic action of neuraminidase of influenza A and B viruses is the hydrolysis of N-acetyl neuraminic acid (sialic acid) residues from specific glycoprotein substrates. Such glycoproteins form specific receptor sites for haemagglutinin on the surface of host cells or erythrocytes.

There is strong evidence that neuraminidase may have a function in mediating the release of virus from the host cell surface receptors after replication and also in allowing virus to penetrate through sialic-acid-containing molecules in the nasal passage. It has also been suggested that viral neuraminidase may play a role in maturation of the virus particle.

In the presence of detergent these mushroom-shaped spikes of the influenza A and B virus neuraminidase are 8 nm long and 4 nm wide with a centrally attached fibre or stem some 10 nm in length. On the removal of detergent the neuraminidase subunits become aggregated, joining together by the tips of their fibre to form characteristic 'cartwheel' structures (Laver and Valentine, 1969). The neuraminidase molecule, like haemagglutinin, appears to possess a hydrophobic end, by which it is anchored to the lipid membrane of the virus, and a hydrophilic end, which contains both the enzymically active site and antigenic determinants. The neuraminidase molecule is orientated rather unusually for a glycoprotein, its N terminus being anchored in the viral membrane. After the three-dimensional structure of influenza neuraminidase was established, the positions on the molecule of the catalytic and antigenic sites could be identified and the binding of antibody or inhibitory chemicals visualized. The molecule has a box-shaped head, with a unique folding pattern. Each monomer has six β-sheets and contains four polypeptide strands. Viewed from above, each monomer has the appearance of a flower with the petals somewhat twisted to resemble a pinwheel. The 'stem' is considered to span the bilayer of the virus with a hydrophobic stretch of amino acids. Of the four carbohydrate chains in crystalline NA, two are on the top and two on the bottom of the box-shaped head. One oligosaccharide is found at a subunit–subunit interface. Both on the NA and on the HA, most of the oligosaccharides are attached on the lateral surfaces of the glycoprotein in positions where the latter might be expected to contact other proteins embedded in the same membrane.

The catalytic site of the NA has been located by difference Fourier analysis of crystals soaked in sialic acid. The site is surrounded by 14 conserved charged residues and contains three hydrophobic residues – Try, Trp and Leu. The new anti-NA drugs bind to eleven of these amino acids (Ives et al., 2001). By examining the sites of amino acid substitutions in a series of 'drifted' field strains and in three antigenic variants selected by growth in the presence of monoclonal antisera, Varghese and colleagues (1983) located antigenic determinants on the NA; they are composed of loops connecting strands of β-sheet and form a nearly continuous surface that encircles the catalytic site on the top of the NA. Although residues in the active site itself are conserved from strain to strain, the proximity of the variable loops to the site suggests that a significant part of the variable loops could interfere with the antibody contacting the active site cavity. Colman and his colleagues (1987) reported the first analysis by X-ray diffraction of the three-dimensional structure of a viral antigen complexed with an antibody Fab fragment. The interaction is not consistent with a rigid 'lock and key' model but rather has been described as a 'handshake'.

THE BIOLOGICAL PROPERTIES OF ANTIBODY TO INFLUENZA HA

Antibody to HA is important in neutralization of virus, complement fixation, prevention of virus attachment and mediation of antibody-dependent cellular cytotoxicity. Resistance to infection is correlated with serum anti-HA antibody levels in human volunteers and by the demonstration of protection against challenge after passive transfer of immune serum in a mouse model. Studies have shown that protection from live virus challenge is associated with local neutralizing antibody and secretory IgA as well as serum anti-HA antibody. In contrast, although antibodies to NA are inefficient in neutralizing influenza viruses, they restrict virus release from infected cells, reduce the intensity of infection and enhance recovery. Antibodies to other structural and non-structural proteins have no known virus-neutralizing activity.

Serum antibody to HA can persist for decades, and retrospective serosurveys suggest that a limited number of influenza subtypes have recycled in humans, namely H1, H2 and H3. Decade-long persistence of immunity was demonstrated dramatically when influenza A (H1N1) strains, similar to viruses that had circulated in 1950, spread throughout the world during 1977/1978. At that time little disease occurred in individuals born before 1950, indicating that substantial immunity remained after almost 50 years. Disease occurred in people <20 years old, however, irrespective of whether they were infected by influenza viruses of the N3N2 subtype. This and numerous observations during the pandemics of 1957 and 1968 suggest that intersubtypic immunity is weak in humans. In contrast, intrasubtypic immunity to influenza A (H3N2) viruses in adults can last for 4–7 years and include two or more variants of the same subtype, depending on the extent of antigenic drift. The fact that a repeat infection with an antigenically related strain boosts the antibody response to the first virus encountered was a unique phenomenon (reviewed by Kilbourne, 1975). This phenomenon, known as 'original antigenic sin', is believed to be a selective anamnestic response orientated toward both the HA and the NA antigens experienced during the original infection. Thus, cross-reactive epitopes will stimulate a predominant secondary antibody response whereas new epitopes in the reinfecting variant induce a primary response. The precise importance of this phenomenon to immunity induced by infection or vaccination is unknown, but observations on original antigenic sin suggest that induced immunity during sequential infections may be biased to older, less relevant strains rather than to the current infection strain. This can be a potential problem in vaccine-induced immunity.

As we have noted above, sequence analysis of the HA genes of monoclonal antibody escape variants, along with field isolates and location of amino acid changes on the three-dimensional structure of the HA, have defined five antibody-combining sites. The relationship of these sites to antibodies produced via natural infection has yet to be determined precisely, but the specificity of the antibody response to HA seems to be limited in humans, particularly in children, and may vary from individual to individual (Natali *et al.*, 1981). A limited range of specificities of anti-HA antibodies in individual mice and rabbits has also been demonstrated (Lambkin and Dimmock, 1995). These findings may have implications for antigenic drift and the epidemiological success of influenza viruses in humans.

Among children with no previous exposure to influenza viruses, vaccination with the live attenuated influenza viruses results in the appearance of serum anti-HA IgM, IgG and IgA antibodies within 2 weeks. IgM and IgA antibodies decline after this time, but the IgG response peaks after about 6 weeks, declines over the next 6 months and then remains relatively stable for 2–3 years. Antiviral IgA, IgG and IgM responses can be detected in nasal secretions in most individuals and persist for several months. A secondary response to infection in primed young adults results in serum IgG and IgA and mucosal IgA in most cases. Serum haemagglutination inhibition (HAI) responses decline initially after infection, but may then remain quite stable for years. Serum antibodies to NA are rarely induced in primary infection and generally occur after re-exposure to NA of the same subtype. Antibodies to type-specific NP and M1 proteins are boosted via reinfection and may help diagnose a recent exposure to virus.

ARE T CELL RESPONSES IMPORTANT IN THE IMMUNE RESPONSE TO INFLUENZA HA?

CD4+ and CD8+ T cell responses to influenza are type specific and are largely cross-reactive among influenza A viruses of different subtypes. T cell recognition of antigen is restricted by the MHC antigens, and the ability to respond to a given viral epitope depends on the HLA phenotype of an individual, a fact that complicates various approaches to vaccination against influenza. In naturally infected humans the CD4+ cell response seems to recognize epitopes on the internal proteins NP and M1 as well as the surface proteins NA and HA. CD8+ cytotoxic T lymphocytes seem to recognise epitopes within the HA, NP, M1 and PB2 proteins. Animal model studies have shown that both CD4+ and CD8+ T cells can contribute to immunity to influenza viruses. Animals that are deficient in both CD4+ and CD8+ T cells do not survive influenza infection, but animals deficient in only CD4+ or CD8+ T cells are able to clear virus.

The protection provided by influenza vaccines is thought to act principally through HAI antibodies, since they are poor at eliciting cytotoxic T lymphocyte responses (McMichael *et al.*, 1981). Production of IgG antibodies by B cells required CD4+ T cell help (Anders *et al.*, 1979), and because CD4+ T cells recognize antigen in association with HLA Class II molecules (Callard and Smith, 1981), polymorphism in these genes has the potential to modulate immune responses to subunit vaccines.

In a recent study (Gelder *et al.*, 2002), we have investigated a cohort of adults who fall into the 'at risk' category for influenza as defined by the Advisory Committee on Immunization Practices. We report an increased frequency of

HLA DRB1*07 and a decreased frequency of DRB1*13 in non-responders to influenza subunit vaccine when compared to matched responders to the same vaccine. The finding that HLA DRB1*07 is over-represented amongst individuals who fail to mount a neutralizing antibody response to influenza is of importance, because it potentially identifies a group of individuals who may not be protected by currently used vaccination strategies.

There may be several possible mechanisms by which HLA Class II might modulate antibody responses to the subunit vaccines. First, the defect may be in the presentation of appropriate antigens to CD4+ T cells: individuals who carry HLA DRB1*07 may fail to recognize peptide epitopes exhibited by the subunit vaccines, because suitable epitopes are either not present or not appropriately processed. By the same mechanisms, DRB1*13 (which was associated with responsiveness) may be particularly efficient at the processing and/or presentation of these antigens. However, it is of note that we have previously investigated CD4* T cell recognition of influenza A/Beijing/32/92 (H3N2) haemagglutinin following both natural infection and subunit vaccination (Gelder *et al.*, 1996). It was shown that HLA DRB1*07 is able to bind synthetic peptides spanning the sequence of this haemagglutinin, and that HLA DRB1*07-restricted CD4+ lymphocytes recognise and proliferate to these peptides.

A second explanation for the lack of an HAI antibody response is that non-responsiveness is a more general phenomenon relating to CD4+ T cell-derived help for B cells necessary for antibody production. The nature of such a defect is unclear, but it is compelling in this regard that HLA DRB1*07 has also been reported to be associated with low responses to hepatitis B vaccine, another highly purified soluble antigen (Alper *et al.*, 1989). Other studies have indicated that the association with non-responsiveness to hepatitis vaccine is with the HLA DRB1*07 DRB1*0202 haplotype (McDermott *et al.*, 1997). Thus it is conceivable that DRB1*07 individuals may have a more general defect in antibody responses to soluble antigens. It will be important in this regard to determine whether non-responders to influenza vaccine respond normally to other vaccines (including whole virus preparation and live vaccines). As a significant minority of DRB1*07 donors mount a normal antibody response to influenza vaccine, the deficit is more likely to be in a gene linked to DRB1*07 than in DRB1*07 itself.

THE PRACTICAL CONSEQUENCES OF ANTIGENIC CHANGE IN HA AND NA: PANDEMICS AND EPIDEMICS

Antigenic variation of influenza has immense practical implications, both in the yearly outbreaks of disease (epidemics) and also the massive worldwide outbreaks (pandemics). Major changes in subtype of influenza, termed **antigenic shift**, are infrequent and occur independently in the haemagglutinin and neuraminidase. Table 4.3 summarizes the old and new designations of these antigens and lists the prototype strains that contain them. The emergence of the pandemic Asian (H2N2) influenza virus in 1957 and its rapid replacement of the previously prevalent pandemic H1N1 virus is an example of the antigenic shift in both the haemagglutinin and the neuraminidase antigens. In contrast, the appearance in 1968 of the pandemic Hong Kong (H3N2) virus, which quickly replaced the H2N2 virus, is an example of antigenic shift in the haemagglutinin alone. All other genes were derived from the preceding H2N2 virus.

Table 4.3 Pandemic viruses: antigenic subtypes of influenza A virus in man during the twentieth century

Antigenic designation haemagglutinin and neuraminidase antigen subtype	Period of prevalence	Type of antigenic variation	Probable mechanism of origin of new virus
H1N1	1918	Shift	Unknown; likely to have emerged in France
	1918–57	Progressive antigenic drift	
	1957	Shift	Genetic recombination between former human H1N1 virus and an influenza A virus from unknown source with surface antigens H2N2. Genes for H2N2 transferred a new pandemic virus
H2H2	1957–68	Progressive antigenic drift	
H3N2	1968	Shift	Genetic recombination between former human H2N2 virus and an influenza A virus from an unknown source with surface antigens H3N2. Gene for H3 transferred to new pandemic virus
	1968–2003	Progressive antigenic drift	
H1N1	1976	Reappearance of swine H1N1 virus	Swine virus related to 1918 pandemic strain infects soldiers at Fort Dix, USA
H1N1	1977	Reappearance of 1950 H1N1 virus	Derived from 1950 H1N1 virus, preserved in nature by unknown mechanism or emerging as a laboratory accident
	1977–2003	Progressive antigenic drift	
H1N1	1989	Reappearance of swine H1N1 virus	Swine virus related to 1918 pandemic virus infects an adult and children at pig fair in USA
H5	1997, 2003	Potential shift	Emerged from chickens
H7	2003	Potential shift	Emerged from poultry

The subtypes of influenza A virus prevalent since 1918 are also shown in Table 4.3. The H1N1 subtype, which probably emerged as a human pandemic virus in 1916–1918 (Oxford et al., 2001, 2002; Phillips and Killingray, 2003), was prevalent for 39 years. In contrast, the H2N2 virus was prevalent for only 11 years, from 1957 to 1968. The H3N2 virus, which, at the time of writing, is still circulating in man, has been prevalent for at least 35 years, and viruses containing N2 neuraminidase have been prevalent since 1957. There is therefore no support whatsoever for the oft-quoted 10-year cycle of pandemic strains of influenza virus. On the contrary, pandemic viruses emerge at unpredictable intervals.

Pandemic disease, however, is not a necessary result of the introduction of a new antigenic subtype of influenza A virus into man. The outbreak associated with the reappearance of swine (H1N1) influenza A virus in man in 1976 was limited in extent and time, although no one under 20 years possessed antibody to this virus. Another entirely unexpected event in the epidemiology of influenza was the reappearance in 1977, after an absence of some 20 years, of H1N1 influenza viruses antigenically and genetically close to the H1N1 strains that circulated in 1950. This newly emerged H1N1 strain caused widespread outbreaks of influenza in children and young adults in many countries between 1977 and 1980 and is still causing disease in 2002; but it did not at first cause illness in persons over the age of 23 years, who were likely to have been naturally infected with H1N1 virus during its earlier period of prevalence up to 1956. The epidemiology of influenza in man since 1977 has been unusually complex, with the simultaneous prevalence of two subtypes of influenza A virus – H1N1 and H3N2 – together with influenza B virus. H1N1 has undergone antigenic drift and older persons are now infected with the new variant. Reassortant viruses with genes derived from both H1N1 and H3N2 subtypes have also been isolated.

At comparatively frequent intervals, minor changes, termed **antigenic drift**, take place in one or both of the HA and NA antigens. Although each episode of 'drift' is in itself minor, the effects are additive and over a period of several years result in strains differing considerably from the original pandemic virus. In antigenic shift virtually all antigenic determinants of the haemagglutinin antigen undergo change, whereas in antigenic drift only certain epitopes change, while others are conserved and enable the haemagglutinin to be identified as belonging to a specific subtype (i.e. H1, N2 or H3). In numerical terms, drift is caused by one or two amino acid substitutions in the HA or NA per year, particularly in and around the major epitopes such as A or B of the HA (Figure 4.2). The frequency of substitutions in HA1 has been about 0.8% per year over the past 31 years. Since five antigenic epitopes have been described in the HA, and as the frequency at which antigenic mutants occur after selection with monoclonal antibodies is between 10^{-3} and 10^{-5}, the possibility of simultaneous changes in all sites is very remote. However, human sera are known to contain antibodies with a very restricted virus-neutralizing repertoire, sometimes to a single epitope (Natali et al., 1981) and may therefore allow emergence of new variants with successive changes in epitopes A–E. The viruses of subtypes H1N1, H2N2 and

H3N2 underwent progressive antigenic drift in their haemagglutinin and neuraminidase antigens during their periods of epidemic activity in man from 1918 to 1956, 1957 to 1967 and 1968 to 2003 respectively.

Influenza vaccines contain the H1N1 and H3N2 virus subtypes and also influenza B. The composition is changed to match antigenic drift. Therefore a view of vaccine reformulation over the years (Figure 4.3) indicates the higher degree of antigenic change in H3N2 viruses compared to H1N1 or influenza B. In contrast, influenza B exists as single subtypes, and the progressive antigenic changes undergone by the prevalent strains since their first isolation in 1940 are attributed entirely to antigenic 'drift'. But the epidemiology of these viruses may also be complex and examination of many influenza B viruses isolated over a period of 40 years showed that serological and genetic relationships may be complicated; e.g. some viruses isolated in 1979 were, serologically, more related to 1970 viruses than to viruses circulating during the previous year.

THE CONFLICTING THEORIES OF THE ORIGIN OF PANDEMIC INFLUENZA

A major unsolved question is the origin of these pandemic influenza A viruses. Two conflicting theories suggest either a recycling phenomenon whereby viruses that caused pandemics in the last century or early in this century return once more to affect mankind (Hope-Simpson, 1992) or, alternatively, that new viruses emerge from the vast influenza ecosystem in birds, swine or horses and infect humans either directly or indirectly after reassorting with an existing human influenza A virus strain (Webby and Webster, 2001). Table 4.4 summarizes the classic HA subtypes and perfectly illustrates the almost boundless ecology of influenza A virus, particularly in birds. At present, the animal or bird theory has most scientific support, but neither hypothesis is proven. The recycling hypothesis presupposes some form of viral latency whereby the virus remains dormant for many years before re-emerging. It might seem unlikely that a virus could cross the strong species barriers from animals and birds to humans, but there is an example in the movement of A/swine and vice-versa and the infection of people in Hong Kong in 1997 by chicken H5 virus (Yuan *et al.*, 1998). The question is not academic because a definitive answer would enable strategic planning of a new pandemic and even stockpiling of vaccines from 'old' human strains or from known animal and bird influenza A viruses (PHLS, 1997).

THE 1918/19 GREAT PANDEMIC OF INFLUENZA A AND SUBSEQUENT OUTBREAKS IN HUMANS CAUSED BY INFLUENZA A/SWINE (H1N1) VIRUS

The years 1918–19 saw the most severe pandemic of influenza yet recorded in human history, the so-called 'Spanish influenza', although there is strong

Table 4.4 Classic HA and NA subtypes of influenza A virus in birds

HA subtype designation	NA subtype designation	Avian influenza A virus
H1	N1	A/duck/Alberta/35/76(H1N1)
	N8	A/duck/Alberta/97/77(H1N8)
H2	N9	A/duck/Germany/1/72(H2N9)
	N9	A/duck/Germany/1/72(H2N9)
H3	N8	A/duck/Ukraine/1/63(H3N8)
	N8	A/duck/England/62(H3N8)
	N2	A/turkey/England/69(H3N2)
H4	N6	A/duck/Czechoslovakia/56(H4N6)
	N3	A/duck/Alberta/300/77(H4N3)
H5	N3	A/tern/South Africa/61(H5N3)
	N9	A/turkey/Ontario/7732/66(H5N9)
	N1	A/chick/Scotland/59(H5N1)
H6	N2	A/turkey/Massachusetts/3740/65(H6N2)
	N8	A/turkey/Canada/63(H6N8)
	N5	A/Shearwater/Australia/72(H6N5)
	N1	A/duck/Germany/1868/68(H6N1)
H7	N7	A/fowl plague virus/Dutch/27(H7N7)
	N1	A/chick/Brescia/1902(H7N1)
	N3	A/turkey/England/63(H7N3)
	N1	A/Fowl plague virus/Rostock/34(H7N1)
H8	N4	A/turkey/Ontario/6118/68(N8N4)
H9	N2	A/turkey/Wisconsin/1/66(H9N2)
	N6	A/duck/Hong Kong/147/77(H9N6)
H10	N7	A/Chick/Germany/N/49(H10N7)
	N8	A/quail/Italy/1117/65(H10N8)
N11	N6	A/duck/England/56(H11N6)
	N9	A/duck/Memphis/546/74(H11N9)
H12	N5	A/duck/Alberta/60/76(H12N5)
N13	N6	A/gull/Maryland/704/77(H13N6)
N14	N4	A/duck/Gurjey/263/83(H14N4)
H15	N8	A/duck/Australia/341/83(H15N8)
	N9	A/shearwater/Australia/2576/83(H15N9)

evidence of emergence in France during the Great War (Oxford et al., 2002). The first large wave in the spring of 1918 was relatively mild, the attack rate being 20–40% in those aged up to 50 years but lower in the elderly. But the second wave, in the autumn of 1918, was exceptionally severe, with huge numbers of deaths in those aged 20–40 years (Figure 4.4). It had been preceded by localized army camp outbreaks in 1916 (Oxford, 2002; Oxford

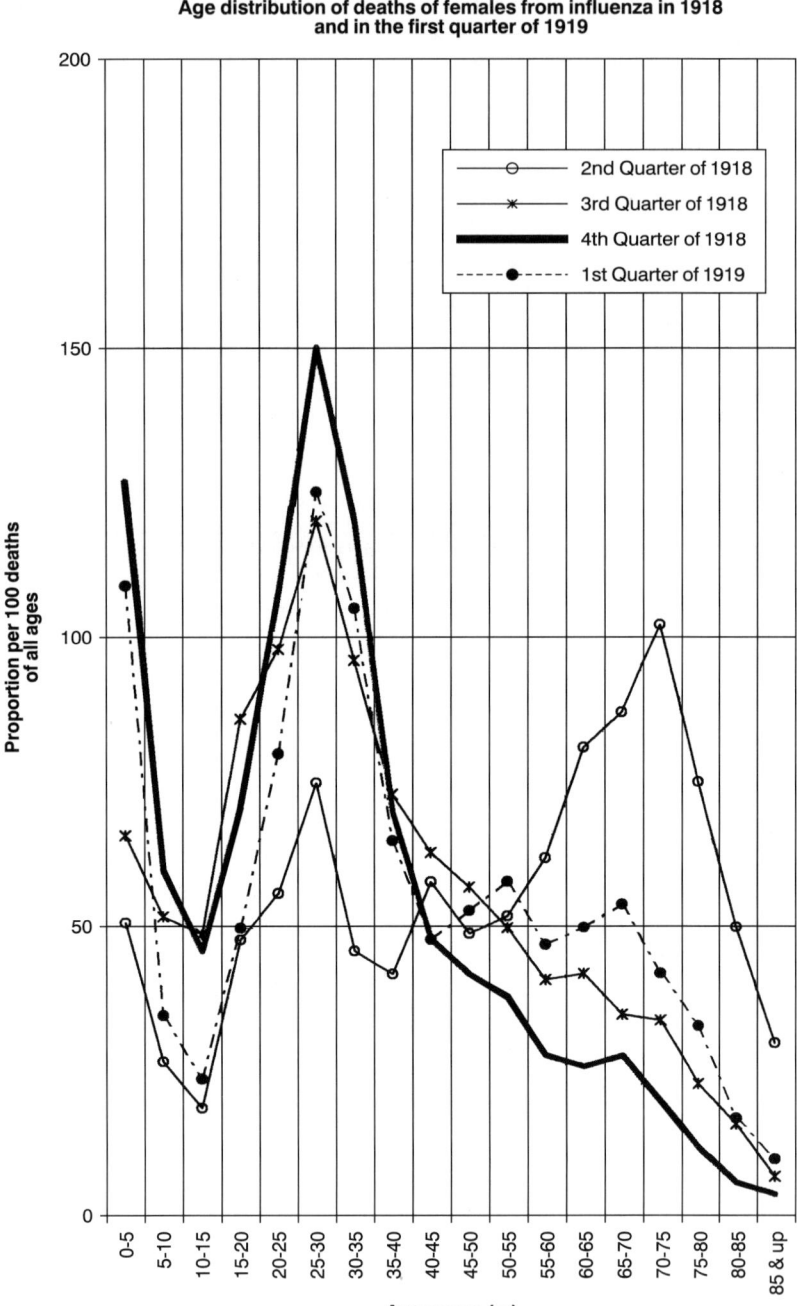

Figure 4.4 Increased mortality in the young compared to the elderly in the Great Pandemic of 1918–1919.

et al., 2002). A less severe outbreak, the third wave (Winternitz *et al.*, 1920; Stuart-Harris *et al.*, 1983; Oxford, 2002), occurred in early 1919. Worldwide, 40 million people are now estimated to have died (Phillips and Killingray, 2003). Retrospective serological evidence and genetic analysis of formalin-fixed and frozen clinical material at the time indicate that this pandemic was caused by the classic (H1N1) influenza virus of a human clade but genetically close to the pig clade (Taubenberger *et al.*, 1997; Davis *et al.*, 2000). We will return to the striking impact of this infection below.

Limited outbreaks caused by a similar influenza A virus strain occurred in military recruits in New Jersey, USA, in January 1976 (Dowdle and Millar, 1978) and in the autumn of 1988 in visitors to a country fair in the USA. There was concern, in the event unjustified, that the first outbreak heralded the onset of a new pandemic era; so in the winter of 1976/1977, 40 million doses of vaccine were administered to persons of all age groups in the USA.

THE 1957 'ASIAN' INFLUENZA PANDEMIC

The Asian (H2N2) influenza virus first appeared in central China in February 1957 (Kilbourne, 1975) and possessed haemagglutinin and neuraminidase antigens that differed completely from the formerly prevalent H1N1 strain and represented the first antigenic shift in the influenza A virus since the pandemic of 1918. The new virus spread rapidly, and virtually every country in the world had been infected by the end of November 1957. The virus had thus behaved in a truly pandemic manner. Some countries, notably Japan and the USSR in early September 1957 and the UK and USSR in early 1958, experienced second waves of infection. The rates of attack by the Asian virus were high during the 1957/58 pandemic period, varying from 20% to 80% in different countries. The highest rates were in the younger age groups and in areas of high population density (reviewed by Stuart-Harris *et al.*, 1983). In the UK and the USA attack rates of about 50% were recorded in the 5–15 year age group. In 1957, before the circulation of the Asian virus, some 30% of serum samples from persons aged 75–85 years contained antibody to the Asian strain. This has been taken to indicate that the Asian virus of 1957 had circulated some 68–70 years previously, and gives some credence to the recycling hypothesis noted above. As already mentioned, influenza pandemics are known to have occurred in the period 1889/90 and might thus have been associated with an earlier pandemic of the H2N2 virus. Between 1958 and 1968 many epidemics of H2N2 infection occurred around the world. By 1967 progressive antigenic 'drift' had occurred in the haemagglutinin and neuraminidase antigens of the prevalent strains, typified by the A/England/12/64 (H2N2) strain in 1965 and the A/Tokyo/3/67 (H2N2) strain in 1967/68. More recent studies have focussed on H2 as a possible re-emergent virus able, potentially, to cause another pandemic.

THE VANISHING PANDEMIC VIRUS

The profile of the percentage of seropositive subjects to A/Singapore/1/57 (H2N2) observed in the 1894–1899 age group, compared to the younger age group, is consistent with the phenomenon of the vanishing subtype, namely a very sudden disappearance of an old pandemic strain rather than a protracted exit from the human community. At present the proportion of the general population exhibiting influenza A (H2N2) antibodies, in comparison to influenza A (H3N2) and (H1N1) antibodies, is low, and there is no serological evidence of recent circulation of influenza A (H2N2) virus in the community (Figure 4.5).

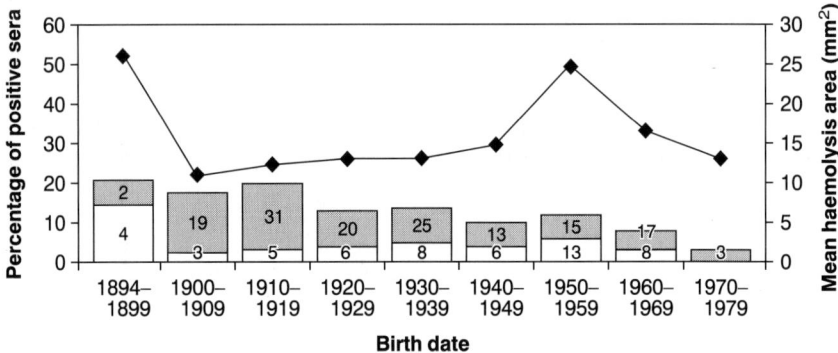

Figure 4.5 Current antibody levels (H1 and SRH) in human sera to influenza A (H2N2) virus from the 1957 pandemic.

The most strongly supported theory on the origin of pandemic influenza suggests a natural animal reservoir for the generation of new human influenza A subtypes. This theory gained added credence when it was found that the Asian (H2N2) strain that appeared in 1957 had retained only five genes of the (H1N1) predecessor, while the remaining genes could have been acquired from avian influenza virus (reviewed by Stuart-Harris et al., 1983). Nevertheless, the possibility still exists that an influenza A virus such as H2N2 could re-emerge from humans, as hypothesized by the alternative theory on the origin of pandemic influenza A viruses, namely recycling of older previous pandemic viruses (Hope-Simpson, 1992). As the number of people already in the community who are seropositive to influenza A (H2N2) virus will decrease as the number of seropositive old people die and the number of people born after 1967 will increase, we would predict that the majority of the population would be susceptible to infection with a re-emerging influenza A (H2N2) virus (Figure 4.5). Obviously a virus such as influenza A (H2N2) could re-emerge as a possible 'new' pandemic virus, from a bird, pig or even human reservoir.

THE 'HONG KONG' VIRUS PANDEMIC AND THE PERIOD 1968-2003

The Hong Kong (H3N2) virus was isolated during influenza outbreaks in Hong Kong in July 1968. The virus spread through India, Japan and Australia by the early autumn (Cockburn *et al.*, 1969). However, only the USA suffered major outbreaks, with increased mortality rates during the winter of 1968/9; in other countries epidemics were mild. The full effect of the Hong Kong pandemic was not felt in Europe, Asia and most other countries until the winter of 1969/70; mortality and morbidity were less severe than in the Asian pandemic in 1957. The antigenic changes in the virus may provide a clue to its epidemiological behaviour and somewhat reduced virulence. Although the haemagglutinin antigen (H3) was completely different from that of the Asian virus (H2), the neuraminidase was of the same antigenic subtype (N2). Thus the Hong Kong virus showed only partial antigenic shift. The common neuraminidase antigen of the Asian and Hong Kong viruses may thus have contributed to a cross-immunity between these strains, which modified the epidemiological impact and initial virulence of the Hong Kong virus.

A CURIOSITY AND A WARNING: REAPPEARANCE OF A INFLUENZA A (H1N1) VIRUS IN 1977

In May 1977, influenza A viruses of the H1N1 subtype were isolated in the northern regions of the People's Republic of China. The outbreak appeared to affect only school children and others under the age of 23 years. In these groups attack rates were high but the illness was generally mild. The infection spread rapidly around the world, but nowhere were there significant numbers of clinical infections in persons older than 25 years, and no country reported increased mortality rates associated with this virus. There is evidence, based on serological investigations made between 1977 and the present, of frequent subclinical infections in older persons with H1N1 virus which has undergone considerable antigenic drift. It is interesting that reassortants that have genes from both H3N2 and H1N1 viruses have been isolated in nature, but there is no evidence that they differ in virulence from the parental viruses.

Conventional antigenic analysis demonstrated a close resemblance between the haemagglutinin and neuraminidase antigens of the 'new' H1N1 strains and those of a virus isolated in the USA in 1950, A/Fort Warren/50 (H1N1). An analysis of the genetic composition of the virus by Nakajima *et al.* (1978), using oligonucleotide mapping of virus RNA, showed that all eight gene segments of the 1977 H1N1 virus had a high degree of sequence homology with those of the A/Fort Warren/50 strains. There is no definitive explanation for the sudden reappearance of this virus. It is possible that the virus represents a genuine 'recycling' of the old strain but, given the sequence homology of the 1950 and 1977 viruses, the two must have been truly latent or genetically 'frozen'. An alternative explanation is that the virus was inadvertently released from a laboratory – not the first time a virus has been thus transmitted. In the first studies with influenza in London in the 1930s a laboratory worker was infected from a ferret, itself given a human virus three days previously.

NUCLEOTIDE SEQUENCE ANALYSIS OF THE INFLUENZA HA GENE FROM SAMPLES FROM THE GREAT 1918 PANDEMIC

The cataclysmic events of the Great Influenza Pandemic in 1918 have provided the impetus for the greater scientific understanding of antigenic structure and correlation of virulence with particular influenza virus genes, particularly the HA gene. Molecular genetics has now allowed a view of the genetic sequence of approximately half of the 1918 influenza genome. The powerful methodologies of reverse genetics now allow a study of the biology and antigenicity of all the structural proteins of this unique virus. To date there is published data on the nucleotide sequence of the HA, NA and NS1 genes (Taubenberger *et al.*, 1997; Reid *et al.*, 1999, 2000; Basker *et al.*, 2001). Retrospective genetic studies of the causative 1918 virus can now be carried out either from pathology museum samples of victims' lung which were wax embedded at the time or from frozen tissue of victims who died in arctic regions (Davis *et al.*, 2000). Taubenberger and his colleagues accessed the pathology collection of the Armed Forces Institute in Washington and examined formalin-fixed lungs from more than 80 victims of the October 1918 outbreak. Sequence data obtained from the HA, NP and NA genes in the initial study and a complete HA sequence from the second study has placed the 1918 virus amongst the influenza A(H1N1) swine or human viruses and this virus appears similar to the first human influenza A viruses isolated, such as A/WSN/33 virus. Our own group has examined pathology samples from victims at the beginning and end of the outbreak in 1919, and these are remarkably similar to the American samples as regards HA sequence (Reid *et al.*, 2003).

A more exhaustive analysis of a subset of representative sequences examined every possible tree, and again the 1918 HAs were placed within the human clade. Analysis of the partial nucleotide sequence of the NP gene also indicated that the gene from the 1918 virus was of the mammalian/swine type rather than avian. This conclusion is consistent with veterinary observation in 1918 of a new respiratory disease in pigs, which later studies showed to be an influenza A virus. Therefore the human influenza A virus appeared to have spread from humans to pigs, where it still resides and indeed re-emerged in 1976/1977 to cause a limited outbreak in Fort Dix. There is a single report which presents computer evidence that the 1918 HA gene is a recombinant (Gibbs *et al.*, 2002). Given the current hypothesis that all human influenza A viruses may have an avian origin the authors concluded that either (i) the progenitor avian strain did exist around 1918, but that it was different from current avian strains, or (ii) the pandemic strain had been adapting in humans before 1918 perhaps between 1900 and 1915.

Previous genetic studies of virulent avian influenza A viruses had identified an inserted sequence of hydrophobic amino acids at the HA1–HA2 junction (Perdue et al., 1997). In addition, loss of a carbohydrate from the HA, which could mask the HA1–HA2 junction, has been shown to be correlated with virulence. For avian influenza A viruses the extra insert extended the spectrum of cellular proteases that were able to cleave HA, so removing the limitation to replication in the respiratory tract. Laboratory studies of avian and other influenza A viruses had also identified a contribution of virulence from NA, M, NSI and NP genes. It is possible that sequence analysis of the entire genome of the 1918 virus may not provide a definitive answer about virulence and nucleotide sequence, particularly if the virulence is multi-genic and involves only few nucleotide changes scattered in several genes. Indeed, in a unique experiment whereby reverse genetics enabled the transposition of the NSI gene of the 1918 virus into another influenza A virus, no increase in virulence for mice was detected, rather the opposite (Basker et al., 2001).

Partial sequence analysis of the HA of a 1919 sample is now complete (A. Reid, J. Taubenberger, A. Elliot, R. Daniels and J. S. Oxford) and shows remarkable similarity to the 1918 specimens. We are now searching The London Hospital archive for formalin-fixed lung samples from influenza victims in 1914, 1915, 1916 and 1917 to obtain a more complete sequence analysis and comparison of HAs before, during and after the main autumn outbreak.

COULD THE HUMAN INFECTIONS IN HONG KONG IN 1997 WITH AVIAN INFLUENZA A VIRUSES WARN OF A NEW PANDEMIC?

Earlier experimental studies in volunteers at the Common Cold Unit in Salisbury had shown that avian influenza viruses, at least of the H7 and H4 subtypes, could break through the species barrier and infect humans. Only mild disease was induced in these volunteer studies and no person-to-person spread of virus was detected.

In contrast, an avian influenza A virus (H5N1) was recovered from a post mortem respiratory sample from a 3-year-old child who died in May 1997 in Hong Kong (Yuan et al., 1998). A similar virus was known to have caused extensive mortality in chickens in the same region in the preceding weeks. There was evidence of a direct link between the child and chickens. Genetic and biological analysis of the virus from the human case established that the HA retained a preference for receptors on avian cells, namely sialic receptors attached to a galactose residue through α, 2, 3 linkage. The HA1 of the H5 human isolate differed in three amino acids compared with the avian viruses and a potentially significant mutation in the human isolate was an absence of carbohydrate at asparagine 156 adjacent to the receptor-binding pocket. Other features of note in the human isolate included basic amino acids at the HA1/HA2 junction. The virus possessed a unique 19 amino acid deletion in the NA. An antigenic similarity to the avian virus was detected by polyclonal antibodies. The human virus isolate retained virulence for chickens and also mice.

Seven months later a second case was detected in a 2-year-old child and then in a further 16 cases. A total of six patients died. Most patients had typical influenza with fever and signs of infection of the respiratory tract. The disease appeared more severe in adults than children. Surveillance of live birds in markets in Hong Kong during this latter period detected influenza H5N1 virus in chickens, ducks and geese. Avian subtypes H9, H6 and H11 were also isolated. The H5 isolates appeared to cause high mortality in chickens but not in ducks or geese. Most significantly there was no evidence of human-to-human transmission of the H5 virus and, as far as can be ascertained, all ill persons had a link with the live chicken markets. All chickens in Hong Kong were slaughtered at the end of December 1997 and no human cases of H5N1 have been reported since, although another avian subtype, H9, now appears to have also spread to humans and there are continuing smaller outbreaks of avian influenza in chickens, whilst H7 has been detected in chicken farmers in The Netherlands.

BIOLOGICAL FACTORS WHICH DRIVE ANTIGENIC DRIFT OF HA

An obvious immunological factor which would be expected to exert strong selective pressure on a quasi-species virus like influenza would be neutralizing antibody or T cells directed towards the HA. It is not uncommon for influenza to infect 10–20% of the population each year and these persons develop neutralizing antibodies and T cell responses directed towards one or more of the four or five epitopes on the HA. However, it must be acknowledged that a conundrum remains. The mutation rate of influenza remains, as noted above, exceptionally high and the virus exists as a quasi-species. Each replication cycle can throw up a mutant for every 10^4 nucleotides transcribed. Given the fact that the entire genome is only 12 k or so nucleotides, it is apparent that mutants of all sorts and all genes will abound. However, the curiosity is that for drift to occur and to allow a drifted virus to infect an individual it would be expected that mutations would be needed in all four epitopes of the HA because individuals would be expected to have antibodies to all four epitopes. But the chances of mutation at four epitopes occurring in a single individual at one time would be remote. So how do new mutants arise and survive? We surmise that sequential pressure in the series of individuals would be required to generate a novel antigenic variant, thus allowing mutations in epitope A in one individual followed by mutation in epitope B in another person and so on. In addition, the surprising finding of the restricted immune response described by Natali *et al.* (1981) could provide an explanation for this.

Within this wider context, Schild and colleagues (1983) published a further surprising finding, namely that biological pressure could force change in antigenic structure of the HA. Essentially, an influenza virus isolate was cultivated in parallel in mammalian MDCK cells and embryonated hen eggs. When the two sets of progeny virus were compared marked antigenic differences were detected in the HA by both monoclonal and polyclonal antibodies. The genetic explanation was found in the amino acid sequence around the receptor-binding site at the tip of the HA molecule. The two viruses differed by single amino acids but since the receptor site bordered an antigenic site, reactivity of the HA with antibodies was altered. Importantly, the reactivity of the egg-grown and mammalian cell-grown virus with post-infection human antibodies was investigated. The MDCK cell-grown virus reacted to higher titres with natural human post-infection antibodies suggesting that, antigenically, this MDCK virus was more related to the natural, uncultivated wild virus with which the volunteer had been infected. Subsequent nucleotide analysis of wild-type virus in throat swabs or direct probing with monoclonal antibodies attached to gold particles showed the presence of a mixture of viruses but in which the MDCK-like genotype predominated.

These studies raised immediate questions about influenza vaccines. It is presumed that the most effective vaccine would have a close antigenic match to the natural virus. However, all vaccines to the present day have been produced using egg-grown virus. In practical terms new mammalian-cell substrates (MDCK and Vero cells) have now been licensed for virus vaccine production, which may allow a new generation of vaccines to be made.

Table 4.5 shows haemagglutinin inhibition (HI) reactions of two influenza B viruses representative of 27 strains, isolated and passaged in eggs or canine kidney cell cultures (MDCK) from specimens collected during a school outbreak

Table 4.5 HI reactions of anti-HA monoclonal antibodies with an influenza B virus isolated on mammalian cells or eggs

Isolation and passage history of B/England/82 viruses		Monoclonal antibodies to:		
		Egg-grown B/Oregon/5/80 virus		MDCK cell-grown B/England/145/82 virus
Specimen no.	Cell substrate for virus isolation and passage	195	238	209
145	M-M	25 600	<	320
	M(x3)	12 800	<	320
	M-egg	38 400	9600	<
	M-egg	12 800	9600	<
	M-egg	12 800	9600	<
222	M-M	>6400	<	160
	M-egg	>6400	6400	<
	Egg	4800	6400	<
	Am-egg-M(x3)	6400	6400	<

Table 4.6 HI antibody in human sera to influenza B virus grown in mammalian cells or eggs

Virus origin and passage history	% of sera with stated HI antibody titres:							
	Children				Young adults			
	>10	>40	>160	GMT	>10	>40	>160	GMT
B/England/222/82 cell-grown	83	76	57	110	100	100	100	590
B/Eng/222/82 egg-grown	20	8	2	>10	20	2	1	<10

GMT, geometric mean HI titre.

of influenza. The virus preparations were tested against anti-HA monoclonal antibodies prepared against influenza B virus cultivated in eggs or MDCK cells. Three patterns of serological reactivity were apparent. Monoclonal antibody 195 reacted to high titres with all the influenza B viruses tested. Monoclonal antibody 209 reacted with all the viruses that had been isolated and subsequently passaged in MDCK cultures but not with those that had been cultivated in eggs at any stage in their passage histories. Antibody 238 reacted only with viruses that had at least one passage in eggs and with none of the viruses that had been passaged in MDCK cultures alone. The mutual exclusivity of the reactions of antibodies 209 and 238 with the virus preparations was striking. The clear antigenic differences between the influenza B viruses described above that differed in their reactions with monoclonal antibodies were also detectable with polyclonal human sera (Table 4.6).

These findings provided the first strong evidence that influenza B virus isolated from clinical specimens and grown in MDCK cultures comprises at least two biologically and antigenically distinct subpopulations. Subsequent studies showed that the cell selection phenomenon was also detectable in influenza A(H1N1) and (H3N2) viruses and equine and avian influenza A viruses (Matrosovich *et al.*, 1997; Ilobi *et al.*, 1998). A minor proportion of the MDCK-grown virus is capable of growth in the allantois of fertile eggs without previous adaptation, and possesses HA that is antigenically distinct from the majority of the virus, which grows poorly or not at all in the allantoic cavity. These studies were not unique in providing evidence of distinct subpopulations of virus within a single strain. Other reports have documented biological or antigenic heterogeneity among populations of influenza A virus cultivated in eggs. However, the studies uncovered for the first time evidence of strong selective pressure for antigenic variants exerted by the cultivation of the MDCK cell/egg systems. It was quickly noted that serological studies of the antigenic characteristics of virus strain for epidemiological purposes or studies of cellular or humoral parameters of immunity in man that are performed exclusively with egg-grown virus may give misleading or at best incomplete data. Importantly these studies also raise questions about the choice of substrate for the preparation of inactivated influenza vaccines. Antigenic differences between egg-grown influenza virus and virus shed by infected persons may be one of the several factors which contribute to the limited efficacy of inactivated influenza vaccines. Finally, non-immune selection of antigenically variant viruses in nature may occur by the selective growth of viruses in different target cell types in the respiratory tract of man and other natural hosts and thereby provide a further potential mechanism for antigenic 'drift'.

WOULD INFLUENZA VACCINE VIRUSES REPLICATED IN MDCK CELLS RATHER THAN EGGS BE MORE IMMUNOGENIC?

To briefly recap, antigenic variation of influenza virus typically results from immune pressure but can also arise by selection pressure from the host cell in which the virus replicates (Schild *et al.*, 1983). The molecular basis for such antigenic differences has been shown to reside in a variety of single amino acid changes in the haemagglutinin (HA) molecules, which are all clustered around a hollow forming the proposed receptor-binding site (RBS) of the HA trimer (Robertson *et al.*, 1987). Furthermore, analysis of antibodies in post-infection human sera has shown that influenza A (H1N1), A (H3N2), and B viruses isolated on a mammalian cell line (MDCK) detect antibody more frequently and to a higher titre than do the corresponding egg-adapted viruses (Schild *et al.*, 1983). A possible interpretation of these results is that the mammalian cell-grown virus is antigenically more related to virus causing infection in humans. An important question was whether these antigenic differences in HA are relevant to influenza vaccine efficacy. Inactivated influenza vaccines were prepared from both an egg- and an MDCK cell-growth isolate of a human influenza A (H1N1) virus that had been previously well characterized serologically with monoclonal and polyclonal antibodies, and biochemically by nucleotide sequence analysis (Robertson *et al.*, 1987). The immunogenicity of the HA component of the vaccines was tested in a number of animal models and, subsequently, intranasal challenge experiments were performed in hamsters with infective homologous and heterologous viruses.

The results of model experiments demonstrated a profound difference in the specificity of antibody stimulated by the two inactivated influenza virus A(H1N1) vaccines where the HAs differ by two amino acids. Vaccine prepared from 157M virus induced equivalent amounts of strain-specific antibody for 157M virus and antibody that cross-reacted well with 157E virus. However, 157E vaccine produced exclusively strain-specific antibody. The differences in antibody specificity could be anticipated in view of the location of residue 189 (Glu in 157M, Lys in 157E) at the tip of the HA molecule adjacent to antigenic site Sb, which is a major antigen on influenza A (H1N1) virus HA. It is possible that the absence of a carbohydrate at residue 163 on 157E virus HA (Asn in 157M, Ser in 157E) also influences immunogenicity, although this is unlikely, as a variant of 157 virus has been selected, which resembles 157M virus antigenically but the only structural difference from 157M virus is the absence of a glycosylation site at residue 163 of the HA (Robertson et al., 1987). In addition to the antigenic changes in HA, the substitutions at positions 189 and 163 could also have an effect on receptor binding. Residue 189 lies on the upper edge of the RBS of the HA molecule and, although residue 163 is not closely associated with the RBS, a carbohydrate attached to Asn-163 may influence the RBS of an adjacent HA monomer.

Subsequent molecular analysis showed that the HA of influenza B virus and also A (H1N1) virus in a clinical sample was in fact identical to that of the MDCK cell-grown virus rather than to the egg-grown virus (Robertson et al., 1991). Gubareva et al. (1994) illustrated the practical consequences of selective pressures in WHO reference viruses, which had, in fact mixtures of influenza viruses.

These selective pressures described for human influenza viruses also apply to equine influenza A viruses (Ilobi et al., 1998) but with the proviso that the egg-grown virus counterpart may be more relevant clinically than the mammalian cell-grown virus. Finally, Gambaryan et al. (1997) have described the application of synthetic sialylglycopolymers to separate influenza viruses into groups which correlate with cultivation on mammalian cells or eggs.

CONCLUSIONS

Nearly 70 years of detailed scientific work has placed the human influenza A virus at the pinnacle of virological knowledge. Compared to any other human virus, more is known about the eight genes of the virus, genetic evolution and antigenic structure. X-ray crystallography has detailed the epitopes of the HA and NA and allowed the design of new antiviral drugs. Reverse genetics now facilitates work into protein and gene structure and biological function such as virulence (Garcia et al., 1996; Solorzano et al., 2000; Basker et al., 2001). It is well established that each of the eight genes contributes to virulence of the virus (reviewed by Burnet, 1979; Oxford et al., 1978; Zambon, 2001). But the HA and NA genes are acknowledged to be the most important contributors to virus pathogenicity.

From a practical viewpoint vaccines have been developed which, in spite of the yearly antigenic changes of the virus HA, offer a 70–80% protective effect against hospitalization, serious illness and death in the 'at risk' groups (Fedson et al., 1993).

But above all looms the spectre of a new global pandemic of influenza (Meltzer, 1999; Oxford, 2002). Stocks of the new anti-neuraminidase antivirals, as well as amantadine, would be of immense importance in combating the first pandemic wave. The new MDCK or Vero cell culture systems (Kistner *et al.*, 1998) could allow one billion doses of monovalent vaccine to be produced in the first 9–12 months before the first wave.

Finally, detailed analysis of the geographical origin in France (Oxford, 2001), the time of first emergence of the virus (Oxford, 2002) and the nucleotide sequence of genes of the 1918 pandemic (Taubenberger *et al.*, 2001) virus has heightened awareness of pandemics and encouraged national countries to make pandemic plans. John Donne's line 'No man is an island' can have no stronger medical relevance than with a virus such as influenza, spread rapidly around the world by aerosol, with symptomless carriers and with a respected and formidable pathogenicity.

ACKNOWLEDGEMENTS

Our genetic study of 1918 influenza is supported by the Wellcome Trust.

REFERENCES

Alper, C. A., Krushall, M. S., Marcus-Bagley, D. *et al.* (1989) *N. Engl. J. Med.* 321, 708–712.
Anders, E. M., Peppard, P. M., Burns, W. H. and White, D. O. (1979) *J. Immunol.* 123, 1356–1361.
Basker, C. F., Reid, A. H., Dybing, J. K., Janczewski, T. A., Fanning, T. G., Zheng, H., Salvatore, M., Perdue, M. L., Swayne, D. E., Garcia-Sastre, A., Palese, P. and Taubenberger, J. K. (2001) *Proc. Natl Acad. Sci. USA* 98, 2746–2751.
Brand, C. M. and Skehel, J. J. (1972) *Nature* 238, 145.
Bullough, P. A., Hughson, F. M., Skehel, J. J. and Wiley, D. C. (1994) *Nature* 371, 37–43.
Burnet, F. M. (1979) *Intervirology* 11, 201.
Callard, R. E. and Smith, C. M. (1981) *Eur. J. Immunol.* 11, 206–212.
Chen, W., Calvo, P. A., Maude, D., Gibbs, J., Schubert, U., Bacik, I., Basta, S., O'Neill, R., Schickli, J., Palese, P., Henklein, P., Bennink, J. R. and Yewdell, J. W. (2002) *Nature Med.* 7, 1306–1310.
Cockburn, W. C., Delon, P. J. and Ferreira, W. (1969) *Bull. WHO* 41, 345–348.
Colman, P. M., Laver, W. G., Varghese, J. N., Baker, A. T., Tulloch, P. A., Air, G. M. and Webster, R. G. (1987) *Nature* 326, 358–363.
Cox, N. J. and Kawaoka, Y. (1998) In *Topley and Wilson's Virology*, Vol. I (eds B. W. J. Mahy and L. Collier), Arnold, London, pp. 385–433.
Crosby, A. W. (1989) *America's Forgotten Pandemic. The Influenza of 1919*. Cambridge University Press, Cambridge.
Davis, J. L., Heginbottom, J. A., Anna, A. P., Daniels, R. S., Berdal, B. P., Bergan, T., Duncan, K. E., Lewin, P. K., Oxford, J. S., Roberts, N., Skehel, J. J. and Smith, C. R. (2000) *J. Forensic Sci.* 20, 68–76.
Dowdle, W. R. and Millar, J. D. (1978) *NY State J. Med.* 62, 1047–1057.

Fedson, D. S., Wajda, A., Nicol, J. P., Hammond, G. W., Kaiser, D. L. and Ross, L. L. (1993) *J. Am. Med. Assoc.* 270, 1956–1961.
Fleming, D. M. (2001) *Phil. Trans. R. Soc. Lond.* 356, 1933–1943.
Francis, T. (1953) *Ann. Intern. Med.* 39, 203.
Gambaryan, A. S., Tuzikov, A. B., Piskarev, V. E., Yamnikova, S. S., Lvov, D. K., Robertson, J. S., Bovin, N. V. and Matrosovich, M. N. (1997) *Virology* 232, 345–350.
Garcia, M., Crawford, J. M., Latimer, J. W., Rivera-Cruz, E. and Perdue, M. L. (1996) *J. Gen. Virol.* 77, 1493–1504.
Gelder, C. M., Lamb, J. R. and Askonas, B. A. (1996) *J. Virol.* 70, 4787–4790.
Gelder, C. M., Lambkin, R., Hart, K. W., Fleming, D., Martin, O. M., Gaughran, F., Bunce, M., Welsh, K. I., Marshall, S. E. and Oxford, J. (2002) *J. Infect. Dis.* 185, 114–117.
Gibbs, M. J., Armstrong, J. S. and Gibbs, A. J. (2002) *Science* 243, 1842–1844.
Gubareva, L. V., Kaiser, L. and Hayden, F. G. (2000) *Lancet* 355, 827–835.
Gubareva, L. V., Wood, J. M., Meyer, W. J., Katz, J. M., Robertson, J. M., Major, D. and Webster, R. G. (1994) *Virology* 199, 89–97.
Hay, A. J., Gregory, V., Douglas, A. R. and Lin, Yi Pu. (2001) *Phil. Trans. R. Soc. Lond.* 356, 1861–1870.
Hayden, F. G., Gubareva, L. V., Monto, A. S., Klein, T. C., Elliott, M. J., Hammond, J. M., Sharp, S. J. and Ossi, M. J. (2000) *N. Engl. J. Med.* 343, 1282–1289.
Hirst, G. K. (1942) *J. Exp. Med.* 75, 47.
Hope-Simpson, R. E. (1992) *The Transmission of Epidemic Influenza.* Plenum Press, New York.
Ilobi, C. P., Henfrey, R., Robertson, J. S., Mumford, J. A., Erasmus, B. J. and Wood, J. M. (1994) *J. Gen. Virol.* 75, 669–673.
Ilobi, C. P., Nicolson, C., Taylor, J., Mumford, J. A., Wood, J. M. and Robertson, J. S. (1998) *Arch. Virol.* 143, 891–901.
Ives, J. A. L., Carr, J. A., Mendel, D. B., Tai, C. Y., Lambkin, R., Kelly, L., Oxford, J. S., Hayden, G. F. and Roberts, N. A. (2002) *Antiviral Research* 55, 307–317.
Kilbourne, E. D. (1975) *The Influenza Viruses and Influenza.* Academic Press, New York.
Kistner, O., Barrett, P. N., Mundt, W., Reiter, M., Schober-Berdixen, S. and Dorner, F. (1998) *Vaccine* 16, 960–968.
Lambkin, R. and Dimmock, N. J. (1995) *J. Gen. Virol.* 76, 889–897.
Laver, W. G. and Valentine, R. C. (1969) *Virology* 38, 105–119.
Laver, W. G. and Webster, R. G. (1972) *Virology* 48, 445–450.
Matrosovich, M. N., Gambaryan, A. S., Teneberg, S., Piskarev, V. E., Yamnikova, S. S., Lvov, D. K., Robertson, J. S. and Karlsson, K-A. (1997) *Virology* 233, 224–234.
McDermott, A. B., Zuckerman, J. N., Sabin, C. A., Marsh, S. E. G. and Madrigal, J. A. (1997) *Tissue Antigens* 50, 8–14.
McMichael, A. J., Gotch, F. M., Cullen, P., Askonas, B. A. and Webster, R. G. (1981) *Clin. Exp. Immunol.* 43, 276–285.
Meltzer, M. I., Cox, N. J. and Fukuda, K. (1999) *Emerg. Infect. Dis.* 5, 659–671.
Meyer, W. J., Wood, J. M., Major, D., Robertson, J. S., Webster, R. G. and Katz, J. M. (1993) *Virology* 196, 130–137.
Monto, A. S., Fleming, D. M., Henry, D., de Groot, R., Makela, M., Klein, T., Elliott, M., Keene, O. N. and Man, C. Y. (1999) *J. Infect. Dis.* 180, 254–261.
Nakajima, K., Desselberger, U. and Palese, P. (1978) *Nature* 274, 334.
Natali, A., Oxford, J. S. and Schild, G. C. (1981) *J. Hyg. (Camb.)* 87, 185.

Nicholson, K. G., Aoki, F. Y., Osterhaus, A. D. M. E., Trottier, S., Carewicz, O., Mercier, C. H., Rode, A., Kinnersley, N. and Ward, P. (2000) *Lancet* 355, 1845–1850.
Oxford, J. S. (2001) *Phil. Trans. R. Soc. Lond.* 356, 1857–1859.
Oxford, J. S. (2002) *Rev. Med. Virol.* 10, 119–133
Oxford, J. S., McGeoch, D. J., Schild, G. C. and Beare, A. S. (1978) *Nature* 273, 778.
Oxford, J. S., Sefton, A., Jackson, R., Innes, W., Daniels, R. S. and Johnson, N. P. A. S. (2002) *Lancet Infect. Dis.* 2, 111–114.
Perdue, M. L., Garcia, M., Senne, D. and Fraire, M. (1997) *Virus Res.* 49, 173–186.
Phillips, H. and Killingray, D. (2003) *The Spanish Influenza Pandemic of 1918–1919: New Perspectives.* Social History of Medicine Series, Routledge, London.
PHLS (1997) Response to a Pandemic of Influenza: An Action Plan. *PHLS Digest* 10, 147–154.
Reid, A. H., Fanning, T. G., Hultin, J. C. and Taubenberger, J. K. (1999) *Proc. Natl Acad. Sci. USA* 96, 1651–1656.
Reid, A. H., Fanning, T. G., Janczewski, T. A. and Taubenberger, J. K. (2000) *Proc. Natl Acad. Sci. USA* 87, 6785–6790.
Reid, A. H., Janczewski, T. A., Greif, R., Elliot, A. J., Daniels, R. S., Berry, C. L., Oxford, J. and Taubenberger, J. K. (2003) *J. Emerg. Dis.* (in press).
Robertson, J. S., Bootman, J. S., Newman, R., Oxford, J. S., Daniels, R. S., Webster, R. G. and Schild, G. C. (1987) *Virology* 160, 31–37.
Robertson, J. S., Nicolson, C., Bootman, J. S., Major, D., Robertson, E. W. and Wood, J. M. (1991) *J. Gen. Virol.* 72, 2671–2677.
Schild, G. C., Oxford, J. S., de Jong, J. C. and Webster, R. G. (1983) *Nature* 303, 706.
Smith, W., Andrewes, C. H. and Laidlaw, P. P. (1933) *Lancet* 225, 66–68.
Solorzano, A., Zheng, H., Fodor, E., Brownlee, G. G., Palese, P. and Garcia-Salestre, A. (2000) *J. Gen. Virol.* 81, 737–742.
Steinhauer, D. A. and Wharton, S. A. (1998) In *Textbook of Influenza* (eds K. G. Nicholson, R. G. Webster and A. J. Hay), Blackwell, Oxford, pp. 54–64.
Stuart-Harris, C. H., Schild, G. C. and Oxford, J. S. (1983) In *Influenza, the Viruses and the Disease*, Edward Arnold, London.
Taubenberger, J. K., Reid, A. H., Janczewski, T. A. and Fanning, T. G. (2001) *Phil. Trans. R. Soc. Lond.* 356, 1829–1839.
Taubenberger, J. K., Reid, A. J., Krafft, A. E., Bijwaard, K. E. and Fanning, T. G. (1997) *Science* 275, 1793–1796.
Treanor, J. J., Hayden, F. G., Vrooman, P. S., Barbarash, R., Bettis, R., Riff, D., Singh, S., Kinnersley, N., Ward, P. and Mills, R. G. (2000) *J. Am. Med. Assoc.* 283, 1016–1024.
Varghese, J. N., Laver, W. G. and Colman, P. M. (1983) *Nature* 303, 35–40.
Webby, R. J. and Webster, R. G. (2001) *Phil. Trans. R. Soc. Lond.* 356, 1817–1828.
Welliver, R., Monto, A. S., Carewicz, O., Schatteman, E., Hassman, M., Hedrick, J., Jackson, H. C., Huson, L., Ward, P. and Oxford, J. S. (2001) *J. Am. Med. Assoc.* 285, 748–754.
Wiley, D. C., Wilson, I. A. and Skehel, J. J. (1981) *Nature* 289, 373–378.
Wilson, L. A., Skehel, J. J. and Wiley, D. C. (1981) *Nature* 289, 366–372.
Winternitz, M. C., Waton, I. M. and McNamara, F. P. (1920) *The Pathology of Influenza.* Yale University Press, New Haven, CT.
Yuan, K. Y., Chan, P. K., Peiris, M. *et al* (1998) *Lancet* 351, 467–471.
Zambon, M. C. (2001) *Rev. Med. Virol.* 11, 227–241.

5

ROTAVIRUS

C. ANTHONY HART AND NIGEL A. CUNLIFFE

Diarrhoeal disease is the fourth commonest cause of death worldwide (Murray and Lopez, 1997). Of the fifty or so pathogens able to cause diarrhoeal disease, rotavirus is undoubtedly the most important. Estimates of the mortality due to rotavirus diarrhoea range from 418 000 to 870 000 deaths each year in children under five years (Institute of Medicine, 1986; Miller and McCann, 2000). Although most of the deaths due to rotavirus are in children in developing countries, the disease burden is the same in developed and developing countries. For example, it is estimated that there were 17 810 hospitalizations due to rotavirus in England and Wales from 1993 to 1994 (Ryan *et al.*, 1996). A community-based study of children under 5 years estimated that there were 762 000 episodes of rotavirus diarrhoea over a five year (1992–1997) period also in England and Wales (Djuretic *et al.*, 1999).

Rotavirus was first described and originally named duovirus by Bishop and colleagues (Bishop *et al.*, 1973), who recognized viral particles in the duodenal mucosa of children admitted to hospital with acute non-bacterial diarrhoea. Subsequently it was found that patients with rotavirus gastroenteritis excrete large numbers of viral particles in their faeces (Flewett *et al.*, 1973). Rotavirus was not established in culture until five years later, when it was realized that proteolytic processing was required for efficient viral replication (Almeida *et al.*, 1978).

VIRUS CLASSIFICATION AND STRUCTURE

Rotavirus is a medium-sized (70–80 nm), unenveloped, round virus with a characteristic wheel-shaped morphology (Figure 5.1). (*Rota* is Latin for a wheel.) Rotavirus is a genus within the family *Reoviridae*, which contains eight separate genera. Its genome consists of eleven segments of double-

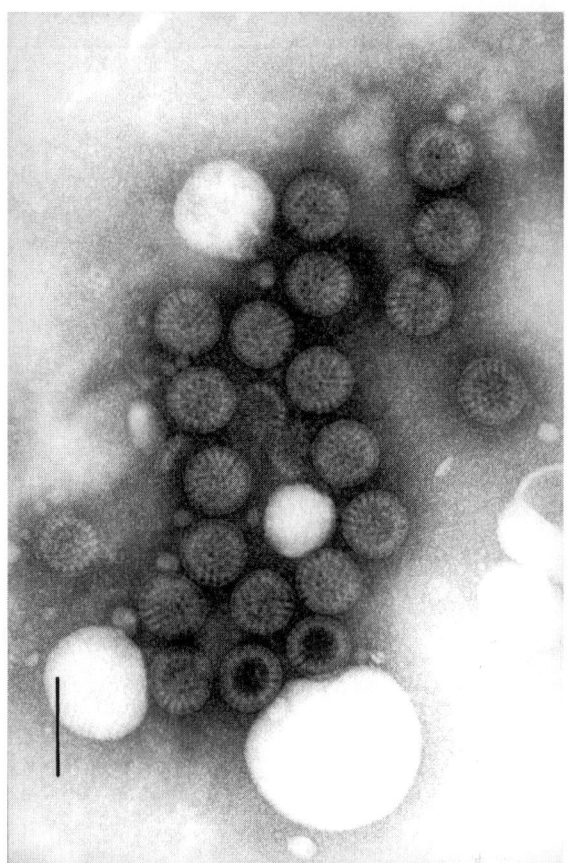

Figure 5.1 Negative stain electron micrograph showing rotavirus particles in a stool sample (Bar = 100 nm).

stranded RNA, which range in size from 0.6 to 3.3 kilobase pairs (kbp). Each segment encodes one or more polypeptides (Table 5.1). The polypeptides produced are either involved in viral replication and not found in the mature virion (non-structural proteins: NSP) or structural proteins making up the virus particles (virus proteins: VP). The mature virion has a trilayered structure (Figure 5.2). The inner layer consists of the 11 dsRNA segments surrounded by VP1, VP2 and VP3. The middle layer, also referred to as the inner capsid, is composed entirely of VP6, and the outer layer or capsid is made up from two proteins, VP7 and VP4. The latter is cleaved by proteolysis to VP5* and VP8*. The mature virion has 60 spikes or knobs extending 120Å from the surface (Estes, 1996). The virion has icosahedral symmetry, with 132 surface capsomers and a triangulation number of T13. There are also 132 large channels that traverse both the inner and the outer capsids.

Antigenic Variation

Table 5.1 Rotavirus genome and gene products

Genome segment	Molecular size (bp)	Gene product	Molecular weight (Kda)	Location in virion	Function
1	3302	VP1	125	Core	RNA polymerase
2	2690	VP2	94	Core	RNA binding
3	2591	VP3	88	Core	Guanylytransferase
4	2362	VP4 (VP5*+VP8*)	88	Outer capsid	Cell attachment; haemagglutinin; neutralizing antigen (P-serotype)
5	1581	NSP1	53	Non-structural	RNA binding (zinc finger)
6	1356	VP6	41	Inner capsid	Group and subgroup antigen
7	1104	NSP3	34	Non-structural	RNA binding
8	1059	NSP2	35	Non-structural	RNA binding
9	1062	VP7	38	Outer capsid	Neutralizing antigen (G-serotype)
10	751	NSP4	28	Non-structural	Virus assembly; enterotoxin
11	667	NSP5	26	Non-structural	RNA binding

ANTIGENS AND EPIDEMIOLOGICAL MARKERS

Rotavirus can be subdivided into a puzzling array of group, subgroup, serotype, genotype, electropherotype and genogroup.

Group

Thus far rotaviruses are split into seven groups based on epitopes on the inner capsid protein VP6 (Table 5.2). Of these, humans are infectable by groups A, B and

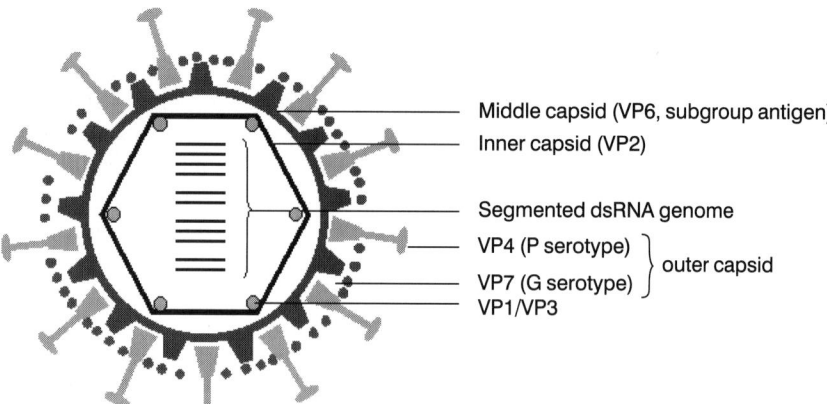

Figure 5.2 The structure of rotavirus.

Table 5.2 Reservoirs of different rotavirus groups

Group	Reservoirs in which infection has been detected
A	Man, primates, horse, sheep, pig, cattle, dog, cat, turkey, chicken, mice
B	Man, pigs, cattle, sheep, rats
C	Man, pigs, cattle, ferret
D	Chicken, turkey
E	Pigs
F	Chicken
G	Chicken

C rotaviruses but group A rotaviruses are responsible for most infections. Group B rotaviruses have been associated with rare but large epidemics of disease in both children and adults (Hung *et al.*, 1984; Krishnan *et al.*, 1999). Group C rotaviruses do cause disease, and approximately 30% of adults have serological evidence of prior exposure (Riepenhoff-Talty *et al.*, 1997). However, in a survey of childhood gastroenteritis in Malawi, only 3.3% of children were excreting group C rotavirus and two-thirds of these were co-infected with group A rotavirus (Cunliffe *et al.*, 2001a). The remaining serogroups have only been detected in other animal species. However, a great diversity of animal species are infectable by Group A rotaviruses, and although they were thought to be host-species-specific, there is increasing evidence that this is not the case. VP6 is the most abundant protein, with 780 molecules per virion. The epitopes imparting group specificity are localized between amino acid residues 48 and 75 (Kohli *et al.*, 1992). Although infection results in production of antibodies directed against group antigen, it is generally thought that they play little role in immunity. However, recent work has shown that in mice, IgA anti-VP6 is protective even though it is non-neutralizing *in vitro* (Burns *et al.*, 1996; Feng *et al.*, 1997).

Subgroup

Two monoclonal antibodies that could immunoprecipitate VP6 but only from group A rotaviruses were produced (Greenberg *et al.*, 1983). These subdivided group A rotaviruses into subgroups I and II. In fact four subgroups are possible – I, II, I and II, and neither. The domain of VP6 carrying subgroup specificity is different from that for group specificity but its location is not known, nor is its role, if any, known.

Serotype (serogroup)

G-types
VP7 is the major neutralization antigen and comprises about 30% of the protein in the mature virion. It forms the smooth surface covering the virion, through

which protrude the 60 knobs of VP4, (VP5*, VP8*). VP7 is a glycoprotein containing N-linked, high-mannose oligosaccharides, which are shortened by losing two mannose residues as the virion matures. It is encoded on the seventh, eight or ninth gene segment depending on the rotavirus strain. The role of VP7 in virus replication is not clear but it was thought to have receptor-binding activity.

Thus far, fourteen VP7 or G (for glycoprotein) serotypes have been delineated by hyperimmune sera and, more recently, monoclonal antibodies (Table 5.3). More recently primers for use in RT-PCR G-typing have been designed (e.g. Gentsch *et al.*, 1996) and the genotype is equivalent to the serotype. Within a particular G-serotype the deduced amino acid sequences are 91–100% similar, but there is much greater variation between serotypes (Green *et al.*, 1987; Green

Table 5.3 Group A rotavirus G-types

Sero(geno)type	Host species
1	Humans, pigs, cattle
2	Humans, pigs
3	Humans, monkeys, dogs, cats, horses, rabbits, mice, goats
4	Humans, pigs
5	Humans, pigs, horses
6	Humans, cattle
7	Birds, cattle
8	Humans, cattle
9	Humans, pigs
10	Humans, pigs, sheep
11	Pigs
12	Humans
13	Horses
14	Horses

and Kapikian, 1992). Most of the divergence is in nine regions of the VP7 amino acid sequence. Of these variable regions (VR), VR5 ($\alpha\alpha 87$–101), VR7 ($\alpha\alpha 142$–152) and VR8 ($\alpha\alpha 208$–221) are the major neutralizing epitopes; these have also been termed antigenic regions A, B and C respectively. Escape mutants with single amino acid changes in these regions occur easily (Kapikian and Chanock, 1996). The extent of glycosylation is also of great importance in the antigenicity of VP7. Infection with one G-serotype virus does not necessarily protect against infection with other rotavirus G-types.

P-types
VP4 forms the 60 spikes with a knob-like end that protrude from the virion surface. It is non-glycosylated and represents only 1.5% of total virion protein. Some rotaviruses are able to agglutinate erythrocytes and this is mediated by

VP4. VP4 is also the ligand by which some rotaviruses bind to enterocytes. For full infectivity VP4 is cleaved to VP5* and VP8* and it has been shown that this trypsin cleavage stabilizes the spikes, which become icosahedrally ordered (Crawford et al., 2001). It is now thought that VP5* forms the body and base of the spikes in beta-pleated sheets and alpha helices respectively and that VP8* forms the spike heads and is in beta-sheets. Both VP5* and VP8* are involved in receptor binding and membrane permeabilization for virus entry into host cells (Tihova et al., 2001). Monoclonal anti-VP4 neutralizes rotavirus infectivity and using such antibodies a number of P- (for protease) serotypes have been defined. However, not all have been characterized this way and a separate P-genotyping system has been developed. Unfortunately, the serotype and genotype numeric assignments do not co-incide and not all genotypes have been defined as serotypes (Table 5.4). It appears that VP8* possesses the neutralizing epitopes

Table 5.4 Rotavirus P-serotypes and genotypes

Genotype	Serotype	Human pathogen	Other animals
[1]	6	−	Cattle, monkeys
[2]		−	Monkeys
[3]	5	+	Monkeys, dogs, cats
[4]	1B	+	
[5]	7	−	Cattle
[6]	2A,2B,2C	+	
[7]	9	−	Pigs
[8]	1A	+	
[9]	3	+	Cats
[10]	4	+	
[11]	8	+	Cattle
[12]		−	Horses
[13]	3B	+	
[14]		+	Pigs
[15]		−	Sheep
[16]	10	−	Mice
[17]		−	Cattle, birds
[18]		−	Horses
[19]		−	Pigs
[20]		−	Mice

that define P-serotype specificity, whereas VP5* has cross-reactive antigens (Larralde and Gorziglia, 1992). There are 20 P-genotypes and some 12 to 15 serotypes (Table 5.4). As with VP7, VP4 also expresses independent neutralizing epitopes, and infection with one P-type does not necessarily prevent infection with another. Thus, in typing group A rotaviruses both G- and P-epitopes are of major importance. For P-types the genotypes are designated by a squared bracket, e.g. P6[1].

Electropherotypes

During acute infection up to 10^{11} virus particles are excreted per gram of faeces. There is so much virus that, following RNA extraction from faeces, the 11 dsRNA genomic segments can be resolved by polyacrylamide gel electrophoresis and visualized by silver-staining. The patterns of migration of the genomic segments provide information for epidemiological purposes (Figure 5.3). In

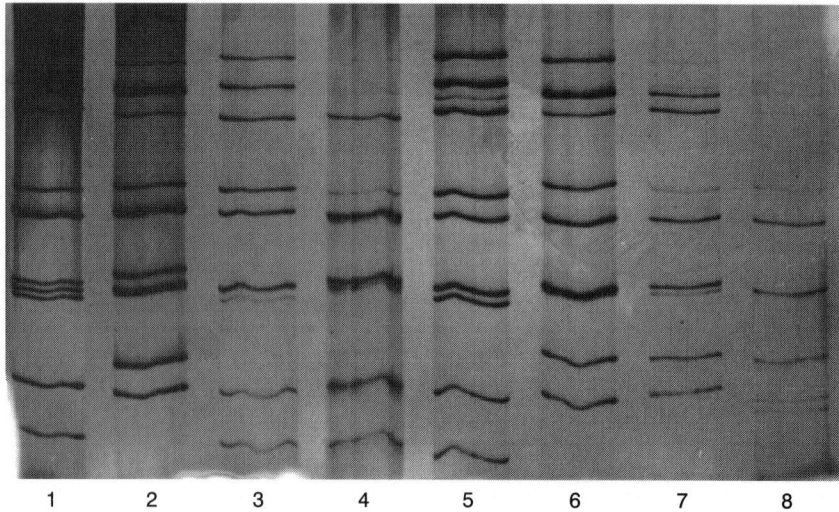

Figure 5.3 Silver-stained PAGE gel of RNA extracted from children with rotavirus diarrhoea showing the 11 genomic dsRNA segments. Lanes 1–2: Standard strains showing long electropherotype profile (strain Wa, lane 1), and short electropherotype profile (strain DS-1, lane 2). Lanes 3–7: Malawi field strains showing long electropherotype profiles (lanes 3–5), and short electropherotype profiles (lanes 6–7). Lane 8: Malawi field strain showing mixed short electropherotype profile.

particular, the rate of migration of the lowest molecular weight segments (in particular segment 11) splits rotaviruses into long, short and super-short electropherotypes. In the short electropherotype the eleventh genomic segment is larger than usual and migrates more slowly to resolve between the ninth and tenth segments. In the super-short pattern the eleventh segment migrates even more slowly. Finally, lower molecular weight tenth and eleventh segments produce the long electropherotype. There can also be minor variability in the migration patterns of the other segments and this variability can be used to compare rotaviruses during epidemic seasons.

Genogroups

By using radiolabelled, positive-sense, single-stranded RNAs transcribed from viral cores as probes at high stringency (Flores *et al.*, 1992), it is possible to split

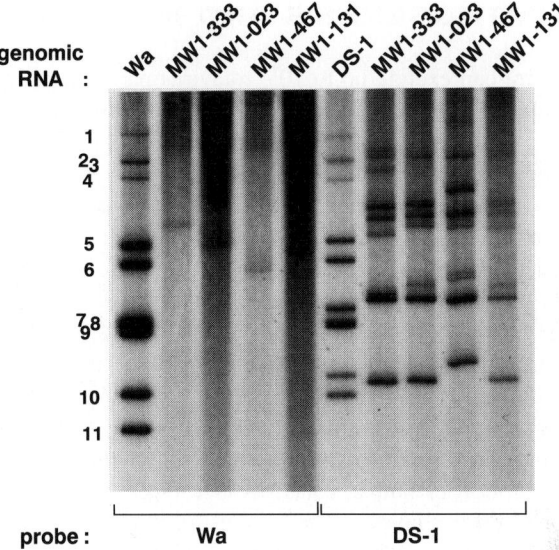

Figure 5.4 Northern blot of rotavirus genomic segments separated by electrophoresis and probed with radiolabelled rotavirus cDNA. These rotaviruses are DS-1 subgroup, as demonstrated by the appearance of bands with the DS-1 probe, and not with the Wa probe.

rotaviruses into genogroups (Figure 5.4). This has revealed two major genogroups of human group A rotaviruses. These are named for the prototype viruses, namely Wa (serotype 1) and DS-1 (serotype 2). The Wa genogroup viruses are generally long electropherotype, subgroup II with VP4 P[6] or P[8] and VP7 G1, G3 or G4 specificities. The DS-1 genogroup are generally of short electropherotype, subgroup I, and VP4 P[4] and VP7 G2 specificities (Table 5.5). In addition, the broad genogroup assignments are matched by sequence analysis of one of the rotavirus non-structural protein (NSP4) genes (Cunliffe *et al.*, 1997).

However, those broad subdivisions into two main genogroups are being eroded away by the tremendous variability of the rotaviruses that infect man and other animals (Cunliffe *et al.*, 2002a).

Table 5.5 Rotavirus genogroups

	Genogroup 1 (Wa-like)	Genogroup 2 (DS-1-like)
Subgroup	II	I
P-type	P[6], P[8]	P[4]
G-type	G1, G3, G4	G2
Electropherotype	Long	Short
NSP4 genetic group	II	I

PATHOGENESIS AND IMMUNITY

An infected infant can excrete up to 10^{11} rotavirus particles per gram of faeces during acute infection and the infective dose is of the order of 10^2. Infection is thus spread very easily and rates of rotavirus infection in children are the same in both developed and developing countries; it is the effects of disease that differ markedly. Infection is acquired by ingestion via the faeco-oral route. However, there is some evidence to suggest that aerosol formation of the extremely fluid stool may also result in short-range transfer and since rotavirus can survive in the environment, transfer via fomites is possible (Abad et al., 1998).

Rotavirus infects the mature villous enterocytes of the small intestine but cannot infect either crypt or colonic enterocytes. Some strains bind to the enterocyte via acetylated sialic acid receptors using VP5* as the ligand. However, recent *in vitro* and *in vivo* studies have shown that sulphated sialyl lipid inhibits the absorption of human rotavirus (serotypes G1, G2, G3 and G4) to MA104 cells and prior administration of the lipid to mice prevented rotavirus diarrhoea (Takahashi et al., 2002). Some rotaviruses enter host cells by receptor-mediated endocytosis but others, especially when trypsin pre-treated, enter directly by membrane permeabilization (Tihova et al., 2001). The next stage, of uncoating and release of genomic dsRNA, is facilitated by low Ca^{2+} concentrations in the cytosol. Replication takes place entirely in the cytoplasm of the host cell. During replication, subviral particles assemble in cytoplasmic viroplasms and bud through the endoplasmic reticulum thus transiently acquiring an envelope. This is lost as the particle moves deeper into the ER, being gradually replaced with a thin layer of protein, which matures to become the outer capsid. Most of the VPs and NSPs are synthesized on free ribosomes but the glycoproteins VP7 and NSP4 are synthesized on ER-associated ribosomes and become inserted into the membrane. It is thought that NSP4 facilitates rotavirus entry into the ER. Virus is released by host cell lysis.

At least four mechanisms have been described by which rotavirus causes gastroenteritis. First, it has been shown that within 12–24 hours of infection small-intestinal brush-border enzymes (sucrase, maltase and lactase) fall to less than a third of normal levels (Batt et al., 1995). These microvillar enzymes have a short half-life and they are continually replaced. It appears that rotavirus interferes with the transport of sucrase–isomaltase from its site of synthesis to the brush border by altering the cytoskeleton (Jourdan et al., 1998). Maltose, sucrose and lactose must be cleaved to monosaccharides to be absorbed. If they are not cleaved, they accumulate in the gut lumen and hold water resulting in a non-absorbable fluid overload and osmotic diarrhoea. Second, NSP4 or even a 22 amino acid peptide (from residues 114–135) when administered intra-orally or intraperitoneally into infant mice causes diarrhoea (Tian et al., 1994, 1996; Ball et al., 1996). It is postulated that NSP4 acts as a viral enterotoxin and causes diarrhoea by opening a cellular cation channel (Perez et al., 1999), which causes an increase in intracellular Ca^{2+} concentration and activation of adenyl cyclase and cAMP-dependent Cl^- secretion. This was thought to be comparable to the mode of action of some bacterial enterotoxins (Ball et al., 1996). Rotavirus is released from enterocytes by cell lysis, and it has been found that increased intracellular Ca^{2+} causes enterocyte death by oncosis (Perez et al., 1998). This leads to a blunting of the small-intestinal villi, as the crypt enterocytes cannot divide quickly enough to replace the mature cells that have died. This results in a great diminution in the intestinal surface area available for absorption. Finally, there is some evidence that stimulation of the enteric neuro-endocrine axis might be involved in the pathogenesis of rotavirus diarrhoea (Lundgren and Svensson, 2001).

Natural infection with rotavirus limits both the risk of subsequent rotavirus infection and rotavirus diarrhoea (Velazquez et al., 1996). The risks of rotavirus infection and rotavirus diarrhoea after one, two or three prior episodes of infection were 0.62, 0.40 and 0.34 for infection and 0.23, 0.17 and 0.08 for symptomatic infection respectively. Furthermore, infection with so-called neonatal strains of rotavirus in neonates is associated with a significantly decreased likelihood of developing rotavirus diarrhoea later in infancy (Bishop et al., 1983). However, it is not clear which are the most important correlates of immunity. In longitudinal studies of infantile gastroenteritis serum virus-specific antibodies (IgG and IgA) have correlated best with protection from re-infection with rotavirus (Matson et al., 1993; Velazquez et al., 2000). However, faecal anti-rotaviral IgA, which is an accurate surrogate of duodenal IgA (Grimwood et al., 1988), also reflects protection against infection (Matson et al., 1993; Coulson et al., 1992). It appears that T cell-mediated immunity is more important in resolution of a current rotavirus infection rather than protection against subsequent infection, at least in mice (Ward, 1996). It is likely that each individual is infected repeatedly with rotavirus throughout their lifetime but it is only on the first few occasions that diarrhoeal disease occurs. An immune response to rotavirus is more important in preventing rotavirus disease and severe disease than in preventing infection at all. The relative importance of antibody to the G- or P-antigens is not known. However, it has been shown in gnotobiotic piglet models that initial rotavirus infection protected against re-infection providing either a VP4 (P) or a VP7 (G) antigen was held in common by the initial and subsequent challenge rotaviruses (Hoshino et al., 1988). However, rotaviruses are in a constant state of genetic and thus antigenic variability, which can allow them to evade immunity.

Genetic and antigenic change

There are four major ways in which the rotavirus genome and thus antigenic composition can alter. These are by mutations resulting in sequence variation (equivalent to antigenic drift), by genome rearrangements, by reassortment (equivalent to antigenic shift) and by the cross-species transmission of animal rotaviruses (Desselberger, 1996; Ramig, 1997; Cunliffe, Bresee et al., 2002; Palombo, 2002).

Sequential point mutations

This is equivalent to the antigenic drift seen in influenza A viruses. The rotaviral genome is replicated by means of a virally encoded, RNA-dependent RNA polymerase (Estes, 1996). This is error-prone and there is no proof-reading capacity. This results in a mutation rate which has been calculated to be of the order of 1×10^5 mutations per nucleotide per replication (equivalent to one mutation per genome replication), which is similar to that of influenza A virus (Blackhall et al., 1996). Sequence variation has been studied most in the gene encoding VP7 (G-glycoprotein) since antibodies to this are important in immunity. In general, viruses of the same G-type exhibited >91% amino acid similarity whereas intertypic variability is much higher (71–86%). However, it is becoming increasingly apparent that there is drift from the original G-types (Palombo, 1999).

Thus far at least four global lineages of human rotavirus G1 VP7 protein have been described (Jin et al., 1996; Maunula and von Bonsdorff, 1998). Analysis of G1 human rotaviruses collected over a six-year period in Melbourne, Australia found two of these lineages (I and II) and that these correlated with monotypes 1a and 1b (based on neutralization in vitro by monoclonal antibodies (Diwakarla and Palombo, 1999). The monotype differences were associated with substitutions at amino acid residue 94 (Asp → Ser/Thr) in a major neutralization epitope (antigenic site A). A study of rotaviruses genotyped as G2 but which could not be serotyped by three separate monoclonal anti-G2 antibodies showed an amino acid substitution at position 96 (Asp → Asn), again in antigenic region A (Iturizza-Gomara et al., 2001a). In each case these rotaviruses were non-neutralizable by standard monoclonal antibody, i.e. mutants that might evade immunity. Although there are fewer studies, other structural and non-structural rotavirus proteins have exhibited lesser variability (up to 14%). Diversity in the VP4 gene has been detected most frequently by failure to P-genotype (Iturizza-Gomara et al., 2000; Cunliffe et al., 2001b), when mutations occurred in the primer-binding sites. P[8] rotaviruses can currently be subdivided into three lineages (Wa-like, F45-like and OP 354-like) on the basis of deduced amino acid substitutions in at least eleven sites on the VP4 protein (Gouvea et al., 1999; Cunliffe et al., 2001b). Similarly, variability in NSP4 and even in the toxic peptide region (Halaihel et al., 2000) is common (Cunliffe et al., 1997; Ciarlet et al., 2000; Lee et al., 2000). The variability in human nongroup A rotaviruses is less well studied (Saif and Jiang, 1994). However, the nucleotide sequence of segment 8 (putative NSP2) of a group B adult diarrhoea rotavirus in Calcutta showed only 77% homology to a murine strain but within human isolates there was 93% identity (Sen *et al.*, 2001). These Calcutta group B rotaviruses showed 8–10% divergence at the nucleotide level from strains obtained from a large Chinese epidemic (which infected over 1 million people) in 1983 (Hung *et al.*, 1984).

Unlike influenza A virus, rotaviruses (except for Group B) do not cause epidemics due to a single strain. In temperate countries there are large upsurges of infection in the winter months and in the tropics, although rotavirus is prevalent throughout the year, more cases occur in the dry seasons (Cunliffe *et al.*, 1998; Hart and Cunliffe, 1999). During each year and in each peak season a number of different rotaviruses of different G- and P-type co-circulate and these vary year on year (Tabassum *et al.*, 1994; Cunliffe *et al.*, 1999; Iturriza-Gomara *et al.*, 2001b). Thus it is difficult to detect sequential changes occurring in the genome of the same rotavirus isolated repeatedly over time. That sequential mutations have accumulated can only be inferred from the great variability within the individual rotavirus genes.

Genome rearrangements

Rotavirus genome rearrangements were detected during investigations of rotaviruses excreted by children with immunodeficiency (Pedley *et al.*, 1984). Children with severe combined immunodeficiency develop chronic infection with rotavirus, which can last for several years (Saulsbury *et al.*, 1980; Chrystie *et al.*, 1982). On PAGE of serial faecal samples from such patients some RNA segments diminished in concentration or were lost and additional bands were found of higher molecular weight (Pedley *et al.*, 1984). On northern blot analysis of the PAGE gels using radiolabelled cDNA probes from bovine rotaviruses, multiple hybridizations with different probes to the atypical higher molecular weight bands were detected (Pedley *et al.*, 1984). These were considered to have arisen by genome rearrangement. A number of these rearranged rotavirus segments have now been sequenced (Desselberger, 1996). In most cases the rearranged segment runs from a normal 5' untranslated region (UTR) through the normal open reading frame (ORF), and at various positions after the stop codon duplicated segments of the ORF, which usually do not include the start codon but have a duplicated stop codon, are added to the normal 3' UTR. Thus these additional genome segments are usually not expressed. This is not always the case and some can produce abnormal proteins, especially when the duplication begins before the stop codon (Tian *et al.*, 1993). Genome rearrangement has now been detected in both human and animal rotaviruses and in rotaviruses from immunocompetent children (Besselaar *et al.*, 1986). They also arise in artificial culture, especially on serial subculture at high multiplicities of infection (Hundley *et al.*, 1985). The importance of the role of genome rearrangements in generating the great diversity of rotaviruses is as yet unclear (Palombo, 2002).

Genome reassortment

The genome of rotavirus is segmented dsRNA and this provides a mechanism for generating antigenic diversity by reassortment. This occurs when two different rotaviruses infect the same host cell simultaneously. This has been shown to occur readily *in vitro* (Ramig, 1997). With 11 genome segments each, two potential strains could theoretically generate 2^{11} genome combinations by random crossover. This has never been detected, probably because to plaque purify and analyse a sufficiently large number of potential reassortants is such an enormous task. It is of course also possible that certain reassortants might be less fit. Genome reassortment has also been demonstrated during experimental co-infection of mice with two different rotaviruses (Gombold and Ramig, 1986). Passive transfer of anti-rotavirus antibody decreased reassortant frequency three-fold, even if both rotaviruses were serologically distinct from that used to generate the antibody (Gombold and Ramig, 1989). This observation suggests that the process of reassortment in animals is likely to be limited, since most animals including humans will have antibody to rotavirus at an early age. However, mixed rotavirus infections have been detected in humans (e.g. Ahmed *et al.*, 1991; Tabassum *et al.*, 1994) and a recent study of 3601 rotavirus strains from children with diarrhoea in the UK found that approximately 2% were reassortant having novel G- and P-type combinations (Iturriza-Gomara *et al.*, 2001b). The authors concluded that reassortment among co-circulating human rotaviruses was not a rare event and contributed to their diversity. Another study in Japan also demonstrated genome segment reassortment between concurrently circulating strains (Watanabe *et al.*, 2001). The consequence of reassortment is that unusual combinations of P- and G-serotype antigens might emerge. In addition, reassortment-mediated genomic alterations can also affect the antigenicity of serotype antigens because of the novel protein–protein interactions introduced (Lazdins *et al.*, 1995).

Interspecies transmission

Although it was originally thought that rotaviruses were species-specific, this is clearly not the case. Challenge experiments have demonstrated that a rotavirus from one species can infect another, for example, the simian rotavirus SA11 infecting mice (e.g. Gombold and Ramig, 1986). Human rotavirus vaccine production has used the Jennerian approach of using either live simian or bovine rotaviruses or live simian/human or bovine/human reassortants (Lynch *et al.*, 2000). Thus there is clear evidence that cross-species transmission can and does occur. The evidence that it occurs naturally is less clear (Palombo, 2002). Most often it is inferred by detecting, for example, an animal-specific VP4 or VP7 in a human rotavirus. This of course will not provide information on whether it is a reassortant animal–human virus or whether it is an animal rotavirus in origin.

Animal group A rotaviruses tend to be of subgroup I and have a long electropherotype. Finding this in a human rotavirus infection is usually taken as preliminary evidence of cross-species transmission, which is then confirmed by genogrouping using radiolabelled cDNA from animal rotavirus. Thus it is possible to determine whether some or all of the genes are from an animal-derived rotavirus. There are nine examples of animal (cat, dog, cattle, or pig) to human, two of animal to animal and one of human to pig transfer (reviewed in Palombo, 2002).

Phylogenetic analysis has been performed on a number of the rotavirus genomic segments (including those encoding VP4, VP7, VP6, NSP1, NSP3 and NSP4) from a variety of human and animal rotaviruses. Such analysis of VP7 from G6 bovine and human isolates detected separate species-specific clusters but there was a separate branch containing one human and one bovine strain (Cooney et al., 2001). This could indicate a point of interspecies transmission. In general VP6 analysis shows clustering by rotavirus group and subgroup. Two clusters are found: one includes subgroup I human, cattle and monkey viruses and the other, subgroup II human and porcine viruses (Mattion et al., 1994). Analysis of NSP1 shows viruses segregating by species except for human and porcine strains, which were related (Dunn et al., 1994). This porcine/human rotavirus relatedness was further shown by analysis of NSP3 and NSP4 genes (Patton et al., 1993; Cunliffe et al., 1997; Kirkwood and Palombo, 1997).

Serotype G5 rotaviruses have been particularly associated with infection of pigs and horses. Such viruses have recently emerged, particularly in Brazil, as important causes of human infection (Gouvea et al., 1994). It appears that at least some of these are derived from porcine strains, but not recently (Racz et al., 2000).

In cattle G6 and G8 rotaviruses are particularly important causes of infection. Some Italian G6 human rotaviruses showed cross-hybridization and sequence homology to bovine G6 strains to all genomic segments except that encoding VP4 (Gerna et al., 1994). This suggests that these were originally bovine strains that have acquired human rotavirus VP4 since this protein is involved in determining host range. Serotype G8 rotavirus has been isolated from humans. Initial strains had a super-short electropherotype (Hasegawa et al., 1984) but subsequently long electropherotype subgroup I strains were found (Gerna et al., 1990). In sub-Saharan Africa, human infection with G8 rotaviruses has been described, and in Malawi over 50% of the isolates were G8 (Cunliffe et al., 1999, 2000, 2001b). These were mostly of short electropherotype with P[6] and P[4]VP4 specificities but two G8 P[6] rotaviruses were of long electropherotype (Cunliffe et al., 2001b). Sequence analysis of the VP7 genes of some of the Malawian G8 strains showed them to have 93.6% and 92.3% identity at the putative amino acid level to bovine rotavirus strains A5 and Cody-1801 respectively (Cunliffe et al., 2000). However, their VP4 genes were most closely related to human strains.

Interspecies rotavirus transmission, especially when associated with reassortant human/animal rotaviruses, appears to be another powerful generator of rotavirus diversity.

Consequences of variability

The constant evolution of rotaviruses means that repeated infections are possible and indeed do occur. Although these reinfections are usually asymptomatic, it is possible for 'new' rotaviruses to emerge that do cause symptomatic disease. Neonatal rotavirus infections are usually asymptomatic (Bishop *et al.*, 1983), and due to viruses with P[6] VP4 gene alleles. Recently, severe outbreaks of rotavirus diarrhoea due to newly emerged G9 P[6] and G8P[6] strains have been described in Holland and Malawi respectively (Widdowson *et al.*, 2000; Cunliffe *et al.*, 2002b). This could perhaps be because the neonate's mother would not have been exposed to G8 or G9 viruses previously and not have antibody to these antigens and thus there would be no transplacentally acquired antibody to protect the neonate.

Secondly, the vaccines developed and in development to prevent rotavirus diarrhoea contain the most important and frequently detected VP7 serotypes (Lynch *et al.*, 2000). In the reassortant rhesus rotavirus tetravalent vaccine (RRV-TV) G1, G2, G3 and G4 were included. It is now clear that others may need to be included, such as G5, G8 and especially G9, which has emerged as a new global serotype. When such vaccines are introduced constant surveillance will be necessary not just for new G- and perhaps P-type viruses but also for homologous G-type escape mutants.

REFERENCES

Abad, F. X., Pinto, R. M. and Bosch, A. (1998) *Appl. Env. Microbiol.* 64, 2392–2396.
Ahmed, M. U., Urasawa, S., Taniguchi, K., Urasawa, T., Kobayashi, N., Wakasugi, F., Islam, A. I. and Sahikh, H. A. (1991) *J. Clin. Microbiol.* 29, 2273–2279.
Almeida, J. D., Hall, T., Banatvala, J. E., Toterdell, B. M. and Chrystie, I. E. (1978) *J. Gen. Virol.* 40, 213–218.
Ball, J. M., Tian, P., Zeng, C. Q-Y., Morris, A. P. and Estes, M. K. (1996) *Science* 272, 101–104.
Batt, R. M., Embaye, H., van de Waal, S., Burgess, D., Edwards, G. B. and Hart, C. A. (1995) *J. Pediatr. Gastroenterol. Nutr.* 20, 326–332.
Besselaar, T. G., Rosenblatt, A. and Kidd, A. H. (1986) *Arch. Virol.* 87. 327–330.
Bishop, R. F., Barnes, G. L., Cipriani, E. and Lund, J. S. (1983) *N. Engl. J. Med.* 309, 72–76.
Bishop, R. F., Davidson, G. P., Holmes, I. H. and Ruck, B. J. (1973) *Lancet* 305, 1281–1283.
Blackhall, J., Fuentes, A. and Magnusson, G. (1996) *Virology* 225, 181–190.
Burns, N. W., Siadat-Pajouh, M., Krishnaney, A. A. and Greenberg, H. B. (1996) *Science* 272, 104–107.
Chrystie, I. L., Booth, I. W., Kidd, A. H., Marshall, W. C. and Banatvala, J. E. (1982) *Lancet* 322, 282
Ciarlet, M., Liprandi, F., Conner, M. F. and Estes, M. K. (2000) *Arch. Virol.* 145, 371–383.

Cooney, K. A., Gorrell, R. J. and Palombo, E. A. (2001) *J. Med. Microbiol.* 59, 462–467.
Coulson, B. S., Grimwood, K., Hudson, I. L., Barnes, G. L. and Bishop, R. F. (1992) *J. Clin. Microbiol.* 30, 1678–1684.
Crawford, S. E., Mukherjee, S. K., Estes, M. K., Lawton, J. A., Shaw, A. L., Ramig, R. F. and Prasad, B. V. V. (2001) *J. Virol.* 75, 6052–6061.
Cunliffe, N. A., Bresee, J. S., Gentsch, J. R., Glass, R. I. and Hart, C. A. (2002a) *Lancet* 359, 640–642.
Cunliffe, N. A., Dove, W., Jiang, B., Thindwa, B. D. M., Broadhead, R. L., Molyneux, M. E. and Hart, C. A. (2001a) *Pediatr. Infect. Dis. J.* 20, 1088–1090.
Cunliffe, N. A., Gentsch, N. R., Kirkwood, C. D., Gondwe, J. S., Dove, W., Nakagomi, O., Nakagomi, T., Hoshino, Y., Bresee, J. S., Glass, R. I., Molyneux, M. E. and Hart, C. A. (2000) *Virology* 274, 309–320.
Cunliffe, N. A., Gondwe, J. S., Broadhead, R. L., Molyneux, M. E., Woods, P. A., Bresee, J. S., Glass, R. I., Gentsch, J. R. and Hart, C. A. (1999) *J. Med. Virol.* 57, 308–312.
Cunliffe, N. A., Gondwe, J. S., Graham, S. M., Thindwa, B. D. M., Dove, W., Broadhead, R. L., Molyneux, M. E. and Hart, C. A. (2001b) *J. Clin. Microbiol.* 39, 836–843.
Cunliffe, N. A., Kilgore, P. E., Bresee, J. S., Steele, A. D., Luo, N., Hart, C. A. and Glass, R. I. (1998) *Bull. WHO* 76, 525–537.
Cunliffe, N. A., Rogerson, S., Thindwa, B. D. M., Greensill, J., Kirkwood, C. D., Broadhead, R. L. and Hart, C. A. (2002b) *J. Clin. Microbiol.* 40, 1534–1537.
Cunliffe, N. A., Wood, P. A., Leite, J. P. G., Das, B. K., Ramachandran, M., Bhan, M. K., Hart, C. A., Glass, R. I. and Gentsch, J. R. (1997) *J. Med. Virol.* 53, 41–50.
Desselberger, U. (1996) *Adv. Virus Res.* 46, 69–95.
Diwakarla, C. S. and Palombo, E. A. (1999) *J. Gen. Virol.* 80, 341–344.
Djuretic, J., Ramsay, M., Gay, N., Wall, P., Ryan, M. and Fleming, D. (1999) *Acta Paediatr.* 426, 38–41.
Dunn, S. J., Cross, T. L. and Greenberg, H. B. (1994) *Virology* 203, 178–183.
Estes, M. K. (1996) In *Fields' Virology* (eds B. N. Fields, D. M. Knipe and P. M. Howley), Lippincott–Raven Press, Philadelphia, pp. 1625–1655.
Feng, N. G., Vo, P. T., Chung, D., Vo, T. V. P., Hoshino, Y. and Greenberg, H. B. (1997) *J. Infect. Dis.* 175, 330–341.
Flewett, T. H., Bryden, A. S. and Davies, H. (1973) *Lancet* 305, 1497.
Flores, J., Greenberg, H. B., Myslinski, J., Kalica, A. J., Wyatt, R. G., Kapikian, A. Z. and Channock, R. M. (1982) *Virology* 121, 288–295.
Gentsch, J. R., Woods, P. A., Ramachandran, K., Das, B. K., Leite, J. P., Alfieri, A., Kumar, R., Bhan, M. K. and Glass, R. I. (1996) *J. Infect. Dis.* 174, S30–S36.
Gerna, G., Sarasini, A., Zentilin, L., Di Matteo, A., Miranda, P., Parea, M., Battaglia, M. and Milanesi, G. (1990) *Arch. Virol.* 112, 27–40.
Gerna, G., Sears, J., Hoshino, Y., Steele, A. D., Nakagomi, O., Sarasini, A. and Flores, J. (1994) *Virology* 200, 66–71.
Gombold, J. L. and Ramig, R. F. (1986) *J. Virol.* 57, 110–116.
Gombold, J. L. and Ramig, R. F. (1989) *J. Virol.* 63, 4525–4532.
Gouvea, V., de Castro, L., Timenetsky, M. D., Greenberg, H. B. and Santos, N. J. (1994) *J. Clin. Microbiol.* 32, 1408–1409.
Gouvea, V., Lima, R. C. C., Linhares, R. E., Clark, H. F., Nosawa, C. M. and Santos, M. (1999) *Virus Res.* 59, 141–147.
Green, K. Y. and Kapikian, A. Z. (1992) *J. Virol.* 66, 548–553.

Green, K. Y., Midthun, K., Gorziglia, M., Hoshino, Y., Kapikian, A. Z. and Chanock, R. M. (1987) *Virology* 6, 153–159.

Greenberg, H. B., McAuliffe, V., Valdesuso, J., Wyatt, R., Flores, J., Kalica, A., Hoshino, Y. and Singh, N. (1983) *Infect. Immun.* 39, 91–99.

Grimwood, K., Lund, J. C., Coulson, B. S., Hudson, I. L., Bishop, R. F. and Barnes, G. L. (1988) *J. Clin. Biol.* 26, 732–738.

Halaihel, N., Lievin, V., Ball, J. M., Estes, M. K., Alvarado, F. and Vasseur, M. (2000) *J. Virol.* 74, 9464–9470.

Hart, C. A. and Cunliffe, N. A. (1999) *Curr. Opin. Infect. Dis.* 12, 447–457.

Hasegawa, A., Inouye, S., Matsuno, S., Yamaoka, K., Eko, R. and Suharyono, W. (1984) *Microbiol. Immunol.* 28, 719–722.

Hoshino, Y., Saif, L. J., Sereno, M. M., Chanock, R. M. and Kapikian, A. Z. (1988) *J. Virol.* 62, 744–748.

Hundley, F., Biryahwaho, B., Gow, M. and Desselberger, U. (1985) *Virology* 143, 88–103.

Hung, T., Chen, G. M., Wang, C. G., Yao, H. L., Fang, Z. Y., Chau, T. X., Chou, Z. Y., Ye, W., Chang, X. J. and Den, S. S. (1984) *Lancet* 326, 1139–1142.

Institute of Medicine (1986) In *New Vaccine Development: Diseases of Importance in Developing Countries*, Vol. 2, National Academy Press, Washington, DC, D13-1–D13-12.

Iturizza-Gomara, M., Cubitt, D. Desselberg, U. and Gray, J. (2001a) *J. Clin. Microbiol.* 29, 3796–3798.

Iturizza-Gomara, M., Green, J., Brown, D. W. G., Desselberger, U. and Gray, J. J. (2000) *J. Clin. Microbiol.* 38, 898–901.

Iturriza-Gomara, M., Isherwood, B., Desselberger, U. and Gray, J. (2001b) *J. Virol.* 75, 3696–3705.

Jin, Q., Ward, R. L., Knowlton, D. R., Gabbay, Y. G., Linhares, A. C., Rappaport, R., Woods, P. A. and Glass, R. I. (1996) *Arch. Virol.* 191, 2057–2076.

Jourdan, N., Brunet, J. P., Sapia, C., Blais, A., Cotte-Laffitte, J. and Forestier, F. (1998) *J. Virol.* 72, 7228–7236.

Kapikian, A. Z. and Chanock, R. M. (1996) In *Fields' Virology* (eds B. N. Fields, D. M. Knipe and P. M. Howley), Lippincott–Raven Press, Philadelphia, pp. 1657–1708.

Kirkwood, C. D. and Palombo, E. A. (1997) *Virology* 236, 258–265.

Kohli, E., Maurice, L., Vautherot, J. F., Bourgeois, C., Bour, J. B., Cohen, J. and Pothier, P. (1992) *J. Gen. Virol.* 73, 907–914.

Krishnan, T., Sen, A., Choudhury, J. S., Das, S., Naik, T. N. and Bhattacharya, S. K. (1999) *Lancet* 353, 380–381.

Larralde, G. and Gorziglia, M. (1992) *J. Virol.* 66, 7438–7443.

Lazdins, I., Coulson, B. S., Kirkwood, C., Dyall-Smith, M., Mosendyez, P. J., Souza, S., and Holmes, I. H. (1995) *Virology* 209, 80–89.

Lee, C-N., Wang, Y-L., Kao, C-L., Zao, C-L., Lee, C-Y. and Chen, H-N. (2000) *J. Clin. Microbiol.* 38, 4471–4477.

Lundgren, O. and Svensson, L. (2001) *Microbes Infect.* 3, 1145–1156.

Lynch, M., Bresee, J. S., Gentsch, J. R. and Glass, R. I. (2000) *Curr. Opin. Infect. Dis.* 13, 495–502.

Matson, D. O., O'Ryan, M. L., Herrera, I., Pickering, L. K. and Estes, M. K. (1993) *J. Infect. Dis.* 167, 577–583.

Mattion, N. M., Cohen, J. and Estes, M. K. (1994) In *Viral Infections of the Gastrointestinal Tract* (ed. A. Z. Kapikian), Marcel Dekker, New York, pp. 169–249.
Maunula, L. and von Bonsdorff, C-H. (1998) *J. Gen. Virol.* 79, 321–332.
Miller, M. A. and McCann, L. (2000) *Health Econ.* 9, 19–35.
Murray, C. J. L. and Lopez, A. D. (1997) *Lancet* 349, 1269–1276.
Palombo, E. A. (1999) *FEMS Microbiol. Lett.* 181, 1–8.
Palombo, E. A. (2002) *Virus Genes* 24, 11–20.
Patton, J. T., Salter-Cid, L., Kalbach, A., Mansell, E. A. and Kattoura, M. (1993) *Virology* 192, 438–446.
Pedley, S., Hundley, F., Chrystie, I., McCrae, A. E. and Deselberger, U. (1984) *J. Gen. Virol.* 65, 1141–1150.
Perez, J. F., Chemello, M. C., Liprandi, F., Ruiz, M. C. and Michelangeli, F. (1998) *Virology* 252, 17–27.
Perez, J. F., Ruiz, M. C., Chemello, M. C. and Michelangeli, F. (1999) *J. Virol.* 73, 2481–2490.
Racz, M. L., Kroeff, S. S., Munford, V., Caruzo, T. A., Durigon, E. L., Hayashi, Y., Gouvea, V. and Palombo, E. A. (2000) *J. Clin. Microbiol.* 38, 2443–2446.
Ramig, R. F. (1997) *Ann. Rev. Microbiol.* 51, 225–255.
Riepenhoff-Talty, M., Morse, K., Wang, C. H., Shapiro, C., Roberts, J., Welter, M., Allen, M., Evans, M. J. and Flanagan, T. D. (1997) *J. Clin. Microbiol.* 35, 486–488.
Ryan, M. J., Ramsay, M., Brown, D., Gay, N. J., Farrington, F. P. and Wall, P. G. (1996) *J. Infect. Dis.* 174, S12–S18.
Saif, L. J. and Jiang, B. (1994) *Curr. Top. Microbiol. Immunol.* 185, 339–371.
Saulsbury, F. T., Winkelstein, J. A. and Yolken, R. J. (1980) *J. Pediatr.* 97, 61–65.
Sen, A., Kobayashi, N., Das, S., Krishnan, T., Bhattacharya, S. K. and Naik, T. N. (2001) *Lancet* 357, 198–199.
Tabassum, S., Shears, P. and Hart, C. A. (1994) *J. Med. Virol.* 43, 50–56.
Takahashi, K., Ohashi, K., Abe, Y., Mori, S., Taniguchi, K., Ebina, T., Nakagomi, O., Terada, M. and Shigeta, S. (2002) *Antimicrob. Agents Chemother.* 46, 420–424.
Tian, P., Ball, J. M., Zeng, C. Q. and Estes, M. K. (1996) *J. Virol.* 70, 6973–6981.
Tian, P., Hu, Y., Schilling, W. P., Lindsay, D. A., Eiden, J. and Estes, M. K. (1994) *J. Virol.* 68, 251–257.
Tian, Y., Tarlow, O., Ballard, A., Desselberger, U. and McCrae, M. A. (1993) *J. Virol.* 67, 6625–6632.
Tihova, M., Dryden, K. A., Bellamy, A. R., Greenberg, H. B. and Yeager, M. (2001) *J. Mol. Biol.* 314, 985–992.
Velazquez, F. R., Matson, D. O., Calva, J. J., Guerrero, M. L., Morrow, A. L., Glass, R. I., Estes, M. K., Pickering, L. K. and Ruiz-Palacios, G. M. (1996) *N. Engl. J. Med.* 335, 1022–1028.
Velazquez, F. R., Matson, D. O., Guerrero, M. L., Shults, J., Calva, J. J., Morrow, A. L., Glass, R. I., Pickering, L. K. and Ruiz-Palacios, G. M. (2000) *J. Infect. Dis.* 182, 1602–1609.
Ward, R. L. (1996) *J. Infect. Dis.* 174, S51–S58.
Watanabe, M., Nakagomi, T., Koshimura, Y. and Nakagomi, O. (2001) *Arch. Virol.* 146, 557–570.
Widdowson, M-A., van Doornum, G. J. J., van der Poel, W. H. M., de Boer, A. S., Mahdi, U. and Koopmans, M. (2000) *Lancet* 356, 1161–1162.

6

HAEMOPHILUS INFLUENZAE

Derek W. Hood and E. Richard Moxon

INTRODUCTION

Haemophilus influenzae is a small Gram-negative, non-spore-forming coccobacillus belonging to the family *Pasteurellaceae*. *H. influenzae* is an obligate commensal of humans, residing predominantly in the upper respiratory tract, but is best known for the occasional associated disease. Capsular strains can cause invasive, bacteraemic infections such as meningitis, septicaemia, epiglottitis, septic arthritis and empyema, particularly in infants. Strains lacking capsule, so-called non-typeable (NT*Hi*) strains, are a common cause of otitis media, sinusitis and conjunctivitis and of acute lower respiratory tract infections which account for many hundreds of thousands of childhood deaths in the Third World. A primary requirement for *H. influenzae* is that it must maintain infectivity through transmission. The occurrence of *H. influenzae* invasive disease is therefore probably an accidental process, as host death does not allow the organism to perpetuate its gene pool.

A critical factor for any microorganism associated with plants or animals is the dynamic nature of the host–microbe interaction. A range of specific molecules that mediate adherence/penetration of host cells, evasion of host defences (innate and acquired), scavenging of nutrients and damage to host tissue are probably required for virulence. To be successful, a microorganism must maintain its fitness in diverse and changing environments and therefore must have evolved mechanisms for responding to such changes. Sensors that monitor environmental changes coupled to a response exerted through a hierarchical system of gene regulation is one genetic mechanism that bacteria have evolved to facilitate acclimation to their host. However, an alternative strategy that has been adopted by some bacteria, including *H. influenzae*, is to increase the effective mutation rate of genes encoding antigens that are critical for interactions with the host. This can generate the diversity necessary for small numbers of infecting organisms to adapt and survive against the plethora of host factors that have evolved to eliminate them. High-frequency, reversible on/off switching of gene expression and function, so-called phase variation, increases the potential of bacteria to adapt to unpredictable host environments. These genetic elements have been termed contingency loci (Moxon et al., 1994).

CAPSULE PHASE VARIATION

The surfaces of pathogenic bacteria display a repertoire of sugar, glycolipid and protein molecules that are accessible to the host immune system. Examples of sugar-containing macromolecules that are prominent to the immune system, by both their abundance and their location, extended from the bacterial cell surface, are capsule and lipopolysaccharide (LPS). Capsular polysaccharide is known to be a critical determinant for invasive systemic disease caused by H. influenzae. Strains of H. influenzae may express one of six serologically distinct capsular types, designated a through f, each of which comprise polymers of distinct sugars in specific linkages. Epidemiological data indicates that strains expressing the polyribosylribitol phosphate-containing serotype b capsule predominate in serious human invasive disease. The amount of type b capsule synthesized has been correlated with virulent infection (Hoiseth et al., 1992; Kroll, 1992). Capsule polysaccharide, in conjunction with LPS, impedes host clearance by complement and phagocytosis (Noel et al., 1996). However, bacteria that produce no capsule may have an advantage in that they will not be recognized by host antibodies to the capsule antigen and may be able to more intimately associate with host cells. Production of capsule polysaccharide represents the less typical mechanism of phase variation found in H. influenzae. Amplification of the 18 kb segment of DNA that is the cap locus involves homologous recombination facilitated by the insertion element IS1016, which flanks the cap locus (Kroll et al., 1991). The entire locus represents a compound transposon. In most clinical type b isolates there exist two copies of the cap locus, although three and up to five copies can be found (Corn et al., 1993). Through recA-dependent events the number of individual copies of the compound transposon can shrink to 0 to produce capsule variants at frequencies of between 1 in 100 and 1 in 1000 per bacterium per generation.

PHASE VARIATION MEDIATED BY SIMPLE NUCLEOTIDE REPEATS

For H. influenzae, the capacity to generate diversity, and in particular antigenic variation of cell-surface-expressed molecules, is largely mediated through specialized DNA sequences, runs of repetitive DNA (microsatellites). This mechanism of phase variation involves changes in the number of nucleotides in simple repeat elements associated with genes. Simple nucleotide repeats, typically iterations of between 2 and 7 bp in *H. influenzae*, are relatively unstable and are hypermutable, allegedly through polymerase slippage (slipped-strand mispairing) during nucleic acid replication (Levinson and Gutman, 1987). This is a *recA*-independent mechanism that results in loss or gain of usually one, but sometimes more, repeat units. Tracts of repetitive DNA are prone to mispair as the complementary strands are separated as DNA 'breathes' during transcription or replication. Whilst the complementary base pairing is temporarily disrupted, slippage may occur so that one strand is displaced relative to the other in either the 5' or the 3' direction (Figure 6.1). This results in one or more repeat units being mispaired when the strands re-align. On the next round of replication, mismatch repair mechanisms correct the incompatibility, such that one or more repeats is gained or lost depending on the direction of slippage. However, as translation depends upon having the correct triplet code, loss or gain of a unit, of say a 4 bp repeat within a reading frame, will cause a frame shift. Thus, if *n* is the

A   ATG...27nt...ATG...73nt...ATGCGTACACTTATCAATCAAT(CAATx12)CAATCAAT....
    TAC...27nt...TAC...73nt...TACGCATGTGAATAGTTAGTTA(GTTAx12)GTTAGTTA.....

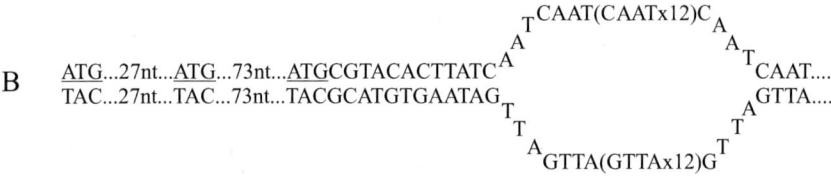

C   ATG...27nt...ATG...73nt...ATGCGTACACTTATCAATCAAT(CAATx12)CAATCAAT....
    TAC...27nt...TAC...73nt...TACGCATGTG   GTTA(GTTAx12)GTTAGTTA....
                                       A   A
                                       A   T
                                       T   T
                                       A   G

D   ATG...27nt...ATG...73nt...ATGCGTACACTTATCAATCAAT(CAATx12)CAAT....
    TAC...27nt...TAC...73nt...TACGCATGTG   GTTA(GTTAx12)GTTAGTTA....
                                       A   A
                                       A   T
                                       T   T
                                       A   G

E   ATG...27nt...ATG...73nt...ATGCGTACACTTATCAATCAAT(CAATx12)CAAT....
    TAC...27nt...TAC...73nt...TACGCATGTGAATAGTTAGTTA(GTTAx12)GTTA....

Figure 6.1 Possible mutational mechanism based on slipped-strand mispairing responsible for phase variation mediated by simple nucleotide repeats in *H. influenzae*. Illustrated is the 5'-CAAT repeats located within the *lic2A* gene; potential initiation codons are underlined. (A), with 16 repeats the *lic2A* gene is expressed; (B) the strands separate during replication; (C) mispairing of 5'-CAAT repeats can occur during reannealling leaving an unpaired copy of 5'-CAAT (asterisks); (D) the unpaired repeat is deleted and the single-strand loop is targeted by excision repair processes; (E) with the number of repeats reduced to 15 the *lic2A* reading frame is not expressed.

number of repeat units that is appropriate for translation, i.e. places the downstream reading frame in frame with the upstream initiation codon, then $n - 1$ or $n + 1$ repeats, resulting from slippage, would cause a frame shift. In turn this would produce either a truncated or an altered peptide depending on whether the frame shift introduced a stop, or merely altered the triplet code. Thus, in this example, repetitive DNA promotes phase-variable expression of genes through the resultant effect on translation of the encoded molecule. This is the common mechanism involving simple nucleotide repeats that promotes phase variation in *H. influenzae*. Phase variation mediated by DNA repeat tracts can also occur in the promoter region of genes. A change in the number or repeat units within the control region of a gene will effect RNA polymerase binding.

PHASE VARIATION BY MODULATION OF TRANSCRIPTION

An example of phase variation in *H. influenzae* that results from altered transcription is the expression of pili (or fimbriae). These long filamentous appendages can act as an adhesin, mediating attachment to respiratory epithelia or other host cell types and are targets for the host immune response (van Ham *et al.*, 1993). During natural infection, nasopharyngeal isolates are often piliated, but when isolated from systemic sites, isolates are almost exclusively non-piliated. The expression of pili is phase variable with rates of transition from 10^{-3} to 10^{-4} per bacterium per generation (Farley *et al.*, 1990). The chromosomal locus encoding pilus determinants (*hifA* to *hifE*) has been studied in detail in type b capsulate strains but the locus is also present in a subset of non-typeable strains (Geluk *et al.*, 1998). The *hifA* (pilus subunit protein) and *hifB* (pilus chaperone protein) genes are transcribed divergently from a common promoter region. Phase variation results from slipped-strand mispairing of dinucleotide (5'-TA) repeats associated with the overlapping promoters of the divergently transcribed genes (Table 6.1). Changes in the number of repeats located within the −10 and −35 regions alter binding of RNA polymerase by repositioning the promoter-binding components (Figure 6.2A). Typically in a type b strain, 10 copies of TA allowed maximal pilus expression, 11 copies reduced expression and 9 copies eliminated expression (van Ham *et al.*, 1993, 1994). A similar finding has been made for the *hif* gene clusters in some non-typeable strains (Geluk *et al.*, 1998).

A further, yet somewhat different example of phase variation mediated by alteration of repeated sequence in the promoter region is demonstrated by the genes encoding the related high molecular weight outer membrane proteins, HMW1 and HMW2. HMW1 and HMW2 are adhesins found exclusively in non-typeable strains (Barenkamp and Leininger, 1992; St Geme *et al.*, 1993) and are predominant targets for the host immune response during infection (Barenkamp and Bodor, 1990). The two proteins interact with distinct host receptor molecules; HMW1 is thought to recognize a sialylated glycoprotein (St Geme *et al.*, 1993).

Table 6.1 Simple sequence repeats and associated phase-variable genes identified in *Haemophilus influenzae*

Gene	Associated repeat 5'-( )$_n$	Location[e]	Gene function	Proven effect upon antigen
hifA/hifB	AT	P	Fimbriae biogenesis	Fimbriae
hxuC	ATT	P	Haem/haemopexin binding	
lic1A	CAAT	O	Phosphocholine synthesis	LPS
lic2A	CAAT	O	Glycosyltransferase	LPS
lic3A	CAAT	O	Sialyltransferase	LPS
lex2A	GCAA	O	Glycosyltransferase	LPS
lgtC	GACA	O	Glycosyltransferase	LPS
HI0635[a], *hgpC, hgbB*	CAAC	O	Haemoglobin/haptoglobin binding	OMP
HI0661, *hgpB, hgbA, hgbC*	CAAC	O	Haemoglobin/haptoglobin binding	OMP
HI0712[a]	CAAC	O	Haemoglobin/haptoglobin binding	OMP
HI1565/HI1567[b]	CAAC	O	Haemoglobin/haptoglobin binding	OMP
Intergenic	CAAC		Unknown	
mod	AGTC	O	Methyltransferase	
yadA[c]	GCAA	O	Outer membrane protein	
HI0687	TTTA	O	Conserved hypothetical	
Intergenic	TTATC		Unknown	
hmw1	ATCTTTC	P	Outer membrane protein	OMP
hmw2	ATCTTTC	P	Outer membrane protein	OMP
Low numbers of other repeats[d]				

[a]HI0712 is very similar to HI0635.
[b]HI1565 and HI1567 are listed in the TIGR database as separate ORFs but are considered as a single protein product in the SWISS-PROT database (Q57408).
[c]The *yadA* reading frame is interrupted by frame shifts in the strain Rd genome sequence.
[d]4 copies of the pentanucleotide 5'-GTCTC, 4 copies of the hexanucleotide 5'-AGCCAG and 3 copies of 5'-GGCAAT and 5'-TTAAAA have been reported for *H. influenzae* (van Belkum *et al.*, 1997). These are located in the intergenic region upstream of HI1278 (*hsdM*), the reading frame of HI0251 (TonB protein) and HI0265 (*hxuC*) and upstream of HI0525 (*pgk*), respectively. Some polymorphism of DNA containing these sequences has been reported for 5'-GTCTC, 5'-AGCCAG and 5'-GGCAAT between strains but not for 5'-TTAAAA (van Belkum *et al.*, 1997). The significance of these findings remains unclear.
[e]The location of the repeats within the relevant gene, P in promoter region, O within ORF.

Changes in the expression of HMW1 and HMW2 involve alteration in the number of heptanucleotide repeats (5'-ATCTTTC) located between alternative promoter sites within the respective gene. For the *hmw2A* gene, an increase in the number of repeat units from 15 to 28 showed an inverse correlation with the amounts of specific RNA transcript and protein product synthesized (Dawid *et al.*, 1999).

PHASE VARIATION MEDIATED BY TETRANUCLEOTIDE REPEATS

With the availability of complete genome sequences of many of the major bacterial pathogens, the presence of tracts of simple nucleotide repeats can be

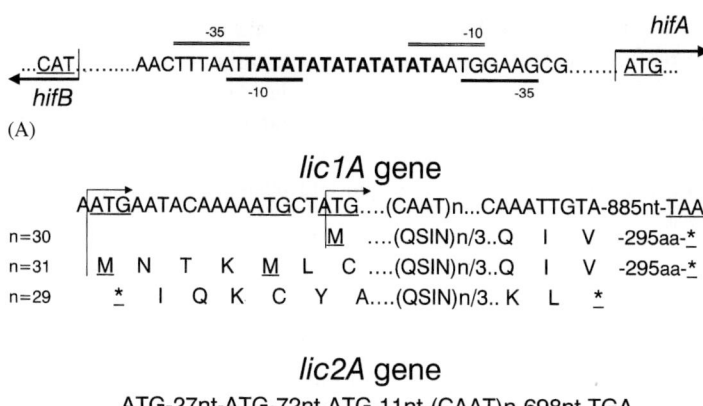

Figure 6.2 Mechanism of phase-variable expression of genes mediated by simple nucleotide repeat tracts in *H. influenzae*. (A) Dinucleotide repeats in the promoter region of the fimbriae biogenesis genes. The start of the divergent reading frames for the *hifA* and *hifB* genes is shown by arrows. (B) Tetranucleotide repeats in the reading frame of the *lic* LPS biogenesis genes. Initiation (ATG) and termination (TAA/TGA/TAG) codons are underlined. Shown in greater detail is the relationship between the number (n) of repeats of the tetranucleotide 5'-CAAT and the reading frame of the *lic1A* gene. The nucleotide sequence and the alternative initiation codons are shown with the amino acid sequence below. When 30 or 31 copies of the repeat are present, correct translation of the reading frame can take place from alternative ATGs in 2 of the 3 reading frames (shown by arrows). With 29 copies of 5'-CAAT present no initiation codon is available for correct translation. The relative spacing of the alternative translational start codons for both *lic2A* and *lic3A* are also illustrated.

used as markers to rapidly identify genes that are likely to be involved with bacterial virulence. A search of the complete genome sequence of *H. influenzae* strain Rd for repetitive DNA sequences identified 12 genes possessing tracts of >5 tetranucleotide repeats. These genes were predicted to code for proteins involved in commensal or virulence functions (Hood *et al.*, 1996b) and included genes required for sequestration of iron-containing compounds, LPS biosynthesis, a putative adhesin and, perhaps surprisingly, components of DNA restriction modification systems (Table 6.1). A similar pattern of genes, but predominantly involving pentanucleotide and mononucleotide repeats, has been described in another pathogen causing a similar pattern of disease, *Neisseria meningitidis* (Saunders *et al.*, 2000). Both the genes for fimbriae biogenesis and the HMW adhesins detailed above are not present in strain Rd. Further analysis of the *H. influenzae* Rd genome sequence identified 2 penta- and 3 hexanucleotide repeat tracts (van Belkum *et al.*, 1997). The variable number of tandem repeats at 18 potentially phase-variable loci were studied by PCR and gel analysis and 14 were found to vary between *H. influenzae* strains (van Belkum *et al.*, 1997). These loci included 11 of the 12 tetranucleotide, 2 pentanucleotide and 1 of the hexanucleotide repeats. Interstrain variation in repeat numbers is a characteristic of phase-variable genes so this data provides some circumstantial evidence for the relevant loci being phase variable.

In the remainder of this chapter we review the phase-variable genes of *H. influenzae* containing the most common and best-characterized type of repeat, tracts of tetranucleotides.

Iron-binding proteins

All living cells require a source of iron but in the human host levels of free iron are very limited, with most iron being bound to specific iron-complexing molecules. A phenotypic trait that distinguishes *H. influenzae* from other organisms is an absolute requirement for an exogenous supply of haem for growth. *H. influenzae* lacks several enzymes necessary for the conversion of δ-aminolevulinic acid to protoporphyrin IX, the intermediate biosynthetic precursor for haem, which is required for aerobic growth (Tatusov *et al.*, 1996). Therefore, *H. influenzae* must procure iron and haem from host iron-carrying molecules such as haem, haemogloboin, haptoglobin, transferrin, and lactoferrin. Invariably this process will involve specific receptor proteins exposed on the bacterial cell surface. The initial screen of the strain Rd genome sequence for loci containing tetranucleotide repeats revealed four loci associated by homology comparison with iron-binding proteins in other organisms (Hood *et al.*, 1996b). Each locus contained tracts of the DNA repeat 5'-CAAC, and a fifth such tract was located in a region of the chromosome with no obvious reading frame with significant homology to any known protein (Table 6.1). The first detailed investigation of one 5'-CAAC-containing gene, *hhuA*, showed that HhuA was an outer membrane protein involved in binding haemoglobin/haptoglobin, although an isogenic *hhuA* mutant was still able to utilize the iron-containing compounds (MacIver *et al.*, 1996). In a NT*Hi* strain, N182, three other ORFs containing 5'-CAAC repeats, *hgbA*, *hgbB* and *hgbC*, which show varying degrees of homology to each other and to *hhuA* were studied. The products of these three genes were shown to bind haemoglobin or haemoglobin/haptoglobin *in vitro* (Cope *et al.*, 2000). Specific monoclonal antibodies have been raised to each of the three proteins and phase-variable expression was demonstrated for two of these proteins, HgbA and HgbB. A translational gene fusion of *hgpA*, encoding a protein in a type b strain related to HhuA, with a β-galactosidase (*lacZ*) reporter has shown that phase-variable gene expression is associated with alteration in the length of the 5'-CAAC repeat tract (Morton *et al.*, 1999). It is apparent that *H. influenzae* contains a repertoire of phase-variable TonB-dependent proteins with varying degrees of homology to each other, which varies between strains and which is induced in the utilization of haem from compounds such as haemoglobin and haemoglobin/haptoglobin. An explanation for the presence of multiple phase-variable genes whose products have overlapping specificity for binding iron-containing compounds may reflect the flux of available iron in changing host compartments or variable expression to allow immune evasion of the host response to these immunogenic but antigenically distinct proteins.

Phase variation of lipopolysaccharide biosynthesis

We will now focus upon those genes that are involved in the biosynthesis of a molecule known to express a complex pattern of antigens, LPS. LPS is the best-characterized system in terms of variation in the number of repeats in multiple genes required for synthesis and the resultant effect of phase variation upon the biology of the bacterium. It illustrates how phenotypic diversity modulated by DNA repeats is biologically relevant to the commensal and virulence behaviour of the organism. LPS is a critical, if not essential, component of the *H. influenzae* cell wall. LPS largely defines the permeability barrier around the cell and influences adherence to host cells, dissemination of bacteria by contiguous or systemic routes and helps mediate inflammation and injury to host tissues. The membrane-anchoring lipid A portion, or endotoxin, is responsible for much of the pathological damage associated with disease whereas the inner and outer core oligosaccharides promote adherence, invasion and extracellular survival of *H. influenzae*. *Haemophilus* LPS lacks the polymerized side chains (O-antigens) distal to the core in many pathogenic Gram-negative species such as members of the *Enterobacteriaceae*. These more truncated glycolipids of *H. influenzae* have been designated lipo-oligosaccharides (LOS) by some investigators; however, we retain the designation LPS. Despite the absence in *H. influenzae* of a true O-antigen, the LPS presents significant structural complexity. It exhibits a significant degree of inter- and intra-strain heterogeneity as detected by reactivity with monoclonal antibodies (mAbs) (Kimura and Hansen, 1986) and staining of LPS with silver after gel electrophoresis (Roche *et al.*, 1994). This antigenic variation of LPS could be generated in one of two ways. Some LPS variation in *H. influenzae* will be as a result of the complex biosynthetic processes that link the sugars and their substituents (phosphate, phosphoethanolamine, pyrophosphoethanolamine, O-acetyl groups, phosphorylcholine or glycine) to complete the tertiary structure of the molecule. Structural variation (microheterogeneity) will occur should not all steps in the biosynthetic process go to completion; all sugar units and substituents may be non-stoichiometric in the absence of any known proof-reading function. However, the major determinant of LPS heterogeneity in *H. influenzae* is phase variation. Phase variation results in multiple LPS structures within a population of cells and the presentation of an array of variant antigens to the host immune system. Indeed, structural heterogeneity has confounded efforts to study the biology of the molecule, its role in virulence and to develop its potential as a common antigen for development of a vaccine to control disease. MAbs specific to LPS provide a necessary tool for detecting variations in LPS structure (Kimura and Hansen, 1986; Gulig *et al.*, 1987), and have been used to study phase variation of LPS epitopes. The molecular basis of LPS phase variation in *H. influenzae* was first characterized by Weiser and colleagues (Weiser *et al.*, 1989a,b). A genomic library from a type b strain, RM7004, was screened for clones that would allow a recipient strain, Rd, to express novel LPS epitopes recognized by murine mAbs (6A2 and 12D9) to outer core structures of *H. influenzae* LPS (Weiser *et al.*, 1989a). Strain Rd does not naturally react with these mAbs, whereas strain RM7004 reacts in a phase-variable manner with each (Figure 6.3) (Patrick *et al.*, 1989). A DNA clone from a genomic library made from strain RM7004 was identified, which conferred some strong reactivity with some of the mAbs. This locus was designated *lic1* (*l*ipopolysaccharide *c*ore epitope 1).

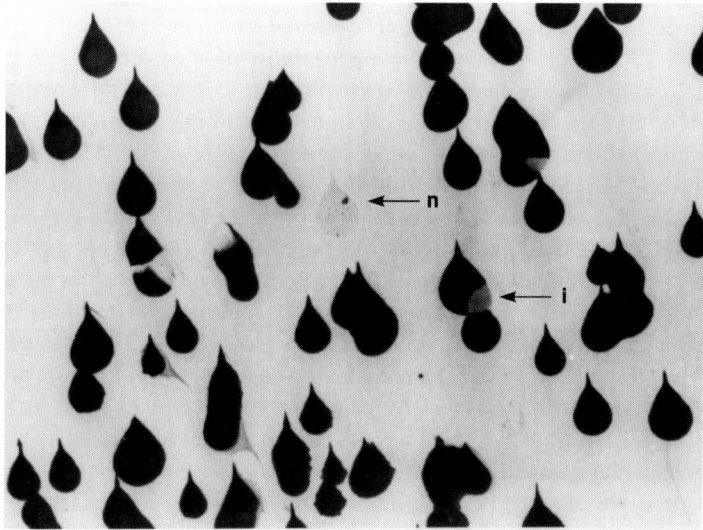

Figure 6.3 Colony immunoblot of *H. influenzae* strain RM7004 stained with an antibody directed towards LPS. The majority of colonies stain strongly; arrows indicate non-reactive (n) and intermediate (i) colonies or colony segments.

A total of five LPS biosynthetic loci relevant to the phase-variable expression of LPS epitopes have currently been identified and characterized in *H. influenzae*. The genes and their relevant details are listed in Table 6.1. Four of these phase-variable genes were identified by classical genetic approaches, similar to that described above for *lic1*, using LPS-specific mAbs to monitor changes in LPS epitope expression patterns. The fifth gene, *lgtC*, was identified from the complete genome sequence of strain Rd (Fleischmann *et al.*, 1995; Hood *et al.*, 1996b). In all cases, the phase-variable genes are involved in synthesis of structures that are part of the outer core (Figure 6.4) of the LPS molecule. Structural details are now available for the LPS from several *H. influenzae* strains (RM153 (Eagan), RM7004, A2, RM118 (Rd),

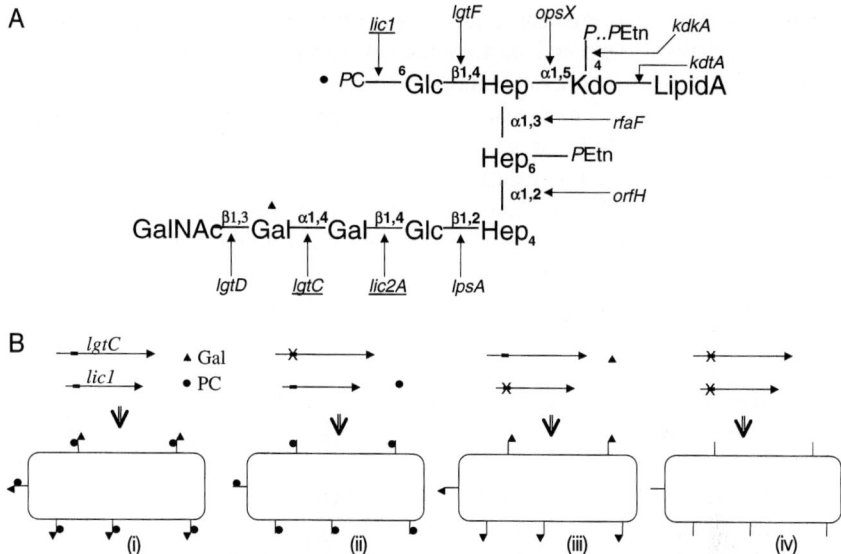

Figure 6.4 The LPS molecule synthesized by *H. influenzae*. (A) A representation of the structure of the LPS from strain RM118 (Rd) showing the genes required for the addition of each of the components. The phase-variable genes are underlined. Represented in the LPS structure: Kdo, 2-keto-3-deoxyoctulosonic acid; Hep, L-glycero-D-manno-heptose; Glc, D-glucose; Gal, D-galactose; GalNAc, N-acetylgalactosamine; *P*Etn, phosphoethanolamine; *P*, phosphate; *PC*, phosphorylcholine. (B) The LPS phenotypes generated by the combined expression patterns of two independent phase-variable genes, *lgtC* adding a galactose (▲) and *lic1* adding phosphorylcholine (●). The repeat tracts in the linear representation of each gene are shown by a box. When the box is overlaid with an X, this indicates an inappropriate number of repeats for expression. (i)–(iv) represent microbial cells expressing the pattern of LPS antigens dependent upon *lic1* and *lgtC* gene expression.

486, 375, 176, 1003, 2019, 9274). However, it was not until the complete genome sequence of strain Rd became available that a comprehensive study of LPS biosynthesis could be carried out. Over 30 genes were identified by homology comparisons with known LPS genes from other organisms (Hood *et al.*, 1996a) and gene function predictions for many have been confirmed by structural analysis of LPS prepared from the relevant mutant strains. All of the major biosynthetic steps for the saccharide portion of the LPS in the *H. influenzae* type d strain, RM118 (Hood *et al.*, 2001b), and the majority for the type b strain RM153 Eagan (Hood *et al.*, 1996a; unpublished data) have been assigned. The details and the contribution of the individual phase-variable genes to the synthesis, heterogeneity and biology of *H. influenzae* LPS are now described.

The **lic1** *locus and phosphorylcholine*
The nucleotide sequence of the *lic1* locus comprises four open reading frames (ORFs) encompassing 3.4 kilobases (kb) of DNA. The ORFs, designated *lic1A*,

lic1B, *lic1C* and *lic1D*, are transcribed in the same direction, each adjacent ORF having overlapping stop and start codons (Weiser *et al.*, 1989a). This suggests that the *lic1* locus is an operon comprising four co-transcribed ORFs. The *lic1A* gene contains a tract of the repeated tetranucleotide 5'-CAAT located within the ORF at its 5' end. Loss or gain of copies of this repeat by slipped-strand mispairing would cause frameshifting with respect to the alternative start codons found in two of the three possible reading frames, resulting in variable gene expression. The *lic1* locus encodes genes for the incorporation of phosphorylcholine (PC) into *H. influenzae* LPS (Weiser *et al.*, 1997) and mutant strains containing deletions encompassing all four ORFs of the *lic1* locus have confirmed their essential role in the expression of PC. Lic1A has similarity to eukaryotic choline kinases, Lic1B to choline permease, Lic1C to a pyrophosphorylase and Lic1D is the predicted diphosphonucleoside choline transferase (Weiser *et al.*, 1997). The function of Lic1D, the phosphorylcholine transferase, is dependent upon an exogenous source of choline, a major component of host lipids, and it is proposed that choline is acquired from the host and assimilated into LPS (Fan *et al.*, 2001). Using colony immunoblotting, three levels of expression of PC, strong (++++), weak (+) and undetectable (–), were detected in the ratio of weak 97%: strong and undetectable 3%, respectively (Weiser *et al.*, 1989a). Oligonucleotides flanking the tetranucleotide region were used to amplify a portion of *lic1A* from chromosomal DNA obtained from colonies exhibiting each of the three levels of PC expression. The most prevalent size of DNA comprising 5'-CAAT repeats found in non-reactive colonies was 184 bp, equivalent to 29 repeats, whereas the size of the repeat for the ++++ reactive colonies was 192 bp (corresponding to 31 repeats) and for + colonies, 188 bp, equivalent to 30 repeats. These results show that the degree of expression of PC in the LPS correlates with translation of *lic1A* from two alternative closely sited ATGs, or no translation (Figure 6.2). The predominant state appears to be 30 repeats under the experimental conditions, but loss or gain of one or more 5'-CAAT repeats results in strong or undetectable mAb binding respectively. Further analysis of *lic1* gene expression in *H. influenzae*, using a *lacZ* reporter gene fused in frame with *lic1D* (Moxon and Maskell, 1992), showed reversible high-frequency variation between three levels of β-galactosidase expression. The results showed some correlation, but were not entirely consistent, with the patterns of PC expression described above.

The transcription and translation of the *lic1* locus is complex but the phenotypic consequences of the phase variation mediated by *lic1A* are that the population of bacteria includes a proportion of individual cells that either express or lack PC on LPS glycoforms. It is evident from experimental studies in animals that the presence or absence of PC may be advantageous in certain host compartments, but not others. Weiser and colleagues have shown that expression of PC was favoured during nasopharyngeal colonization and carriage in the infant rat, an animal model of *H. influenzae* infection, and that bacteria expressing PC are prevalent in cultures from human respiratory secretions

(Weiser and Pan, 1998; Weiser et al., 1998). In a different model of H. influenzae infection, the induction of otitis media in chinchillas, it has been shown that expression of PC greatly favours survival of NTHi strains in the nasopharynx and subsequent infection of the middle ear (Tong et al., 2000). This raises the important issue of how the variable presence of PC on LPS glycoforms modulates its biological functions.

H. influenzae adheres to and subsequently invades human bronchial epithelial cells through an interaction of LPS with platelet adherence factor receptor (PAF-R) (Swords et al., 2000). PAF-R is found on both human epithelial and endothelial cells and is able to promote a series of host signalling events. It has been proposed following inhibition studies that PC on the LPS is a key component of this interaction. It has also been shown that PC-expressing bacteria are more susceptible to the bactericidal activity of human serum. This increase in susceptibility occurs through the binding of C-reactive protein (CRP) (Weiser et al., 1998) and subsequent activation of the classical pathway of complement. However, CRP (and complement) can be found in the upper respiratory tract and might sterically interfere with the interaction of PC with PAF-R. An additional proposed mechanism through which PC may affect bacterial survival is to confer relative resistance to antimicrobial peptides (Lysenko et al., 2000a). The complexity of these potential host–microbe interactions may have been a decisive factor in the evolution of the phase-variable expression of PC. It is presumed to have the potential to afford an adaptive mechanism through the natural selection of variants best suited to the differing host microniches and provides flexibility over the time course of colonization or infection.

Distinct from the phase variation of PC within a clonal population, the precise location of PC on LPS glycoforms of genetically distinct strains can be different. The altered molecular environment of PC seems to be a factor in determining its accessibility, for example for binding of CRP and its subsequent effector role in mediating susceptibility to complement-mediated killing (Lysenko et al., 2000b). Allelic differences in Lic1D (the PC transferase) apparently preferentially direct the addition of PC to different glycoforms (Lysenko et al., 2000b). Analyses of LPS structure from H. influenzae strains has shown that, depending on the strain, each of the three oligosaccharide extensions of the triheptosyl backbone (Figure 6.4) of the inner core can be substituted with PC.

lic2A *and* lgtC *synthesize a phase-variable digalactoside epitope of* H. influenzae *LPS*

When digested chromosomal DNA from strain RM7004 was hybridized against an oligonucleotide composed of 5'-CAAT repeats, it resulted in the discovery of two loci additional to *lic1*, designated *lic2* and *lic3*. When first characterized, the *lic2* locus comprised four ORFs, only one of which apparently had relevance to LPS synthesis (High et al., 1993). The first ORF of the *lic2* locus, designated *lic2A*, contained multiple copies of 5'-CAAT within its 5' end and when mutated resulted in loss of reactivity with LPS specific mAbs 5G8 and 4C4 (Weiser *et al.*, 1990a). These mAbs bind to a digalactoside, the αGal(1–4)βGal epitope. This antigen is present on the LPS of other bacteria (Virji *et al.*, 1990) and is a component of the globoseries glycolipids expressed by some human epithelial cells. The digalactoside can act as receptor for the P fimbriae expressed by *Escherichia coli* and for the B subunit of shiga toxin. A potential biological role for the αGal(1–4)βGal structure as a virulence determinant was hypothesized based upon its potential to mimic host structures. *lic2A* contains two possible start codons (ATGx and ATGy) upstream of the repeats in one frame and a third potential start codon (ATGz) closer to the repeats in another frame (Figure 6.2). Three levels of mAb reactivity ('on', 'intermediate' and 'off') could be detected upon colony immunoblotting of *H. influenzae* type b strains using the specific mAbs (High *et al.*, 1993). These occurred at different relative frequencies in the strains. Direct sequencing of DNA amplified by PCR from immunostained colonies of each strain showed a correlation, albeit imperfect, between the number of 5'-CAAT repeats in *lic2A* and the mAb reactivity observed (High *et al.*, 1993). An identical gene, *lex1*, has been described independently in another *H. influenzae* type b strain (Cope *et al.*, 1991). Much of this incomplete correlation of Lic2A expression, digalactoside incorporation and mAb reactivity was made apparent with the discovery of a further LPS phase-variable gene, *lgtC*. *lgtC* was one of nine novel loci with multiple tandem tetranucleotide repeats revealed after analysis of the complete genome sequence of strain Rd (Hood *et al.*, 1996b). *lgtC* is a homologue of a gene encoding a glycosyl transferase implicated in LPS biosynthesis in *Neisseria* and has multiple copies of the tetranucleotide 5'-GACA located just within its 5' end and just 3' to a single ATG initiation codon. Deletion of *lgtC* in strain RM7004 showed loss of reactivity to mAbs specific for the digalactoside and the isolated LPS had a simpler electrophoretic profile when compared to wild type. Mass spectrometric analysis of the LPS isolated from a number of *lgtC* mutant strains has indicated in each case that the terminal αGal residue of the αGal(1–4)βGal epitope was absent (Figure 6.4) (Hood *et al.*, 1996a, 2001b). Similar experiments have confirmed Lic2A as a β-galactosyltransferase which adds the proximal residue of the digalactoside epitope (Hood *et al.*, 2001b). Biochemical analysis of LgtC has confirmed that it has α-galactosyltransferase activity (Hood *et al.*, 2001b). The importance of the digalactoside epitope to LPS biology is underlined by the evidence that two apparently independent phase-variable genes, *lgtC* and *lic2A*, contribute to its biosynthesis. The apparent contradictions observed in experiments to monitor LPS gene repeat numbers and phenotype (as described above for *lic2A*), are probably due to multiple phase-variable genes whose contributions to LPS phenotype are interdependent and capable of producing multiple reactive epitopes within the LPS molecule.

Phase-variable expression of the digalactoside is likely to have a role in evasion of host immune defences. Weiser and colleagues demonstrated in human patients that greater than 99% of cells isolated from the CSF of patients newly diagnosed with meningitis reacted with the appropriately specific mAbs (Weiser et al., 1989b; Weiser and Pan, 1998). In contrast, among H. influenzae cells isolated from the respiratory tract of people colonized with the organism, <0.1% were reactive. This would be consistent with the occurrence of phase variation in vivo. Expression of a digalactoside favours the survival of the organism against the killing effect of human serum (Kimura et al., 1987; Weiser et al., 1990; Cope et al., 1991; Hood et al., 1996b; Weiser and Pan, 1998). The data from these experiments would indicate that the presence or absence of the digalactoside structure might influence translocation from the respiratory tract to the bloodstream and subsequent survival in the blood. The role of the digalactoside in serum resistance is presumed to be by presentation of a self-antigen (molecular mimic) to the host immune system. In another respiratory pathogen, Moraxella catarrhalis, the digalactoside is similarly a factor in resistance to bactericidal activity (Zaleski et al., 2000). It is reasonable to propose that expression of this epitope in H. influenzae is simply not required, or is even disadvantageous, in alternative niches such as the nasopharynx.

The fifth LPS-associated phase-variable locus characterized for H. influenzae is lex2. A genomic library was constructed from the type b strain, DL42, which reacts with the LPS-specific mAb, 5G8. Clones from this library were used to transform another type b strain, DL180, which is non-reactive with mAb 5G8 (Jarosik and Hansen, 1994). A clone that conferred mAb 5G8 reactivity was identified and this was sequenced to reveal two contiguous ORFs. The first ORF, designated lex2A, contained the tetranucleotide 5'-GCAA repeated 18 times within its 5' end (Jarosik and Hansen, 1994). Transposon mutagenesis indicated that the second ORF, lex2B, is required for expression of the mAb 5G8-reactive LPS epitope. Lex2B has recently been shown in our laboratory to encode a β-glucosyltransferase in type b strains (R. Aubrey, D. Hood, and E. R. Moxon, unpublished). This transferase permits further oligosaccharide chain extension, including a digalactoside epitope, to occur in the LPS of these strains. Thus three loci, each containing a gene with tetranucleotide repeats, that can be relevant to digalactoside expression in the LPS of H. influenzae have been identified.

Lic3A, a phase-variable sialyl transferase mediating serum resistance

DNA sequence analysis of cloned DNA of the third of the lic loci, lic3, identified four ORFs (Maskell et al., 1991). The first ORF, lic3A, contains 5'-CAAT repeats just within its 5' end. Potential start codons are located 1 and 15 bp upstream of the repeats (Figure 6.2). In the original investigation, mutation of lic3A had no detectable effect on LPS structure and so gene expression could not be correlated to a phase-varying phenotype. Recently, based on its homology to a sialyl transferase in Campylobacter jejuni (Gilbert et al., 2000), lic3A has been shown to encode an α-2,3-sialyltransferase required for the phase-variable addition of

Neu5Ac to a lactose structure in *H. influenzae* LPS (Hood *et al.*, 2001a). Immediately downstream of *lic3A* is an ORF predicted to make a protein that has 56% identity with UDP-galactose-4-epimerase, designated *galE*. Downstream of *galE*, the third and fourth ORFs of a possible *lic3* transcriptional unit are proteins with homology to enterobacterial AmpG and adenylate kinase (Adk). By inserting the *lacZ* gene downstream of the repeats in *lic3A* as a translational reporter, some correlation between the number of 5'-CAAT repeats and LacZ activity was made (Szabo *et al.*, 1992).

Based on *in vitro* studies, sialic acid is likely to be critical in conferring resistance of *H. influenzae* to complement-mediated killing and phagocytic ingestion. Incorporation of sialic acid as a terminal epitope in LPS has been shown to be particularly relevant for resistance of capsule-deficient or NT*Hi* strains to the killing effect of human serum (Hood *et al.*, 1999, 2001a). This may be important to the pathogenesis of a number of diseases caused by *H. influenzae*. Cell surface sialylation is a characteristic of many host cells and prevents complement activation by binding the serum regulatory protein, factor H (Varki, 1993). Further, addition of sialic acid can mask existing antigens in the LPS and indeed has the potential to create new ones that could themselves be mimics of host cell surface structures. A majority, if not all, NT*Hi* strains display LPS sialylation (Bauer *et al.*, 2001) and analysis of the strain Rd genome (a strain now known to be capable of synthesizing sialylated glycoforms) and detailed metabolic studies (Vimr *et al.*, 2000) indicate that *H. influenzae* must obtain sialic acid from an external source. Since *H. influenzae* is an obligate commensal/pathogen of man, the source of sialic acid must be from the human host environment. However, sialic acid metabolism is evidently a balance between catabolic pathways and activation for incorporation into macromolecules. Furthermore, in addition to Lic3A that adds Neu5Ac in an α2,3-linkage to a lactosyl residue, there are other enzymes involved in the incorporation of sialic acid into *H. influenzae* LPS (Hood *et al.*, 2001a; Jones *et al.*, 2002). Recent experiments in a chinchilla model of acute otitis media suggest that LPS sialylation may be crucial in the pathogenesis of disease (Bouchet, Hood, Li *et al.*, unpublished results). As described above for the phase-variable expression of PC, *H. influenzae* can modulate its surface sialylation and this may be important for optimizing survival in different host compartments such as the nasopharynx, the Eustachian tube and the middle ear. For example, in another mucosal pathogen, *Neisseria gonorrhoeae*, LPS sialylation downregulates interactions with host epithelial cells, thus modulating microbial–host interactions (van Putten, 1993).

The localization of *lic3A* and the other genes in the *lic3* locus may be more than simple coincidence. The *galE* gene helps control the balance of activated galactose available for LPS synthesis and other cellular processes. It is possible that the clustering of these genes in the *lic3* locus may promote a regulatory mechanism linking the biosynthesis of LPS antigens (Lic3A and GalE) with the energy status of the cell (Adk) through a transducing mechanism involving AmpG. However, to date there is little evidence to support this hypothesis.

Phase variation and infection

To study the impact on virulence of interdependent genes involved in expression of phase-variable epitopes that may switch spontaneously during an experimental infection poses some problem. Hosking and colleagues (Hosking *et al.*, 1999) monitored the number of repeats present at each of the three *lic* loci for organisms isolated from the nasopharynx, blood and CSF of the infant rat following infection. This indicated that each compartment was colonized by a heterogeneous population of organisms probably expressing different combinations of the *lic* genes. The predominant number of repeats did vary between the populations in each of the compartments and the inoculum used. In a subsequent experiment, Weiser and Pan inoculated infant rats with organisms predominantly (> 98%) not expressing either PC or the digalactoside (Weiser and Pan, 1998). Ten days following intranasal challenge, half of the colonies obtained in nasal washes had reverted to a phenotype expressing PC, with no apparent change in the number expressing the digalactoside. The number of repeats present in *lic1*, and the *lic2A/lgtC* genes and expression of the respective LPS epitopes, PC and the digalactoside has been investigated in clinical samples (Weiser and Pan, 1998)

One approach to simplify interpretation upon experiment is to delete the tetranucleotide repeats from any gene so as to result in constitutive translation in the 'on' phase. A mutant of *lic2A*, lacking the 5'-CAAT repeats, was constructed in strain RM7004 (High *et al.*, 1996). The LPS from this strain was identical to the wild-type strain as assessed by colony immunoblotting and gel fractionation of purified LPS. Deletion of the 5'-CAAT repeats reduced the rate of phase variation, but did not abolish it, confirming the involvement of multiple genes in the synthesis of the relevant epitope. Using phase locked 'on' mutants, there was enhanced survival in infant rats of the *H. influenzae* strain able to express two, as compared to one, digalactoside in its LPS (R. Aubrey, D. Hood, E.R. Moxon, unpublished).

FACTORS INFLUENCING THE RATE OF PHASE VARIATION

For *H. influenzae*, an organism containing multiple phase-variable genes, the potential number of patterns of expression for the relevant antigens is enormous. A crucial factor underlying changes in the precise pattern of antigens elaborated will be the rate of switching relevant to each gene. Therefore it is necessary to have knowledge of the factors that influence and direct the switching rates. Some investigation of the factors influencing the process, and in particular the rate, of phase variation mediated by tetranucleotide repeats in *H. influenzae* has been undertaken. The *mod* gene of *H. influenzae* has homology to methyltransferases of type III restriction/modification systems in other organisms and in the Rd genome sequence contains 40 tetranucleotides (5'-AGTC) within the reading frame (Hood et al., 1996b). The gene is found in the vast majority of H. influenzae strains but can contain an altered tetranucleotide motif (Hood et al., 1996b) and indeed is associated with repeats in only about half of the strains tested (De Bolle et al., 2000). A derivative of strain Rd was constructed by fusing the β-galactosidase gene in frame with the 5' end of the *mod* gene distal to the repeats (De Bolle et al., 2000). As described above for the LPS genes, in the resultant strain LacZ expression, monitored by colorimetric changes in substrate (X-gal), serves as a reporter of gene expression and therefore can be used to determine concomitant changes in repeat numbers which directly alter translation. Phase variation was observed at high frequency (4 $\times 10^{-3}$ variants per colony) in the strain with the wild-type number of repeats. Further strains were constructed containing genetically altered numbers of repeats ranging from 6 to 38 copies (De Bolle et al., 2000). The mutation rate of the repeat tract increased linearly with tract length over the range 17 to 38 repeat units. Below these lengths the experimental determination of phase variation was made difficult by the inherent low rates. PCR amplification and sequencing of the DNA containing the repeat tracts indicated that the majority of tract alterations were insertions or deletions of one repeat unit. Thus, the number of repeats in a tract is an important determinant of tetranucleotide-mediated phase variation in this organism.

It is known that mutation of tetranucleotide repeat tract length is independent of *recA*-mediated recombination (Dawid et al., 1999). A study of other *trans*-acting factors influencing mutation rates of tetranucleotide repeat tracts in *H. influenzae* has been carried out using the *mod* gene reporter system described above (Bayliss et al., 2002). Mutation of only one of a number of genes involved in mismatch repair and DNA metabolism, *polI*, destabilized 5'-AGTC tetranucleotide repeat tracts. A reduction in the number of repeats within a tract was predominant in *polI* mutants, this was proposed to be due to incorrect Okazaki fragment processing during DNA replication. In contrast, mutation of a mismatch repair gene, *mutS*, destabilized dinucleotide repeats of the sequence (5'-AT) found in the fimbriae genes. Studies in another organism, *N. meningitidis*, have likewise shown that rates of variation of short repeats, in this case mononucleotides, are dependent upon both tract length and mismatch repair (Richardson and Stojilkovic, 2001). Mutation of mismatch repair genes in the same organism gives rise to 'mutator' phenotypes with greatly enhanced global mutation rates. It is thought that localized hypermutation through changes in the number of tetranucleotide repeats in *H. influenzae* confers a fitness advantage whilst avoiding any deleterious effects associated with 'mutator' phenotypes in other organisms (Bayliss et al., 2002).

In summary, *H. influenzae* has made a considerable investment in the process of phase variation mediated largely through tetranucleotide repeat tracts. Each of the phase-variable genes identified in *H. influenzae* can be described as components of a mechanism required to generate complex repertoires of variant antigens allowing these bacteria to adapt to the differing microenvironments of the host and evade immune responses. If one considers the exemplar of LPS, the potential combinatorial effects of on or off expression of five independent phase-variable biosynthesis genes on the heterogeneity of the molecule is significant. Although the phenotypes resulting from expression of LPS genes are not always independent, the 32 possible expression patterns of the phase-variable genes could result in substantial structural and antigenic variation of the LPS molecules present in the cells of a population. If one extrapolates from the example of LPS to include the products of the other phase-variable genes that are exposed on the bacterial cell surface, the enormous potential for diversity between organisms inherent to this mode of gene regulation is evident.

REFERENCES

Barenkamp, S. and Bodor, F. F. (1990) *Paediatr. Infect. Dis. J.* 9, 333–339.
Barenkamp, S. and Leininger, E. (1992) *Infect. Immun.* 60, 1302–1313.
Bauer, S. H. J., Månsson, M., Hood, D. W., Richards, J. C., Moxon, E. R. and Schweda, E. K. H. (2001) *Carbohydrate Res.* 335, 251–260.
Bayliss, C. D., van de Ven, T. and Moxon, E. R. (2002) *EMBO J.* 21, 1465–1476.
Cope, L. D., Hrkal, Z. and Hansen, E. J. (2000) *Infect. Immun.* 68, 4092–4101.
Cope, L. D., Yogev, R., Mertsola, J., Latimer, L. J., Hanson, M. S., McCracken, G. H. and Hansen, E. J. (1991) *Mol. Microbiol.* 5, 1113–1124.
Corn, P. G., Anders, J., Takala, A. K., Kayhty, H. and Hoiseth, S. K. (1993) *J. Infect. Dis.* 167, 356–364.
Dawid, S., Barenkamp, S. and St Geme, J. W. III (1999) *Proc . Natl Acad. Sci. USA* 96, 1077–1082.
De Bolle, X., Bayliss, C. D., Field, D., van de Ven, T., Saunders, N., Hood, D. W. and Moxon, E. R. (2000) *Mol. Microbiol.* 35, 211–222.
Fan, X., Goldfine, H., Lysenko, E. and Weiser, J. N. (2001) *Mol. Microbiol.* 41, 1029–1036.
Farley, M. M., Stephens, D. S., Kaplan, S. L. and Mason, E. O. (1990) *J. Infect. Dis.* 154, 752–759.
Fleischmann, R. D., Adams, M. D., White, O., Clayton, R. A., Kirkness, E. F., Kerlavage, A. R., Butt, C. J., Tomb, J-F., Dougherty, B. A., Merrick, J. M., McKenney, K., Sutton, G., FitzHugh, W., Fields, C., Gocayne, J. D., Scott, J., Shirley, R., Liu, L-I., Glodek, A., Kelley, J. M., Weidman, J. F., Phillips, C. A., Spriggs, T., Hedblom, E., Cotton, M. D., Utterback, T. R., Hanna, M. C., Nguyen, D. T., Saudek, D. M., Brandon, R. C., Fine, L. D., Fritchman, J. L., Fuhrmann, J. L., Geoghagen, N. S. M., Gnehm, C. L., McDonald, L. A., Small, K. V., Fraser, C. M., Smith, H. O. and Venter, J. C. (1995) *Science* 269, 496–512.
Geluk, F., Eijk, P. P., van Ham, S. M., Jansen, H. M. and van Alphen, L. (1998) *Infect. Immun.* 66, 406–417.
Gilbert, M., Brisson, J-R., Karwaski, M-F., Michniewicz, J. J., Cunningham, A. M., Wu, Y., Young, N. M. and Wakarchuk, W. W. (2000) *J. Biol. Chem.* 275, 3896–3906.
Gulig, P. A., Patrick, C. C., Hermanstorfer, L., McCracken, G. H. Jr and Hansen, E. J. (1987) *Infect. Immun.* 55, 513–520.
High, N. J., Deadman, M. E. and Moxon, E. R. (1993) *Mol. Microbiol.* 9, 1275–1282.

High, N. J., Jennings, M. P. and Moxon, E. R. (1996) *Mol. Microbiol.* 20, 165–174.
Hoiseth, S. K., Corn, P. G. and Anders, J. (1992) *J. Infect. Dis.* 165, S114.
Hood, D. W., Cox, A. D., Gilbert, M., Makepeace, K., Walsh, S., Deadman, M. E., Cody, A., Martin, A., Brisson, J-R., Richards, J. C. and Wakarchuk, W. W. (2001a) *Mol. Microbiol.* 39, 341–351.
Hood, D. W., Cox, A. D., Schweda, E. K. H., Walsh, S., Deadman, M. E., Martin, A., Moxon, E. R. and Richards, J. C. (2001b) *Glycobiology* 11, 957–967.
Hood, D. W., Deadman, M. E., Allen, T., Masoud, H., Martin, A., Brisson, J. R., Fleischmann, R., Venter, J. C., Richards, J. C. and Moxon E. R. (1996a) *Mol. Microbiol.* 22, 951–965.
Hood, D. W., Deadman, M. E., Jennings, M. P., Bisceric, M., Fleischmann, R. D., Venter, J. C. and Moxon, E. R. (1996b) *Proc. Natl Acad. Sci. USA* 93, 11121–11125.
Hood, D. W., Makepeace, K., Deadman, M. E., Rest, R. F., Thibault, P., Martin, A., Richards, J. C. and Moxon, E. R. (1999) *Mol. Microbiol.* 33, 679–692.
Hosking, S. L., Craig, J. E. and High, N. J. (1999) *Microbiol.* 145, 3005–3011.
Jarosik, G. P. and Hansen, E. J. (1994) *Infect. Immun.* 62, 4861–4867.
Jones, P. A., Samuels, N. M., Phillips, N. J., Munson, R. S., Bozue, J. A., Arseneau, J. A., Nichols, W. A., Zaleski, A., Gibson, B. W. and Apicella, M. A. (2002) *J. Biol. Chem.* 277, 14598–14611.
Kimura, A. and Hansen, E. J. (1986) *Infect. Immun.* 51, 60–79.
Kimura, A., Patrick, C. C., Miller, E. E., Cope, L. D., McCracken, G. H., Jr and Hansen, E. J. (1987) *Infect. Immun.* 55, 1979–1986.
Kroll, J. S. (1992) *J. Infect. Dis.* 165, S93–S96.
Kroll, J. S., Loynds, B. M. and Moxon, E. R. (1991) *Mol. Microbiol.* 5, 1549–1560.
Levinson, G. and Gutman, G. A. (1987) *Mol. Biol. Evol.* 4, 203–221.
Lysenko, E. S., Gould, J., Bals, R., Wilson, J. M. and Weiser J. N. (2000a) *Infect. Immun.* 68, 1664–71.
Lysenko, E., Richards, J. C., Cox, A. D., Stewart, A., Martin, A., Kapoor, M. and Weiser, J. N. (2000b) *Mol. Microbiol.* 35, 234–245.
MacIver, I., Latimer, J. L., Liem, H. H., Muller-Eberhard, U., Hrkal, Z. and Hansen, E. J. (1996) *Infect. Immun.* 64, 3703–3712.
Maskell, D. J., Szabo, M. J., Butler, P. D., Williams, A. E. and Moxon, E. R. (1991) *Mol. Microbiol.* 5, 1013–1022.
Morton, D. J., Whitby, P. W., Jin, H., Ren, Z. and Stull, T. L. (1999) *Infect. Immun.* 67, 2729–2739.
Moxon, E. R. and Maskell, D. J. (1992) In *Molecular Biology of Bacterial Infection, Current Status and Future Perspectives, Society for General Microbiology Symposium 49* (eds C. E. Hormaeche, C. W. Penn and C. J. Smythe), Cambridge University Press, Cambridge, pp. 75–96.
Moxon, E. R., Rainey, P. B., Nowak, M. A. and Lenski, R. E. (1994) *Curr. Biol.* 4, 24–33.
Noel, G. J., Brittingham, A., Granato, A. A. and Mosser, D. M. (1996) *Infect. Immun.* 64, 4769–4775.
Patrick, C. C., Pelzel, S. E., Miller, E. E., Haanes-Fritz, E., Ruolf, J. D., Gulig, P. A., McCracken, G. H. and Hansen, E. J. (1989) *Infect. Immun.* 57, 1971–1978.
Richardson, A. R. and Stojilkovic, I. (2001) *Mol. Microbiol.* 40, 645–655.
Roche, R. J., High N. J. and Moxon, E. R. (1994) *FEMS Microbiol. Lett.* 120, 279–284.

Saunders, N. J., Jeffries, A. C., Peden, J. F., Hood, D. W., Tettelin, H., Rappuoli, R. and Moxon, E. R. (2000) *Mol. Microbiol.* 37, 207–215.

St Geme, J. W., Falkow, S. and Barenkamp, S. J. (1993) *Proc. Natl Acad. Sci. USA* 90, 2875–2879.

Swords, W. E., Buscher, B. A., Ver Steeg, I. K., Preston, A., Nichols, W. A. and Weiser, J. N. (2000) *Mol. Microbiol.* 37, 13–27.

Szabo, M., Maskell, D., Butler, P., Love, J. and Moxon, E. R. (1992) *J. Bacteriol.* 174, 7245–7252.

Tatusov, R. L., Mushegian, A. R., Bork, P., Brown, N. P., Hayes, W. S., Borodovsky, M., Rudd, K. E. and Koonin, E. V. (1996) *Curr. Biol.* 6, 279–291.

Tong, H. H., Blue, L. E., James, M. A., Chen, Y. P. and DeMaria, T. F. (2000) *Infect. Immun.* 68, 4593–4597.

Van Belkum, A., Scherer, S., van Leeuwen, W., Willemse, D., van Alphen, L. and Verbrugh, H. A. (1997) *Infect. Immun.* 65, 5017–5027.

Van Ham, S. M., van Alphen, L., Mool, F. R. and van Putten, J. P. M. (1993) *Cell* 73, 1187–1196.

Van Ham, S. M., van Alphen, L., Mool, F. R. and van Putten, J. P. M. (1994) *Mol. Microbiol.* 13, 673–684.

Van Putten, J. P. M. (1993) *EMBO J.* 12, 4043–4051.

Varki, A. (1993) *Glycobiology* 3, 97–130.

Vimr, E., Lichtensteiger, C. and Steenbergen, S. (2000) *Mol. Microbiol.* 36, 1113–1123.

Virij, M., Weiser, J. N., Lindberg, A. A. and Moxon, E. R. (1990) *Microb. Pathog.* 9, 441–450

Weiser, J. N. and Pan, N. (1998) *Mol. Microbiol.* 30, 767–775

Weiser, J. N., Lindberg, A. A., Manning, E. J., Hansen, E. J. and Moxon, E. R. (1989a) *Infect. Immun.* 57, 3045–3052.

Weiser, J. N., Love, J. M. and Moxon, E. R. (1989b) *Cell* 59, 657–665.

Weiser, J. N., Maskell, D. J., Butler, P. D., Lindberg, A. A. and Moxon, E. R. (1990a) *J. Bacteriol.* 172, 3304–3309.

Weiser, J. N., Pan, N., McGowan, K. L., Musher, D., Martin, A. and Richards, J. (1998) *J. Exp. Med.* 187, 631–640.

Weiser, J. N., Shchepetov, M. and Chong, S. T. H. (1997) *Infect. Immun.* 65, 943–950.

Weiser, J. N., Williams, A. and Moxon, E. R. (1990b) *Infect. Immun.* 58, 3455–3457.

Zaleski, A., Scheffler, N. K., Densen, P., Lee, F. K. N., Campagnari, A. A., Gibson, B. W. and Apicella, M. A. (2000) *Infect. Immun.* 68, 5261–5268.

7

PHASE VARIATION IN *HELICOBACTER PYLORI* LIPOPOLYSACCHARIDE

BEN J. APPELMELK AND
CHRISTINA M. J. E. VANDENBROUCK-GRAULS

INTRODUCTION

The human Gram-negative gastric bacterium *Helicobacter pylori* is a very successful pathogen: it persists for a lifetime in the stomach of as much as 50% of the world population. *H. pylori* causes a variety of diseases, i.e. gastric and duodenal ulcers, mucosa-associated lymphoid tissue (MALT) lymphoma and adenocarcinoma, a disease which ranks second only to lung cancer with regard to mortality worldwide (Kuipers and Appelmelk, 1997). The type of disease that follows after *H. pylori* infection is strongly influenced by host genetic factors. *H. pylori*-infected patients with certain gene polymorphisms in IL-1 and IL-1 receptor antagonist (IL-1RA) that predispose to a pro-inflammatory environment in the stomach are particularly prone to developing gastric cancer. That the bacterium is able to colonize so many people chronically is most likely due to an ability to adapt well to the gastric environment. The finding that *H. pylori* is genetically extremely heterogeneous presumably reflects this ability (Suerbaum, 2000). Bacterial genetic diversity within a single patient has also been reported (Israel *et al.*, 2001). Analysis of the genome of the two *H. pylori* strains that have been sequenced gives clues as to how variability may be obtained. Many genes contain long homopolymeric repeats (Tomb *et al.*, 1997; Saunders *et al.*, 1998; Alm *et al.*, 1999), including those whose products are involved in lipopolysaccharide synthesis (glycosyltransferases or synthetases) (Berg *et al.*, 1997). These C-tracts, encountered also in other species like *Neisseria* spp. (Jennings, 1999) and *Campylobacter jejuni* (Linton *et al.*, 2000; Parkhill *et al.*, 2000), contribute to phenotypic diversity through a process called slipped-strand mispairing. During replication, DNA polymerase slippage in the tracts yields daughter DNA strands with either one C lacking or one extra C. This results in a high-frequency, reversible frameshifting; in one frame a complete gene product (glycosyltransferase) will be formed but when frameshifted, downstream either 'non-sense' polypeptide will be synthesized or a stop-codon may appear leading to a truncated enzyme (Figure 7.1). Either way, this leads to a high-frequency activation–inactivation of enzyme activity and hence to a high-frequency switching of LPS phenotype. Phase variation is defined as the random switching of phenotype at frequencies that are much higher (up to 1% or higher) than classical mutation rates. This process results in reversible loss and gain of certain LPS epitopes and results in a bacterial population that is heterogeneous with regard to LPS expression.

ATGTTCCAACCCCTATTAGACGCCTTTATAGAAAGCGCTTCCATTGAAAAAATGGCCTCTAAATCT
CCCCCCCCCCCCCTAAAAATCGCTGTGGCGAATTGGTGGGGAGATGAAGAAATTAAAGA
ATTTAAAAAGAGCGTTCTTTATTTTATCCTAAGCCAACGCTACGCAATCACCCTCCACCAAAACCCC
AATGAATTTTCAGATCTAGTTTTTAGCAATCCTCTTGGAGCGGCTAGAAAGATTTTATCTTATCAAA
ACACTAAACGAGTGTTTTACACCGGTGAAAACGAATCACCTAATTTCAACCTCTTTGATTACGCCAT
AGGCTTTGATGAATTGGATTTTAATGATCGTTATTTGAGAATGCCTTTGTATTATGCCCATTTGCAC
TATAAAGCCGAGCTTGTTAATGACACCACTGCGCCCTACAAACTCAAAGACAACAGCCTTTATGCT
TTAAAAAAAACCCTCTCATCATTTTAAAGAAAACCACCCTAATTTGTGCGCAGTAGTGAATGATGAG
AGCGATCTTTTAAAAAGAGGGTTTGCCAGTTTTGTAGCGAGCAACGCTAACGCTCCTATGAGGAAC
GCTTTTTATGACGCTCTAAATTCCATAGAGCCAGTTACTGGGGGAGGAAGTGTGAGAAACACTTTA
GGCTATAAGGTTGGAAACAAAAGCGAGTTTTTAAGCCAATACAAGTTCAATCTCTGTTTTGAAAAC
TCGCAAGGTTATGGCTATGTAACCGAAAAAATCCTTGATGCGTATTTTAGCCATACCATTCCTATTT
ATTGGGGGAGTCCCAGCGTGGCGAAAGATTTTAACCCTAAAAGTTTTGTGAATGTGCATGATTTCA
ACAACTTTGATGAAGCGATTGATTATATCAAATACCTGCACACGCACCCAAACGCTTATTTAGACA
TGCTCTATGAAAACCCTTTAAACACCCTTGATGGGAAAGCTTACTTTTACCAAGATTTGAGTTTTAA
AAAAATCCTAGATTTTTTTAAAACGATTTTAGAAAACGATACGATTTATCACAAATTCTCAACATCT
TTCATGTGGGAGTACGATCTGCATAAGCCGTTAGTATCCATTGATGATTTGAGGGTTAATTATGATG
ATTTGAGGGTTAATTATGACCGGCTTTTACAAAACGCTTCGCCTTTATTAGAACTCTCTCAAAACAC
CACTTTTAAAATCTATCGCAAAGCTTATCAAAAATCCTTGCCTTTGTTGCGCGCGGTGAGAAAGTTG
GTTAAAAAATTGGGTTTG

Figure 7.1 Role of C-tracts in phase variation. Shown is the sequence of an α3-*fucT* gene. Note the C13 stretch near the 5' end of the gene, which allows a high-frequency phase variation based on DNA slippage during replication. A C13 tract leads to biosynthesis of an active full-length gene product, but C12 will lead to a TAA stop codon right after the tract.

Phase variation contributes to virulence by generating heterogeneity, following which host pressure selects those bacteria that express the best-adapted phenotype. Thus, the process which generates the variants is not controlled *per se* but random, with the best-adapted organisms surviving to replicate. Based on DNA sequence, 27 phase-variable genes were identified in *H. pylori*, three of which are fucosyltransferase (*fucT*) genes required for LPS biosynthesis; C-tracts are also present in β3-galactosyl transferase (β3-*galT*) (Appelmelk *et al.*, 2000) and heptosyltransferase genes or in as yet unidentified glycosyltransferases (Logan, 2001). Hence, at any single moment a population of *H. pylori* bacteria could potentially comprise 2^{27} ($>10^8$) different phenotypes! As shown below, from eight different LPS phenotypes that can theoretically be formed due to phase variation in the three *fucT* genes, several could indeed be isolated.

Bacteria that express the phenotype that is adapted best to the environment (host) may outcompete the others. For example, *Neisseria gonorrhoeae* phase-variably expresses LPS that is either sialylated or not (van Putten, 1993). Variants expressing non-sialylated LPS are invasive but serum-sensitive; variants expressing sialylated LPS are not invasive but serum-resistant. Thus, bacteria expressing the non-sialylated phenotype invade the host, and these are killed by serum, while the sialylated variants that will necessarily have formed after invasion survive and expand clonally. Consequently, phase variation increases adaptation and survival of the pathogen.

In this chapter we will discuss the molecular mechanisms of phase variation in *H. pylori* LPS and its biological role.

H. PYLORI LPS EXPRESSES BLOOD GROUP ANTIGENS

For a limited number of *H. pylori* strains, the LPS structure has been determined chemically (methylation analysis, mass spectrometry). Most of them express Lewis blood group antigens (Table 7.1) (Simoons-Smit *et al.*, 1996; Wirth *et al.*, 1996; Heneghan *et al.*, 1997). Many strains express Lewis × (Le^x) and/or Le^y, but strains expressing H type I have also been found frequently; less often, strains expressing Le^a, Le^b, i-antigen or sialyl-Le^x or even blood group A have been encountered (Aspinall *et al.*, 1996; Monteiro, 2001). Often, strains express more than one Lewis antigen (Monteiro *et al.*, 1998). For example, strain NCTC 11637 (= ATCC 43504) expresses polymeric Le^x with up to eight or nine residues, which is substituted terminally in non-stoichiometric amounts with Le^y or H type 1. Table 7.1 shows the structure of various Lewis and related blood groups, the glycosyltransferases required for biosynthesis of those structures, as well the structure of a few characteristic *H. pylori* LPS. Like any LPS, *H. pylori* LPS comprise lipid A, a core and the O-antigen. A few examples: the O-antigen of strain P466 is a poly-lactosamine (equivalent to the i-antigen expressed by fetal erythrocytes), α-3 fucosylated to form poly Le^x and terminally substituted with $\alpha 2$-fucose to form Le^y; the structure of strain NCTC 11637 is similar to that of strain P466 but expresses less Le^y; strains O6 and MO19 form a single Le^y (polymeric Le^y does not exist). To form these fucosylated polylactosamines several glycosyltransferases are required: GalT and GlcNAcT to form the backbone, plus $\alpha 3$-FucT to form the Le^x; for synthesis of Le^y, an additional $\alpha 2$-FucT is needed. Of course, for *H. pylori* O-antigen biosynthesis additional enzymes are required, like the enzymes for conversion of GDP-mannose to GDP-fucose (Jarvinen *et al.*, 2001).

For O-antigen characterization of large collections of *H. pylori* strains, serological typing methods (Elisa or immunoblot) with specific monoclonal antibodies (Mabs) or carbohydrate-reactive lectins have been used (Hynes *et al.*, 1999). Serology has the advantage of being much faster and easier to perform than structural analysis; it allows mass-screening, and it is more sensitive (see below).

Table 7.1 Lewis blood group antigens expressed by *H. pylori*, glycosyltransferases required for biosynthesis and structure of representative *H. pylori* LPS

Blood group structure	Name	Glycosyltransferases required
Fucα1→2Galβ1→3GlcNAc	H type 1	α2FucT, β3GalT
Fucα1→2Galβ1→4GlcNAc	H type 2*	α2FucT, β4GalT
Galβ1→4GlcNAc 3 ↑ Fucα1	Lewis x (Le^x)	α3FucT, β4GalT
Fucα1→2Galβ1→4GlcNAc 3 ↑ Fucα1	Lewis y (Le^y)	α2FucT, α3FucT, β4GalT
Galβ1→3GlcNAc 4 ↑ Fucα1	Lewis a (Le^a)	α4FucT, β3GalT
Fucα1→2Galβ1→3GlcNAc 4 ↑ Fucα1	Lewis b (Le^b)	α2FucT, α4FucT, β3GalT
→3 (Galβ1→4GlcNAcβ1→)$_n$	i-antigen	β4GalT
NANA2→3 Galβ1→4GlcNAc 3 ↑ Fucα1	sialyl Lewis x (sLe^x)	α3FucT, β4GalT, sialylT
Fucα1→2Galβ1→3GlcNAc 3 ↑ GalNAcα1	Blood group A†	α2FucT, β3GalT, α3GalNAcT

H. pylori LPS structure

Le^y, (Le^x)$_n$-core-lipid A (strain P466)
Le^y-core-lipid A (strains MO19, O6)
i-Ag,H$_I$-core-lipid A (strain J223)

Abbreviations: Gal, D-galactose; Fuc, L-fucose; GlcNAc, *N*-acetyl-D-glucosamine; NANA, N-acetylneuraminic (= sialic) acid. α2/3/4FucT = α2/3/4 fucosyltransferase; β3/4GalT = b3/4 galactosyltransferase; GalNAcT, *N*-acetyl-D-galactosaminyltransferase; GlcNAcT, *N*-acetyl-D-glucosaminyltransferase. For biosynthesis of all blood group antigens shown an active β3-GlcNAcT is also required.
*Structural chemical data to show the presence of H type 2 in *H. pylori* LPS are lacking.
†*Helicobacter mustelae* expresses blood group A; data for *H. pylori* are based on serology only.

On the other hand, use of Mabs may yield false-negative results, and it was demonstrated that some *H. pylori* strains that do not react with anti-Le^x Mabs, by structural analysis were shown to express Le^x (Knirel *et al.*, 1999). False-positive results may occur too: a strain that reacted strongly with an anti-Le^b Mab did not express Le^b but only its Fucα1.4GlcNAc partial structure (Monteiro *et al.*, 1998). Hence, the success of serology depends on the use of well-characterized Mabs that have been tested for binding with series of glycoconjugates. Using such well-defined Mabs, serology was found to be predictive for LPS structure (Appelmelk et al., 1999a), and in fact the i-antigen, H type I, sialyl-Le^x and the presence of small amounts of Le^y in strain NCTC 11637 were first detected by serology and only confirmed later by chemical structure analysis (Appelmelk et al., 1998; Monteiro et al., 2000). There is consensus that at least 80% of *H. pylori* strains worldwide express Lewis blood group antigens; for a few strains the lack of Lewis antigen expression has been confirmed by structure analysis (Knirel et al.,1999; Monteiro et al., 2001). Apart from *H. pylori*, only two other parasites, *Schistosoma mansoni* and *Dictyocaulus viviparus*, express Le^x (van Dam et al., 1996; Haslam et al., 2000).

Thus, Lewis antigen expression in *H. pylori* is highly conserved. This restricted diversity in O-antigen structure is striking, as for instance, *Escherichia coli* or *Salmonella* spp. express >100 different O-antigens. Apart from *Helicobacter mustelae* (blood group A; Monteiro et al., 1997) and *Helicobacter acynonychis* (Le^y, unpublished) other species of the genus *Helicobacter* do not express blood group antigens (B. J. Appelmelk, unpublished). Hence, the question arises whether *Helicobacter pylori* Lewis antigens play a role in pathogenesis. A comparable situation is found in *N. gonorrhoeae* LPS, where highly conserved lactosamine epitopes in LPS bind to the asialoglycoprotein receptor, a process that starts bacterial invasion (Harvey et al., 2001).To answer the question on the biological role of *H. pylori* Lewis antigens becomes even more urgent now it is known that these antigens display phase variation.

PHASE VARIATION IN *H. PYLORI* LPS LEWIS ANTIGENS

Phase variants can be detected by colony blotting with Mabs specific for Lewis antigens (Appelmelk et al., 1998). For example, Figure 7.2a shows a colony blot where *H. pylori* strain NCTC 11637 was probed with a Mab specific for H type 1. Three types of colonies were present. First are those that are completely reactive with the Mab (dark colonies); the bacteria forming this colony originate from a single bacterial cell expressing H type 1, with no switching-off to the H type 1-negative phenotype occurring during multiplication. Likewise, non-reactive colonies (white) originate from a bacterial cell with a switched-off phenotype. Colonies with dark sector(s) originate from a cell with a switched-off phenotype that switched on (often in more than one independent event per colony); clonal outgrowth of a switched-on variant gives rise to the sectored patterns. When switch frequencies are low, (almost) no sectored colonies are found (Figure 7.2b). By colony blotting with several different Mabs we were able to isolate many (>10) serologically different LPS phase variants from a single strain (strain NCTC 11637). The variants were subsequently serotyped in Elisa, analysed by SDS-PAGE/immunoblotting, and finally their LPS structures were determined by chemical methods (Figure 7.3, Tables 7.2 and 7.3). The frequency

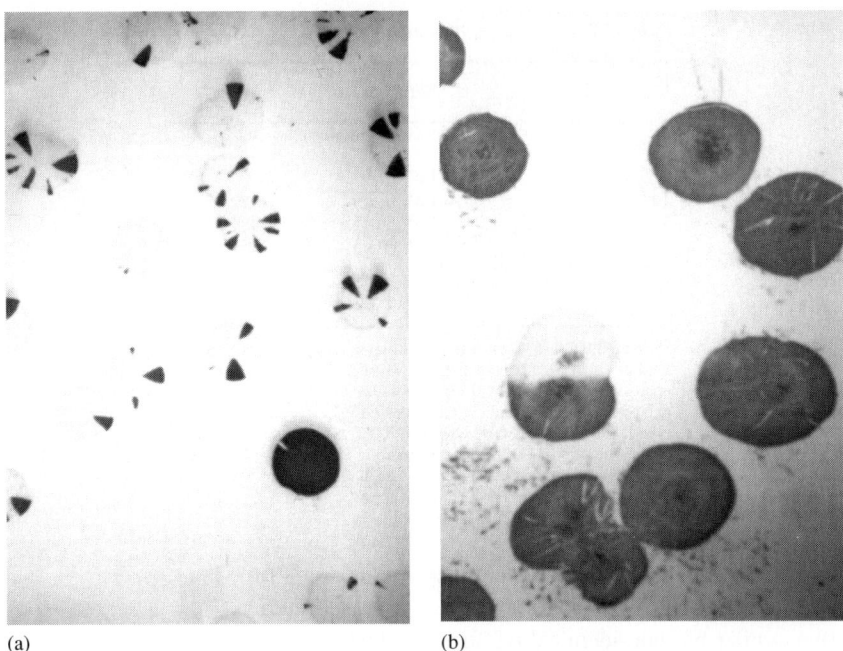

(a)   (b)

Figure 7.2 LPS phase variation in *H. pylori* LPS. (a) *H. pylori* strain B1.5 was colony blotted with an anti-H type 1 Mab and immunostained; evident are the white, the black and the sectored colonies; (b) low-frequency phase variation, not yielding sectored colonies.

of LPS phase variation was in the range of 0.5–1% (Appelmelk *et al.*, 1998) and in many cases reversibility of switching was demonstrated. Phase variation is not restricted to laboratory strains but also occurs in clinical isolates (data not shown), and even generates LPS diversity in a single patient (see below).

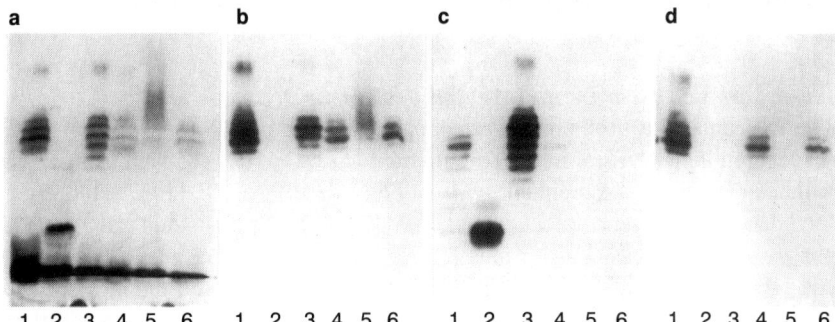

Figure 7.3 Characterization of LPS phase variants by SDS-PAGE/silverstain (a) and immunoblot with anti-Le^x Mab (b), anti-Le^y Mab (c) and anti-H type 1 (d). Lane 1, NCTC 11637; lane 2, variant 1b; lane 3, variant 1c; lane 4, variant 2b; lane 5, variant 3a; lane 6, variant 3c. The low MW fraction in (a) represents the core-lipid A part. (Reproduced from *Trends in Microbiology*, vol. 8, issue 12, Appelmelk, B. J., pp. 565–570. Copyright © 2000, with permission from Elsevier.)

Table 7.2 LPS phase variants of strain NCTC 11637*

Phase variant	Serotype
1b	Le^y
1c	Le^y, $(Le^x)_n$
2b	$Le^y \ldots (Le^x)_n$, H type 1
3a	Le^a, $(Le^x)_n$, ...H type I
K4.1	i-Ag, H type 1
H11	Le^y,$(Le^x)_n$, i -Ag

*Strain NCTC 11637 and variant 2b expresses identical serotypes, i.e. $Le^y \ldots (Le^x)_n$, H type 1. $Le^y \ldots (Le^x)_n$ designates a weak expression of Le^y and a strong expression of Le^x; in addition both express H type I. Variant 3a expresses Le^a, $(Le^x)_n$, and weakly H type I. Most colonies isolated from NCTC 11637 express the 2b phenotype. Le^y, $(Le^x)_n$ designates a strong expression of both Le^x and Le^y (variant 1c).

Molecular mechanisms of LPS phase variation: α3-fucT

Two similar but non-identical α3-fucosyltransferase (α3-*fucT*) genes have been identified, HP0379 and HP0651, in strain 26695 (JHP 1002 and 0596 in strain J99); the names *futA* and *futB* have been proposed for HP0379 and HP0651, respectively (Appelmelk *et al.*, 1999a; Wang *et al.*, 2000). Functional studies with the cloned and expressed gene products show that both FucT enzymes are able to form Le^x from lactosamine acceptors (Martin *et al.*, 1997). However, insertional mutagenesis, together with serotyping with well-defined anti-Le^x Mabs and structure analysis, has shown that the two α3-FucT enzymes differ in fine-specificity (Appelmelk *et al.*, 1999a). FutA has a preference for internal GlcNAc residues and yields polymeric Le^x, while FutB has a preference for GlcNAc residues at the non-reducing terminus and forms mono/oligomeric Le^x.

Phase variation from Le^x to i-Ag and back to Le^x

The molecular basis of phase variation in *futA* and *futB* was investigated by sequencing their C-tracts in the parent strain (NCTC 11637) as well as in phase variants (Table 7.3) (Appelmelk *et al.*, 1999a). In the NCTC 11637, HP0651 is 'off' due to the presence of a C9 tract; HP0379 is 'on' in this strain (C10). In the phase variant expressing i-Ag plus H type 1 (variant K4.1) both HP0651 (C9) and HP0379 (C11) are off, which explains the lack of Le^x and the biosynthesis of non-fucosylated polylactosamine (= i-Antigen) in strain K4.1. In addition, K4.1 expresses H type 1; thus, parent and variant express an active FutC (α2FucT); no C-tract sequencing was done for this gene. Strain K5.1, a variant with a serotype identical to that of NCTC 11637 parent, was isolated by colony blotting K4.1 with a Le^x Mab, and hence K5.1 represents the switch-back to the

Table 7.3 Molecular mechanism of phase variation due to α3-fucosyltransferase gene C-tract length changes

Strain	Specificity of monoclonal antibody used (Mab code)						α3-*fucT* C-tract length	
	mono-Le^x (6H3)^a	poly-Le^x (54.1F6A)^a	i-antigen	H type 1	H type 2	Le^y	HP0651	HP0379
NCTC 11637	−	+++^{b}	−	+++	+	+	C9 (off)	C10 (on)
K5.1	−	++	−	++	Not done	+	C9 (off)	C10 (on)
2b	+	+++	−	+++	Not done	+	C9 (off)	C10 (on)
K4.1	−	−	+++	+++	+	+	C9 (off)	C11 (off)
1c	+	+++	−	+	+	+++	C10 (on)	C10 (on)
1b	−	−	−	−	Not done	+++	C9 (off)	C10 (on)
4187E	+++	+++^{c}	−	++	−	+++^{d}	C10 (on)	C10 (on)
4187E-KO651	+	+++	−	+++	−	+	ND	ND
4187E-KO379	+++	+++	+	+	−	+++	ND	ND
4187E-KO379/651	−	−	+++	+++	−	+	ND	ND

^{a}Mabs 6H3 and 54.1F6A recognize monomeric and polymeric Le^x, respectively.
^{b}−, OD$_{492}$ <0.3; +, 0.3<OD$_{492}$ <1.3; ++, 1.3<OD$_{492}$<2.3; +++, OD$_{492}$>2.3.
^{c}When titrated this Mab reacted 128 times better with 4187E-KO651 than with strain 4187E and 64 times less with 4187E-KO379 than with 4187E parent.
^{d}When titrated, Mab Hp 151 (anti-Le^y) reacted equally well with strains 4187E and 4187E-KO379.

parental phenotype. In K5.1, HP0379 is 'on' again (C10). Thus, the reversible phase variation from Le^x to i-Ag and back to Le^x can be understood at the molecular level through reversible length changes in the C-tract of α3-*fucT* gene HP0379, from C10 to C11 and back to C10. A HP0379/HP0651 double knockout of strain 4187E (4187E -KO379/651) expresses a serotype identical to that of strain K4.1, i.e. i-antigen and H type 1. Clinical isolate J233 expresses H type 1 plus i-Ag as determined by structural chemistry (Monteiro *et al.*, 1998) and by serology, and in that strain also both α3-*fucT* genes are off (B. J. Appelmelk, unpublished). We conclude that the LPS serotype is determined by the on/off status of α3-*fucT*.

Phase variation from Le^x to Le^x plus Le^y
While strain NCTC 11637 expresses polymeric Le^x, H type 1 and a little Le^y, phase variant 1c strongly expresses both Le^x and Le^y. C-tract analysis shows that both HP0379 and HP0651 are 'on' in strain 1c. Knockout studies in strain 4187E also show that the presence of an intact HP0651 is associated with a stronger Le^y expression and with reactivity with Mab 6H3 that recognizes monomeric Le^x. We conclude that HP0651 FucT preferentially fucosylates GlcNAc at the non-reducing terminus, thus forming an efficient acceptor for α2-FucT to form Le^y. In contrast, HP0379 α3-FucT prefers internal GlcNAc, thus forming polymeric Le^x from the inside out, a structure which evidently is a less efficient acceptor. Consequently, as compared to variant 1c, less Le^y is formed in the parent strain.

Molecular mechanisms of LPS phase variation: α2-fucT

HP0093/94 (JHP0086) is an α2-*fucT*. This gene, also named *futC*, is required for biosynthesis of both Le^y and H type 1 (Wang *et al.*, 1999a,b; Appelmelk *et al.*, 2000). H type 2 epitopes have not been encountered yet in *H. pylori* LPS, and knocking out both α3 -*fucT* genes in a strain that expresses Le$^{x/y}$ yields LPS that expresses i-antigen but no H type 2 (Appelmelk *et al.*, 1999a).This implies that α3-fucosylation has to precede α2-fucosylation. This was confirmed in enzyme assays with recombinant FutC, where the enzyme was shown to form Le^y from synthetic Le^x but not H type 2 from Galα1.4GlcNAc (Wang *et al.*, 1999a). FutC forms H type 1 with a Galα1.3GlcNac acceptor (Wang *et al.*, 1999a); thus, α2-fucosylation can precede α4-fucosylation.

Variant 3a was isolated by colony blotting of NCTC 11637 as a Le^y-negative colony. This strain strongly expresses polymeric Le^x and weakly Le^a (Appelmelk *et al.*, 2000) and H type I. Hence, as compared to the parent, a simultaneous loss of Le^y and H type I has taken place. This phenotype can be explained by phase variation in α2-*fucT*, (on in NCTC 11637 and off in variant 3a) and indeed insertional inactivation of this gene in NCTC 11637 yields a mutant with a serotype indistinguishable from that of strain 3a. The α2-*fucT* gene also contains a C-tract and hence phase variation may occur along the lines sketched above for α3-*fucT*. However, phase variation in *futC* is more complicated than in *futA* and *futB*, as a second mechanism for phase variation was observed in *futC* which can overrule variation due to C-tract changes. *futC* contains a sequence (AAAAAAG) that allows mRNA slippage at the translational level (Wang *et al.*, 1999b).The result of this slippage is a –1 frame shift. The mechanism is as follows: the anticodon for lysine is both UUU and CUU. However, from the whole genome sequence it is known that *H. pylori* codes only for a tRNA^{Lys} with the UUU anticodon, while tRNA^{Lys} with the CUU anticodon is missing. Hence, when AAG is encountered in the mRNA of α2*fucT*, the amino-acyl-loaded transfer RNA molecule, tRNA^{Lys} (anti-codon:UUU), slips one base back to allow the stronger interaction with AAA. This second mechanism may therefore compensate for +1 frameshifting due to C-tracts. These two mechanisms operate in the genome strain 26695. While this strain expresses Le^y (Monteiro *et al.*, 2000), its α2-*fucT* gene is frameshifted (+1) due to the C-tract (Tomb *et al.*, 1997) and theoretically would yield an inactive α2FucT. However, presence of the translational –1 frame shift cassette AAAAAAG causes a –1 shift in reading frame; an active enzyme is formed and Le^y synthesis takes place. The mechanism of –1 slippage has been well investigated for the *dnaX* gene of *E. coli* (Tsuchihashi and Brown, 1992; Farabaugh, 1996). We have not sequenced C-tracts in *futC*, but the role of this gene in generating the 3a phenotype was demonstrated by insertional inactivation of this gene in mutant KO 0093/94, which has a phenotype similar to that of 3a. Finally, 3a was shown to switch back to the parental phenotype, which again proves the reversibility of phase variation. Interestingly, variant 3a and the KO 093/94 also express Le^a, and a similar variant (variant LeA) can be isolated after colony blotting of NCTC 11637 with a Le^a-specific Mab. Insertional inactivation of *futA* in variant 3a leads to a mutant (3a KO 0379) that does not express Le^a; this and other data (Rasko *et al.*, 2000a,b) prove that *futA* is bi-functional and is both an α3-*fucT* and an α4-*fucT*.

Molecular mechanisms of LPS phase variation: the GlcNAcT gene

Variant 1b was isolated from NCTC 11637 after colony blotting with a Le^y-reactive Mab and strongly expresses Le^y but no other Lewis antigen; this serotype is similar to that of strains MO19 and O6. SDS-PAGE demonstrates that variant 1b has lost the polymeric chain and expresses a truncated LPS, which consists of lipid A, core and a single Le^y (Figure 7.3; Table 7.2); the frequency of back-switching to the parental phenotype is very low. Enzymatic analysis showed that the 1b variant lacks GlcNAcT activity (Appelmelk *et al.*, 1998). The serotype of this strain can be understood as follows: we assume that there are two different GlcNAcT in *H. pylori*, i.e. one that recognizes the core and puts on the first GlcNAc and a second one that recognizes Gal and thus is responsible for chain elongation. This situation is identical to that in mammals, where two GlcNAcT are also present. The lack of GlcNAcT enzymatic activity in variant 1b is interpreted by us to indicate the absence of the second, elongating enzyme. Thus, the core plus a single GlcNAc is formed first in this variant. HP0379 is 'on' in variant 1b, therefore terminal Le^x is formed; α2-FucT then forms Le^y. Although GlcNAcT genes have been identified in other species (Blixt et al., 1999), these do not show significant homology with H. pylori ORFs, and so this gene has not been identified yet, let alone the mechanism of phase variation.

Molecular mechanisms of LPS phase variation: β3-galT

Variant H11 was isolated from NCTC 11637 after colony blotting with 4D2, a H type I-specific Mab. This variant expresses Le^x, Le^y but no H type 1; phase variation in β3-*galT* (Appelmelk et al., 2000) can explain the phenotype. Several genes with sequence homology to glycosyltransferases have been identified in the two strains sequenced. One of them (HP0619 in 26695 and JHP563 in J99) has strong sequence homology to *lic2b*, a gene necessary for LPS biosynthesis in *Haemophilus influenzae*. Both JHP563 and HP0619 contain C-tracts. Insertional inactivation of this gene in NCTC 11637 yields a mutant that expresses a phenotype similar to that of variant H11, which provided evidence that JHP563 codes for a β3-GalT. Inactivation of this gene in strain G27 led to loss of both Le^a and H type I epitopes, which shows that indeed this β3-GalT is needed for biosynthesis of the Galβ1.3GlcNAc structure present in both epitopes. By colony blot of H11 with a H type I Mab, a back-switch variant (H111) that expressed the parental phenotype was isolated. C-tract analysis showed NCTC 11637 and H111 to have a C14 tract (full-length gene product) while in the H type I-negative variant H111 a C13 tract that yields a truncated, inactive product was found. In conclusion, reversible phase variation in β3-*galT* can be explained at the molecular level.

Phase variation in sialyl-Le^x.

Initially, strain P466 LPS was chemically characterized as expressing polymeric Le^x, terminating in Le^y. Subsequent serological studies showed this strain to react with a sialyl-Le^x-reactive Mab. It is unclear why sialyl-Le^x was not detected initially but further research showed phase variation in this epitope to occur and both sialylated and non-sialylated variants of strain P466 were isolated. Chemical analysis of one of the variants that expressed sialyl-Le^x serologically demonstrated this structure indeed to substitute about 30% of the poly-Le^x chain (Monteiro et al., 2000). The genetic basis of phase variation in sialyl-Le^x has not yet been clarified but homologues of *neuA* and *neuB* (HP0178), genes essential for biosynthesis of the sialyl-group, were reported to be present in the two sequenced strains; by southern analysis we found *neuB* to be present in three additional strains, including strain NCTC 11637 (unpublished data); *neuB* contains a C6-tract (strains 26695 and J99). It is unclear why only so few *H. pylori* strains seem to express sialyl-Le^x.

A double switch

Variant D1.1 expresses a truncated LPS and does not react with any anti-Lewis Mab. This variant arose through phase variation from a K4.1-like variant that expresses i-antigen plus H type I only, through subsequent loss of the elongating GlcNAcT (Appelmelk *et al.*, 1998).

BIOLOGICAL ROLE OF LPS PHASE VARIATION

From a single strain many LPS phase variants can be isolated, each with a phenotype corresponding to that of other, independent strains, including clinical isolates. Potentially, a single strain may even be able to express *all* serotypes encountered in *H. pylori* LPS. It is puzzling why in certain strains (MO19 and O6), the 1b-phenotype is the most prevailing type while in strain J223 the K4.1 phenotype prevails, and still other strains express the 3a phenotype (Monteiro, 2001). One hypothesis is that, depending on the host, a different LPS-phenotype is selected. We isolated multiple *H. pylori* colonies ($n = 30$) from a single patient and found that 20% of the colonies expressed Le$^{x/y}$, while 80% express the i-Ag (Appelmelk *et al.*, 1999b). By molecular typing combined with C-tract sequencing it was demonstrated that they are phase variants of the same strain. We concluded that LPS phase variation contributes to strain diversity *in vivo*. In another study, bacterial Le^y-expression increased in strains from older patients (Munoz *et al.*, 2001) or after growth at low pH (Moran *et al.*, 2002), but whether phase variation is involved was not determined. Which factor(s) determine the actual serotype expressed by a strain isolated from a clinical sample, or the distribution of serotypes of multiple isolates obtained from a single patient? It is possible that host selection does not play a role but that the distribution of phenotypes is determined by the inherent switch-rates in the various glycosyltransferases A forward switch (for example from C13 to C14) does not necessarily have the same rate as the corresponding back-switch (C14 to C13). Likewise, depending on the nucleotides surrounding a C-tract, a C14 tract in one gene may exhibit a switch frequency different from that of a C14 tract in another gene. In theory, the most prevalent phenotype may be determined by the switch frequencies inherent to the various C-tracts and at present no single environmental or host factor has been identified that causes a change in LPS phenotype through selection of LPS phase variants. However, in analogy with other species, much attention has been given to host-induced antibodies as selectors of LPS-phenotype.

LEWIS ANTIGEN MIMICRY AND IMMUNE EVASION

More than 80% of all *H. pylori* strains express Lewis antigens, most often $Le^{x/y}$. $Le^{x/y}$ antigens are also expressed by the gastric epithelial tissue. The phenomenon that pathogens express epitopes similar to those of the host is called molecular mimicry. It is also seen in *Neisseria* and *H. influenzae*. Strikingly, the ferret-specific *Helicobacter*, called *H. mustelae*, expresses blood group A on LPS, in mimicry with blood group A of ferret gastric mucosa (Monteiro *et al.*, 1997). These findings suggest that phase variation in Le^x and Le^y may serve immune evasion, i.e. the escape from immune attack by expressing a Lewis antigen that exactly mimics the gastric niche. In this concept, phase variation and immune evasion would contribute to persistence, a mechanism recently put forward also for the gut mucosal pathogen *Bacteroides fragilis* (Krinos *et al.*, 2001). It can be hypothesized that, in analogy with ABO blood group antigen expression, a host that expresses Le^x forms anti-Le^y but not anti-Le^x antibodies. Hence, a Le^x-positive *H. pylori* strain that infects a Le^x-positive host would escape immune attack and persists, while a Le^y-positive strain would be eradicated by anti-Le^y antibodies. The same applies to persistence of *H. mustelae* in the ferret, both host and pathogen expressing blood group A. Experimental infection in Rhesus monkeys confirms this concept: an *H. pylori* strain isolated from Le^y-positive animals (in gastric mucosa) expresses more Le^y than Le^x; the *same* strain expresses more Le^x than Le^y when isolated after colonization of Le^x-positive animals (Wirth *et al.*, 1998). Thus, the expression of *H. pylori* $Le^{x/y}$ epitopes depends on the host. It is conceivable that *in vivo* outgrowth of Le^y-expressing *H. pylori* variants is favoured because variants expressing Le^x are suppressed in Le^y-positive hosts, which form anti-Le^x but not anti-Le^y antibodies. However, whether the two variants are phase variants of the same strain was not investigated, nor was it shown that the animals formed serum antibodies to $Le^{x/y}$. In agreement with the 'adaptation' hypothesis is the finding that *H. pylori* strains isolated from Chinese patients more often express Le^a or Le^b as compared to strains isolated in Western countries (Zheng *et al.*, 2000). Other studies in humans provide far less consistent results and in two out of three studies no correlation between the Lewis phenotypes of host and pathogen is found (Wirth *et al.*, 1997; Heneghan *et al.*, 2000). In addition, strains expressing Le^x and strains expressing Le^y can be isolated from a single patient, which argues against adaptation based on Lewis antigens (Wirth *et al.*, 1999). Finally, selection and outgrowth of *H. pylori* $Le^{x/y}$ LPS variants would be driven by anti-LPS antibodies, specifically those directed to $Le^{x/y}$. However, there is some debate whether *H. pylori* infection induces anti-$Le^{x/y}$ antibodies. Although infection with *H. pylori* induces a high anti-LPS antibody response, in direct Elisa with synthetic Le^x as coating antigen, an anti-$Le^{x/y}$ response was not detectable (Claeys *et al.*, 1998); in a subpopulation of non-secretors (patients that do not express gastric Le^b) a detectable but very low response was observed (Kurtenkov et al., 1999). In fact, anti-Le$^{x/y}$ antibodies occur naturally in sera from persons not infected by *H. pylori* (Chmiela et al., 1999). However, with a competitive immunoassay with larger amounts of synthetic Le^x in solution, Moran and colleagues have recently shown that *H. pylori* indeed induces anti-Le$^{x/y}$ in a majority of patients (Heneghan et al., 2001). No relationship between anti-Le^x antibody titre and host Le^x expression was observed, a finding contraindicative of evasion through mimicry. Yet, the possibility of immune evasion cannot be completely excluded. It remains possible that gastric anti-Le^x response is locally diverse within a single patient and LPS phase variation may help evasion from this local response. This mechanism is in agreement with our finding that LPS phase variation drives variability *in vivo* (Appelmelk et al., 1999b).

In summary, the hypothesis that phase variation in *H. pylori* LPS serves bacterial persistence through evasion of immune response is yet unproven.

H. PYLORI LPS PHASE VARIATION AND BACTERIAL ADHESION

Recent data suggest that LPS phase variation may play a role in transmission, adherence and colonization. LPS phase variants of a single strain have not yet been tested for their ability to colonize. However, several colonization studies with mutants knocked out in LPS biosynthesis genes have been performed. The expression of Le$^{x/y}$ proved to be crucial for *in vivo* colonization of mice. β1,4-GalT was inactivated in strain SS-1, which expresses Le$^{x/y}$. The mutant does not express Le$^{x/y}$, and as compared to the parent, the mutant colonizes less well (Logan et al., 2000). Mutants in another LPS biosynthesis gene, *galE* (HP0360, UDP-galactose-4-epimerase), express a shorter LPS devoid of Lewis antigens (Kwon et al., 1998; Moran et al., 2000) and also colonize mice less well (Moran et al., 2000). However, a decreased colonization does by necessity mean that Lewis antigens *per se* are involved: from other Gram-negative pathogens it is known that shortening of LPS will lead to a decrease in virulence. Strains with a shorter LPS are simply more sensitive to the lytic action of serum, or are more easily phagocytosed and killed. Thus, the lesser colonization of the β1,4-GalT and *galE* knockouts could simply be due to enhanced killing. To circumvent this we created knockouts that do not express Le$^{x/y}$ yet have a long LPS. Inactivation of both α3-*fucT* genes yielded a mutant that expresses i-antigen and H type 1 (see Table 7.3) and with a length identical to that of the parent strain (Appelmelk et al., 1999a). While the parent strain (Le$^{x/y}$ positive) colonizes mice well, the mutant does not (Figure 7.4), which demonstrates that Le$^{x/y}$ antigens are essential for colonization (Martin et al., 2000), at least in the mouse.

Current data suggest that the effect of LPS structure on colonization is due to its role in mediating bacterial adhesion to host tissue. A Mab specific for *H.*

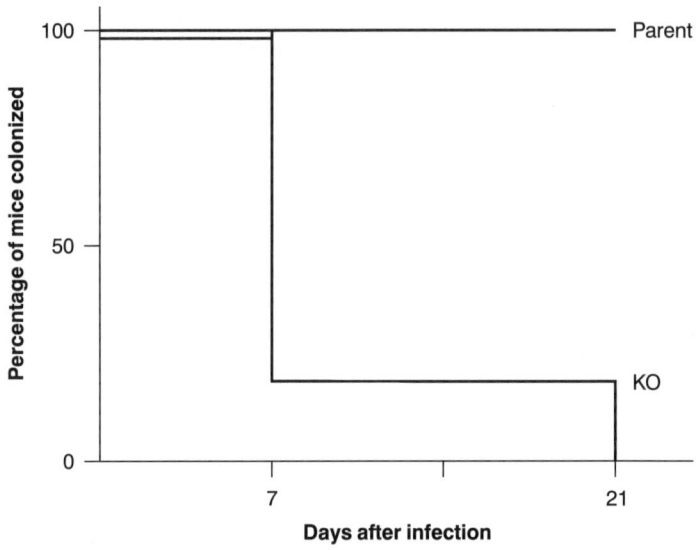

Figure 7.4 Role of Le^x in colonization of mice. Strain 4187 expresses Le^x and colonizes well; the α3-*fucT* double knockout (KO) colonizes much less well. (Reproduced from *Trends in Microbiology*, vol. 8, issue 12, Appelmelk, B. J., pp. 565–570. Copyright © 2000, with permission from Elsevier.)

pylori LPS (Le^x) inhibits adhesion of bacteria to gastric epithelial cells (Osaki *et al.*, 1998); *H. pylori* LPS itself was also inhibitory. In another study, synthetic Le^x was found to inhibit bacterial adhesion (Campbell *et al.*, 1999). Mutants in *rfbM* (HP0043, GDP-mannose pyrophosphorylase) express a long, fucose-lacking i-antigen (Edwards *et al.*, 2000). *galE* and *rfbM* mutants did not adhere to gastric sections, while the parent (strain NCTC 11637, Le$^{x/y}$ positive) adhered well (Edwards *et al.*, 2000) (Figure 7.5); this demonstrates the role of Le^x in adhesion. In addition, synthetic polyvalent Le^x, either free (Campbell *et al.*,1999) or bound to 1-micron-sized polystyrene beads (Edwards *et al.*, 2000), bound to human

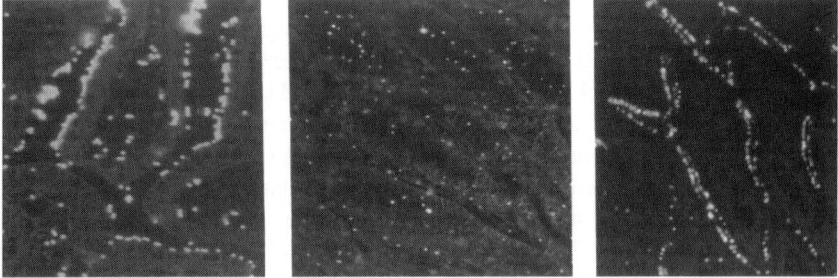

Figure 7.5 Role of Le^x in adhesion of *H. pylori* to gastric mucosal epithelium: (*left*) adhesion of Le^x-positive parent; (*middle*) adhesion of Le^x-negative mutant; (*right*) adhesion of Le^x-polystyrene beads. (Reproduced from Edwards *et al.* 2000 with permission from Blackwell Publishing.)

gastric epithelial cells; in the latter case, the binding pattern of the laden beads was similar to that of Le^x-expressing bacteria. However, until now a direct role for LPS in adhesion has been demonstrated only for strain NCTC 11637 through comparison with isogenic mutants. In addition, only biopsy material from a single patient was tested. Clearly, the role of LPS in *in situ* adherence needs confirmation. Indirectly, results from clinical studies also suggest a role for Le$^{x/y}$ in adhesion: as compared to *H. pylori* strains that express Le$^{x/y}$ weakly (Heneghan *et al.*, 2000), strains that expressed Le$^{x/y}$ strongly cause a higher colonization density in the stomach of infected patients.

The above studies suggest that *H. pylori* LPS might be a ligand for host lectins which mediate bacterial adhesion. A role for Le^x as a ligand is not unexpected as a Le^x–lectin interaction has been shown in several studies to be of biological significance in different systems (Table 7.4). Which host lectin mediates LPS-dependent adhesion of *H. pylori* to gastric mucosa? Several host lectins are

Table 7.4 Lewis × as a ligand for lectins

Sperm expresses Le^x-binding lectins;[a] lectin unknown
Le^x-binding lectins play a role corneal cell–cell adhesion and differentiation; lectin unknown[b]
Le^x-selectin binding relevant to interaction of *S. mansoni* to eosinophils or endothelial cells[c]
Le^x on pulmonary mucin is a ligand for a *Pseudomonas aeruginosa* flagellum[d]
Le^x–Le^x homotypic interaction is crucial in murine embryogenesis[e]
C-type lectins (selectins, mannose-binding protein, surfactant protein D, mannose receptor) may recognize Lewis antigens[f]

[a] Yoshitani *et al.*, 2001
[b] Cao *et al.*, 2001
[c] Lejoly-Boisseau *et al.*, 1999; Nutten *et al.*, 1999
[d] Scharfman *et al.*, 2001
[e] Eggens *et al.*, 1989; Fenderson *et al.*, 1984
[f] Larsen *et al.*, 1990; Zhang *et al.*, 1998

already known to interact with host Lewis antigens (Table 7.4). For example, selectins bind to Le^x and, in particular, sialylLe^x (Larsen *et al.*, 1990); however, our attempts to demonstrate binding of recombinant selectins to *H. pylori* LPS failed (unpublished data). Several other C-type (calcium-dependent) lectins are known to interact specifically with mannose and fucose. Examples are mannose-binding protein (MBP), surfactant protein D (SPD) and macrophage mannose receptor (MR) (Weis *et al.*, 1998). Hence, it is likely that poly-fucosylated *H. pylori* LPS can interact with C-type host lectins. Indeed, it has been shown that surfactant protein D, which is also expressed in the stomach epithelium (Fisher and Mason, 1995; Murray *et al.*, 2002), is able to bind *H. pylori* LPS, and LPS variants not able to do so have been isolated (Moran, 2001). In immunoblot, *H. pylori* LPS and Le^x have been found to bind to gastric polypeptides of about 16–29 kDa (Campbell *et al.*, 1999), 100 kDa (Edwards, 2000). Despite intensive investigations in several research groups, the identity of gastric Le^x-binding proteins remains unknown.

A role for LPS/Le$^{x/y}$ in adherence seems likely but this role is certainly not absolute: Le$^{x/y}$-negative mutants adhere as strongly as their Le$^{x/y}$-positive parents when the strain expresses the Le^b-binding lectin BabA and when the host expresses Le^b (J. Mahdavi, unpublished data). In addition, Le$^{x/y}$-negative strains colonize human hosts well (Rasko et al., 2000c). Thus a Le$^{x/y}$–lectin interaction may contribute to adhesion only for *H. pylori* strains that do not express BabA, or for strains that colonize non-secretors. This is in contrast with the situation in mice, which do not express gastric Le^b (Guruge et al., 1998), and here colonization might require the presence of Le^x-binding lectins in the gastric mucosa and the presence of Le^x on *H. pylori* LPS. Alternatively, gastric Le^x-binding lectins may be localized more luminally in the stomach as compared to gastric Le^b mucins, and thus provide the initial contact between *H. pylori* and gastric mucosa, also in Le^b-positive patients. In a later stage, the bacteria might move more inward to interact through babA with Le^b on mucin or epithelial cells. In this model, phase variation fulfils a biological role by allowing detachment of bacteria not expressing Le$^{x/y}$ and hence transmission to another host; subsequently, switch-back variants expressing Le$^{x/y}$ adhere and colonize a new host.

CONCLUSION

The molecular mechanisms of *H. pylori* LPS phase variation are known in some detail, involving the action of an array of enzymes on sugar structures. However, the biological role of phase variation in evasion of immune response or adhesion and transmission remains to be elucidated.

REFERENCES

Alm, R. A., Ling, L. S., Moir, D. T., King, B. L., Brown, E. D., Doig, P. C., Smith, D. R., Noonan, B., Guild, B. C., deJonge, B. L., Carmel, G., Tummino, P. J., Caruso, A., Uria-Nickelsen, M., Mills, D. M., Ives, C., Gibson, R., Merberg, D., Mills, S. D., Jiang, Q., Taylor, D. E., Vovis, G. F. and Trust, T. J. (1999) *Nature* 397, 176–180.

Appelmelk, B. J., Martin, S. L., Monteiro, M. A., Clayton, C. A., McColm, A. A., Zheng, P. Y., Verboom, T., Maaskant, J. J., van den Eijnden, D. H., Hokke, C. H., Perry, M. B., Vandenbroucke-Grauls, C. M. J. E. and Kusters, J. G. (1999a) *Infect. Immun.* 67, 5361–5366.

Appelmelk, B. J., Martino, M. C., Veenhof, E., Monteiro, M. A., Maaskant, J. J., Negrini, R., Lindh, F., Perry, M., Del Giudice, G. and Vandenbroucke-Grauls, C. M. (2000) *Infect. Immun.* 68, 5928–5932.

Appelmelk, B. J., Shiberu, B., Trinks, C., Tapsi, N., Zheng, P. Y., Verboom, T., Maaskant, J., Hokke, C. H., Schiphorst, W. E., Blanchard, D., Simoons-Smit, I. M., van den Eijnden, D. H. and Vandenbroucke-Grauls, C. M. (1998) *Infect. Immun.* 66, 70–76.

Appelmelk, B. J., Wirth, H. P., Lansbergen, R., Schilders, I., Martin, S. L., Verboom, T. and Vandenbroucke-Grauls, C. M. J. E. (1999b) *Gut* 45, A23–A24.

Aspinall, G. O., Monteiro, M. A., Pang, H., Walsh, E. J. and Moran, A. P. (1996) *Biochemistry* 35, 2489–2497.
Berg, D. E., Hoffman, P. S., Appelmelk, B. J. and Kusters, J. G. (1997) *Trends Microbiol.* 5, 468–474.
Blixt, O., van Die, I., Norberg, T. and van den Eijnden, D. H. (1999) *Glycobiology* 9, 1061–1071.
Campbell, B. J., Rogerson, K. A. and Rhodes, J. M. (1999) *Gastroenterology* 116, G0565.
Cao, Z., Zhao, Z., Mohan, R., Alroy, J., Stanley, P. and Panjwani, N. (2001) *J. Biol. Chem.* 276, 21714–21723.
Chmiela, M., Jurkiewicz, M., Wisniewska, M., Czkwianianc, E., Planeta-Malecka, I., Rechcinski, T. and Rudnicka, W. (1999) *Acta Microbiol. Pol.* 48, 277–281.
Claeys, D., Faller, G., Appelmelk, B. J., Negrini, R. and Kirchner, T. (1998) *Gastroenterology* 115, 340–347.
Edwards, N. J. (2000) Data presented at the 4th International Workshop on Pathogenesis and Host Response in *Helicobacter* Infections, July 6–9, Helsingor, Denmark.
Edwards, N. J., Monteiro, M. A., Faller, G., Walsh, E. J., Moran, A. P., Roberts, I. S. and High, N. J. (2000) *Mol. Microbiol.* 35, 1530–1539.
Eggens, I., Fenderson, B., Toyokuni, T., Dean, B., Stroud, M. and Hakomori, S. (1989) *J. Biol. Chem.* 264, 9476–9484.
Farabaugh, P. J. (1996) *Annu. Rev. Genet.* 30, 507–528.
Fenderson, B. A., Zehavi, U. and Hakomori, S. (1984) *J. Exp. Med.* 160, 1591–1596.
Fisher, J. H. and Mason, R. (1995) *Am. J. Resp. Cell Mol. Biol.* 12, 13–18.
Guruge, J. L., Falk, P. G., Lorenz, R. G., Dans, M., Wirth, H. P., Blaser, M. J., Berg, D. E. and Gordon, J. I. (1998) *Proc. Natl Acad. Sci. USA* 95, 3925–3930.
Harvey, H. A., Jennings, M. P., Campbell, C. A., Williams, R. and Apicella, M. A. (2001) *Mol. Microbiol.* 42, 659–672.
Haslam, S. M., Coles, G. C., Morris, H. R. and Dell, A. (2000) *Glycobiology* 10, 223–229.
Heneghan, M. A., McCarthy, C. F., Janulaityte, D. and Moran, A. P. (2001) *Infect. Immun.* 69, 4774–4781.
Heneghan, M. A., McCarthy, C. F. and Moran, A. P. (1997) *Gut* 41, A107–A108.
Heneghan, M. A., McCarthy, C. F. and Moran, A. P. (2000) *Infect. Immun.* 68, 937–941.
Hynes, S. O., Hirmo, S., Wadstrom, T. and Moran, A. P. (1999) *J. Clin. Microbiol.* 37, 1994–1998.
Israel, D. A., Salama, N., Krishna, U., Rieger, U. M., Atherton, J. C., Falkow, S. and Peek, R. M., Jr (2001) *Proc. Natl Acad. Sci. USA* 98, 14625–14630.
Jarvinen, N., Maki, M., Rabina, J., Roos, C., Mattila, P. and Renkonen, R. (2001) *Eur. J. Biochem.* 268, 6458–6464.
Jennings, M. P., Srikhanta, Y. N., Moxon, E. R., Kramer, M., Poolman, J. T., Kuipers, B. and van der Ley, L. P. (1999) *Microbiology* 145, 3013–3021.
Knirel, Y. A., Kocharova, N. A., Hynes, S. O., Widmalm, G., Andersen, L. P., Jansson, P. E. and Moran, A. P. (1999) *Eur. J. Biochem.* 266, 123–131.
Krinos, C. M., Coyne, M. J., Weinacht, K. G., Tzianabos, A. O., Kasper, D. L. and Comstock, L. E. (2001) *Nature* 414, 555–558.
Kuipers, E. J. and Appelmelk, B. J. (1997) *Biomed. Pharmacother.* 51, 150–155.
Kurtenkov, O., Klaamas, K., Miljukhina, L., Shljapnikova, L., Ellamaa, M., Bovin, N. and Wadstrom, T. (1999) *FEMS Immunol. Med. Microbiol.* 24, 227–232.

Kwon, D. H., Woo, J. S., Perng, C. L., Go, M. F., Graham, D. Y. and El Zaatari, F. A. (1998) *Curr. Microbiol.* 37, 144–148.

Larsen, E., Palabrica, T., Sajer, S., Gilbert, G. E., Wagner, D. D., Furie, B. C. and Furie, B. (1990) *Cell* 63, 467–474.

Lejoly-Boisseau, H., Appriou, M., Seigneur, M., Pruvost, A., Tribouley-Duret, J. and Tribouley, J. (1999) *Exp. Parasitol.* 91, 20–29.

Linton, D., Gilbert, M., Hitchen, P. G., Dell, A., Morris, H. R., Wakarchuk, W. W., Gregson, N. A. and Wren, B. W. (2000) *Mol. Microbiol.* 37, 501–514.

Logan, S. M. (2001) Data presented at the 11th International Workshop on *Campylobacter, Helicobacter* and related organisms, September 1–5, Freiburg.

Logan, S. M., Conlan, J. W., Monteiro, M. A., Wakarchuk, W. W. and Altman, E. (2000) *Mol. Microbiol.* 35, 1156–1167.

Martin, S. L., Edbrooke, M. R., Hodgman, T. C., van den Eijnden, D. H. and Bird, M. I. (1997) *J. Biol. Chem.* 272, 21349–21356.

Martin, S. L., McColm, A. A. and Appelmelk, B. J. (2000) *Gastroenterology* 119, 1414–1416.

Monteiro, M. A. (2001) *Adv. Carbohydr. Chem. Biochem.* 57, 99–158.

Monteiro, M. A., Appelmelk, B. J., Rasko, D. A., Moran, A. P., Hynes, S. O., MacLean, L. L., Chan, K. H., St Michael, F., Logan, S. M., O'Rourke, J., Lee, A., Taylor, D. E. and Perry, M. B. (2000) *Eur. J. Biochem.* 267, 305–320.

Monteiro, M. A., Chan, K. H., Rasko, D. A., Taylor, D. E., Zheng, P. Y., Appelmelk, B. J., Wirth, H. P., Yang, M., Blaser, M. J., Hynes, S. O., Moran, A. P. and Perry, M. B. (1998) *J. Biol. Chem.* 273, 11533–11543.

Monteiro, M. A., St Michael, F., Rasko, D. A., Taylor, D. E., Conlan, J. W., Chan, K. H., Logan, S. M., Appelmelk, B. J. and Perry, M. B. (2001) *Biochem. Cell Biol.* 79, 449–459.

Monteiro, M. A., Zheng, P. Y., Appelmelk, B. J. and Perry, M. B. (1997) *FEMS Microbiol. Lett.* 154, 103–109.

Moran, A. (2001) Data presented at the 11th International Workshop on *Campylobacter, Helicobacter* and related organisms; September 1–5, Freiburg.

Moran, A. P., Knirel, Y. A., Senchenkova, S. N., Widmalm, G., Hynes, S. O. and Jansson, P. E. (2002) *J. Biol. Chem.* 277, 5785–5795.

Moran, A. P., Sturegard, E., Sjunnesson, H., Wadstrom, T. and Hynes, S. O. (2000) *FEMS Immunol. Med. Microbiol.* 29, 263–270.

Munoz, L., Gonzalez-Valencia, G., Perez-Perez, G. I., Giono-Cerezo, S., Munoz, O. and Torres, J. (2001) *J. Infect. Dis.* 183, 1147–1151.

Murray, E., Khamri, W., Walker, M. M., Eggleton, P., Moran, A. P., Ferris, J. A., Knapp, S., Karim, Q. N., Worku, M., Strong, P., Reid, K. B. and Thursz, M. R. (2002) *Infect. Immun.* 70, 1481–1487.

Nutten, S., Papin, J. P., Woerly, G., Dunne, D. W., MacGregor, J., Trottein, F. and Capron, M. (1999) *Eur. J. Immunol.* 29, 799–808.

Osaki, T., Yamaguchi, H., Taguchi, H., Fukuda, M., Kawakami, H., Hirano, H., Watanabe, S., Takagi, A. and Kamiya, S. (1998) *J. Med. Microbiol.* 47, 505–512.

Parkhill, J., Wren, B. W., Mungall, K., Ketley, J. M., Churcher, C., Basham, D., Chillingworth, T., Davies, R. M., Feltwell, T., Holroyd, S., Jagels, K., Karlyshev, A. V., Moule, S., Pallen, M. J., Penn, C. W., Quail, M. A., Rajandream, M. A., Rutherford, K. M., van Vliet, A. H., Whitehead, S. and Barrell, B. G. (2000) *Nature* 403, 665–668.

Rasko, D. A., Wang, G., Monteiro, M. A., Palcic, M. M. and Taylor, D. E. (2000a) *Eur. J. Biochem.* 267, 6059–6066.
Rasko, D. A., Wang, G., Palcic, M. M. and Taylor, D. E. (2000b) *J. Biol. Chem.* 275, 4988–4994.
Rasko, D. A., Wilson, T. J., Zopf, D. and Taylor, D. E. (2000c) *J. Infect. Dis.* 181, 1089–1095.
Saunders, N. J., Peden, J. F., Hood, D. W. and Moxon, E. R. (1998) *Mol. Microbiol.* 27, 1091–1098.
Scharfman, A., Arora, S. K., Delmotte, P., Van Brussel, E., Mazurier, J., Ramphal, R. and Roussel, P. (2001) *Infect. Immun.* 69, 5243–5248.
Simoons-Smit, I. M., Appelmelk, B. J., Verboom, T., Negrini, R., Penner, J. L., Aspinall, G. O., Moran, A. P., Fei, S. F., Shi, B. S., Rudnica, W., Savio, A. and de Graaff, J. (1996) *J. Clin. Microbiol.* 34, 2196–2200.
Suerbaum, S. (2000) *Int. J. Med. Microbiol.* 290, 175–181.
Tomb, J. F., White, O., Kerlavage, A. R., Clayton, R. A., Sutton, G. G., Fleischmann, R. D., Ketchum, K. A., Klenk, H. P., Gill, S., Dougherty, B. A., Nelson, K., Quackenbush, J., Zhou, L., Kirkness, E. F., Peterson, S., Loftus, B., Richardson, D., Dodson, R., Khalak, H. G., Glodek, A., McKenney, K., Fitzegerald, L. M., Lee, N., Adams, M. D. and Venter, J. C. (1997) *Nature* 388, 539–547.
Tsuchihashi, Z. and Brown, P. O. (1992) *Genes Dev.* 6, 511–519.
Van Dam, G. J., Claas, F. H., Yazdanbakhsh, M., Kruize, Y. C., van Keulen, A. C., Ferreira, S. T., Rotmans, J. P. and Deelder, A. M. (1996) *Blood* 88, 4246–4251.
Van Putten, J. P. (1993) *EMBO J.* 12, 4043–4051.
Wang, G., Boulton, P. G., Chan, N. W., Palcic, M. M. and Taylor, D. E. (1999a) *Microbiology* 145 (Pt 11), 3245–3253.
Wang, G., Ge, Z., Rasko, D. A. and Taylor, D. E. (2000) *Mol. Microbiol.* 36, 1187–1196.
Wang, G., Rasko, D. A., Sherburne, R. and Taylor, D. E. (1999b) *Mol. Microbiol.* 31, 1265–1274.
Weis, W. I., Taylor, M. E. and Drickamer, K. (1998) *Immunol Rev.* 163, 19–34.
Wirth, H. P., Karita, M., Yang, M. and Blaser, M. J. (1996) *Gastroenterology* 110, A296.
Wirth, H. P., Yang, M., Dubois, A., Berg, D. E. and Blaser, M. J. (1998) *Gut* 43, A26.
Wirth, H. P., Yang, M., Peek, R. M., Jr, Hook-Nikanne, J., Fried, M. and Blaser, M. J. (1999) *J. Lab. Clin. Med.* 133, 488–500.
Wirth, H. P., Yang, M., Peek, R. M., Jr, Tham, K. T. and Blaser, M. J. (1997) *Gastroenterology* 113, 1091–1098.
Yoshitani, N., Mori, E. and Takasaki, S. (2001) *Glycobiology* 11, 313–320.
Zhang, K., Chuluyan, H. E., Hardie, D., Shen, D. C., Larsen, R. and Issekutz, A. (1998) *Hybridoma* 17, 445–456.
Zheng, P. Y., Hua, J., Yeoh, K. G. and Ho, B. (2000) *Gut* 47, 18–22.

8

GENETIC VARIATION IN THE PATHOGENIC *NEISSERIA* SPECIES

THOMAS F. MEYER AND STUART A. HILL

INTRODUCTION

The *Neisseriae* are Gram-negative diplococci and form a group that includes a wide variety of commensal species besides the two pathogenic species *Neisseria gonorrhoeae* (gonococci) and *Neisseria meningitidis* (meningococci), which cause the diseases gonorrhoea and meningitis, respectively. Meningococci and gonococci, as well as some of their commensal relatives, are unusual in that they are only able to infect humans as their natural host and as such represent typical mucosal colonizers. Generally, localized uncomplicated gonorrheal infections involve considerable tissue damage, whereas the frequently occurring localized meningococcal infections (e.g. of the nasopharynx) of normal human individuals cause little or no damage and are asymptomatic. In this respect, meningococcal infections are therefore somewhat reminiscent of the mucosal colonization by the commensal *Neisseria* species. Under rare, yet still undefined conditions, the pathogenic *Neisseria* can disseminate to cause severe or even life-threatening diseases, including meningitis, bacteraemia, pelvic inflammatory disease or septic arthritis.

Localized neisserial infections invariably involve a series of receptor-mediated interactions between the bacteria and their primary target cells. The formation of pili is a prerequisite for neisserial attachment to the mucosal epithelial cell surface. Often surface-attached bacteria will penetrate epithelial cells and – whilst contained within a membranous vacuole – transcytose towards sub-epithelial tissues. Furthermore, the pathogenic *Neisseriae* strongly interact with phagocytic cells, such as neutrophils and macrophages, which seem to provide them with a protected intracellular environment.

Neisserial pathogenesis is characterized by a strong inflammatory response to the infection rather than by the action of distinct bacterial toxins. Consequently, neisserial infections can be regarded as multifactorial processes. With respect to their putative virulence attributes, meningococci and gonococci share many important features, with one prominent difference being the presence of a polysaccharide capsule in the meningococcus. Therefore, most of the topics discussed in this review are relevant to both pathogenic species, despite their different disease spectra.

GENETIC DIVERSITY AND EVOLUTIONARY ADAPTABILITY OF *NEISSERIA* SPECIES

Horizontal genetic exchange plays a dominant role in the evolution and epidemiology of *Neisseria* species. A continuous horizontal gene flow affects the chromosomal composition of not only the pathogenic *Neisseria* species, but also of many of the commensal species. In fact, considerable circumstantial evidence exists for horizontal exchange of genes between commensal and pathogenic *Neisseria* species, between meningococci and gonococci, as well as within species, with the outcome being the generation of mosaic gene structures (Halter *et al.*, 1989; Manning *et al.*, 1991; Feavers *et al.*, 1992; Spratt *et al.*, 1992; Zhou and Spratt, 1992). Despite the fact that it is difficult to give precise estimates on the frequencies at which horizontal genetic exchanges occur in nature, several examples exist where it appears that such events may have occurred in recent years (Achtman, 1994).

Conceptually, horizontal genetic exchange can be envisioned as a long-term adaptive mechanism that is suitable for responding to gross environmental changes as well as a means by which the genetic flexibility of the various *Neisseria* species can be secured. However, evolution involving such dramatic chromosome homogenization through the exchange of a common neisserial gene pool suggests that perhaps *Neisseria* species should be regarded as a collective group rather than as a set of seemingly independent distinct species.

Natural competence for transformation

The ability of *N. gonorrhoeae* to engage in DNA transformation under natural conditions has long been recognized (Catlin and Cunningham, 1961; Sparling, 1966) and is the only natural process for the horizontal exchange of chromosomal DNA between the *Neisseriae*. Bacteriophages, in particular transducing phages, have not been identified and, although many conjugative plasmids do exist, conjugative (Hfr-like) mobilization of chromosomal determinants has not been reported under natural conditions. Thus, the only known mechanism that can account for the observed horizontal exchange among the *Neisseriae* is DNA transformation. Horizontal exchange of chromosomal markers via transformation is readily observed by co-cultivation of different *Neisseria* strains *in vitro* (Frosch and Meyer, 1992; Zhang *et al.*, 1992). In a typical experiment, the efficiency of transfer of chromosomal markers between two gonococcal strains following 1 hour of co-cultivation is in the order of 10^{-5} per cell per genetic locus. Exchange is completely abrogated when DNase is present within the culture medium, indicating that, despite the apparent viability of the cultured cells, there is substantial release of DNase-accessible DNA into the medium. How such large quantities of chromosomal DNA are released has not yet been elucidated. While there may well be specific export mechanisms (Dorward *et al.*, 1989), spontaneous autolysis is, however, a common observation during cultivation of gonococci (Hebeler and Young, 1975).

Natural transformation competence differs markedly from artificial transformation protocols using recombinant systems (e.g. transformation of *Escherichia coli*). Unlike these artificial systems, transforming donor DNA is taken up into *Neisseria* as a linear molecule and requires RecA protein and homologous DNA sequences within the recipient cell in order for either the re-circularization of a plasmid DNA to occur or the incorporation of the donor DNA into the chromosome (Biswas *et al.*, 1986; Koomey *et al.*, 1987). Exclusion of foreign DNA occurs at two levels: a DNA homology requirement, which has been shown to be an effective mechanism of protection against bacterial transformation with unrelated DNA in other naturally competent species (Harris-Warrick and Lederberg, 1978), as well as through a requirement for an 8 bp genus-specific uptake sequence located within the donor DNA (Goodman and Scocca, 1988). The requirement for the presence of an uptake sequence is not absolute (Boyle-Vavra and Seifert, 1996). However, because the uptake signal is usually part of a transcriptional terminator (where it constitutes the palindromic stem loop) it occurs at high frequency within neisserial DNA. Whether the uptake signal primarily serves for DNA recognition or whether it also represents a site for the DNA linearization and/or a signal for the direction of the DNA transport is currently unknown. Likewise, the role of DNA restriction/methylation systems, which are highly abundant among the *Neisseria* species, in the transformation-mediated exchange is currently not well understood, and is also puzzling given the fact that no viruses are known to infect the *Neisseriae* (Sullivan and Saunders, 1989; Gunn *et al.*, 1992; Gunn and Stein, 1993).

Generation of *Neisseria* mutants has allowed the dissection of the gonococcal transformation process into: (i) uptake and conversion of the transforming DNA into a DNase-resistant state; and (ii) subsequent processing events (Biswas *et al.*, 1989). One such DNA uptake-deficient mutant (*dud-1*) has been characterized (Dorward and Garon, 1989), providing a biochemical basis to the early observations that suggested a role for gonococcal pili (which belong to the type IV or N-methyl-Phe class of pili) in transformation competence (Wolfgang *et al.*, 1998a). An interesting outcome of these studies is the observation that the major pilus subunit, PilE (encoded by the *pilE* locus), rather than intact pili, is required for DNA uptake (Gibbs *et al.*, 1989). One intriguing aspect of gonococcal biology is that organisms can spontaneously shut down PilE synthesis (and pilus formation) by irreversible deletion of the *pilE* locus. This not only results in a transformation defect (in the order of 5–6 logs) but also has considerable implications with respect to the infection process (see below).

In addition to a requirement for PilE polypeptide, the minor pilus-associated protein PilC (Jonsson *et al.*, 1991), in conjunction with other type IV pilus assembly factors (Tonjum and Koomey, 1997; Wolfgang *et al.*, 1998b) is also required for DNA uptake (Hobbs and Mattick, 1993; Rudel *et al.*, 1995a). The involvement of multiple pilus-associated factors in neisserial transformation competence is not unexpected since DNA uptake (as well as type IV pilus assembly) shares common mechanistic features with similar systems reported in other bacteria (Hobbs and Mattick, 1993). Consequently, many neisserial competence defects show pleiotrophic effects primarily because many of the competence gene products are also involved in other essential cellular processes. An example of a non-essential competence determinant is ComA, which is present in all competent *Neisseria* species. ComA appears to be a typical inner membrane protein and is involved in a transformation step subsequent to the initial DNA uptake (Facius and Meyer, 1993), more importantly, however, its distribution among all neisserial species and the presence of autologues in even Gram-positive species (Dubnau, 1997) suggests a common transformation mechanism. Hopefully, future studies on natural transformation will shed more light on the significance of this interesting process for the genetic evolution and the pathogenic properties of the *Neisseria* species.

RAPID MICRO-ENVIRONMENTAL ADAPTATION

Genetic variation versus gene regulation

Microbial populations need to be able to adapt to gross, long-term environmental changes and to rapidly respond to recurrent changes within a micro-environment. For the pathogenic *Neisseria*, environmental changes often occur during the course of an infection. In order to respond to such recurrent changes, microorganisms maintain retrievable genetic programs. There are two principle types of adaptive programs that are used by most microorganisms: (i) genetic variation through the accumulation of temporally random spontaneous mutations within chromosomal DNA; and (ii) controlled gene regulation of specific gene products. Changes occurring through genetic variation where spontaneous mutations accumulate at spatially distinct loci allows for the inheritance of these mutations by subsequent progeny, ultimately leading to the synthesis of altered gene products within a subsection of the population. Consequently, this type of event generates heterogeneous populations, with only a minor fraction of the population likely to show improved micro-environmental adaptation over time.

In contrast, controlled gene regulation of specific gene products in response to environmental change is a determined adaptive process that typically affects the whole population. Despite this apparent difference in the benefits to the population as a whole, phenotypic changes that occur through either spontaneous mutation or controlled gene regulation need not necessarily be exclusive phenomena and often are interconnected. For example, the frequency or the direction of a genetic switch may be influenced by environmental effectors, and conversely, a phase-variable regulator protein (whose expression is effected through spontaneous mutation) may control the expression of genes that respond to the changes within the environment (Robertson and Meyer, 1992).

An interesting feature of some (but not all) variable surface proteins in the *Neisseria* is that they are often represented within the genome by gene families rather than by individual genes. This applies for at least three variable surface proteins that have essential functions in the infection process (i.e. PilE, PilC and the Opa proteins) (Robertson and Meyer, 1992), while other variable factors, such as the meningococcal Opc (Olyhoek *et al.*, 1991) and class I proteins (Barlow *et al.*, 1989), are encoded by single copy genes. Mechanistically, genetic changes within the *Neisseria* fall into two broad categories: illegitimate events (RecA-independent), which account primarily for on/off switching, and homologous recombination schemes (RecA-dependent), which involve the physical exchange of nucleotide sequences between the various genes. Furthermore, due to its natural competence, *Neisseria* can also use both intra- and intercellular recombination routes to vary their chromosome compositions. Thus, this remarkable capacity for variation of the neisserial cell surface (which occurs within the same order of magnitude as observed for the variability of antibodies and T cell receptors of the immune system) allows the pathogenic *Neisseriae* to effectively evade the human immune response.

Genetic variation via homologous recombination

The most intensely studied system for gene variation via homologous recombination in *Neisseria* (primarily through studies with the gonococcus) is the mechanism involved in the antigenic variation of PilE polypeptide. PilE polypeptide is encoded by the *pilE* gene, and when polymerized, is the major structural component of the pilus organelle (reviewed by Swanson and Koomey, 1989; Seifert, 1996). Different PilE variants arise through the genetic alteration of *pilE* and occur at high frequency. On plates, pilin variants are spawned at the rate of approximately 10^{-3} variants per colony-forming unit (Serkin and Seifert, 1998). Similar high switching frequencies are also indicated *in vivo* as different PilE variants can be identified within 24 hours following the initiation of an experimental infection (Swanson *et al.*, 1987, 1992). Moreover, it is estimated that gonococci possess the genetic capacity to produce approximately 10^7 variant PilE proteins (Haas *et al.*, 1992). Rearrangement of *pilE* occurs predominantly through an *intra*-cellular (Swanson *et al.*, 1990; Zhang *et al.*, 1992; Facius and Meyer, 1993), RecA-dependent (Koomey *et al.*, 1987), nonreciprocal, gene-conversion-like mechanism; sequences located within the multiple, silent, variant *pil* genes (*pilS*) are exchanged for the corresponding gene sequences residing at the *pilE* locus (Haas and Meyer, 1986; Swanson *et al.*, 1986). Pilin variants can also arise occasionally through *pilE* deletion events (Segal *et al.*, 1985; Hill *et al.*, 1990; Manning *et al.*, 1991; Hill and Grant, 2002), as well as by an *inter*-cellular DNA transformation-mediated route (Seifert *et al.*, 1988; Gibbs *et al.*, 1989). The contribution of transformation to the generation of pilin variants remains controversial. However, when compared to *intra*-cellular recombination events where discrete segments of *pil* DNA are exchanged between *pilE* and *pilS*, genetic variation via the horizontal route tends to be restricted to the exchange of intact *pilE* alleles, which strongly indicates a prominent role for flanking homology in driving *pilE* allelic exchange (Hill, 1996).

In contrast to the PilE system, *opa* recombination events are primarily of the illegitimate kind (Swanson *et al.*, 1992), in that Opa surface variation is mostly confined to on/off switching. Yet, limited *opa* gene variation can occur via the homologous route and arises predominantly, though not exclusively, through *inter*-cellular recombination (Bhat *et al.*, 1992). Consequently, homogenization of *opa* gene sequences seems to occur via the *in vivo* horizontal spread of *opa* alleles throughout a population via the DNA transformation route (Hobbs *et al.*, 1994, 1998).

Genetic requirements for homologous recombination

A major obstacle for the study of PilE variation is the intractability of neisserial genetics. Consequently, experiments need to be continually tempered to the unique characteristics of the neisserial systems. In other microorganisms (e.g. *E. coli*, *Salmonella typhimurium* and *Saccharomyces cerevisiae*), insight into the mechanics of genetic recombination has relied heavily on the use of defined *rec* mutants (reviewed in Kowalczykowski *et al.*, 1994). In *E. coli*, numerous *rec* gene products have been identified, leading to the identification of three distinct recombination pathways (the RecBCD, RecF and RecE pathways), with participation in any particular pathway apparently dependent upon the genetic composition of the cell (Kowalczykowski *et al.*, 1994; Myers and Stahl, 1994).

In *N. gonorrhoeae*, initial efforts focused on demonstrating a requirement for RecA protein in *pilE* gene rearrangements and in DNA transformation (Koomey *et al.*, 1987; Koomey and Falkow, 1987). Accordingly, in the absence of RecA protein, gross *pilE* gene rearrangements are not seen, and, likewise, DNA transformation is abrogated. However, RecA involvement in pilin variation occurs at the synaptic stage of recombination. Therefore, no insight was gained into what initiates a *pilE/pilS* recombination event, nor any subsequent molecular event.

With the availability of the various neisseria genome sequences (Parkhill et al., 2000; Tettelin et al., 2000; Oklahoma genome project), other rec genes could now be assessed for their effects on PilE variation. In E. coli, the RecBCD pathway is the primary recombination pathway and involves the RecBCD holoenzyme (a ubiquitous ATP-dependent exonuclease found in all Gram-negative bacteria) (reviewed in Kowalczkowski, 1994; Myers and Stahl, 1994), which works in concert with RecA protein. E. coli recB or recC mutants are severely impaired for genetic recombination in the absence of suppressor mutations. Surprisingly, neither N. gonorrhoeae recB nor N. gonorrhoeae recC mutants showed any defect in PilE antigenic variation, with only a minor defect being observed in their DNA repair capacity (Mehr and Seifert, 1998). Consequently, the only pathway that seemed to be compromised by a recB or recC mutation in Neisseria was DNA transformation (Mehr and Seifert, 1998). Attention then switched to the analysis of recD mutations, which proved to be more controversial. E. coli recD mutants are hyper-recombinogenic (Myers and Stahl, 1994). A recD mutant was constructed on an N. gonorrhoeae MS11 background, and displayed elevated recombination rates, at which non-parental pilin variants were spawned independently of DNA transformation (Chaussee et al., 1999). DNA sequencing of the recombinant pilE alleles, as well as immunoblotting for the PilE polypeptide, established that pilE recombination events in the recD mutants were typical and resulted from pilE/pilS exchanges. Interestingly, the majority of recombination events were pilus+ to pilus++ antigenic switches, with pilus– to pilus+ switches being rarely seen (S.A.H., unpublished observations). Consequently, those observations suggested that pilE gene variation in strain MS11 somehow involved the RecBCD holoenzyme. An analogous study using a different gonococcal strain (FA1090) challenged this viewpoint, however (Mehr and Seifert, 1998). In this study, using a less well-characterized gonococcal recD mutant, inactivation of the RecD subunit slightly depressed pilin variation in this particular strain background, providing support for the previous contention that the RecBCD pathway played no role in PilE antigenic variation (Mehr and Seifert, 1998).

Analysis of other FA1090 rec mutations supported the contention that PilE variation was mediated via a 'RecF-like' pathway. The RecF pathway, which in E. coli is activated in a recBC sbcBC genetic background, involves many different genes (e.g. recQ, RecO, recJ, recR, recF, recN, ruvA, ruvB, ruvC and recG) (Kowalczkowski et al., 1994). The Neisseria lack a recF gene as well as the sbcB suppressor allele that encodes Exonuclease I. Yet, despite these deficiencies, recO and recQ mutations showed approximately a 10-fold reduction in the generation of PilE variants (Mehr and Seifert, 1998). Inactivation of other 'RecF-like' genes led to less robust effects on PilE variation: inactivation of RecJ caused a three-fold depression in switching rates, whereas FA1090 recN mutants showed wild-type switching frequencies (Skaar et al., 2002). In strain MS11, qualitatively similar observations were made in pilin switching rates when MS11 ΔrecJ mutants were assessed microscopically (Hill, 2000; S.A.H., unpublished observations). A more prominent role for RecJ protein was revealed when switching was mediated through *pilE/pilS* templated deletion events (Hill and Grant, 2002). Here inactivation of the MS11 *recJ* gene suppressed *pilE* deletion events, which was in direct contrast to MS11 *recD* mutants, where the *pilE* locus seemed to become hyper-deletionogenic (Hill and Grant, 2002).

Data from the above synopsis of the genetic requirements for pilin switching in *Neisseria* seem to indicate that the various genetic requirements may well be strain-specific as well as *pilE* allele-specific. It is therefore possible that no specific 'pathway' actually exists for PilE variation. This is in keeping with the data from studies in *E. coli* (Sawitze and Stahl, 1994), where little specificity seems to be exhibited by any Rec protein regarding its participation in any particular pathway. Indeed it may be the availability of recombination functions, and the occurrence of specific DNA substrates within the neisserial cell, that actually dictates the route by which a *pilE* recombination event may proceed.

MOLECULAR MODELS FOR PILE VARIATION

The vast majority of data on *pilE* gene variation fits within a gene conversion concept (Haas and Meyer, 1986; Swanson and Koomey, 1989). Information flow is overwhelmingly unidirectional (i.e. always from *pilS* to *pilE*; in a single instance, a reciprocal exchange was documented between *pilE* and *pilS6*) (Gibbs *et al.*, 1989), with the 'donor' *pilS* gene remaining genetically unchanged following conversion of the 'acceptor' *pilE* gene. However, just how this is achieved currently remains a mystery. Over the years, two *intra*-cellular models have been proposed and are briefly summarized below.

The 'mini-cassette exchange' model

The 'mini-cassette exchange' model (Figure 8.1) was proposed to account for many of the early observations (Hagblom *et al.*, 1985; Haas and Meyer; 1986) and gained considerable favour because of its simplicity. Subsequently, the model found considerable support from the overwhelming majority of studies on pilin variation that were performed in other laboratories (e.g. by Swanson and colleagues; Swanson *et al.*, 1985, 1986, 1987).

The basis of this model is the unique structure of the various *pil* alleles (see Figure 8.1), with the salient features being the following:

1. within a *pilE* gene there are six variable gene segments (mini-cassettes, mc) of *pil* sequence that are located within the 3' 2/3's of the gene, with each variable segment being flanked by short conserved regions (Meyer *et al.*, 1984; Hagblom *et al.*, 1985; Haas and Meyer, 1986; Haas *et al.*, 1992);
2. the 150 bp at the 5' end of *pilE* is conserved between all *pilE* genes and is not found at any *pilS* locus (Hagblom *et al.*, 1985; Haas *et al.*, 1987);

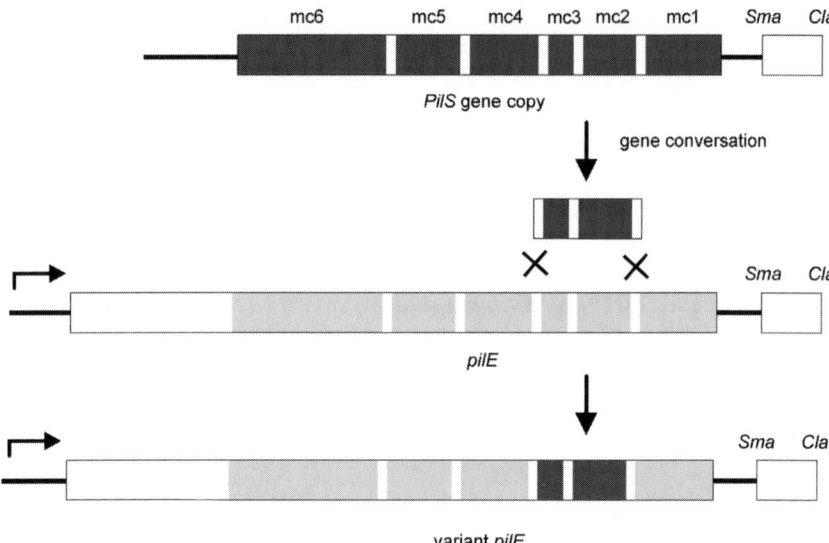

Figure 8.1 'Mini-cassette exchange' model for pilin variation, adapted from Haas and Meyer (1986). Pilin variation occurs through the exchange of 'mini-cassettes' (mc) of variable *pil* sequence from a *pilS* gene copy to *pilE*. In the example provided, a variant *pilE* gene is created through the exchange of mini-cassettes mc3 and mc2 at the *pilE* locus. During conversion, the resident *pilE* gene sequences are lost, whereas the *pilS* gene copy remains unchanged (note: only a single *pilS* gene copy is included in the figure). Unshaded boxes represent the conserved gene segments found in all *pil* genes; the unshaded segment towards the 5' end of the *pilE* genes represents the conserved sequence found in all *pilE* genes. The shaded boxes represent the variable gene segments. The relative position of the *Sma/Cla* repeat is also shown. Pilin switching is abrogated in Ngo mutants where the *Sma/Cla* repeat is deleted from the *pilE* locus (Wainwright *et al.*, 1994).

3 most of the *pil* gene copies that are located within the various *pilS* loci also contain the six variable segments of *pil* sequence as well as being flanked by the short conserved regions (Haas and Meyer, 1986; Haas *et al.*, 1992);

4 immediately downstream of *pilE* and downstream of each *pilS* locus resides a short repeat called the *Sma/Cla* repeat (Meyer *et al.*, 1984; Haas and Meyer, 1986; Haas *et al.*, 1992).

Therefore, according to this model, PilE variation occurs through the exchange of the 'mini-cassettes'-worth of variable sequence between *pilS* and *pilE* with 'crossing-over' predicted to occur within the conserved *pil* segments that flank the variable gene segments (see Figure 8.1). Whether *pilE/pilS* exchanges are confined to events that occur within the variable portions of the *pilE* coding sequence has recently been questioned following the observation that when genetic markers were placed within the *Sma/Cla* repeat located downstream of *pilE* they were found to co-convert back to wild-type at high frequency during a PilE switching event (Howell-Adams and Seifert, 1999; Hill and Grant, 2002). What these observations suggest is that if homology is present between *pilE* and a *pilS* gene copy then incorporation of a recombination tract will proceed until sequence non-identity is encountered to terminate the event. In most cases, this will occur within the *pilE* open reading frame because *pilE/pilS* recombination is a classic example of homologous recombination (i.e. recombination between homeologous gene sequences that contain considerable stretches of sequence non-identity). Consequently, recombination between the non-identical *pil* alleles will provide frequent opportunities for sequence non-identities to be encountered, which will preferentially terminate most recombination tracts within the *pilE* open reading frame.

The 'looping-out' model

A different gene conversion model for PilE variation was proposed by Seifert and collegues (see Figure 8.2) (Howell-Adams *et al.*, 1996; Howell-Adams and Seifert, 1999, 2000). The molecular basis of this model came from studies that initially identified a RecA-dependent fusion of *pilE* with a modified *pilS* locus (the *pilS* modification was the insertion of a large promoter-less antibiotic gene cassette) (Howell-Adams *et al.*, 1996). Subsequent experiments then led to the proposal of a 'looping-out' model for PilE variation (Figure 8.2) (Howell-Adams and Seifert, 1999, 2000). Briefly, *pilE* recombines with a *pilS* locus, producing a *pilE/pilS* hybrid fusion product, whose formation causes extensive tracts of *pilE* flanking DNA to be duplicated. As a consequence of the large chromosomal duplication within the vicinity of *pilE*, the repeated DNA tracts are then able to pair, which then allows for the excision of a *pilE/pilS* fusion product in a closed-circular form. The excised circle carrying the *pilE/pilS* fusion then recombines with another *pilE* gene located within the same cell, which then results in the formation of novel *pilE* recombinants. Unfortunately, there has been no actual demonstration of novel PilE variants following recombination of a hybrid circle carrying a *pilE::pilS* fusion with a resident *pilE* gene to date. Closed-circular molecules do form readily as a consequence of normal chromosome dynamics in gonococci (Howell-Adams and Seifert, 2000; Barten and Meyer, 2001). However, closed-circular molecules containing *pilE/pilS* fusions have yet to be demonstrated in a wild-type background and their demonstration would seem to be a crucial feature of the 'looping-out' model. None the less, circles carrying *pilS* genes have been identified (R. Barten and T.F.M., unpublished data) as well as circles carrying *opa* genes, so it seems that intracellular circle formation may be an important mechanism by which the *Neisseria* vary their chromosome composition (Barten and Meyer, 2001), and this relatively uncomplicated version of the circle formation model may indeed constitute the basis of the directional *pilE* recombination. Complicating matters though is the fact that different PilE antigenic variants seem to arise through multiple *pilE/pilS* recombination events (Swanson *et al.*, 1987). This observation demands multiple *pilE/pilS* fusion events coupled with

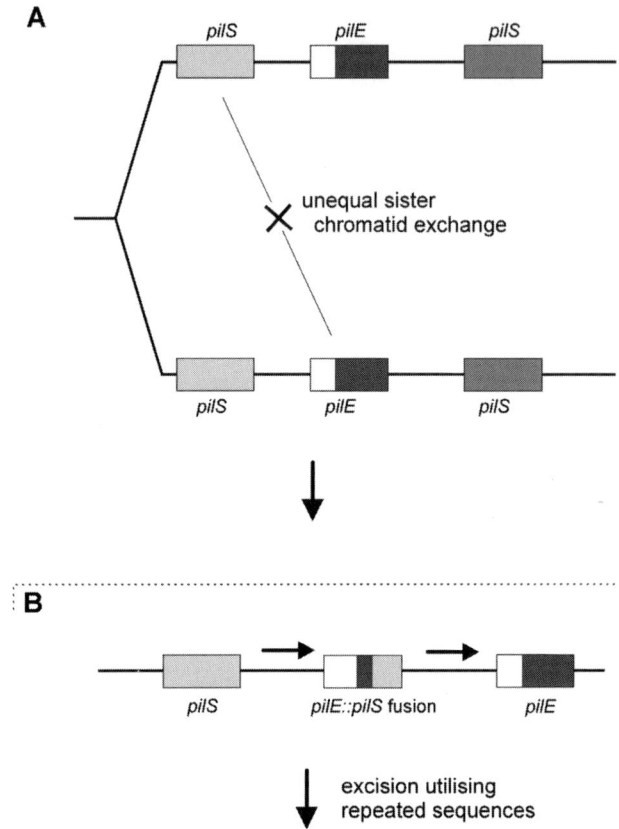

Figure 8.2 'Looping-out' model for pilin variation (adapted from Howell-Adams and Seifert, 1999). (A) Following DNA replication, *pilE* engages in unequal sister chromatid exchange with a *pilS* locus. (B) Unequal sister chromatid exchange results in a *pilE::pilS* fusion such that the chromosomal sequence immediately upstream of the *pilE* gene is duplicated (as indicated by the horizontal arrows).

multiple excision events in the formation of each PilE variant. Therefore, perhaps a specific system exists that promotes the formation and excision of the *pilE/pilS* fusion product.

GENETIC VARIATION VIA THE ILLEGITIMATE ROUTE

By definition, illegitimate recombination occurs without the assistance of RecA protein, and, for the *Neisseria*, appears to be a major route by which the antigenic content of the cell surface changes. Many neisserial genes phase vary

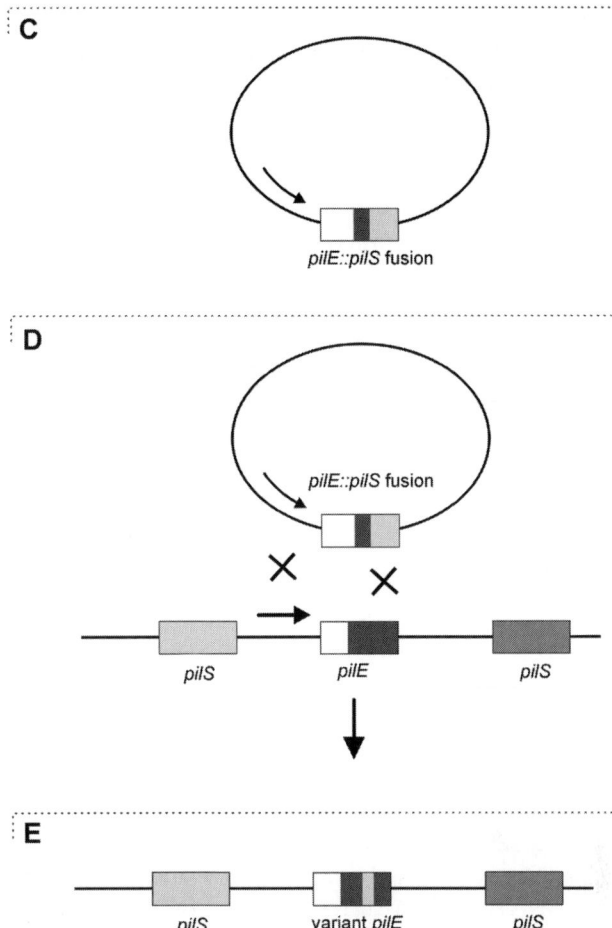

Figure 8.2 (*Continued*) (C) Owing to the duplication of sequence, the duplicated sequences are now able to pair, allowing the *pilE::pilS* fusion to excise from the chromosome as a closed-circular molecule. (D) The *pilE::pilS* fusion contained on the closed-circular molecule then recombines with a conventional *pilE* gene located within the same cell. (E) Following recombination, a variant *pilE* gene is produced.

(i.e. switch on and off) through ill-defined frameshifting mechanisms, and involve either the gain or the loss of bases within homopolymeric nucleotide tracts (Koomey *et al.*, 1987; Rudel *et al.*, 1992, 1995a; Sarkari *et al.*, 1994; Jennings *et al.*, 1995; van der Ende *et al.*, 1995;; Hammerschmidt *et al.*, 1996; Yang and Gotschlich, 1996), or alternatively, the gain or loss of short oligonucleotide repeats (Stern *et al.*, 1986). Yet, despite the high prevalence of this type of antigenic switching event in the *Neisseria*, very few genetic requirements have been identified.

Classical studies on frameshifting in *E. coli* have indicated a prominent role for DNA polymerase in the generation of frame shifts within regions of repeated base sequences (Streisinger *et al.*, 1966; Streisinger and Owen, 1985). Frameshifting can also occur through sister chromatid interactions, which may be of importance in the *Neisseria* when one considers *opa* repeat expansion (Levinson and Gutman, 1987; Bzymek *et al.*, 1999). Recently, the genetic requirement for frameshifting within repeats was studied in *Haemophilus influenzae* (Bayliss *et al.*, 2002), where *polI* mutations were found to destabilise tetrameric repeat stretches whereas a *mutS* mutation destabilized dinucleotide repeat tracts.

The genetic requirements for phase variation in the *Neisseria* are less clear, with the *opa* system being perhaps the best understood. Phase variation of the *opa* genes occurs through slipped-strand mispairing of –CTCTT– pentameric repeats, which takes an individual *opa* gene in and out of frame (Stern *et al.*, 1986; Belland *et al.*, 1989; Murphy *et al.*, 1989). The number of –CTCTT– repeats varies greatly between the various *opa* genes found both within a cell as well as between strains. Switching frequencies tend to reflect the number of repeats within an *opa* gene: the higher the number of repeats the higher the switching frequency (Swanson, 1994). However, a threshhold that seems to depend upon the individual *opa* gene can be reached (Swanson, 1994). When *opa::phoA* fusions were examined in *E. coli*, *opa* genes with five repeat units were found to be stable and were unable to phase-vary (Belland *et al.*, 1989). This also is apparently the case when a five repeat *opa* gene resides within the neisserial chromosomal setting (J. Swanson, personal communication). Thus, these latter observations suggest that some special structural feature may be involved in promoting a switching event. Along these lines, the presence of the –CTCTT– repeats within the chromosome creates an atypical nucleotide strand bias, in that, one DNA strand consists of a large homopyrimidine tract, whereas the complementary strand is a homopurine tract. Physically, such DNA strand bias can result in a non-B-DNA configuration known as H-DNA (Belland, 1991). H-DNA is a triple-stranded structure with a nuclease-sensitive single strand being left unpaired. Whether this unique DNA structure actually contributes to frameshifting is unclear.

Studies on the contribution of mismatch repair to phase variation within the *Neisseria* are currently in their infancy. Inactivation of either *mutS* or *mutL* led to between 250- and 500-fold increases in the phase variation of haemoglobin receptors in *N. meningitidis* (Richardson and Stojiljkovic, 2001), with this increased frequency being accompanied by an overall elevation in the spontaneous mutation rate. Similar observations were made in the gonococcus, where inactivation of *mutS* also causes a similar increase in switching frequency of the haemoglobin receptor, yet has no effect on *opa* phase variation (J. Cannon, personal communication). Based on these limited studies, it seems that neisserial mismatch repair systems may influence frameshifting within reiterated base tracts, with perhaps a different genetic system modulating frameshifting of repeated elements.

FUNCTIONAL SIGNIFICANCE OF GENETIC VARIATION

The neisserial pili

Pili, which are fine, hair-like organelles that protrude from the bacterial cell surface, probably represent the most variable structure produced by the pathogenic *Neisseria* spp. Their presence on the cell surface is an absolute requirement for the initiation of an infection (Kellogg *et al.*, 1968) in that they confer the attachment of the bacteria to epithelial cells (McGee *et al.*, 1981; Stephens and McGee, 1981). Pilus binding appears to be specific for human cells and therefore represents a major determinant of the neisserial species tropism. Owing to their exposed location, pili are also strong targets of an antibody response that could interfere with receptor recognition. Efforts in generating a pilus-based vaccine have thus far failed due to the enormous variability of the pilus structure, most notably in the chemical composition of PilE polypeptide (Johnson *et al.*, 1991). It therefore remains a crucial question as to how pili deceive the immune system yet still fulfil their function as an adhesin. This problem certainly bears important practical and theoretical implications that are not necessarily restricted to the *Neisseria* model.

Recently, considerable insight has been gained into the nature of the pilus components that are actually involved in receptor recognition. Early studies suggested a direct role of PilE (the major polypeptide subunit in the pilus organelle) in promoting adherence to various cellular substrates (Schoolnik *et al.*, 1984; Virji and Heckels, 1984; Rothbard *et al.*, 1985). While this possibility still remains, recent studies demonstrated that the different adherence properties of the pili need to be carefully differentiated because there are at least three distinct *Neisseria* adherence specificities: one for epithelial cells, one for erythrocytes and one allowing inter-gonococcal adherence. Each of these properties can be distinguished genetically or otherwise (Rudel *et al.*, 1992). Interestingly, adherence of *Neisseria* to epithelial cells, and interbacterial aggregation, have been shown to be influenced by PilE variation whereas pilus-dependent haemagglutination appears to be unaffected (Lambden *et al.*, 1980; Rudel *et al.*, 1992; Nassif *et al.*, 1993). PilE polypeptide can also be modified through phosphorylation and/or glycosylation (Robertson *et al.*, 1977; Schoolnik *et al.*, 1984; Virji *et al.*, 1993a), and it has recently been demonstrated that glycosylation of specific variant pilins can also influence receptor recognition for gonococci (Virji *et al.*, 1993a). However, the observed effect of PilE variation on adherence may be indirect and does not preclude a potential role of the minor pilus subunits.

Of the minor protein components located within the gonococcal pilus (Muir *et al.*, 1988; Parge *et al.*, 1990), the PilC proteins have been strongly implicated in pilus biogenesis (Jonsson *et al.*, 1991) and as such have been afforded a role in transformation competence (Rudel *et al.*, 1995a). None the less, pili can be assembled in the absence of either of the two known PilC1 and PilC2 proteins (Rudel *et al.*, 1992, 1995a,b). Such modified pili are still able to haemagglutinate but they lack the potential to adhere to epithelial cells (Rudel *et al.*, 1992), indicating a role for PilC in epithelial cell adherence. Rudel *et al.* (1995b) have succeeded in purifying gonococcal PilC protein to homogeneity from a *pilE* deletion mutant. Using antisera raised against the purified PilC protein, it was possible to firmly establish that the PilC proteins actually represent the epithelial cell-specific pilus-associated adhesins. Importantly, purified PilC protein could effectively compete with the binding of both gonococci and meningococci (which expressed different PilC and PilE polypeptides) to epithelial cells. Furthermore, an additional polypeptide, PilV, has recently been implicated in adherence by promoting the functional display of PilC protein within the context of the pilus fibre (Winther-Larsen *et al.*, 2001). In conclusion, these studies led us to the intriguing conclusion that the pili of the pathogenic *Neisseria* species, irrespective of their structural variability, recognize identical, or a closely related group of, receptor(s).

Cell surface structural variability allows us to also distinguish between two principal functions of genetic variation, i.e. as an escape mechanism as well as as an adaptive function, examples of which seem to be provided by the *pil* and the *opa* systems. A typical escape mechanism is antigenic variation. In the context of the *Neisseria* infection, this term describes the competition between two variable systems (i.e. the host immune system and the bacteria themselves) that are operating with similar mechanisms, and which, when viewed from the bacterial perspective, has the purpose of avoiding molecular interaction with antibody. The opposite applies for the adaptive variation model. Here structural change is tailored to enhance productive molecular interactions, in contrast to stochastic mechanisms that are likely to cause unnecessary bacterial extinction due to generating excessive molecular variation.

The extreme variability of PilE (the major pilus polypeptide subunit) classifies it as an escape factor. The mechanism used for PilE variation, the more or less random intragenic recombination, lowers the chance that novel recombinants with a specific molecular configuration will be produced. However, functional integrity must be maintained on at least two levels: PilE polymerization and pilus interaction with a conserved host receptor. The polymerization function is less problematic because it involves the conserved hydrophobic regions of the PilE polypeptide, which are neither surface exposed nor immuno-susceptible. The dilemma, however, is how to accommodate conserved receptor-binding function in the highly variable context of the pili? To solve this problem, the organism utilizes the less variable pilus-associated adhesin, PilC, in conjunction with PilV. In conclusion, PilE variation primarily serves to protect pili from interacting with immunoglobulins but is probably inadequate to modulate receptor specificity.

Cell tropisms conferred by the Opa proteins

Opa proteins are major outer membrane constituents of primarily the pathogenic *Neisseria* spp. Computer predictions suggest Opa proteins fold into a β-pleated sheet structure (which is typical of many outer membrane proteins), whereby three variable loop regions and a fourth conserved loop are oriented outwards from the bacterial cell surface (Meyer et al., 1986; Barritt et al., 1987; Bhat et al., 1992). Gonococci possess the most variant *opa* genes (~11), which is considerably higher than the number found in meningococci (3–4) or *Neisseria lactamica* (~2). Interestingly, independent *N. gonorrhoeae* isolates rarely possess *opa* genes of the same sequence (Haas et al., 1992), indicating that the repertoire of variant *opa* genes within the gonococcal population is substantially larger than that of a single strain.

The role of Opa proteins in the various adherence functions, such as interbacterial adhesion and interaction with human epithelial and phagocytic cells, has long been recognized (King and Swanson, 1978; Lambden et al., 1979; Virji and Heckels, 1986; Fischer and Rest, 1988; Makino et al., 1991; Belland et al., 1992; Virji et al., 1993b). Recent work suggests that Opa proteins not only cause bacterial adherence but also trigger important cellular functions. Invasion of *N. gonorrhoeae* MS11 into human epithelial cells was shown to be dependent on the expression of one distinct Opa protein (Makino et al., 1991; Weel et al., 1991). Using a reverse genetics approach, the remainder of the Opa proteins from this strain were subsequently shown to confer binding to human polymorphonuclear cells (PMNs) but not to epithelial cells (Kupsch et al., 1993), with PMN chemoluminescense only being induced if certain Opa proteins were expressed (Belland et al., 1992). However, when human peripheral blood monocytes (PMBs) were assessed, uptake and chemoluminescence was triggered by the expression of a different set of Opa proteins (including the one that facilitates uptake into epithelial cells). Interestingly, for all the Opa proteins of strain MS11, at least one binding specificity has been observed, indicating that each variant gene encodes a functional protein. Furthermore, the specific binding properties of Opa proteins are maintained if the genes are cloned and expressed in a different neisserial host (Kupsch et al., 1993). In the much smaller meningococcal Opa protein repertoire, Opa proteins have been identified that are required for epithelial cell invasion (Virji et al., 1993b) and PMN stimulation. Notably, this species also often carries a copy of the phase-variable *opc* gene, whose product, although structurally unrelated, has a function similar to the epithelial cell-specific Opa proteins (Virji et al., 1992). Thus, it is evident that the variable Opa and Opc adhesins represent important cell tropism determinants of *N. gonorrhoeae* and *N. meningitidis* and that the variability of these proteins allows for multiple cellular interactions. More recent work provided a wealth of information on the molecular interactions of Opa proteins with human cells, indicating the binding of variant Opa proteins to two principal receptor classes, i.e. the heparan-sulphate proteoglycans and the carcinoembryonic cell adhesion molecules (reviewed in Dehio et al., 1998).

In order to conform to the adaptive variation model outlined above, the underlying genetic mechanism giving rise to the different *opa* alleles needs to avoid unnecessary mutation and recombination (which could lead to non-productive phenotypes), and, instead, maintain a selected set of pre-existing functional genes. Indeed, this is true for the *opa* gene system, which exhibits limited phenotypic variability. Recombination, bearing the risk of generating non-productive hybrids, is a rare event among the *opa* genes (Stern *et al.*, 1986; Connell *et al.*, 1988). Furthermore, all native *opa* genes encode functional Opa proteins capable of recognizing distinct cellular receptors (Kupsch *et al.*, 1993), which is in direct contrast to *in vitro*-engineered hybrid *opa* genes, which often encode non-functional Opa proteins (T.F.M., unpublished information). It is therefore evident that the mechanism underlying Opa variation favours productive interaction with target cells and sacrifices immune evasion by minimizing unnecessary variation.

If this reasoning applies, then why is it so essential for the pili rather than for the Opa proteins to vary antigenically? Furthermore, why is it that many other surface proteins of *Neisseria*, such as the major outer membrane protein P.I (Judd, 1989), are able to withstand host immunity without undergoing any intra-strain variation at all? These questions lead us to consider yet another neisserial variation system, the variable lipopolysacharide (LPS) system, because it seems that neisserial membrane-associated proteins, in contrast to the highly exposed pili, are efficiently protected against host immunity through a variable carbohydrate mantle produced by the bacteria.

Optional resistance to the extracellular environment: LPS variation and capsule

The *Neisseriae* produce a short type of lipopolysacharide (LPS or LOS) which lacks any repetitive O-side chains. None the less, LPS preparations from *in vitro*-cultured bacteria reveal multiple size classes indicating a structural heterogeneity of the neisserial LPS (Schneider *et al.*, 1988). Antibodies can be raised against distinct LPS types, and these are useful to demonstrate sectored colonies on plates (Mandrell and Apicella, 1993). Several lines of evidence suggest that LPS variation occurs *in vivo*. In meningococcal carriers, the majority (70%) of bacteria isolated from the nasopharynx are unencapsulated and preferentially express a short LPS species (Broome, 1986), whereas in the diseased state, 97% of the blood and CSF isolates are encapsulated and express a long LPS species (Jones *et al.*, 1992). Likewise, experimental gonococcal infection in human volunteers indicates that bacteria isolated early in the infection have a short LPS, whereas after the development of an inflammatory response, a different phenotype emerges, where a long LPS species predominates (Schneider *et al.*, 1991).

A major difference between the variant LPS molecules is the presence of additional terminal galactose residues in the longer LPS forms. Interestingly, these terminal galactose residues can be externally modified by a membrane-associated bacterial sialyltransferase in the presence of soluble CMP-NANA. While the gonococcus depends on scavenging CMP-NANA from the host bodily fluids, the meningococcus is able to synthesize its own CMP-NANA (Smith, 1991; Mandrell and Apicella, 1993; van Putten, 1993). Consequently, given the presence of CMP-NANA, LPS may or may not be sialylated, leading to LPS variation. The functional relevance of the LPS phase transitions has recently been elucidated (van Putten, 1993), and LPS function appears to be dependent on the amount of sialic acid that is incorporated in the different LPS forms. A low sialylation phenotype, such as found early in an infection (Schneider et al., 1991), enables entry of the bacteria into mucosal cells, but makes them more susceptible to bactericidal activity. In contrast, highly sialylated bacteria are incapable of entering epithelial cells but are resistant to phagocytosis and killing by antibodies and complement, thus allowing an infection to persist. Consequently, depending upon the degree of sialylation, bacteria are either adapted to an extracellular environment capable of resisting humoral immune mechanisms or, when they are more sensitive to bacteriocidal activities, better adapted to enter an intracellular milieu via Opa-mediated cellular interactions.

The function of highly sialylated LPS has many similarities with that of polysialic capsules produced by *N. meningitidis* and many of the commensal *Neisseria* spp. in that the capsule protects extracellular bacteria against both specific and non-specific host responses (Frosch et al., 1989). Although phase variation of capsule expression has been observed, this structure appears to be primarily regulated by environmental factors (Brener et al., 1981). None the less, loss of capsule expression favours Opa- and Opc-mediated interactions of *N. meningitidis* with target cells and thus seems to be a prerequisite for the cellular invasion (Virji et al., 1993a). Thus, unencapsulated meningococci are phenotypically similar to non-sialylated gonococci, and the recent advances made in the analysis of capsular (*cps*) gene organization and function (Frosch et al., 1989) are likely to be important for our understanding of the conditions favouring invasive meningococcal infections.

CONCLUSION

Antigenic variation in *Neisseria* is extreme, with several surface antigens changing either by on/off switching (e.g. *opa* gene expression and the turning on and off of certain genes that are involved in lipooligosaccharide (LOS) biosynthesis), or, alternatively, by *bona fide* genetic rearrangement (e.g. the diversification of *pil* gene sequences). Mechanistically these events fall into two broad categories, i.e. illegitimate events (RecA-independent), which account primarily for on/off switching, and homologous recombination schemes (RecA-dependent), which involve the physical exchange of nucleotide sequences between the various genes. Furthermore, because of their natural competence, horizontal movement of DNA can also contribute to genetic variation.

At the molecular level, many fundamental mechanistic questions about how *Neisseria* vary the genetic content of their chromosomes remain unanswered.

Several questions waiting to be addressed include: (1) What are the factors that dictate whether *Neisseria* uses *inter-* or *intra-*cellular pathways to shuffle its genetic content? (2) Is pilin variation random or programmed? (3) How is recombination initiated between *pilE* and *pilS*? (4) How is the neisserial chromosome stably maintained when multiple opportunities exist for gross rearrangement (deletions or inversions) between the various *pil* and *opa* alleles? (5) What is the fate of flanking markers, and do they exchange or become converted as a consequence of recombination events at *pilE*? (6) Why are classical *intra-*cellular gene-conversion-like events apparently suppressed between the multiple *opa* genes in the chromosome, in contrast to the more promiscuous pilin system? (7) What are the genetic systems that promote phase variation through illegitimate recombination?

In contrast to the mechanistic questions, the functional consequences of antigenic variation in the *Neisseria* are becoming much clearer. As discussed in this chapter, antigenic variation can provide either an escape or an adaptive function as exemplified by the analysis of the *pil* and the *opa* systems. As more advances are made in connection to cellular interactions then some of the structural nuances that determine cellular uptake are likely to be revealed. In this regard, definitive host cell receptor interactions need to be defined, as well as convincing molecular scenarios as to what prompts the gonococcus to turn off pilus production during epithelial cell uptake, and likewise, what prompts the meningococcus to turn off capsule production in conjunction with pilus production during its invasive strategy.

Finally, the *Neisseria* are clearly capable of residing in diverse molecular environments, ranging from the inhospitable environs of an inflamed mucosa, to the hardly more congenial chemical surroundings of life within a vacuole inside an epithelial or phagocytic cell. The fact that humans are the sole host for *Neisseria* spp. places immense evolutionary pressure on the various members of the genus. Consequently, *Neisseria* spp. must adapt rapidly to their changing environments as well as adapt to continual molecular onslaught in the form of the host immune system. Although much is known about how the *Neisseria* spp. confuse the immune system, less is known about how they microadapt to their changing environments. The presence of so many phase-variable frameshifting systems in the *Neisseria* genomes perhaps provides a clue as to how microadaptation is achieved; stochastic frameshifting will allow multiple phenotypes to be present within a population without necessarily increasing the mutation load. Consequently, multiple phenotypes can be presented yet the strict adaptive functions will still be maintained within the gene pool. This strategy allows micropopulations that facilitate immediate as well as evolutionary survival to emerge.

REFERENCES

Achtman, M. (1994) *Mol. Microbiol.* 11, 15–22.
Barlow, A. K., Heckels, J. E. and Clarke, I. N. (1989) *Mol. Microbiol.* 3, 131–139.

Barritt, D. S., Schwalbe, R. S., Klapper, D. G. and Cannon, J. G. (1987) *Infect. Immun.* 55, 2026–2031.
Barten, R. and Meyer T. F. (2001) *Mol. Gen. Genet.* 264, 691–701.
Bayliss, C. D., van de Ven, T. and Moxon, E. R. (2002) *EMBO J.* 21, 1465–1476.
Belland, R. J. (1991) *Mol. Microbiol.* 5, 2351–2360.
Belland, R. J., Chen, T., Swanson, J. and Fischer, S. H. (1992) *Mol. Microbiol.* 6, 1729–1737.
Belland, R. J., Morrison, S. G., van der Ley, P. and Swanson, J. (1989) *Mol. Microbiol.* 3, 777–786.
Bhat, K. S., Gibbs, C. P., Barrera, O., Morrison, S. G., Jahnig, F., Stern, A., Kupsch, E. M., Meyer, T. F. and Swanson, J. (1992) *Mol. Microbiol.* 5, 1889–1901.
Biswas, G. D., Burnstein, K. L. and Sparling, P. F. (1986) *J. Bacteriol.* 168, 756–761.
Biswas, G. D., Lacks, S. A. and Sparling, P. F. (1989) *J. Bacteriol.* 171, 657–664.
Boyle-Vavra, S. and Seifert, H. S. (1996) *Microbiology* 142, 2839–2845.
Brener, D., DeVoe, I. W. and Holbein, B. E. (1981) *Infect. Immun.* 33, 59–66.
Broome, C. V. (1986) *J. Antimicrob. Chemother.* 18 Suppl A, 25–34.
Bzymek, M., Saveson, C. J., Feschenko, V. V. and Lovett, S. T. (1999) *J. Bacteriol.* 181, 477–482.
Catlin, B. W. and Cunningham, L. S. (1961) *J. Gen. Microbiol.* 26, 303–312.
Chaussee, M. S., Wilson, J. and Hill, S. A. (1999) *Microbiology* 145, 389–400.
Connell, T. D., Black, W. J., Kawula, T. H., Barritt, D. S., Dempsey, J. A., Kverneland, K., Jr, Stephenson, A., Schepart, B. S., Murphy, G. L. and Cannon, J. G. (1988) *Mol. Microbiol.* 2, 227–236.
Dehio, C., Gray-Owen, S. D. and Meyer, T. F. (1998) *Trends Microbiol.* 6, 489–495.
Dorward, D. W. and Garon, C. F. (1989) *J. Bacteriol.* 171, 4196–4201.
Dorward, D. W., Garon, C. F. and Judd, R. C. (1989) *J. Bacteriol.* 171, 2499–2505.
Dubnau, D. (1997) *Gene* 192, 191–198.
Facius, D. and Meyer, T. F. (1993) *Mol. Microbiol.* 10, 699–712.
Feavers, I. M., Heath, A. B., Bygraves, J. A. and Maiden M. C. J. (1992) *Mol. Microbiol.* 6 (4), 489–495.
Fischer, S. H. and Rest, R. F. (1988) *Infect. Immun.* 56, 1574–1579.
Frosch, M. and Meyer, T. F. (1992) *FEMS Microbiol. Lett.* 79, 345–349.
Frosch, M., Weisgerber, C. and Meyer, T. F. (1989) *Proc. Natl Acad. Sci. USA* 86, 1669–1673.
Gibbs, C. P., Reimann, B.-Y., Schulz, E., Kaufmann, A., Haas, R. and Meyer, T. F. (1989) *Nature* 338, 651–652.
Goodman, S. D. and Scocca, J. J. (1988) *Proc. Natl Acad. Sci. USA* 85, 6982–6986.
Gunn, J. S. and Stein, D. C. (1993) *Gene* 132, 15–20.
Gunn, J. S., Piekarowicz, A., Chien, R. and Stein, D. C. (1992) *J. Bacteriol.* 174, 5654–5660.
Haas, R. and Meyer, T. F. (1986) *Cell* 44, 107–115.
Haas, R., Schwarz, H. and Meyer, T. F. (1987) *Proc. Natl Acad. Sci. USA* 84, 9079–9083.
Haas, R., Veit, S. and Meyer, T. F. (1992) *Mol. Microbiol.* 6, 197–208.
Hagblom, P., Segal, E., Billyard, E. and So, M. (1985) *Nature* 315, 156–158.
Halter, R., Pohlner, J. and Meyer, T. F. (1989) *EMBO J.* 8 (9), 2737–2744.
Hammerschmidt, S., Muller, A., Sillmann, H., Muhlenhoff, M., Borrow, R., Fox, A., van Putten, J., Zollinger, W. D., Gerardy-Schahn, R. and Frosch, M. (1996) *Mol. Microbiol.* 20, 1211–1220.

Harris-Warrick, R. M. and Lederberg, J. (1978) *J. Bacteriol.* 133, 1237–1245.
Hebeler, B. H. and Young, F. E. (1975) *J. Bacteriol.* 122, 385–392.
Hill, S. A. (1996) *Mol. Microbiol.* 20, 507–518.
Hill, S. A. (2000) *Mol. Gen. Genet.* 264, 268–275.
Hill, S. A. and Grant, C. C. R. (2002) *Mol. Gen. Genom.* 266, 962–972.
Hill, S. A., Morrison, S. G. and Swanson, J. (1990) *Mol. Microbiol.* 4, 1341–1352.
Hobbs, M. and Mattick, J. S. (1993) *Mol. Microbiol.* 10, 233–243.
Hobbs, M. M., Malorny, B., Prasad, P., Morelli, G., Kusecek, B., Heckels, J. E., Cannon, J. G. and Achtman, M. (1998) *Microbiology* 144, 157–166.
Hobbs, M. M., Seiler, A., Achtman, M. and Cannon, J. G. (1994) *Mol. Microbiol.* 12, 171–180.
Howell-Adams, B. and Seifert, H. S. (1999) *J. Bacteriol.* 181, 6133–6141.
Howell-Adams, B. and Seifert, H. S. (2000) *Mol. Microbiol.* 37, 1146–1158.
Howell-Adams, B, Wainwright, L. A. and Seifert, H. S. (1996) *Mol. Microbiol.* 22, 509–522.
Jennings, M. P., Hood, D. W., Peak, I. R., Virji, M. and Moxon, E. R. (1995) *Mol. Microbiol.* 18, 729–740.
Johnson, S. C., Chung, R. C., Deal, C. D., Boslego, J. W., Sadoff, J. C., Wood, S. W., Brinton, C. C., Jr and Tramont, E. C. (1991) *J. Infect. Dis.* 163, 128–134.
Jones, D. M., Borrow, R., Fox, A. J., Gray, S., Cartwright, K. A. and Poolman, J. T. (1992) *Microb. Pathog.* 13, 219–224.
Jonsson, A. B., Nyberg, G. and Normark, S. (1991) *EMBO J.* 10, 477–488.
Judd, R. C. (1989) *Clin. Microbiol. Rev.* 2 Suppl, S41–S48
Kellogg, D. S., Jr, Cohen, I. R., Norins, L. C., Schroeter, A. L. and Reising, G. (1968) *J. Bacteriol.* 96, 596–605.
King, G. J. and Swanson, J. (1978) *Infect. Immun.* 21, 575–584.
Koomey, J. M. and Falkow, S. (1987) *J. Bacteriol.* 169, 790–795.
Koomey, M., Gotschlich, E. C., Robbins, K., Bergstrom, S. and Swanson, J. (1987) *Genetics* 117, 391–398.
Kowalczykowski, S. C., Dixon, D. A., Eggleston, A. K., Lauder, S. D. and Rehrauer, W. M. (1994) *Microbiol. Rev.* 58, 401–465.
Kupsch, E. M., Knepper, B., Kuroki, T., Heuer, I. and Meyer, T. F. (1993) *EMBO J.* 12, 641–650.
Lambden, P. R., Heckels, J. E., James, L. T. and Watt, P. J. (1979) *J. Gen. Microbiol.* 114, 305–312.
Lambden, P. R., Robertson, J. N. and Watt, P. J. (1980) *J. Bacteriol.* 141, 393–396.
Levinson, G. and Gutman, G. A. (1987) *Mol. Biol. Evol.* 4, 203–221.
Makino, S., van Putten, J. P. and Meyer, T. F. (1991) *EMBO J.* 10, 1307–1315.
Mandrell, R. E. and Apicella, M. A. (1993) *Immunobiology* 187, 382–402.
Manning, P. A., Kaufmann, A., Roll, U., Pohlner, J., Meyer, T. F. and Haas, R. (1991) *Mol. Microbiol.* 5, 917–926.
McGee, Z. A., Johnson, A. P. and Taylor-Robinson, D. (1981) *J. Infect. Dis.* 143, 413–422.
Mehr, I. J. and Seifert, H. S. (1998) *Mol. Microbiol.* 30, 697–710.
Meyer, T. F., Billyard, E., Haas, R., Storzbach, S. and So, M. (1984) *Proc. Natl Acad. Sci. USA* 81, 6110–6114.
Meyer, T. F., Haas, R., Stern, A., Fiedler, H., Frosch, M., Jahnig, F., Muralidharan, K. and Veit, S. (1986) *Ann. Sclavo. Collana. Monogr.* 3, 407–414.
Muir, L. L., Strugnell, R. A. and Davies, J. K. (1988) *Infect. Immun.* 56, 1743–1747.

Murphy, G. L., Connell, T. D., Barritt, D. S., Koomey, M. and Cannon, J. G. (1989) *Cell* 24, 539–547.
Myers, R. S. and Stahl, F. W. (1994) *Ann. Rev. Genet.* 28, 49–70.
Nassif, X., Lowy, J., Stenberg, P., O'Gaora, P., Ganji, A. and So, M. (1993) *Mol. Microbiol.* 8, 719–725.
Olyhoek, A. J., Sarkari, J., Bopp, M., Morelli, G. and Achtman, M. (1991) *Microbiol. Pathog.* 11, 249–257.
Parge, H. E., Bernstein, S. L., Deal, C. D., McRee, D. E., Christensen, D., Capozza, M. A., Kays, B. W., Fieser, T. M., Draper, D., So, M. et al. (1990) *J. Biol. Chem.* 265, 2278–2285.
Parkhill, J., Achtman, M., James, K. D., Bentley, S. D., Churcher, C., Klee, S. R., Morelli, G., Basham, D., Brown, D., Chillingworth, T., Davies, R. M., Davis, P., Devlin, K., Feltwell, T., Hamlin, N., Holroyd, S., Jagels, K., Leather, S., Moule, S., Mungall, K., Quail, M. A., Rajandream, M-A., Rutherford, K. M., Simmonds, M., Skelton, J., Whitehead, S., Spratt, B. G. and Barrell, B. G. (2000) *Nature* 404, 502–506.
Richardson, A. R. and Stojiljkovic, I. (2001) *Mol. Microbiol.* 40, 645–655.
Robertson, B. D. and Meyer, T. F. (1992) *Trends Genet.* 8, 422–427.
Robertson, J. N., Vincent, P. and Ward, M. E. (1977) *J. Gen. Microbiol.* 102, 169–177.
Rothbard, J. B., Fernandez, R., Wang, L., Teng, N. N. and Schoolnik, G. K. (1985) *Proc. Natl Acad. Sci. USA* 82, 915–919.
Rudel, T., Facius, D., Barten, R., Scheuerpflug, I., Nonnenmacher, E. and Meyer, T. F. (1995a) *Proc. Natl Acad. Sci. USA* 92, 7986–7990.
Rudel, T., Scheurerpflug, I. and Meyer, T. F. (1995b) *Nature* 373, 357–359.
Rudel, T., van Putten, J. P., Gibbs, C. P., Haas, R. and Meyer, T. F. (1992) *Mol. Microbiol.* 6, 3439–3450.
Sarkari, J., Pandit, N., Moxon, E. R. and Achtman, M. (1994) *Mol. Microbiol.* 13, 207–217.
Sawitzke, J. A. and Stahl, F. W. (1994) *J. Bacteriol.* 176, 6730–6737.
Schneider, H., Griffiss, J. M., Boslego, J. W., Hitchcock, P. J., Zahos, K. M. and Apicella, M. A. (1991) *J. Exp. Med.* 174, 1601–1605.
Schneider, H., Hammack, C. A., Apicella, M. A. and Griffiss, J. M. (1988) *Infect. Immun.* 56, 942–946.
Schoolnik, G. K., Fernandez, R., Tai, J. Y., Rothbard, J. and Gotschlich, E. C. (1984) *J. Exp. Med.* 159, 1351–1370.
Segal, E., Billyard, E., So, M., Storzbach, S. and Meyer, T. F. (1985) *Cell* 40, 293–300.
Seifert, H. S. (1996) *Mol. Microbiol.* 21, 433–440.
Seifert, H. S, Ajioka, R. S., Marchal, C., Sparling, P. F. and So, M. (1988) *Nature* 24, 392–395.
Serkin, C. D. and Seifert, H. S. (1998) *J. Bacteriol.* 180, 1955–1958.
Skaar, E. P., Lazio, M. P. and Seifert, H. S. (2002) *J. Bacteriol.* 184, 919–927.
Smith, H. (1991) *Proc. R. Soc. Lond. B Biol. Sci.* 246, 97–105.
Sparling, P. F. (1966) *J. Bacteriol.* 92, 1364–1371.
Spratt, B. G., Bowler, L. D., Zhang, Q. Y., Zhou, J. and Smith, J. M. (1992) *J. Mol. Evol.* 34, 115–125.
Stephens, D. S. and McGee, Z. A. (1981) *J. Infect. Dis.* 143, 525–532.
Stern, A., Brown, M., Nickel, P. and Meyer, T. F. (1986) *Cell* 47, 61–71.
Streisinger, G. and Owen, J. (1985) *Genetics* 109, 633–659.
Streisinger, G., Okada, Y., Emrich, J., Newton, J., Tsugita, A., Terzaghi, E. and Inouye, M. (1966) *Cold Spring Harbor Symp. Quant. Biol.* 31, 77–84.

Sullivan, K. M. and Saunders, J. R. (1989) *Mol. Gen. Genet.* 216, 380–387.
Swanson, J. (1994) In *Pathobiology and Immunobiology of Neisseriaceae* (eds Conde-Glez, C. J., Morse, S., Rice, P., Sparling, F. and Calderon, E.), Instituto Nacional de Salud Publica, Cuernavaca, Mexico.
Swanson, J. and Koomey, J. M. (1989) In *Mobile DNA* (eds Berg, D. E. and Howe, M. M.), American Society for Microbiology, Washington, DC, pp. 743–761.
Swanson, J., Belland, R. J. and Hill, S. A. (1992) *Curr. Opin. Genet. Dev.* 2, 805–811.
Swanson, J., Bergstrom, S., Barrera, O., Robbins, K. and Corwin, D. (1985) *J. Exp. Med.* 162, 729–744.
Swanson, J., Bergstrom, S., Robbins, K., Barrera, O., Corwin, D. and Koomey, J. M. (1986) *Cell* 47, 267–276.
Swanson, J., Morrison, S., Barrera, O. and Hill, S. (1990) *J. Exp. Med.* 171, 2131–2139.
Swanson, J., Robbins, K., Barrera, O., Corwin, D., Boslego, J., Ciak, J., Blake, M. and Koomey, J. M. (1987) *J. Exp. Med.* 165, 1344–1357.
Tettelin, H., Saunders, N. J., Heidelberg, J., Jeffries, A. C., Nelson, K. E., Eisen, J. A., Ketchum, K. A., Hood, D. W., Peden, J. F., Dodson, R. J., Nelson, W. C., Gwinn, M. L., DeBoy, R., Peterson, J. D., Hickey, E. K., Haft, D. H., Salzberg, S. L., White, O., Fleischmann, R. D., Dougherty, B. A., Mason, T., Ciecko, A., Parksey, D. S., Blair, E., Cittone, H., Clark, E. B., Cotton, M. D., Utterback, T. R., Khouri, H., Qin, H., Vamathevan, J., Gill, J., Scarlato, V., Masignani, V., Pizza, M., Grandi, G., Sun, L., Smith, H. O., Fraser, C. M., Moxon, E. R., Rappuoli, R. and Venter, J. C. (2000) *Science* 287, 1809–1815.
Tonjum, T. and Koomey, M. (1997) *Gene* 192, 155–163.
Van der Ende, A., Hopman, C. T., Zaat, S., Essink, B. B., Berkhout, B. and Dankert, J. (1995) *J. Bacteriol.* 177, 2475–2480.
van Putten, J. P. (1993) *EMBO J.* 12, 4043–4051.
Virji, M. and Heckels, J. E. (1984) *J. Gen. Microbiol.* 130 (Pt 5), 1089–1095.
Virji, M. and Heckels, J. E. (1986) *J. Gen. Microbiol.* 132 (Pt 2), 503–512.
Virji, M., Makepeace, K., Ferguson, D. J., Achtman, M. and Moxon, E. R. (1993b) *Mol. Microbiol.* 10, 499–510.
Virji, M., Makepeace, K., Ferguson, D. J., Achtman, M., Sarkari, J. and Moxon, E. R. (1992) *Mol. Microbiol.* 6, 2785–2795.
Virji, M., Saunders, J. R., Sims, G., Makepeace, K., Maskell, D. and Ferguson, D. J. (1993a) *Mol. Microbiol.* 10, 1013–1028.
Wainwright, L. A., Pritchard, K. H. and Seifert, H. S. (1994) *Mol. Microbiol.* 13, 75–87.
Weel, J. F., Hopman, C. T. and van Putten, J. P. (1991) *J. Exp. Med.* 173, 1395–1405.
Winther-Larsen, H. C., Hegge, F. T., Wolfgang, M., Hayes, S. F., van Putten, J. P. and Koomey M. (2001) *Proc. Natl Acad. Sci. USA* 98, 15276–15281.
Wolfgang, M., Lauer, P., Park, H. S., Brossay, L., Hebert, J. and Koomey, M. (1998a) *Mol. Microbiol.* 29, 321–330.
Wolfgang, M., Park, H. S., Hayes, S. F., van Putten, J. P. and Koomey M. (1998b) *Proc. Natl Acad. Sci. USA* 95, 14973–14978.
Yang, Q. L. and Gotschlich, E. C. (1996) *J. Exp. Med.* 183, 323–327.
Zhang, Q. Y., DeRyckere, D., Lauer, P. and Koomey, M. (1992) *Proc. Natl Acad. Sci. USA* 89, 5366–5370.
Zhou, J. and Spratt, B. G. (1992) *Mol. Microbiol.* 6, 2135–2146.

9

CANDIDA ALBICANS

DAVID R. SOLL

INTRODUCTION

A variety of microorganisms, most notably pathogens, have developed mechanisms for generating phenotypic variability for rapid adaptation in response to environmental challenges, such as an immune response by the host. In the majority of cases, the result of rapid adaptation is a change in antigenic state, and the molecular mechanism involves DNA reorganization. In the case of *Salmonella typhimurium*, a conserved, reversible DNA inversion leads to the reorientation of a promoter and expression of an alternative flagellin protein (Andrews, 1922; Glasgow *et al.*, 1989). In the case of *Escherichia coli*, a similar mechanism leads to expression of type 1 fimbriae (Branton, 1959; Glasgow *et al.*, 1989). In *Borrelia hermsii*, recombinational events result in a variety of serotypes, the basis of relapsing fever (Barbour, 1989), and in *Trypanosoma brucei*, DNA reorganization also results in a variety of serotypes, the basis of African sleeping sickness (Donelsen, 1989). The list of systems that generate antigenic variation in both pathogenic and nonpathogenic microorganisms is now quite extensive, and in the majority of cases has been demonstrated to involve the reorganization of DNA and transcription of alternative genes, usually in a family encoding two or more surface molecules. The advantage of such a strategy is obvious in the case of pathogens. Each infecting population expresses a dominant phenotype, which is successful for the expressed host conditions, but each population also harbours minor variants, which provide an advantage under a different set of host conditions. In response to an environmental challenge, such as a host immune response, a change in anatomical location, drug therapy or a fundamental change in host physiology, a minor variant more adapted to that change enriches while the previously dominant phenotype is cleared.

Antigenic or phase variation in the form of alternative yeast mating types has been the focus of intensive investigation for years in the budding yeast

Antigenic Variation
ISBN: 0–12–194851-X

Saccharomyces cerevisiae and the fission yeast *Schizosaccharomyces pombe* (Klar, 1989). Even though these yeasts are not closely related evolutionarily, interconversion between mating types in both cases results from the transfer of DNA sequence information from silent cassette loci to an expressed mating type locus. The transfer results in changes in cell physiology in addition to antigenic differences related to fusion. In both cases, switching is reversible. Phenotypic switching and phase transitions have also been identified in several species of infectious fungi, including *Candida albicans* (Pomes *et al.*, 1985; Slutsky *et al.*, 1985, 1987; Soll *et al.*, 1987; Soll, 1989, 1992a), *Candida glabrata* (Lachke *et al.*, 2000) and *Cryptococcus neoformans* (Goldman *et al.*, 1998; Fries *et al.*, 1999). In all three organisms, switching affects putative virulence traits and appears, therefore, to play a role in pathogenesis. Switching in *C. albicans* has been the most carefully studied in the pathogenic yeasts, and has proved unique among pathogens because of its extraordinary pleiotrophic effect on phenotype (Soll, 1992a). Switching in *C. glabrata* has also proved unique in its graded effect on a variety of phenotypic characteristics (Lachke *et al.*, 2000). It is the intent of this chapter to review the discoveries of these systems, the progress that has been made in elucidating the mechanisms involved in the switching process and the regulation of phase-specific genes, and experiments and observations that implicate switching in pathogenesis.

THE DISCOVERY OF SWITCHING IN *C. ALBICANS*

The discovery of switching was based on observations that clonal populations included variants of colony morphology that reverted back to the dominant colony morphology. Negroni (1935) first observed the formation of variant rough colony morphologies and Mackinnon (1940) subsequently reported the appearance of variant colonies with a 'spiky' morphology. Brown-Thomsen (1968) then described 15 different colony morphologies, and demonstrated differences between them in sugar-assimilation patterns and the reduction of tetrazolium salts. However, experiments were not performed in these early studies to test the exact frequencies of switching or, most importantly, reversibility. During the subsequent 17 years, there was little interest in the origin or role of variants in colonizing populations of *C. albicans*. However, in 1985, Pomes *et al.* reported reversible switching between a 'smooth' and a 'rough' colony morphology in *C. albicans* strain 1001 and Slutsky *et al.* (1985) discovered reversible switching between several colony morphologies in *C. albicans* strain 3153A. Pomes *et al.* (1985) demonstrated that the frequency of rough colony formation in clonally derived smooth colony populations was $<10^{-4}$, but could be stimulated more than 30-fold by low doses of ultraviolet (UV) irradiation. They found that the frequency of smooth colony formation in clonally derived rough colony populations was $<9 \times 10^{-4}$, but the frequency of rough colonies with smooth sectors was 4×10^{-3} (Pomes *et al.*, 1985). Pomes *et al.* demonstrated that switching in their strain was reversible at high frequency. Slutsky et al. (1985) demonstrated that cells of strain 3153A switched between eight colony types (original smooth, star, ring, irregular wrinkle, stippled, hat, fuzzy and revertant smooth) (Figure 9.1A–H), and that the frequency of switching could be stimulated with a low dose of UV. They demonstrated that cells exhibiting any of the phenotypes other than original white were in a high-frequency mode of switching (Slutsky et al., 1985) (Figure 9.1J). While the original smooth phenotype switched spontaneously to variant phenotypes at a frequency of $<10^{-4}$, variants in turn switched at frequencies ranging between 3×10^{-2} and 5×10^{-4} (Figure 9.1J). Again, Slutsky et al. (1985) demonstrated interconvertability (i.e. reversibility) at high frequency. In Figure 9.1I an example is presented of a spontaneous switch from ring to star in strain 3153A.

It was clear that although switching in strains 1001 and 3153A shared general characteristics, the switch phenotypes differed. In 1987 two additional reports

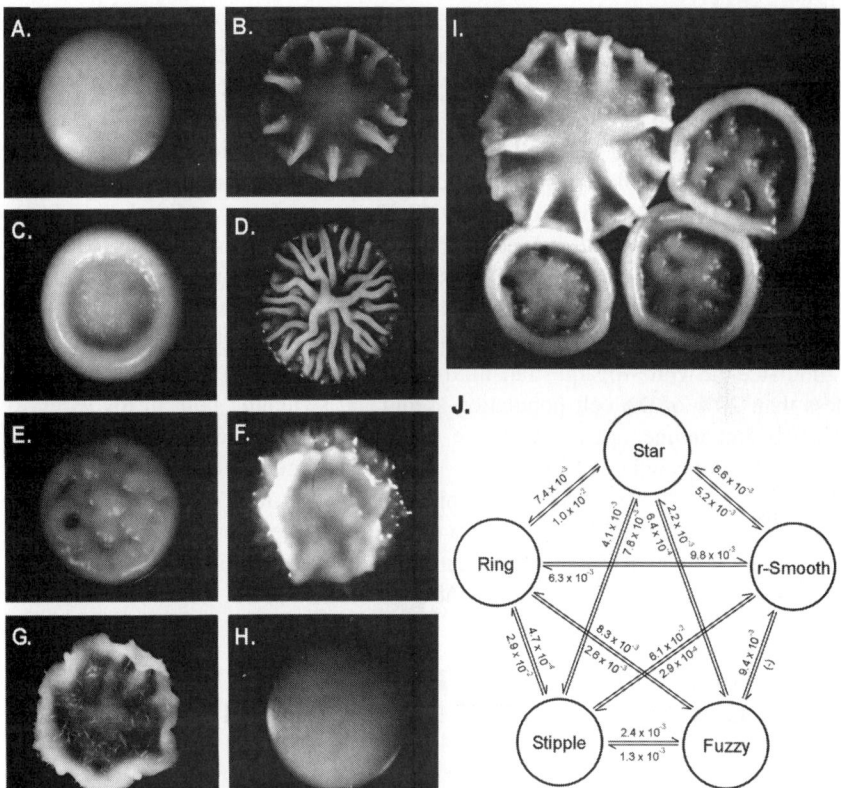

Figure 9.1 Switching in *Candida albicans* strain 3153A: (A) 'original smooth'; (B) 'star'; (C) 'ring'; (D) 'irregular wrinkle'; (E) 'stippled'; (F) 'hat'; (G) 'fuzzy'; (H) 'revertant smooth'; (I) an example of a switch from ring to star; (J) frequencies of switching. (For details, see Slutsky et al. (1985) *Science* 230, 666–669, from which the figure is reproduced with permission.)

supported the conclusion that different strains of *C. albicans* differed in the colony phenotypes composing their switching repertoire. First, Soll *et al.* (1987) found that the predominant switching system amongst independent isolates collected from vaginitis patients involved a reversible transition between a heavy myceliated and unmyceliated colony phenotype. Second, Slutsky *et al.* (1987) identified a switching system in strain WO-1 that involved a reversible phase transition between a smooth hemispheric white ('white') and a smooth flat grey ('opaque') colony phenotype, which was named the 'white–opaque transition' (Figure 9.2). This latter switching system, which is expressed in approximately 3% of *C. albicans* isolates, has become the major experimental system for investigating the molecular basis of switching (Soll *et al.*, 1991; Soll, 1997, 2002a,b).

THE WHITE–OPAQUE TRANSITION

Strain WO-1 was cultured from the blood of a bone marrow transplant patient at the University of Iowa Hospitals and Clinics (Slutsky *et al.*, 1987). The patient died as a result of this systemic infection. Strain WO-1 exhibited the white and opaque phenotype on modified Lee's medium (Lee *et al.*, 1975; Bedell and Soll, 1979), as well as other tested media, and could be distinguished by colour on agar containing phloxine B (Anderson and Soll, 1987). While white colonies and white sectors stained very light pink, opaque-phase colonies and opaque sectors stained bright red (Figure 9.2) (Anderson and Soll, 1987). The frequency of white-phase colony-forming units in clonally derived opaque-phase populations was approximately 10^{-3}, and the frequency of opaque-phase colony-forming units in clonally derived white-phase populations was approximately 10^{-3} (Anderson and Soll, 1987; Slutsky *et al.*, 1987). As in the case of strains 1001 and 3153A switching, UV stimulated the white–opaque transition (Morrow *et al.*, 1989). UV doses that killed less than 20% of the cell population stimulated switching both in the white to opaque and in the opaque to white direction. The UV effect was heritable. Increased frequencies of switching lasted many generations after treatment (Morrow *et al.*, 1989). Because the white–opaque transition can be discriminated on a variety of agar media, involves a single reversible transition and, as will be described, involves a dramatic change in cell morphology, this system was selected to pursue the molecular basis of switching.

THE WHITE–OPAQUE TRANSITION AFFECTS THE BUDDING YEAST CELL MORPHOLOGY

Switching in strain 3153A (Slutsky *et al.*, 1985) and in strains exhibiting the 'unmyceliated–heavy myceliated' transition (Soll *et al.*, 1987) affects the constraints on the bud–hypha transition in colony domes. Indeed, the variant colony morphologies in these strains arise in large measure as a result of differences in the spatial distributions and temporal dynamics of budding yeast,

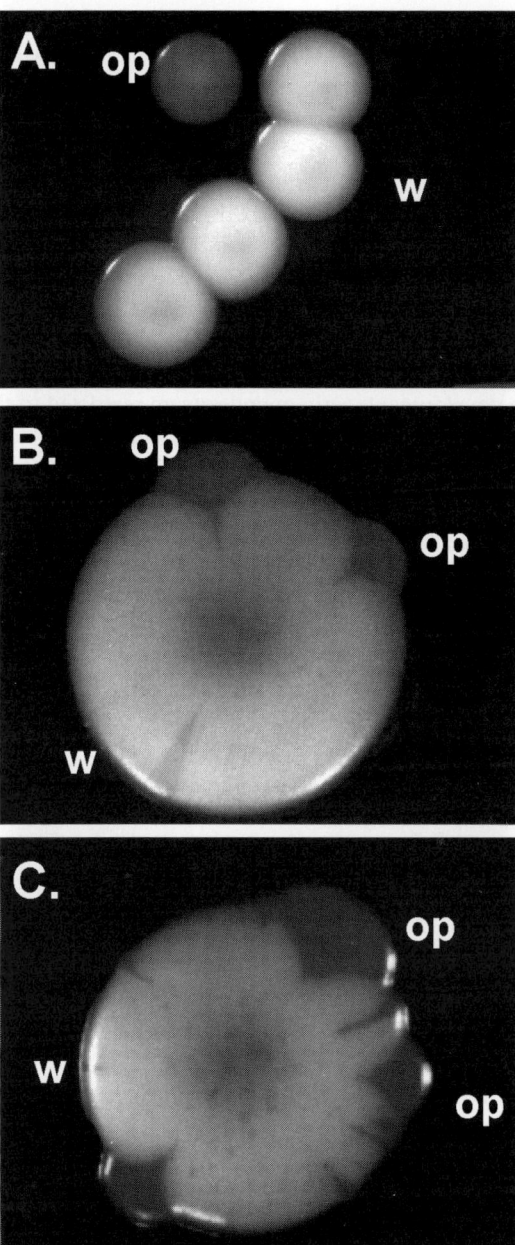

Figure 9.2 The white–opaque transition in strain WO-1: (A) cells from a white colony plated on agar at low density; (B,C) examples of opaque sectors in white colonies aged on dishes wrapped with parafilm to block gas exchange. In all cases cells were plated on agar containing phloxine B, which stains opaque-phase cells red (darker sectors and colonies in black and white micrograph); op, opaque; wh, white. (For details, see Slutsky *et al.* (1987) and Anderson and Soll (1987).)

pseudohypha and hypha formation. In these systems, the basic budding yeast cell morphologies are similar. In the white–opaque transition, however, the yeast cell morphology differs between the two phases (Anderson and Soll, 1987; Slutsky *et al.*, 1987; Rikkerink *et al.*, 1988; Anderson *et al.*, 1990; Soll, 1992a). In the white phase, budding cells are round to ovoid (Figure 9.3A), and exhibit the budding pattern of diploid *S. cerevisiae* strains (Slutsky *et al.*, 1987). In contrast, in the opaque phase, yeast cells are bean-shaped and asymmetric (Figure 9.3B), are approximately two to three times larger than white-phase budding cells, possess twice the mass of white-phase cells, and exhibit a significant proportion of bipolar budding (Slutsky *et al.*, 1987). F-actin staining with rhodamine-conjugated phalloidin revealed that early in opaque-phase cell development actin granules were distributed throughout the cortex, like a white-phase budding yeast cell, but later in development, actin granules were restricted to the daughter cell apex, like a white-phase developing hypha (Anderson and Soll, 1987). These results suggested that budding opaque-phase cells exhibit characteristics of both budding cells and hypha-forming cells.

Scanning electron microscopy (SEM) and transmission electron microscopy (TEM) demonstrated further differences in cell wall morphology and ultrastructure (Anderson and Soll, 1987; Anderson *et al.*, 1990). SEMs of opaque-phase cells revealed unique cell wall 'pimples' (Figure 9.3D) absent in walls of white-phase cells (Figure 9.3C). These pimples were evident only on mother cells and daughter cells that had attained lengths greater than half that of their mother cells (Anderson and Soll, 1987). The average number of pimples on the surface of a mature opaque-phase cell was 141 (Anderson *et al.*, 1990). The density of pimples on a mature opaque-phase cell was 1.33 pimples per μm^2 (Anderson *et al.*, 1990). Transmission electron micrographs revealed thickenings in the walls of opaque-phase cells corresponding to pimples (Figure 9.4B). Such thickenings were absent in the walls of white-phase cells (Figure 9.4A). In some cases, channels were visible traversing the wall through the centres of pimples. In many cases, these channels terminated extracellularly with membrane blebs (Anderson *et al.*, 1990). Although there is no data on the nature of these blebs, it is likely that they represent plasma membrane that has been drawn through channels as a result of fixation and may, therefore, represent ultrastructural artefacts.

TEM also revealed a large vacuole in opaque-phase cells containing vesicles (Anderson *et al.*, 1990) (Figure 9.4B, C). Similar large vacuoles were not evident in thin sections of white-phase cells (Figure 9.4A). In a majority of cases, the vesicles appeared collapsed (e.g., Figure 9.4B), presumably a result of fixation.

ANTIGENIC CHANGES IN THE WHITE–OPAQUE TRANSITION

To test for antigenic changes in the white–opaque transition, polyclonal antiserum was raised against opaque-phase cells by injecting heat-killed cells

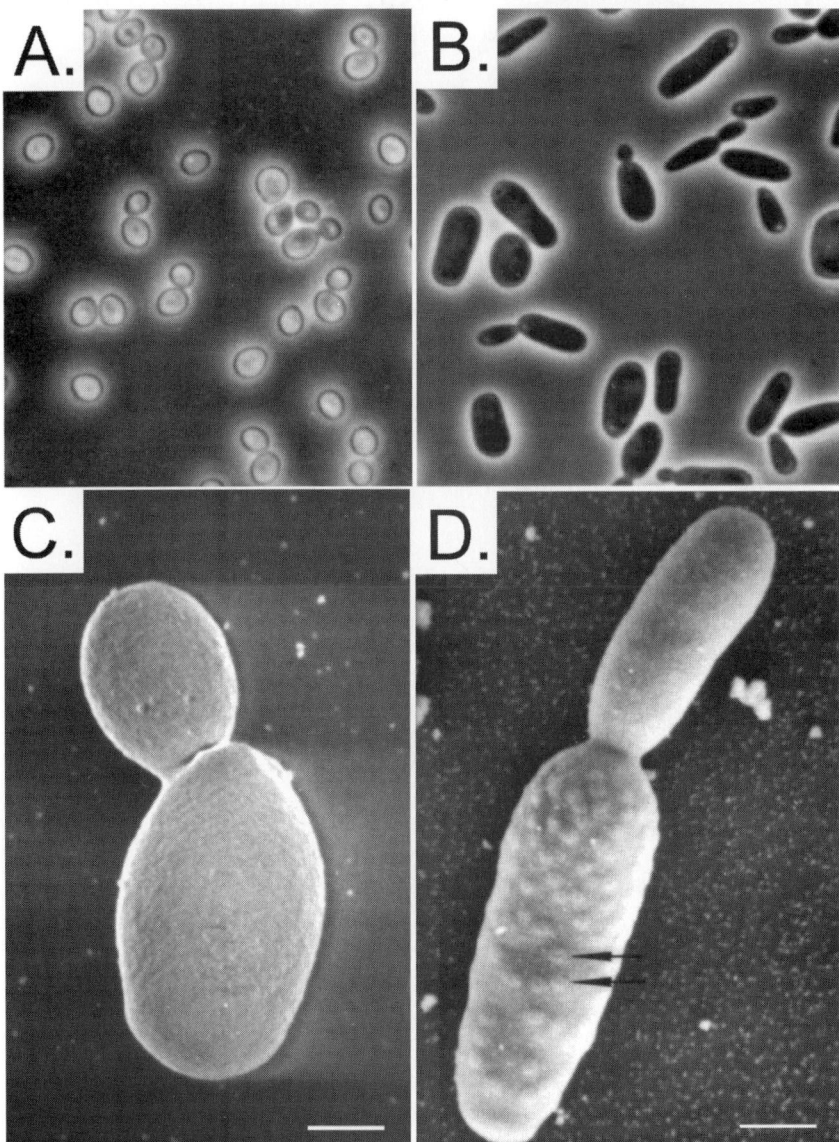

Figure 9.3 White- and opaque-phase cells exhibit different morphologies in the budding yeast form of growth: (A) phase contrast images of white-phase budding yeast cells; (B) phase contrast images of opaque-phase budding cells; (C) scanning electron micrograph (SEM) of white-phase budding yeast cell; (D) SEM of opaque-phase budding cell (pimples are pointed to with thin black arrows). Scale bars in panels C and D represent 0.5 μm.

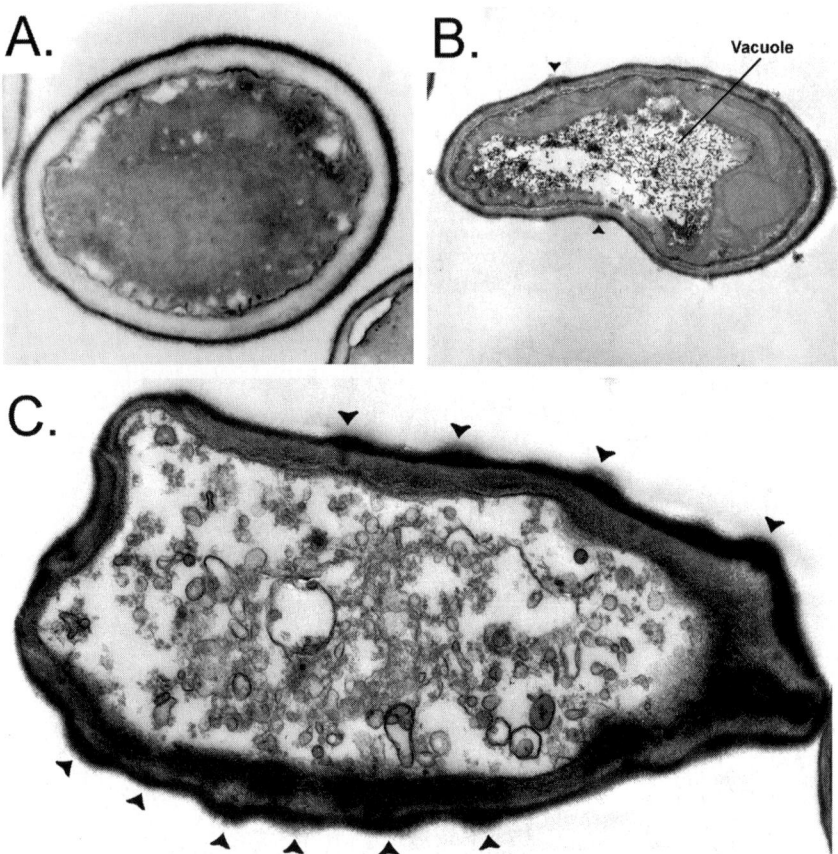

Figure 9.4 Opaque-phase cells exhibit unique cell wall pimples: (A) transmission electron micrograph (TEM) of white-phase yeast cell; (B) TEM of opaque-phase yeast cell (pimples pointed to with arrow heads; note collapsed vesicles in vacuole); (C) TEM of opaque-phase yeast cell (pimples pointed to with arrow heads; note vesicles in vacuole are not collapsed). (For details, see Anderson *et al.* (1990) *J. Bacteriol.* 172, 224–235, from which the figure is reproduced with permission.)

into rabbits (Anderson and Soll, 1987). The resulting antiserum uniformly stained the surface of heat-fixed white-phase budding cells, white-phase cells forming hyphae and opaque-phase budding cells (Anderson and Soll, 1987, Anderson *et al.*, 1990). When the antiserum was absorbed repeatedly with heat-fixed white-phase budding yeast cells, it no longer stained heat-fixed white-phase budding yeast cells (Figure 9.5A), but it did stain hyphae formed by white-phase cells, beginning at the first septum formed along the tube (Figure 9.5B), and it stained opaque-phase cells, but in a punctate fashion (Figure 9.5C). To test whether the antiserum contained antibodies against opaque-specific antigens, the original antiserum was absorbed with white budding yeast cells induced to form

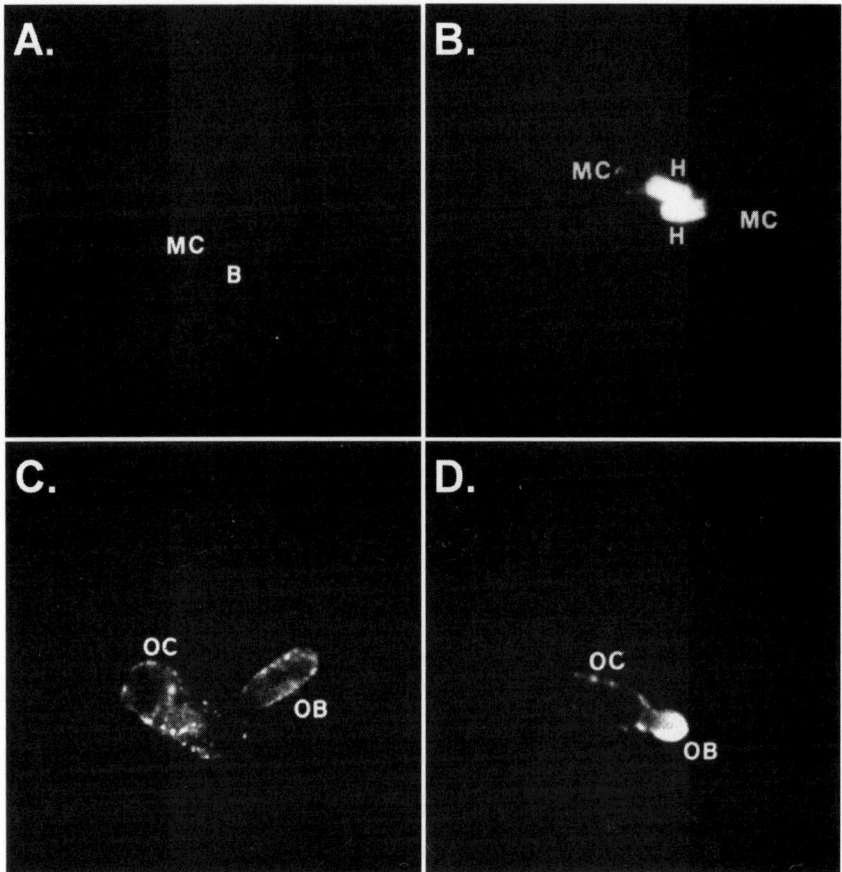

Figure 9.5 Demonstration of antigenic differences between white- and opaque-phase cells. Polyclonal antiserum generated against heat-fixed opaque-phase cells was absorbed with heat-fixed white-phase cells and used to stain heat-fixed white-phase cells (A), heat-fixed white-phase cells forming hyphae (B) and opaque-phase cells (C). When the antiserum was absorbed with hypha-forming white-phase cells, it still stained opaque-phase cells in a punctate fashion (D), demonstrating the presence of opaque-phase-specific antigens. (For details, see Anderson and Soll (1987) *J. Bacteriol.* 169, 5579–5588, from which the figure is reproduced with permission.)

germ tubes, thus removing antibodies to budding yeast cells and hyphae. This absorbed antiserum stained neither white budding nor white hypha-forming cells, but still stained opaque budding yeast cells (Figure 9.5D), demonstrating the presence of opaque-specific antigens (Anderson and Soll, 1987).

When western blots of white-phase cell extract and opaque-phase cell extracts were stained with the absorbed antiserum, three opaque-phase-specific antigens were identified at 31, 21 and 14.5 kDa (Figure 9.6B) (Anderson *et al.*, 1990). To test which of these were on the cell surface, the antiserum was absorbed with non-permeabalized opaque-phase cells to remove those antibodies reacting with antigens on the cell surface. The absorbed antiserum was then used to stain a western blot of white- and opaque-phase cell extracts. Only the antibodies against the 14.5 kDa antigen were removed (Figure 9.6C), suggesting that the 21

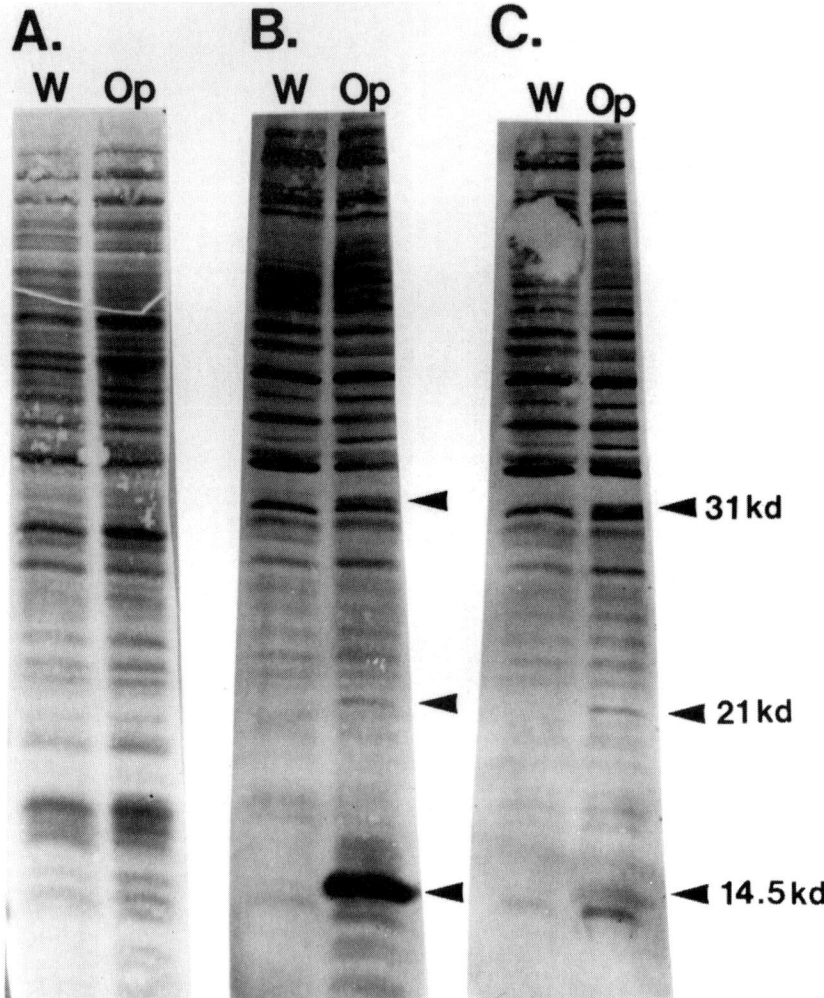

Figure 9.6 Demonstrations that two opaque-phase-specific antigens (31 kDa and 21 kDa) are intracellular and one (14.5 kDa) is extracellular by western blot analysis: (A) amido black-stained western blot of white (W)- and opaque (Op)-phase cell extracts; (B) western blot stained with antiserum generated against opaque-phase cells revealed three opaque-phase-specific antigens; (C) western blot stained with antiserum absorbed with non-permeabilized opaque-phase cells demonstrates that only the 14.5 kDa protein was removed and the latter, therefore, is on the opaque-phase cell wall. (For details, see Anderson *et al.* (1990) *J. Bacteriol.* 172, 224–235, from which the figure is reproduced with permission.)

and 31 kDa antigens were intracellular, while the 14.5 kDa antigen was on the cell surface (Anderson *et al.*, 1990). Since absorbed antiserum stained the surface of non-permeabilized opaque-phase cells in a punctate fashion (Figure 9.5C), immunogold labelling was used to test whether the 14.5 kDa antigen was localized to the pimples (Anderson *et al.*, 1990). TEMs of the immunostained cells revealed that the 14.5 kDa antigen was indeed localized to the surface of opaque-phase cell wall pimples (Figure 9.7).

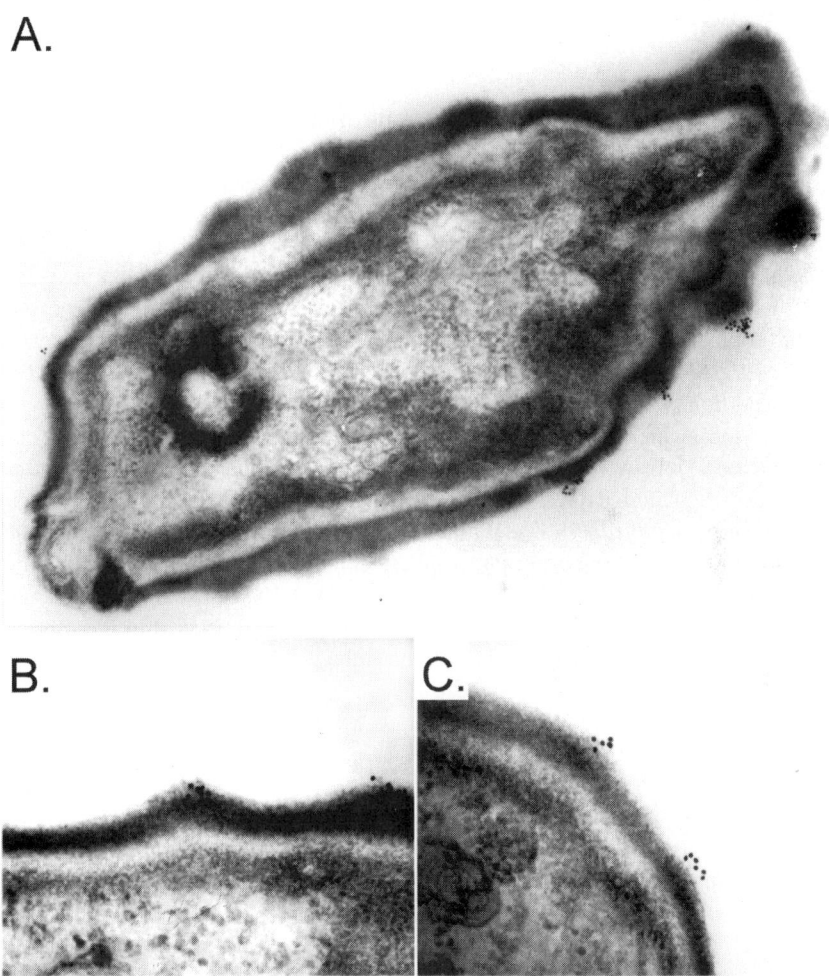

Figure 9.7 Indirect immunogold staining of fixed, non-permeabilized opaque-phase cells with antiserum generated against opaque-phase cells demonstrates localization of the 14.5 kDA protein at the cell wall pimples. (For details, see Anderson *et al.* (1990) *J. Bacteriol.* 172, 224–235, from which the figure is reproduced with permission.)

SWITCHING INVOLVES DIFFERENTIAL GENE EXPRESSION

The demonstration of antigenic variation in the late 1980s (Anderson and Soll, 1987; Anderson et al., 1990) suggested that switching involved differential gene expression. In 1992 Morrow et al. used a differential hybridization screen of an opaque-phase cDNA library to identify the first phase-specific gene *PEP1*, which encoded a secreted aspartyl proteinase. Since 1992, the list of phase-specific genes has continued to grow.

SAP1 and *SAP3*

PEP1 proved to be a member of a family of aspartyl proteinases, and was subsequently renamed SAP1 (White et al., 1993; Hube et al., 1994). It is generally believed that one of the functions of the secreted aspartyl proteinases is in tissue penetration. Naglik et al. (1999), using a reverse transcription PCR protocol, demonstrated that *SAP1* and *SAP3* were expressed in the saliva of individuals only during oral infections, and *SAP3*, which had not been observed to be expressed *in vitro* by the laboratory strains used by these researchers, was found to be expressed *in vivo*. Schaller et al. (2000) demonstrated that the *SAP*s were expressed during disease progression in the order *SAP1* and *SAP3*, *SAP8*, *SAP6* and finally *SAP3*. It has been demonstrated by northern analysis that *SAP1* is expressed at high levels in the opaque phase and at negligible levels in the white phase (Morrow et al., 1992; White et al., 1993; Hube et al., 1994), and that this phase-regulated expression pattern correlated with extracellular protease activity (Morrow et al., 1992). While the addition of serum to growth medium enhances *SAP1* expression in opaque-phase cells, it does not activate transcription in a white-phase-specific manner in strain WO-1 (Morrow et al., 1992). It has also been demonstrated that *SAP3* is differentially expressed in the opaque phase of strain WO-1 (White et al., 1993; Hube et al., 1994).

OP4

Again using a differential hybridization screen of an opaque-phase cDNA library, the second phase-specific gene, *OP4*, was identified (Morrow et al., 1993). *OP4* encodes a putative protein of 402 amino acids with unusually high proportions of alanine, leucine and serine, the three accounting for 38% of the protein mass. The initial amino-terminal 26 amino acids are hydrophobic, and extensive enough to span a membrane. Neither the DNA nor the deduced amino acid sequence of *OP4*, however, showed homology to respective sequences in databases (Morrow et al., 1993). Thus *SAP1* and *OP4* are coordinately regulated in the white–opaque transition.

WH11

Using a differential hybridization screen of a white-phase cDNA library subtracted with opaque-phase cDNA, the white-phase-specific gene *WH11* was cloned (Srikantha and Soll, 1993). *WH11* is differentially expressed exclusively in white-phase budding yeast cells. It is not expressed in opaque-phase cells nor in white-phase hyphae (Srikantha and Soll, 1993). The WH11 protein is homologous to the glucose/lipid-regulated protein Glp1p of *S. cerevisiae* (Stone *et al.*, 1990), which was subsequently demonstrated to be Hsp12p, a heat-shock protein (Praekelt and Meacock, 1990). A homologue of *WH11* was subsequently identified in *Aspergillus nidulans* (Dutton *et al.*, 1997). A polyclonal antibody against recombinant Wh11p stained white-phase cells throughout the cytoplasm, but did not stain vesicles, nuclei or the cell surface (Schröppel *et al.*, 1996). This antibody did not stain opaque-phase cells (Schröppel *et al.*, 1996).

CDR3 and *CDR4*

Two genes, *CDR3* and *CDR4*, that encode ABC transporter proteins are regulated by the white–opaque transition. *CDR3* is opaque phase-specific (Balan *et al.*, 1997) and *CDR4* white phase-specific (Sanglard *et al.*, 1999). It is curious that these two genes, which are involved in the development of drug resistance, exhibit alternate expression in the white–opaque transition.

EFG1

EFG1 encodes a trans-acting factor with homologues in *S. cerevisiae* (Gimeno and Fink, 1994), *A. nidulans* (Miller *et al.*, 1992) and *Neurospora crassa* (Aramayo *et al.*, 1996). In *C. albicans*, *EFG1* has been demonstrated to play a role in the regulation of hypha formation (Lo *et al.*, 1997; Stoldt *et al.*, 1997) and in the process of switching (Sonneborn *et al.*, 1999; Srikantha *et al.*, 2000). *EFG1* is transcribed as a 3.2 kb mRNA in white-phase cells and as a far less abundant 2.2 kb mRNA in opaque-phase cells (Srikantha *et al.*, 2000). The alternative mRNAs are initiated at different transcription start sites (Srikantha *et al.*, 2000).

HOS3

HOS3 encodes a histone deacetylase with a homologue in *S. cerevisiae* (Srikantha *et al.*, 2001). In *C. albicans*, *HOS3* is transcribed as a 2.5 kb mRNA in white-phase cells and as a 2.3 kb mRNA in opaque-phase cells (Srikantha *et al.*, 2001).

Other phase-regulated genes

A number of additional genes have recently been demonstrated to be regulated by the white–opaque transition. The two component hybrid kinase regulator gene *NIK1* (Srikantha *et al.*, 1998), the nitric oxide synthase gene *NOS3* (Srikantha and Soll, unpublished observation), and the phosphofructokinase gene *PFK2* (Lockhart and Soll, unpublished observations) are expressed in the opaque phase at levels far higher than in the white phase. The phosphoglycerate mutase gene *GPM* (Lockhart and Soll, unpublished observation) and the universal regulator gene *MCM1* (Srikantha *et al.*, 2001; Zhao *et al.*, 2002) are expressed in the white phase at levels far higher than in the opaque phase.

In summary, the white–opaque transition involves the differential expression of a variety of phase-specific or phase-enriched genes (Figure 9.8). Some of these genes have been implicated in pathogenesis. The list of phase-specific genes has

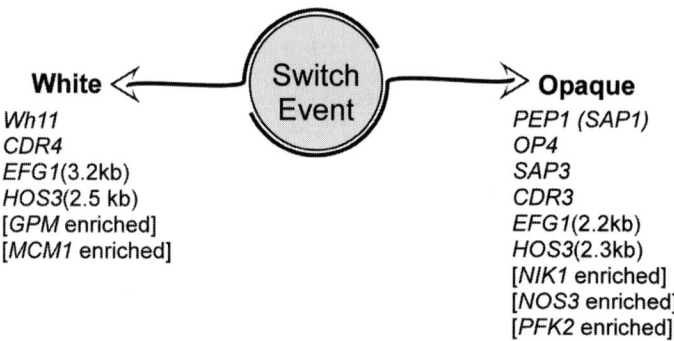

Figure 9.8 Synopsis of genes regulated by the white–opaque transition in strain WO-1.

grown rapidly in the past several years, and will no doubt increase dramatically when results from microarrays are reported. The number and variety of genes regulated by the white–opaque transition contrasts markedly with the majority of switching systems in other pathogens, which usually involve variation in a single protein species expressed at the cell surface. Although phase-specific genes have been identified almost exclusively in the white–opaque transition, and the expression patterns of these genes analysed almost exclusively in this experimental model, the expression of *WH11*, *OP4* and *SAP1* was also analysed in the 3153A switching system (Morrow *et al.*, 1994). It was demonstrated that *WH11* was constitutively expressed in the original smooth (o-s), irregular wrinkle (iw), revertant smooth (r-s) and star (st) switch phenotypes. However, *OP4* was expressed in iw, r-s and st cells, but not in o-s cells, and *SAP1* was expressed only in st cells (Morrow *et al.*, 1994). These results demonstrate that *OP4* and *SAP1* are expressed only in variant phenotypes of the 3153A switching system, and switching-regulated expression of the two phase-specific genes can be dissociated. They also suggest that in the more complex switching systems in *C. albicans* phase-specific genes are combinatorially expressed in the different variant phenotypes.

A CONTEXTUAL FRAMEWORK FOR INVESTIGATING THE MECHANISMS OF GENE REGULATION DURING SWITCHING

In order to understand how the switch event is regulated and in turn how phase-specific genes are regulated, one must formulate plausible models that are predictive, testable and alterable, and which provide a contextual framework for interpreting experimental results, especially those collected with mutants. Perhaps, the simplest switching model involves a switch at a master locus resulting in the synthesis of alternate *trans*-acting factors, each involved in the activation of genes in the alternate phases (Figure 9.9A). In this simple model, a switch from the white to opaque phase turns off expression of a white-phase-specific *trans*-acting activator, and turns on an opaque-phase-specific *trans*-acting activator, and a switch from opaque to white results in converse events (Figure 9.9A). Other simple models can be developed. For instance, the switch from white to opaque may simply turn on expression of a white-phase gene repressor–opaque-phase gene activator and a switch from opaque to white simply turns this *trans*-acting factor off. In this model, white-phase genes are constitutively expressed in the white phase. In the opaque phase, white-phase genes are repressed and opaque-phase genes are activated through expression of the single regulatory factor. Alternatively, a switch event may result in a sequence of events that amplify the number of phase-specific activators and/or repressors (Figure 9.9C). In the latter more complex models, the switch event results in multiple activators and/or repressors in a single phase, which interact with different *cis*-acting sequences, defining subsets of coordinately regulated phase-specific genes. In these more complex models, the number of steps between the switch event and downstream gene regulation is expanded and the number of *trans*-acting factors increases in each phase.

Developing the kinds of models presented in Figure 9.6 highlights several points one must keep in mind when analysing the mechanisms involved in phase-specific gene expression. First, a gene that is differentially expressed in a particular phase may be involved in an amplification step immediately downstream of the switch event or may represent a structural molecule or enzyme that functions as a component of the terminal phenotype (i.e. white or opaque). It is highly likely that the number of genes involved increases in each downstream step. Therefore, the majority of genes regulated by switching will not be involved in gene regulation, but rather will be individual terminal components involved in expression of the final phenotype. Terminal genes may include the ABC transporter genes *CDR*3 and *CDR4*, the secreted aspartyl

180 Antigenic Variation

A. Simple Model 1

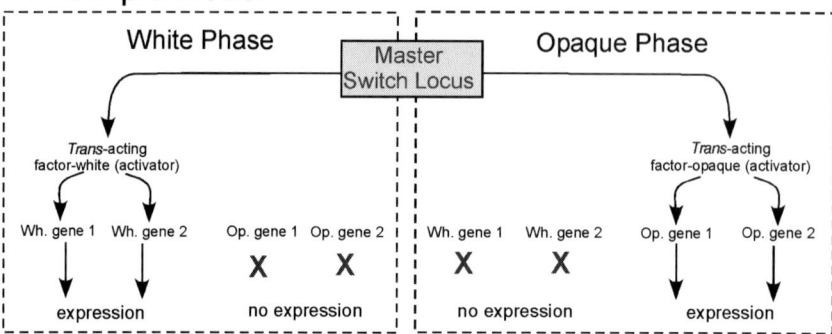

B. Simple Model 2

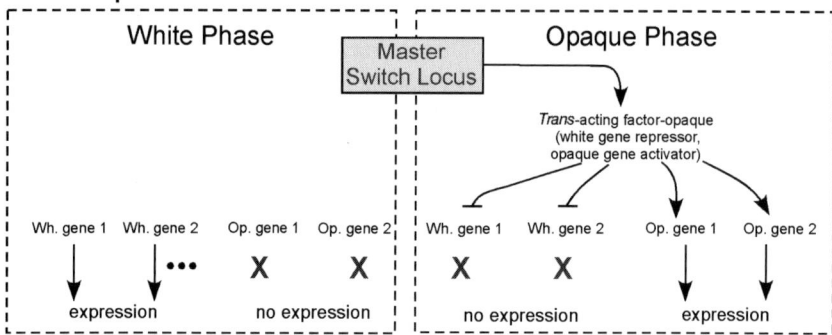

C. Complex Model

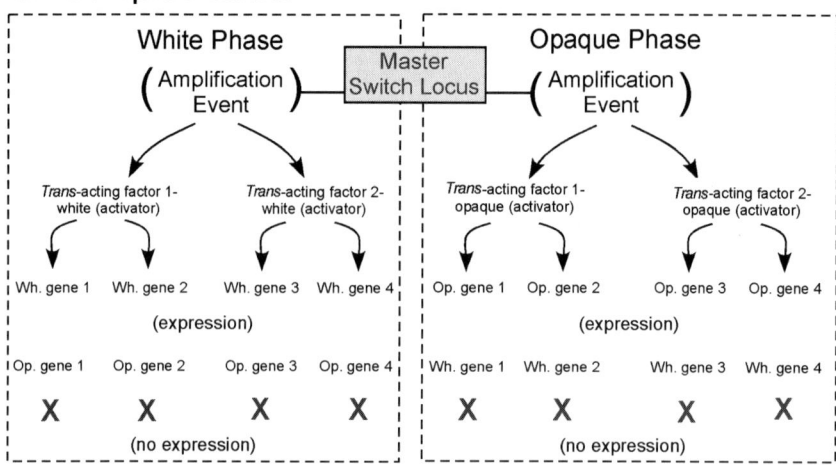

Figure 9.9 Examples of simple and complex models for the regulation of phase-specific gene expression in the white–opaque transition.

proteinases *SAP1* and *SAP2* and enzymes such as *NOS3*, *PFK* and *GPM*. On the other hand, regulatory proteins such as *EFG1*, *MCM1* and *HOS3* may play roles in the regulation of gene expression.

A second point that must be kept in mind is that the switch event(s) probably occurs transiently, in a very short period of time, and may not be identifiable through characterization of established alternative states. The phenotypes of the alternative states (white versus opaque) represent 'steady states' established by the switch event. A third point that must be kept in mind is that a mutation that blocks a phenotypic transition need not occur in a gene specifically involved in the switch event. If a gene is essential for the downstream expression of the opaque phenotype, a mutation in that gene may block expression of the terminal phenotype or a portion of it, but not the switch event itself. Plating assays that simply distinguish between terminal phenotypes cannot distinguish between these alternatives.

Functional analyses of phase-specific gene promoters

In order to understand how phase-specific genes are coordinately regulated, one must first test whether they share common *cis*-acting sequences. To do this one must functionally characterize their promoters by analysing the expression of a reporter gene fused to deletion derivatives of the promoter. Unfortunately, one of the most commonly used reporters, the firefly luciferase, was used in developing the deletion derivatives in the first functional analysis of a phase-specific gene promoter of *C. albicans* (Srikantha *et al.*, 1995). Unfortunately the firefly luciferase gene was not expressed in *C. albicans* because of altered codon usage (Santos *et al.*, 1990, 1993; Ohama *et al.*, 1993). *C. albicans* and several related species use a tRNA with a CAG anticodon to decode the codon CUG as serine, rather than leucine. Therefore, reporters with leucines in functional domains, as in the case of the firefly luciferase (Tatsumi *et al.*, 1992), will not be expressed as a functional protein in *C. albicans*. The deletion derivatives of the promoter of the white-phase-specific gene *WH11* were characterized using a relatively slow and insensitive transcription assay (Srikantha *et al.*, 1995). Subsequently, a highly sensitive luciferase reporter that was compatible with the altered codon usage of *C. albicans* was identified, the sea pansy *Renilla reniformus* luciferase (Lorenz *et al.*, 1991). The functional characterizations of the promoters of several phase-specific genes with this reporter system (Srikantha *et al.*, 1996) suggest a more complex mechanism of coordinate regulation involving more than one phase-specific *trans*-acting factor in each phase.

Promoters of white-phase-specific genes

The white-phase-specific gene *WH11* is expressed only in budding yeast cells in the white-phase (Srikantha and Soll, 1993). First, using northern blot analysis as a method for measuring the strength and developmental regulation of promoter constructs, Srikantha *et al.* (1995) demonstrated that *WH11* expression was regulated through two *cis*-activation sequences, a distal sequence (wDAS) between –475 and –388 bp, and a proximal sequence (wPAS) between –307 and –203 bp upstream of the transcription start site (Figure 9.10A). These activation sequences were confirmed and their strengths quantified using the *R. reniformis* reporter (Srikantha *et al.*, 1997). Ishii *et al.* (1997) subsequently demonstrated that wPAS contained an Rbf1p-binding site. Although no repressor sites were revealed

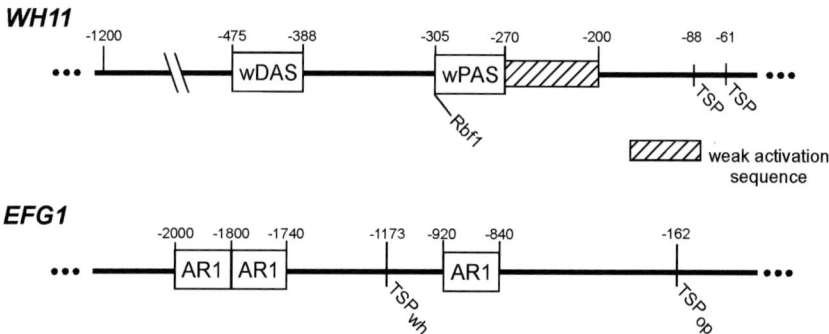

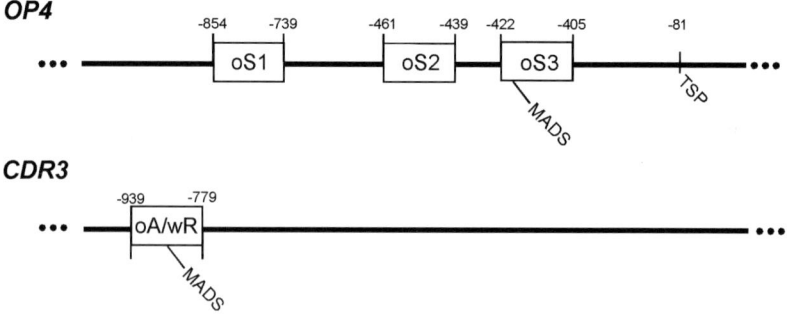

Figure 9.10 The regulatory *cis*-acting sequences in the promoters of phase-regulated genes. (A) wDAS, wPAS, distal and proximal white-phase-specific activation sequences, respectively, in the *WH11* promoter (Srikantha *et al.*, 1995, 1997). AR1, AR2 and AR3 white-phase-specific activation sequences in the *EFG1* promoter (Lachke *et al.*, 2002a). TSP, transcription start point; TSP_{wh}, TSP_{op}, transcription start points for the white-phase- and opaque-phase-specific transcripts of *EFG1*. (B) oS1, oS2, enhancer-like regions in the *OP4* promoter. oS3, phase-specific activator sequence in the OP4 promoter (Lockhart *et al.*, 1998). oA/ wR, the opaque activation/white repressor sequence in the CDR3 promoter (Lockhart *et al.*, 2003).

by this functional analysis, the physical demonstration that Rbf1p, a known repressor, binds to wPAS suggests that repression may also occur at that site.

The second white-phase-specific promoter examined was that of *EFG1*. *EFG1* is transcribed as a more abundant 3.2 kb mRNA in the white phase and a less abundant 2.2 kb mRNA in the opaque phase (Srikantha *et al.*, 2000). The transcription start site for the white-phase transcript is at -1173 and the transcription start site for the opaque-phase transcript is at -162 bp (Srikantha *et al.*, 2000). The *EFG1* promoter contains three white-phase-specific activation sequences between -2000 and -1800 bp (AR1), between -1800 and -1740 bp (AR2) and between -920 and -840 bp (AR3) (Figure 9.10A) (Lachke *et al.*, 2002b). AR3 is downstream of the transcription start point for the white-phase transcript (Figure 9.10A). Therefore, AR3 is within the leader sequence of the white-phase *EFG1* transcript. Interestingly, there are no activation sequences for the opaque-phase transcript (Lachke *et al.*, 2003). An analysis of deletion derivatives revealed that opaque-phase promoter activity decreased proportionally to deletion lengths between -2000 and -74 bp. Together, these results demonstrate overlapping white- and opaque-phase promoters for the differential expression of the 3.2 and 2.2 kb *EFG1* transcripts.

Both *WH11* and *EFG1* (3.2 kb transcript) are coordinately expressed in the white phase, but downregulated in the opaque phase, and both contain white-phase-specific activation sequences in their promoters. If coordination is through the same phase-specific DNA binding protein, then wDAS and wPAS of *WH11* should exhibit homology with the activation sequences of *EFG1*. Sequence comparisons between the four activation sequences revealed no homologies, suggesting that the simplest models involving common DNA binding proteins (e.g. Figure 9.9A or 9.9B) are not compatible. However, these results do not exclude a simple model in which a single phase-specific *trans*-acting factor confers phase-specificity to a number of different activation and/or repression complexes that differ in DNA binding proteins.

Promoters of opaque-phase-specific genes

The opaque-phase-specific gene *OP4* is expressed exclusively in opaque-phase budding cells (Morrow *et al.*, 1993). A functional analysis of the *OP4* promoter using the *R. reniformis* luciferase reporter revealed three sequences involved in activation: oS1 between -854 and -739 bp, oS2 between -469 and -439 bp, and oS3 between -422 and -405 bp (Figure 9.10B) (Lockhart *et al.*, 1998). Deletion of oS1 reduced opaque-phase-specific expression to 50% of the full promoter, deletion of oS2 reduced opaque-phase-specific expression to 20%, and deletion of oS3 eliminated expression in the opaque phase (Lockhart *et al.*, 1998). When a polylinker was substituted for a portion of oS3, leaving oS1 and oS2 intact, opaque-phase-specific expression was completely eliminated. Therefore, while oS1 and oS2 appear to enhance opaque-phase-specific expression, oS3 is essential for opaque-phase-specific expression. We have tentatively interpreted oS1 and oS2 to represent enhancer sequences and oS3 to represent a phase-specific activator sequence.

oS3 contains a MADS box consensus-binding site similar to the Mcm1-binding sites in *S. cervisiae* (Lockhart *et al.*, 1998) (Figure 9.11A). Mutations in the *S. cerevisiae* Mcm1-binding site that cause loss of function (Acton *et al.*, 1997) also cause dramatic decreases in activity in the MADS box-binding site in the *OP4* promoter (Lockhart *et al.*, 1998). Together, these results suggest that a Mcm1-like protein plays a regulatory role in opaque-phase-specific gene expression. To make the story even more provocative, we discovered in recharacterizing the OP4 promoter sequence that an α2-binding site resides 5 bases upstream of the *OP4* MADS box-binding site, and an α1-binding site resides immediately downstream (Lockhart *et al.*, 2003) (Figure 9.11A), a highly unlikely coincidence.

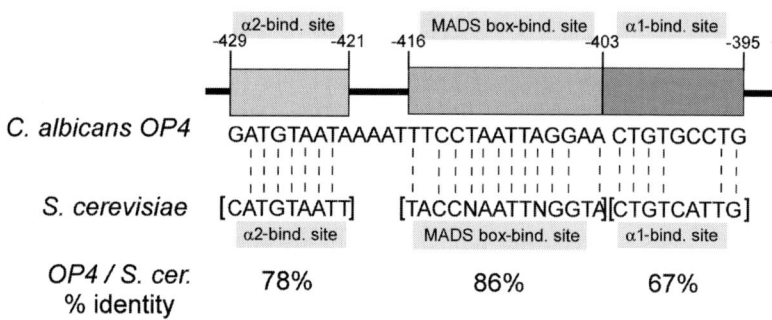

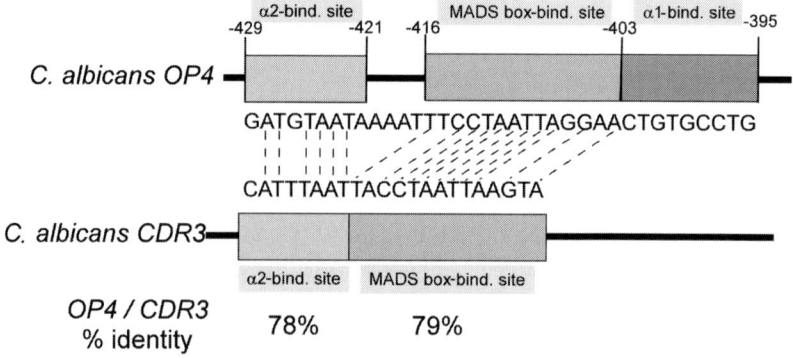

Figure 9.11 (A) Homologies between the *C. albicans OP4* promoter activation site and *S. cerevisiae* sites for the MADS box-binding site, α2-binding site and α1-binding site. (B) Homologies between the *C. albicans OP4* and *CDR3* promoters for the MADS box-binding sites and α2-binding sites.

If opaque-phase genes are regulated by a common DNA binding protein and hence a common *cis*-acting sequence, then other opaque-phase-specific genes should be regulated by a MADS box consensus-binding site. The *PEP1* (*SAP1*) promoter contains a MADS box consensus-binding site sequence 60% identical to that in the *OP4* promoter, but this sequence is outside the activation region (Lockhart *et al.*, 2003). When this sequence was substituted for the endogenous MADS box-binding site sequence in the *OP4* promoter, it did not support activation (Lockhart *et al.*, 1998). Therefore, *OP4* and *PEP1* (*SAP1*), although coordinately expressed, are not regulated by common *cis*-acting sequences. In contrast, the promoter of *CDR3* contains a MADS box consensus-binding site in its single phase-specific regulatory sequence (Lockhart *et al.*, 2003). This sequence functions in both activation of *CDR3* transcription in the opaque phase and repression of *CDR3* transcription in the white phase (Lockhart *et al.*, 2003). This sequence is 79% identical to the *OP4* MADS box-binding site sequence and is bordered upstream by an α2-binding site that is 78% identical to the α2-binding site upstream of the *OP4* MADS box-binding site (Figure 9.11B). No sequence comparable to an α1-binding site was found downstream (Figure 9.11B). These results are the first to suggest that at least a subset of opaque-phase-specific genes may be coordinately regulated by the same *cis*-acting sequences and hence by the same DNA binding proteins in activation complexes. They further suggest that Mcm1p plays a role in the regulation of phase-specific gene expression. The proximity of the α2-binding site to the MADS box-binding sequences in the two genes also suggests a role for α2. However, the selective presence of the α1-binding site downstream of the MADS box-binding sequence suggests differential regulation as well.

PHASE-SPECIFIC *TRANS*-ACTING FACTORS

The results obtained so far demonstrate that phase-specific genes are regulated by phase-specific *trans*-acting factors that function through *cis*-acting sequences as both activators and repressors. Gel mobility shift assays using the wDAS and wPAS sequences from the *WH11* promoter and white- and opaque-phase cell extracts provided the first direct evidence of phase-specific factors. One white-phase-specific complex was identified with wDAS and two with wPAS (Srikantha *et al.*, 1995). No opaque-phase-specific complexes were identified in these experiments (Srikantha *et al.*, 1995). Gel mobility shift assays using the AR2 sequence of the *EFG1* promoter, and white- and opaque-phase cell extracts have revealed a white-phase-specific activation complex (Lachke *et al.*, 2003). The next step in this reverse genetics approach will be to identify the phase-specific *trans*-acting factors that interact with the identified *cis*-acting sequences.

At least two *trans*-acting factors that are regulated by switching have already been identified. First it has been demonstrated that the gene encoding the *trans-*

acting factor Mcm1p is expressed at a far higher level in white-phase cells than in opaque-phase cells (Srikantha et al., 2001). Second, it has been demonstrated that the gene encoding the *trans*-acting factor Efg1p is expressed as a more abundant 3.2 kb transcript in the white phase, and as a less abundant 2.2 kb transcript in the opaque phase (Srikantha et al., 2000). To test the role of *EFG1* in the white–opaque transition, a deletion mutant was generated in *C. albicans* strain WO-1. The deletion mutant could switch between the white and opaque phases and turn white- and opaque-phase genes on and off in the appropriate manner (Srikantha et al., 2000). However, cells in the white phase exhibited the elongate bean shape of cells in the opaque phase (Srikantha et al., 2000). These cells did not, however, form pimples on the cell wall, demonstrating that the aberration was solely in cell shape. These results demonstrate that *EFG1* plays a role in the downstream expression of a portion of the white-phase phenotype, namely the genesis of a round cell shape. The white-phase-specific genes under the regulation of Efg1p that are responsible for generating the round shape of a white-phase yeast cell have not yet been identified.

POSSIBLE SWITCHING MECHANISMS

Although the discovery of switching occurred more than 15 years ago, the exact molecular basis of the switch event has not been elucidated. The slow progress in elucidating the switch mechanism has in large part been due to the absence of a genetic system in *C. albicans*. However, mutational analyses and a reverse genetics approach (Soll, 1997) have provided some insights into mechanisms involved in or which influence the switch event. There are, basically, two obvious mechanisms that may be the basis of the white–opaque transition. The first involves a change in DNA organization. Examples of switching systems that involve DNA reorganization include *S. typhimurium* (Andrews, 1922; Glasgow et al., 1989), *E. coli* (Branton, 1959; Glasgow et al., 1989), *B. hermsii* (Barbour, 1989), *Neisseria gonorrhoeae* (Swanson and Koomey, 1989), *T. brucei* (Donelsen, 1989), mating type interconversion in *S. cerevisiae* and *S. pombe* (Klar, 1989), and class switching of the immunoglobulin heavy chain (Lutsker and Alt, 1989). The mechanisms of DNA reorganization that have evolved in the different organisms are diverse, the subject of the majority of chapters in this volume.

The second switch mechanism involves chromatin reorganization and is best exemplified in experiments performed in *S. cerevisiae* that demonstrate positional effects of gene expression caused by sub-telomeric regions of the chromosome (Pilus and Rhine, 1989; Gottschling et al., 1990; Aparacio et al., 1991). When the *ADE2* gene of *S. cerevisiae* was moved close to a sub-telomeric region, it switched spontaneously between an expressed and an unexpressed state at a frequency of 10^{-3} (Gottschling et al., 1990). If the gene *SIR2*, which is involved in gene silencing (Gartenberg, 2000), is deleted in these strains, *ADE2* is constitutively expressed (Gottschling *et al.*, 1990, Aparacio *et al.*, 1991). These experiments have been interpreted as follows. Sub-telomeric regions cause metastable changes in chromatin state in associated chromosome regions. The chromatin of genes within an associated region switch between inactive and active states. These states may include the acetylation and deacetylation, respectively, of chromatin. Such a mechanism is compatible with the white–opaque transition. As will become clear in the discussion that follows, evidence has accumulated which is consistent with, but which does not definitively prove, such a mechanism for the white–opaque transition.

GROSS CHROMOSOMAL REARRANGEMENTS ARE NOT THE BASIS OF SWITCHING

Switching in *C. albicans* strain WO-1 is perfectly reversible. Cells that have passed through 50 sequential transitions still switch at the same frequency as the original clone. In addition, cell lineages that have been monitored through several white–opaque transitions have the same karyotype (Soll, 1992a). In *C. albicans* strain 3153, it has been demonstrated that variant phenotypes in a high-frequency mode of switching undergo frequent changes in karyotype, restricted almost entirely to those chromosomes harbouring the repeated ribosomal RNA cistrons (Ramsey *et al.*, 1994). However, the specific changes observed in switch lineages do not correlate with the precise phenotypes, demonstrating that cells switching frequently also undergo high-frequency rDNA reorganization (Ramsey *et al.*, 1994), in a fashion similar to *S. cerevisiae* null mutants of the transcription-silencing gene *SIR2* (Gottlieb and Esposito, 1989).

THE DEACETYLASES PLAY A ROLE IN THE SUPPRESSION OF SWITCHING

Chu *et al.* (1992) provided the first evidence that genes exist that suppress switching. They fused an auxotrophic strain of WO-1, which switched between white and opaque, with two unrelated strains, neither of which underwent the white–opaque transition. By correlating switching, or the lack of switching, between white and opaque with chromosome retention after heat shocking fusants, they concluded that genes essential for the white–opaque transition resided on chromosome 3, and that repressors of switching were located on chromosome 2, 5 or 6.

The suppression of gene expression in yeast and other eukaryotes has been demonstrated to involve silencing genes, the most prominent of which is *SIR2* (Gartenberg, 2000), and the deacetylases, which mediate the deacetylation of histones H3 and H4 (Workman and Kingston, 1998; Ayer, 1999). Iwai *et al.*

(2000) demonstrated that Sir2p is in fact an NAD-dependent deacetylase. Perez-Martin *et al.* (1999) first deleted a *SIR2* homologue in *C. albicans* strain CAI4, which undergoes 3153A-like switching (Srikantha *et al.*, 2001). The null mutant exhibited an increase in 3153A-like phenotypic switching, spontaneous hypha formation at 30 °C and an increase in chromosomal reorganization (Perez-Martin *et al.*, 1999). Pujol and Soll (2003) subsequently deleted *SIR2* from *C. albicans* strain WO-1 and observed no effect on the basic frequency of the white–opaque transition. They did, however, observe that the sir2⁻ mutant of strain WO-1 exhibited an increase in switching between phenotypes analogous to the 3153A-type switching system, suggesting that strain WO-1 underwent the predominant white–opaque transition, and at very low frequency the 3153A-like switching system, and only the latter was suppressed by Sir2p.

The major deacetylases in *S. cerevisiae* include *RPD3* (Vidal and Gaber, 1991), *HDA1* (Rundlett *et al.*, 1996), and *HOS1*, *HOS2* and *HOS3* (Rundlett *et al.*, 1996). Homologues to all of these genes have been cloned from *C. albicans* (Srikantha *et al.*, 2001). In an initial test of the role of these deacetylases in switching, Klar *et al.* (2001) tested the effects of trichostatin A, a histone deacetylase inhibitor (Yoshida *et al.*, 1995; Ekwall *et al.*, 1997), on the white–opaque transition. Trichostatin A selectively caused an increase in the frequency of switching from the white to opaque phase, but had no effect on the frequency of switching from the opaque to white phase (Klar *et al.*, 2001). Since the effect of trichostatin A is most severe on Hda1p (Carmen *et al.*, 1996), *HDA1* was deleted from *C. albicans* strain WO-1 and switching tested in the mutant. The frequency of switching from white to opaque was increased in the mutant, but the frequency from opaque to white was unaffected, a phenocopy of trichostatin A-treated cells (Klar *et al.*, 2001; Srikantha *et al.*, 2001). Deletion of *RPD3* also affected the frequency of switching, but in this case, the mutant exhibited increased frequencies in both the white to opaque and the opaque to white direction (Srikantha *et al.*, 2001). Deletion of *HOS3* had no effect on switching in either direction (Yong *et al.*, 2003). Deletion of *HDA1* resulted in reduced white-phase-specific expression of the *EFG1* (3.2 kb) transcript, and deletion of *RPD3* resulted in reduced opaque-phase-specific expression of *OP4*, *SAP1* and *SAP3* (Srikantha *et al.*, 2001).

The characterization of the deacetylase mutants demonstrated that HDA1 and RPD3 play distinct roles in the suppression of switching, and in the expression of phase-specific genes. Since the deacetylases play a fundamental role in the suppression of gene transcription, we must first consider the possibility that they function directly to suppress switching by deacetylating histones 3 and 4 at the site of the switch. A switch, therefore, would involve acetylation through a histone acetyltransferase. However, we must also consider the possibility that suppression by the deacetylases is effected at gene loci encoding *trans*-activators of switching, and not at a master switch locus. In the case of phase-specific gene regulation, the deacetylases uniformly play roles in gene activation. This suggests that the deacetylases function to suppress other silencing genes, like those encoding the Sir proteins. In *S. cerevisiae*, deletion of *RPD3* results in increased repression of the mating type loci, telomeres and ribosomal DNA (DeRubertis *et al.*, 1996; Rundlett *et al.*, 1996; Hassig *et al.*, 1998), presumably through upregulation of Sir proteins, which function as transcription repressors.

HEAT-INDUCED MASS CONVERSION AND A MODEL FOR SWITCHING

In all of the mutant studies described in previous sections, two characteristics of the switching process were assessed. First, mutants were tested for their capacity to switch. Second, the steady-state phenotypes of the alternate phases (white versus opaque), including cell morphology, cell ultrastructure, cell physiology and phase-specific gene expression, were compared. These analyses did not include a biochemical analysis of the actual switch event. To achieve this, one must be able to induce mass synchronous conversion from one phase to the other, so that a population undergoing synchronous conversion can be sampled up to, at and after the switch event, and the population, because of its synchrony, can then be considered at each time point representative of the single cell. Heat-induced mass conversion from the opaque to white phase affords us with such a dynamic preparation. When a homogenous population of opaque-phase cells is shifted from 25 °C to >36 °C (e.g. 42 °C), synchronous mass conversion to the white phase is initiated (Slutsky *et al.*, 1987; Rikkerink *et al.*, 1988; Morrow *et al.*, 1993; Srikantha and Soll, 1993). By then returning cells to 25 °C at time intervals, the precise time of the switch from opaque to white can be defined. Prior to the switch event, or point of 'phenotypic commitment', cells transferred back to 25 °C will continue multiplying in the opaque-phase phenotype, but after the switch event, cells transferred back to 25 °C will then multiply in the white-phase phenotype. In Figure 9.12A, cell concentration and percentage white-phase cells are plotted as a function of the time after a shift of opaque-phase cell from 25 °C to 42 °C. After a 1.5 hour lag, cellular divisions occur semisynchronously (completion of each cell doubling is noted by a numbered arrow in Figure 9.12A). While 0% of the population had switched to the white phase through the first cell doubling, 70% had switched through the second cell doubling, and 83% had switched through the third cell doubling (Figure 9.12A). While *WH11* was not expressed through the first cell division, it began to be expressed at the time of the second cell doubling (Srikantha and Soll, 1993), and while both *OP4* and *SAP1* expression could be induced prior to the second cell doubling, it could not be induced after the second cell doubling (Morrow *et al.*, 1993) (Figure 9.12A).

In Figure 9.12B, a model that is consistent with these data is presented. In the model, the two alleles of a master switch locus, located on the two homologous chromosomes 'a' and 'b', can exist in either of two states, the white state (Wh)

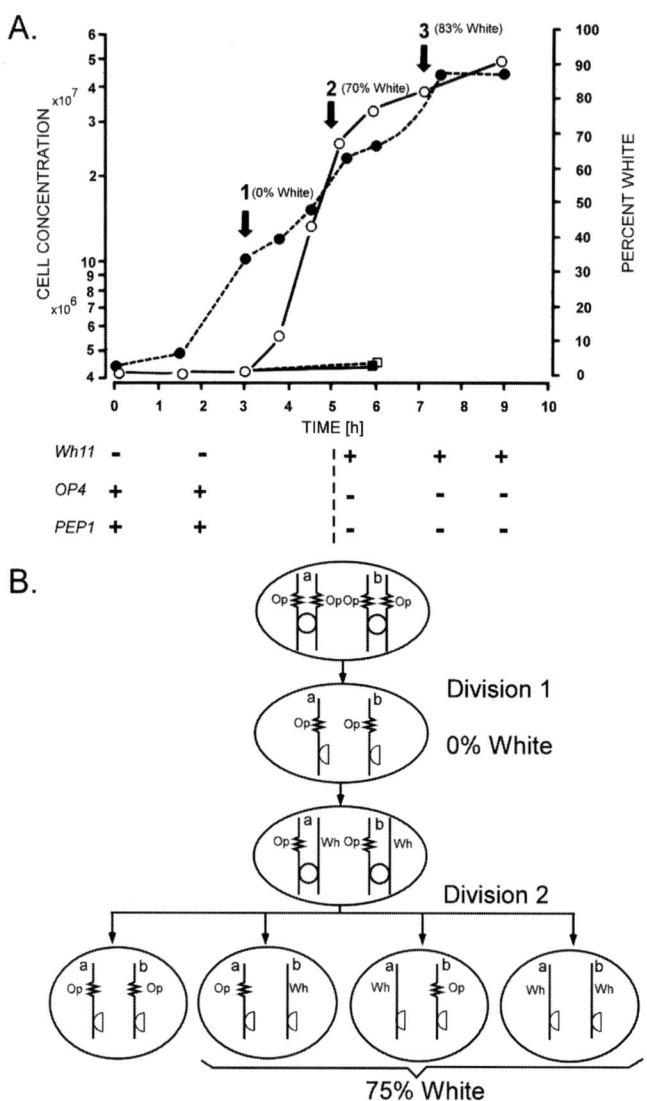

Figure 9.12 A model of switching based on data obtained from semisynchronous mass conversion from opaque to white. (A) A mid-log-phase population of opaque-phase cells was transferred to 42 °C. After a 1.5 hour lag, cells progressed through three semisynchronous doublings (numbered arrows). At time points, cells were removed from the 42 °C culture and plated at 25 °C to assess phenotype through the assessment of colony morphology. The percentage white-phase cells is coplotted with cell concentration, and the percentage white-phase cells after each cell doubling is presented in parentheses (see Srikantha and Soll, 1993). (B) In the model, the switch locus is represented in the recessive opaque state (Op) by a zigzagged line and in the dominant white state (Wh) by a straight line on each chromatid. A sphere between chromatids and hemisphere on a single chromatid represent the centromere. The predicted percentage Wh is based on the assumption that the Wh state is dominant to the Op state at the master switch locus. Note the estimates obtained from the model in panel B are similar to the results obtained in the experiment in panel A.

represented as a straight line or the opaque state (Op) represented as a zigzagged line. The white state is dominant. Therefore, a cell must possess both alleles of the master switch locus in the Op state to be opaque, but need only possess one of the two alleles in the Wh state to be white. In the model, synchronized cells in the opaque-phase population just prior to the first cell division contain both replicated alleles of the master switch locus in the Op state. Therefore, after the first cell division all daughter cells contain both alleles in the Op state. However, after chromatid duplication, all newly replicated versions of the master switch locus are in the Wh state. During the second cell division, random segregation of the chromatids of the homologues a and b leads to four allelic combinations: aOp:bOp, aOp:bWh, aWh:bOp and aWh:bWh (Figure 9.12B). Since Wh is dominant, 75% of the daughter cells will be white. In the third cell division, 96% of the daughter cells will be white. The percentage white-phase cells predicted from the model in Figure 9.12B after the first, second and third cell doublings would, therefore, be 0, 75 and 96%, respectively, and the observed percentage white-phase cells obtained in the experiment in Figure 9.12A was 0, 70, and 83%, respectively.

If the white state of the master switch locus is dominant, then the transition from opaque to white should entail a single allelic change, but the transition from white to opaque should entail two allelic changes. Therefore, the transition from opaque to white should involve only one step, but the transition from white to opaque should involve two steps, and should pass through a 'heterozygous' intermediate. Interestingly, Bergen *et al.* (1990) demonstrated in single cell analyses of the white–opaque transition over a decade ago that opaque-phase cells directly formed white-phase daughter cells, while white-phase cells formed an elongate, pseudohypha-like intermediate in the transition to opaque, which is consistent with the model.

THE ROLE OF *C. ALBICANS* SWITCHING IN PATHOGENESIS

When high-frequency, reversible switching in *C. albicans* strain 3153A was first reported in the journal *Science* by Slutsky *et al.* (1985), it was suggested by the authors that the process played a role in pathogenesis. This obvious suggestion was deemed controversial, and letters were written to the editor arguing against this suggestion, even in the face of the growing body of evidence demonstrating the role of antigenic variability and switching in the virulence of other pathogens. Since that time, a number of experiments and studies have been performed and observations made that support a role for switching in pathogenesis. First, a study was performed in which cells were plated at low density directly from the vaginal canals of 11 vaginitis patients infected with *C. albicans*, so that the colony phenotypes of individual cells in the vagina could be immediately assessed (Soll *et al.*, 1987). Four of the 11 tested samples contained cells with multiple colony phenotypes. In all cases, DNA fingerprinting with the complex probe Ca3 (Pujol *et al.*, 1997; Soll, 2000) demonstrated that the multiple phenotypes represented the same strain, and that switching, therefore, occurred at the site of infection. Subsequent plating experiments verified that cells switched reversibly and at high frequency between the observed phenotypes (Soll *et al.*, 1987). It was subsequently demonstrated that in a sequence of three recurrent vaginitis episodes caused by *C. albicans*, cells switched to a different morphological phenotype in each episode following drug therapy (Soll *et al.*, 1989). DNA fingerprinting and plating experiments again demonstrated that the different phenotypes represented switch phenotypes of the same strain. In a comparison of primary isolates from the oral cavities of healthy individuals and individuals suffering from oral candidiasis, the data suggested that infecting strains switched on average at higher frequencies than commensal strains (Hellstein *et al.*, 1993). Jones *et al.* (1994) performed a comparison of switching in isolates from deep mycoses and superficial mycoses, and found that the former switched on average at higher frequencies than the latter. Finally, Vargas *et al.* (2000) performed a comparison of isolates from the oral cavity of healthy and HIV-positive individuals, and found that isolates from the latter individuals switched on average at higher frequencies. In this study (Vargas *et al.*, 2000), the first convincing evidence was presented of the dramatic effects switching had on both aspartyl proteinase secretion and drug resistance of fresh clinical isolates.

The results reported by Vargas *et al.* (2000) on the impact of switching on virulence traits reinforced a number of observations that had previously been made over the past 14 years on the effects of the white–opaque transition on putative virulence factors. It had been reported that the white–opaque transition affected (1) cell adhesion and cohesion (Kennedy *et al.*, 1988; Vargas *et al.*, 1994); (2) proteinase secretion (Morrow *et al.*, 1992; White *et al.*, 1993; Hube *et al.*, 1994); (3) drug susceptibility (Soll *et al.*, 1991); (4) antigenicity (Anderson and Soll, 1987; Anderson *et al.*, 1990); (5) restraints on the bud–hypha transition (Anderson *et al.*, 1989); and (6) sensitivity to white blood cells and oxidants (Kolotila and Diamond, 1990). Therefore, the capacity to switch between a limited number of general phenotypes, each expressing a different combination of virulence traits, represents a strategy for phenotypic variability which could provide the basis for rapid adaptation (Soll, 1992b; Odds, 1997). Rapid adaptation in the form of enrichment of a minor switch variant in a population could occur in response to drug therapy, the host immune response, a change in body location, entrance into a biofilm, a change in host physiology and a change in the normal competing microflora.

Finally, the use of animal models supports a role for switching in pathogenesis. In the very commonly used mouse model for systemic infections, white-phase cells are far more virulent that opaque-phase cells (Kvaal *et al.*, 1997, 1999). In a standard tail injection experiment, while 50% death occurs at approximately 6 days in white-phase-cell-injected animals, 50% death occurs after 17 days in opaque-phase-cell-injected animals (Kvaal *et al.*, 1999). The delay in the latter animals correlates with the conversion of opaque-phase cells to the white phase, suggesting white is highly virulent and opaque nonvirulent (Kvaal *et al.*, 1997). In contrast, in a mouse model of cutaneous infection, opaque-phase cells are far more successful at colonizing skin than white-phase cells (Kvaal *et al.*, 1999), providing proof for the first time that alternative switch phenotypes exhibit alternative pathogenic specialization. In addition, the misexpression of phase-specific genes provided some insight into the roles of these genes in virulence. Misexpression of the white-phase-specific gene *WH11* in the opaque phase led to increased opaque-phase instability and a dramatic increase in the frequency of switching from the opaque to white phase (Kvaal *et al.*, 1997). *WH11* misexpression mutants in the opaque phase were far more virulent in the systemic mouse model than wild-type cells in the opaque phase by virtue of the increased frequency of switching to the more virulent white phase (Kvaal *et al.*, 1997). On the other hand, misexpression of the opaque-phase-specific aspartyl proteinase gene *SAP1* in the white phase had no effect on the already high level of virulence of white-phase cells in the mouse systemic model, but it did increase dramatically the capacity of white-phase cells to colonize skin (Kvaal *et al.*, 1999). In this latter case, the misexpression of a single phase-specific gene conferred virulence in a particular model, suggesting that phase-specific virulence in this case was dependent on the expression of a single phase-specific gene.

SWITCHING IN *C. GLABRATA*

C. glabrata has emerged as a major commensal in the oral cavity of elderly people (Lockhart *et al.*, 1999) and as the second most common yeast bloodstream infection (Pfaller *et al.*, 1998). Although a member of the genus *Candida*, *C. glabrata* is far more closely related to *S. cerevisiae* than to *C. albicans* (Barnes *et al.*, 1991; Pesole *et al.*, 1995). The most troublesome aspect of *C. glabrata* pathogenicity is its natural drug resistance. Bloodstream infections and chronic infections, such as recurrent vaginitis, caused by *C. glabrata* are commonly refractory to treatment. Up until 1999, *C. glabrata* was somewhat of an enigma. Although it was second to *C. albicans* in most aspects of infection and commensalism (Odds, 1988; Pfaller *et al.*, 1998), it appeared to undergo neither the bud–hypha transition nor high-frequency phenotypic switching, two developmental programs that contribute to the pathogenic success of *C. albicans*. Hence it should have been no surprise to learn recently that *C. glabrata* does indeed undergo the bud–pseudohypha transition (Csank and Haynes, 2000; Lachke *et al.*, 2002b), forms invasive tubes (Lachke *et al.*, 2002b) and switches reversibly and frequently between a number of colony phenotypes (Lachke *et al.*, 2000, 2002b).

Cells of *C. glabrata* strain 35B11 switch reversibly and at high frequency between a white (Wh), light brown (LB), dark brown (DB) and very dark brown (vDB) colony phenotype that can be visualized on agar containing 1mM

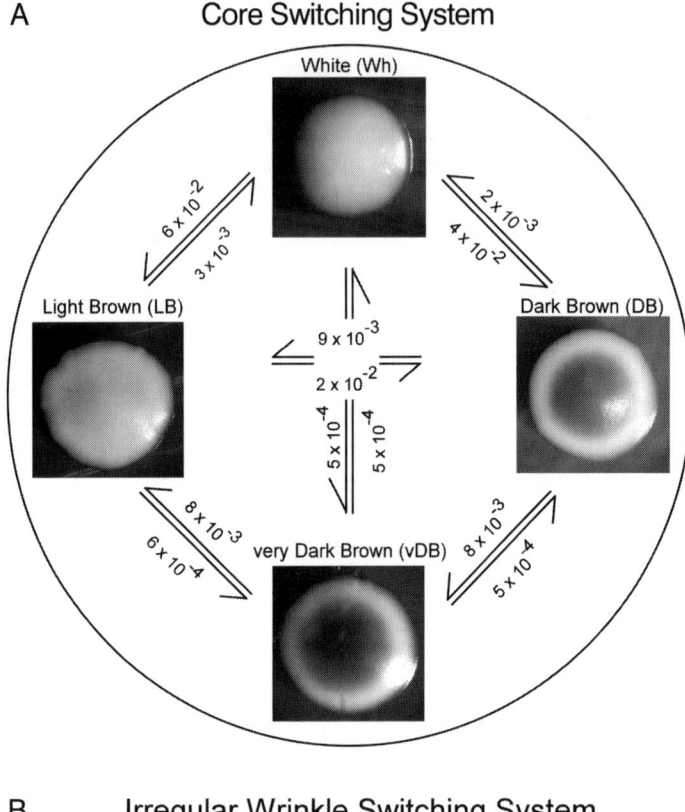

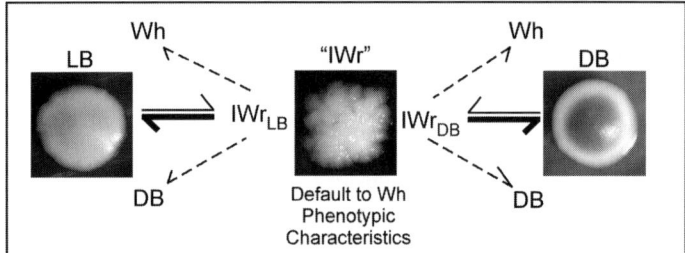

Figure 9.13 The two switching systems in *C. glabrata*. (A) The core switching system based on graded brown coloration on agar containing $CuSO_4$ (here shown as graded levels of grey). The value above each arrow refers to the frequency of the phenotype pointed to by each arrow in a clonal population of the phenotype from which the arrow emanates. (B) In the irregular wrinkle (IWr) switching system, light brown (LB) and dark brown (DB) cells switch at low frequency to irregular wrinkle, IWr_{LB} and IWr_{DB}, respectively. IWr can switch to core phenotypes, but has a propensity (i.e. exhibits the highest frequency and represented by thick arrow) to switch back to the core phenotype from which it emanates, hence IWr_{LB} and IWr_{DB}. An IWr colony is composed almost exclusively of pseudohyphal cells that exhibit characteristics of the core phenotype Wh. (For details, see Lachke *et al.* (2000) and Lachke *et al.* (2002b) *Microbiol.*, from which the figure is reproduced with permission.)

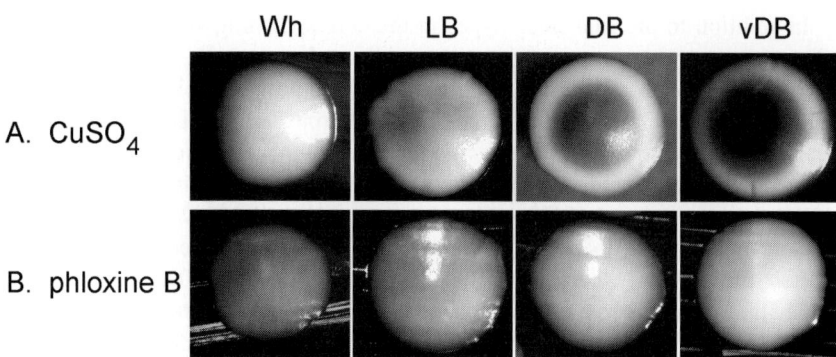

Figure 9.14 The core switch phenotypes of *C. glabrata* exhibit graded brown coloration (here shown as graded levels of grey) on agar containing CuSO$_4$ according to the intensity hierarchy Wh<LB<DB<vDB. When cells from each core phenotype are then grown on agar containing phloxine B, but lacking CuSO$_4$, they exhibit graded pink coloration according to the intensity hierarchy Wh>LB>DB>vDB, the reverse of that on CuSO$_4$. (For details, see Lachke *et al.* (2002b) *Microbiol.*, from which the figure is reproduced with permission.)

CuSO$_4$ (Lachke *et al.*, 2000). This core switching system has been demonstrated in the majority of tested *C. glabrata* isolates (45 out of 55), (Lachke *et al.*, 2002b). It is shown diagrammatically in Figure 9.13A. It should be immediately evident that the phenotypes in the core switching system are graded according to coloration caused by reduction of CuSO$_4$ to Cu$_2$S. If cells from the four core phenotypes are plated on agar containing phloxine B but lacking CuSO$_4$, each phenotype exhibits a different colour intensity from white to red, but the gradation of phloxine B coloration is opposite that of CuSO$_4$ coloration (i.e. Wh>LB>DB>vDB) (Lachke *et al.*, 2002b) (Figure 9.14). At the cellular level, colonies of the Wh, LB, DB and vDB phenotypes at 6 days contain comparable proportions of buds, pseudohyphae and tubes (Lachke *et al.*, 2002b). However, the switch phenotypes express select switching-regulated genes in a graded fashion. For example, cells of the four core switch phenotypes express the metallothionein gene *MTII* in the hierarchy Wh<LB<DB≤vDB, while expressing a variety of constitutively expressed genes (e.g. *TRP1*, *MTI*, *AMT1*, *EPA1*, *PDH1*, *H1S3*) at the same levels (Lachke *et al.*, 2002b). The frequency of switching also shows gradation in the hierarchy Wh>LB>DB>vDB (Figure 9.15).

CuSO$_4$ coloration	Wh < LB < DB < vDB
Phloxine B coloration	Wh > LB > DB > vDB
Expression MT-II	Wh < LB < DB ≤ vDB
Frequency of switching	Wh > LB > DB > vDB

Figure 9.15 A synopsis of graded characteristics of the phenotypes expressed by *C. glabrata* in the core switching system (Lachke *et al.*, 2000, 2002b).

In addition to the four phenotypes in the core switching system, *C. glabrata* also switches reversibly to at least one additional phenotype, irregular wrinkle (IWr) (Lachke *et al.*, 2002b). Switches to IWr have now been obtained in LB and DB population. The IWr phenotypes obtained from LB and DB are phenotypically indistinguishable. Both colony types are composed almost exclusively of pseudohyphal cells and both exhibited the coloration, switching frequency and *MHII* expression level of Wh cells. However, IWr cells derived from LB cells switch preferentially back to LB cells, and IWr cells derived from DB cells switch preferentially back to DB cells (Lachke *et al.*, 2002b). Hence, IWr emanating from LB retains, or 'remembers', its LB origin, and IWr emanating from DB retains, or remembers, its DB origin. These results demonstrate that a switch to irregular wrinkle results in expression of the Wh phenotype in addition to a restriction in cellular phenotype, but that IWr cells retain a core phenotype identity. These results suggest that switching to IWr represents a second switching system, which occurs independently of the core switching system, can occur while cells are expressing any of the core phenotypes, and can impact the terminal phenotypic characteristics under the control of the core switching system.

Because switching has only recently been demonstrated in *C. glabrata*, virtually nothing is known concerning its role in pathogenesis. There should be little doubt, however, that it plays a role, especially when one considers the fact that it provides reversible variability in colonizing populations in both gene expression and invasive phenotypes (Lachke *et al.*, 2000, 2002b). Interest in *C. glabrata* switching will no doubt be stimulated by the unique experimental opportunities this system affords in pursuing the mechanism of switching. First, because it is so closely related to *S. cerevisiae* one may be able to draw from the wealth of information on the regulatory circuitry of *S. cerevisiae* in predicting the molecular pathways involved in *C. glabrata* switching. Second, because it is haploid, mutagenesis strategies such as 'restriction enzyme mediated integration' (REMI) can be used in conjunction with colony morphology screens to identify genes that are involved in switching. Such strategies are far less effective when applied to diploid organisms such as *C. albicans* and have, therefore, not been used in the search for genes involved in *C. albicans* switching. Third, the gradation of switching-regulated phenotypic characteristics in either a Wh<LB<DB<vDB hierarchy or a Wh>LB>DB>vDB hierarchy is unique, with few obvious counterparts in the switching systems of other organisms. The mechanisms regulating gene expression in this system may, therefore, be distinct from those regulating gene expression in other yeast switching systems, including those now being analysed in *C. albicans*. In the switching systems in *C. albicans*, genes are turned either on or off in the different phases. Finally, *C. glabrata* provides us with the first apparent example of an organism that possesses two interacting but distinct switching systems (i.e. the core switching system and the IWr switching system). Although different switching systems have been demonstrated in the species *C. albicans*, they have been assigned to different strains. In the strain *WO-1*, there is a hint of multiple switching systems, the white–opaque transition and a 3153A-like switching system, but the latter appears to occur at low frequency, and no data has been obtained which proves interaction between the two switching systems.

PHASE VARIATION IN THE FUNGI: GENERAL LESSONS AND SUMMARY

The infectious bacteria have taught us that phase transitions provide a pathogen with mechanisms for rapid adaptation in response to environmental, therapeutic and host challenges. Each population contains variants that can enrich in a population when cells expressing the dominant phenotype are cleared. That means that different phenotypes either provide different adaptive advantages or represent escape mechanisms from a specific immune response. When one compares the mechanisms that have evolved to generate antigenic variability in *S. typhimurium*, *B. hermsii*, *N. gonorrhoeae* and the trypanosomes, it is evident that different molecular solutions have evolved to effect the same end. In *S. cerevisiae* and *S. pombi*, elegant systems have evolved for changing mating types, which involve conserved DNA reorganizational events. In these latter systems, the reorganization of one locus has pleiotrophic effects on a variety of phenotypic characteristics. Again, the mechanisms involved in the mating type transitions have no exact counterparts in the bacteria. In *C. albicans*, switching appears to play a role similar to that in bacterial pathogens, but like the mating type systems in yeast, the effect of a switch is pleiotropic. In addition, switching in the pathogenic yeast can occur between a number of phenotypes (e.g. seven in strain 3153A (Slutsky *et al.*, 1985) or just two (e.g. the white–opaque transition in strain WO-1) (Slutsky *et al.*, 1987) and one strain can possess two switching systems. Based on recent characterization of the role of the deacetylases in switching, it will not be surprising to find that the different switching systems involve different switch loci and different molecular mechanisms. In *C. glabrata*, graded switching appears to be relatively distinct from switching in *C. albicans* and undoubtedly involves a unique graded mechanism for the transcription of developmentally regulated genes. Switching to IWr in *C. glabrata* appears to be distinct from the core switching system, but interacts with it, providing for the first time an experimental model for delving into the interactions of distinct switching systems. It is clear from the preceding review of switching in *Candida* species that the focus of current research is first on the mechanisms involved in the regulation of phase-specific or switching-regulated genes, and second on the elucidation of the switch event. It seems apparent that elucidation of the mechanism basic to one system may not be applicable to other systems. The mechanism of switching in *C. albicans*, the oldest identified switching system in the infectious fungi, has still not been elucidated, primarily because of the diploid nature of this organism and the absence of a classical genetic system for the analysis of mutants. Reverse genetic approaches are under way, but they are by nature slow. *C. glabrata* on the other hand is haploid and allows us for the first time the use of strategies such as REMI mutagenesis combined with screens of colony morphology to identify genes involved in the switching process.

ACKNOWLEDGEMENTS

The author is indebted to Dr T. Srikantha, Dr S. Lockhart, Dr C. Pujol, Mr R. Zhao and Mr S. Lachke for sharing unpublished data, to Dr K. Daniels in developing the figures, and to Rebecca Brincks for help in assembling the manuscript. The recent work from the Soll laboratory was supported by NIH grant AI2392.

REFERENCES

Acton, T., Zhang H. and Vershon, A. (1997) *Mol. Cell Biol.* 17, 1881–1889.
Anderson, J. and Soll, D. R. (1987) *J. Bacteriol.* 169, 5579–5588.
Anderson, J., Cundiff, L., Schnars, B., Gao, M., Mackenzie, I. and Soll, D. R. (1989) *Infect. Immun.* 57, 458–467.
Anderson, J., Mihalik, R. and Soll, D. R. (1990) *J. Bacteriol.* 172, 224–235.
Andrews, F. W. (1922) *J. Pathol. Bacteriol.* 25, 1509–1514.
Aparacio, O. M., Billington, B. C. and Gottschling, D. E. (1991) *Cell* 66, 1279–1287.
Aramayo, R., Peleg, Y., Addison, R. and Mutzenberg, R. (1996) *Genetics* 144, 991–1003.
Ayer, D. E. (1999) *Trends in Cell Biol.* 9, 193–198.
Balan, I., Alareo, A. and Raymond, M. (1997) *J. Bacteriol.* 179, 7210–7218.
Barbour, A. (1989) In *Mobile DNA* (eds D. E. Berg and M. M. Howe), American Society of Microbiology, Washington, DC, pp. 783–790.
Barnes, S. M., Lane, D. J., Sogin, M. L., Biblau, C. and Weisburg, W.G. (1991) *J. Bacteriol.* 173, 2250–2255.
Bedell, G. and Soll, D .R. (1979) *Infect. Immun.* 26, 348–354.
Bergen, M., Voss, E. and Soll, D. R. (1990) *J. Gen. Microbiol.* 136, 1925–1936.
Branton, C. C. (1959) *Nature* 183, 782–786.
Brown-Thomsen, J. (1968) (Robin) Berkhous. *Hereditas* 60, 335–398.
Carmen, A. A., Rundlett, S. E. and Grunstein, M. (1996) *J. Biol. Chem.* 271, 15837–15844.
Chu, W-S., Rikerink, E. H. A. and Magee, P. T. (1992) *J. Bacteriol.* 174, 2951–2957.
Csank, C. and Haynes, K. (2000) *FEMS Microbiol. Lett.* 189, 115–120.
DeRubertis, F., Kadosh, D., Henchoz, S., Pauli, D., Reuter, G., Struhl, K. and Spierer, P. (1996) *Nature* 384, 589–591.
Donelsen, J. E. (1989) In *Mobile DNA* (eds D. E. Berg and M. M. Howe), American Society of Microbiology, Washington, DC, pp. 763–782.
Dutton, J. R., Johns, S. and Miller, B. L. (1997) *EMBO J.* 16, 5710–5721.
Ekwall, K., Olsson, T., Turner, B. M., Cranson, G. and Allshire, R. C. (1997) *Cell* 91, 1021–1032.

Fries, B. C., Goldman, D. C., Cherniak, R., Ju, R. and Casadevall, A. (1999) *Infect. Immun.* 67, 6076–6083.

Gartenberg, M. R. (2000) *Curr. Opin. Microbiol.* 3, 132–137.

Gimeno, C. J. and Fink, G. R. (1994) *Mol. Cell. Biol.* 14, 2100–2112.

Glasgow, A. C., Hughs, K. T. and Simon, M. I. (1989) In *Mobile DNA* (eds D. E. Berg and M. M. Howe), American Society of Microbiology, Washington, DC, pp. 637–659.

Goldman, D., Fries, B., Franzot, S., Montella, L. and Casadevall, A. (1998) *Proc. Natl Acad. Sci. USA* 95, 14967–14972.

Gottlieb, S. and Esposito, S. (1989) *Cell* 56, 771–776.

Gottschling, D. E., Asparacio, O. M., Billington, B. L. and Zakian, V. A. (1990) *Cell* 63, 751–762.

Hassig, C. A., Tong, J. K., Fleischer, T. C., Owa, T., Grable, P. G., Ayer, D. E. and Schreiber, S. L. (1998) *Proc. Natl. Acad. Sci. USA* 95, 3519–3524.

Hellstein, J., Vawter-Hugart, H., Fotos, P., Schmid, J. and Soll, D. R. (1993) *J. Clin. Microbiol.* 31, 3190–3199.

Hube, B., Monod, M., Schofield, D., Brown, A. and Gow, N. (1994) *Mol. Microbiol.* 14, 87–99.

Ishii, N., Yamanato, F., Yoshiro, M., Arisawa, M. and Aoki, Y. (1997) *Microbiol.* 143, 429–435.

Iwai, S., Armstrong, C. M., Kaeberlein, M. and Gaurente, L. (2000) *Nature* 403, 795–800.

Jones, S., White, G. and Hunter, P. G. (1994) *J. Clin. Microbiol.* 32, 2869–2870.

Kennedy, M. J., Rogers, A. C., Hanselman, L. A., Soll, D. R. and Yancey, R. J. (1988) *Mycopath.* 102, 149–156.

Klar, A. J. S. (1989) In *Mobile DNA* (eds D. E. Berg and M. M. Howe), American Society of Microbiology, Washington, DC, pp. 671–691.

Klar, A. J. S., Srikantha, T. and Soll, D. R. (2001) *Genet.* 158, 919–924.

Kolotila, M. P. and Diamond, R. D. (1990) *Infect. Immun.* 58, 1174–1179.

Kvaal, C., Lachke, S. A., Srikantha, T., Daniels, K., McCoy, J. and Soll, D. R. (1999) *Infect. Immun.* 67, 6652–6662.

Kvaal, C., Srikantha, T. and Soll, D. R. (1997) *Infect. Immun.* 65, 4468–4475.

Lachke, S., Joly, S., Simonson, P., Daniels, K. and Soll, D. R. (2003) *Microbiol.* 148, 2661–2674.

Lachke, S., Srikantha, T. and Soll, D. R. (2003) *Mol. Micro.* 48, 523–536.

Lachke, S., Srikantha, T., Tsai, L., Daniels, K. and Soll, D. R. (2000) *Infect. Immun.* 68, 884–895.

Lee, K. L., Buckley, H. R. and Campbell, C. C. (1975) *Sabouraud.* 13, 148–153.

Lo, H. J., Kohler, J. R., DiDomenico B., Loebenberg D., Cacciapuoti, A. and Fink, G. R. (1997) *Cell* 90, 939–949.

Lockhart, S., Radke, J., Stonebraker, C. and Soll, D. R. (2003) In preparation.

Lockhart, S. R., Joly, S., Vargas, K., Swails-Wenger, J., Enger, L. and Soll, D. R. (1999) *J. Dent. Res.* 78, 857–868.

Lockhart, S. R., Nguyen, M., Srikantha, T. and Soll, D. R. (1998) *J. Bacteriol.* 180, 6607–6616.

Lorenz, W. W., McCann, R. O., Langiaru, M. and Cormier, M. J. (1991) *Proc. Natl Acad. Sci. USA* 88, 4438–4442.

Lutsker, S. G. and Alt, F. W. (1989) In *Mobile DNA* (eds D. E. Berg and M. M. Howe), American Society of Microbiology, Washington, DC, pp. 693–714.

Mackinnon, J. E. (1940) *J. Infect. Dis.* 66/67, 59–77.
Miller, K. Y., Wu, J. and Miller, B. L. (1992) *Genes Dev.* 6, 1770–1782.
Morrow, B., Anderson, J., Wilson, E. and Soll, D. R. (1989) *J. Gen. Microbiol.* 135, 1201–1208.
Morrow, B., Ramsey, H. and Soll, D. R. (1994) *J. Med. Vet. Mycol.* 32, 287–294.
Morrow, B., Srikantha, T., Anderson, J. and Soll, D. R. (1993) *Infect. Immun.* 61, 1823–1828.
Morrow, B., Srikantha, T. and Soll. D. R. (1992) *Mol. Cell. Biol.* 12, 2997–3005.
Naglik, J. R., Newport, G., White, T. C., Fernandes-Naglik, L. L., Greenspan, J. S., Greenspan, D., Sweet, S. P., Challacombe, S. J. and Agabian, N. (1999) *Infect. Immun.* 67, 2482–2490.
Negroni, P. (1935) *Rev. Soc. Argent. Biol.* 11, 449–453.
Odds, F. C. (1988) *Candida and Candidiasis.* Bailliere Tindall, London.
Odds, F. C. (1997) *Mycoses* 40(2), 9–12.
Ohama, T., Suzuki, T., Mori, M., Osawa, S., Ueda, T., Watanabe, K. and Nakace T. (1993) *Nucl. Acid Res.* 21, 4039–4045.
Perez-Martin, J., Uria, J. A. and Johnson, A. D. (1999) *EMBO J.* 18, 2580–2592.
Pesole, G., Lotti, M., Alberghina, L. and Saccone, C. (1995) *Genet.* 141, 903–907.
Pfaller, M., Jones, R. N., Messer, S. A., Edmond, M. B., Wenzel, R. P. and Group S. P. (1998) *Diag. Micro. Infect. Dis.* 30, 121–129.
Pilus, L. and Rine, J. (1989) *Cell* 59, 637–647.
Pomes, R., Gil, C. and Nombela, C. (1985) *J. Gen. Microbiol.* 131, 2107–2113.
Praekelt, U. M. and Meacock, P. A. (1990) *Mol. Gen. Genet.* 223, 97–106.
Pujol, C. and Soll, D. R. (2003) In preparation.
Pujol, C., Lockhart, S. R., Noel, S., Tibayrenc, M. and Soll, D. R. (1997) *J. Clin. Microbiol.* 35, 2348–2358.
Ramsey, H., Morrow, B. and Soll, D. R. (1994) *Microbiol.* 140, 1525–1531.
Rikkerink, E. H. A., Magee, B. B. and Magee, P. T. (1988) *J. Bacteriol.* 170, 895–899.
Rundlett, S. E., Carmen, A. A., Kobajashi, R., Bavykin, S., Turner, B. M. and Grunstein, M. (1996) *Proc. Natl Acad. Sci. USA* 93, 14503–14508.
Sanglard, D., Ischer, F., Monod, M., Dogra, S., Prasad, R. and Bille, J. (1999) ASM Conference on Candida and Candidiasis, Charleston, SC, ASM C27.
Santos, M., Colthurst, D. R., Wills, N., McLauglin, C. S. and Tuite, M. F. (1990) *Curr. Genet.* 17, 487–491.
Santos, M. A., Keith, G. and Tuite, M. F. (1993) *EMBO J.* 12, 607–616.
Schaller, M., Schackert, C., Korting, H. C., Januschke, E. and Hube, B. (2000) *J. Invest. Dermatol.* 114, 712–717.
Schröppel, K., Srikantha, T., Wessels, D., DeCock, M., Lockhart, S. R. and Soll, D. R. (1996) *Microbiololgy* 142, 2245–2254.
Slutsky, B., Buffo, J. and Soll, D. R. (1985) *Science* 230, 666–669.
Slutsky, B., Staebell, M., Anderson, J., Risen, L., Pfaller, M. and Soll, D. R. (1987) *J. Bacteriol.* 169, 189–197.
Soll, D. R. (1989) In *Mobile DNA* (eds D. E. Berg and M. M. Howe), American Society of Microbiology, Washington, DC, pp. 791–798.
Soll, D. R. (1992a) *Clin. Microbiol. Rev.* 5, 183–203.
Soll, D. R. (1992b) In *New Fungal Strategies* (eds. J. E. Bennett, R. J. Hay and P. K. Peterson), Churchill Livingston, Edinburgh, pp. 156–172.
Soll, D. R. (1997) *Microbiology* 143, 279–288.

Soll, D. R. (2000) *Clin. Microbiol. Rev.* 13, 332–370.
Soll, D. R. (2002a) In *Candida and Candidiasis* (ed. R. A. Calderone), ASM Press, Washington, DC, pp. 123–142.
Soll, D. R. (2002b) In *Fungal Pathogenesis* (eds R. A. Calderone and R. L. Cihlar), Marcel Dekker, New York, pp. 161–182.
Soll, D. R., Anderson, J. and Bergen M. (1991) In *Candida albicans* (ed. R. Prasad), Cellular and Molecular Biology, Springer Verlag, pp. 20–45.
Soll, D. R., Galask, R., Isley, S., Rao, T. V. G., Stone, D., Hicks, J., Schmid, J., Mac, K. and Hanna, C. (1989) *J. Clin. Microbiol.* 27, 681–690.
Soll, D. R., Langtimm, C. J., McDowell, J., Hicks, J. and Galask, R. (1987) *J. Clin. Microbiol.* 25, 1611–1622.
Sonneborn, A., Tebarth, B. and Ernst, J. F. (1999) *Infect. Immun.* 67, 4655–4660.
Srikantha, T. and Soll, D. R. (1993) *Gene* 131, 53–60.
Srikantha, T., Chandrasekhar, A. and Soll, D. R. (1995) *Mol. Cell Biol.* 15, 1797–1805.
Srikantha, R., Klopach, A., Lorenz, W. W., Tsai, L. K., Laughlin, L. A., Gorman, J. A. and Soll, D. R. (1996) *J. Bacteriol.* 178, 121–129.
Srikantha, T., Tsai, L., Daniels, K., Enger, L., Highley, K. and Soll, D. R. (1998) *Microbiology* 14, 2715–2729.
Srikantha, T., Tsai, L., Daniels, K., Klar, A. and Soll D. R. (2001) *J. Bacteriol.* 183, 4614–4625.
Srikantha, T., Tsai, L., Daniels, K. and Soll, D. R. (2000) *J. Bacteriol.* 182, 1580–1591.
Srikantha, T., Tsai, L. and Soll, D. R. (1997) *J. Bacteriol.* 179, 3837–3844.
Stoldt, V. R., Sonneborn A., Leuker, C. E. and Ernst, J. F. (1997) *EMBO J.* 16, 1982–1991.
Stone, R. L., Matarese, V., Magee, B. B., Magee, P. T. and Bernlohr, D. A. (1990) *Gene* 96, 171–176.
Swanson, J. and Koomey, J. M. (1989) In *Mobile DNA* (eds D. E. Berg and M. M. Howe), American Society of Microbiology, Washington, DC, pp. 743–762.
Tatsumi, H., Kajiyama, N. and Nakano, E. (1992) *Biochem. Biophys. Acta* 1131, 161–165.
Vargas, K., Messer, S. A., Pfaller, M., Lockhart, S. R., Stapleton, J. T., Hellstein, J. and Soll, D. R. (2000) *J. Clin. Microbiol.* 38, 3595–3607.
Vargas, K., Wertz, P. W., Drake, D., Morrow, B. and Soll, D. R. (1994) *Infect. Immun.* 62, 1328–1335.
Vidal, M. and Gaber, R. F. (1991) *Mol. Cell Biol.* 11, 6317–6327.
White, T. C., Miyasaki, S. H. and Agabian, N. (1993) *J. Bacteriol.* 175, 6126–6135.
Workman, J. C. and Kingston, R. E. (1998) *Ann. Rev. Biochem.* 67, 545–579.
Yong, Y., Srikantha, T. and Soll, D. R. (2003) In preparation.
Yoshida, M. S., Horinouchi, S. and Bippu, T. (1995) *Bioessays* 17, 423–430.
Zhao, R., Lockhart, S. R., Daniels, K. and Soll, D. R. (2002) *Euc. Cell* 1, 353–365.

10

THE MSG GENE FAMILY AND ANTIGENIC VARIATION IN THE FUNGUS *PNEUMOCYSTIS CARINII*

JAMES R. STRINGER

INTRODUCTION

The ability of a microbe to rapidly vary surface antigens is usually detected as an unusually high number of surface phenotypes in a population recently derived from a single predecessor. Therefore, studies on the genes that confer the 'antigenic variation' phenotype usually come after the phenotype has been demonstrated. However, it is possible to approach the issue of antigenic variation in the opposite sense by first studying the genes and genomic structures that would appear to confer the capacity to vary surface antigens at high frequency. In the case of *Pneumocystis carinii*, focusing on genes has been necessary because *P. carinii* does not propagate well in culture, and animal infection models have not been sufficiently refined. These two technical problems make studies on surface phenotypes very difficult. Nevertheless, evidence for antigenic variation in *P. carinii* is becoming available, thanks to the insights and tools provided by molecular genetic approaches.

P. CARINII IS ONE SPECIES IN A LARGE FUNGAL GENUS

At the outset of a discourse on *Pneumocystis* organisms it is useful to clarify the phylogenetic position of the genus *Pneumocystis*, which definitely belongs among the fungi, and the complexity of the genus, which contains many species in addition to *P. carinii*. Although members of the *Pneumocystis* genus have been known to science for nearly a century, and one member has been a significant human pathogen since the beginning of the AIDS pandemic, basic knowledge about this genus has been slow to be acquired, primarily due to the lack of a robust culture system. What is more, although a much more detailed picture of the genus has emerged over the past decade, dissemination of new information has been slow. For instance, nearly all sources in the literature still call the organism that causes human PCP *P. carinii*. However, *P. carinii*, which is found in rats, causes pneumocystis pneumonia (PCP) only in rats, not in other species. PCP in humans, which is a frequent problem in AIDS patients, is caused by *Pneumocystis jiroveci*. As far as is known, *P. jiroveci* infects only humans.

The persistence in the literature of obsolete names for members of the genus *Pneumocystis* is understandable given that the nomenclature pertaining to the genus has been in flux for nearly a decade. Nevertheless, *P. jiroveci* and *P. carinii* have had official species status since 1999 (Frenkel, 1999). Here we use the most current, and presumably the final, names for these organisms. This nomenclature corresponds to that used in previous literature as follows. *P. jiroveci* and *P. carinii* were previously called special forms of *P. carinii*, i.e. *P. carinii* formae specialis *jiroveci* and *P. carinii* formae specialis *carinii*, respectively. The latter two names have been superceded because it is now clear that *P. jiroveci* and *P. carinii* are distinct species, not merely special forms of a single species (Smulian, 2001). While it is possible to study *P. jiroveci* by obtaining samples from patients, most of what is known about the genus has been learned from *P. carinii*. Hence, it is the focus of this chapter.

P. CARINII AND LABORATORY RATS

The study of *P. carinii* has relied heavily on infected rats because culture of the microbe has been problematic. Most reports pertaining to attempts to maintain *P. carinii* in continuous culture have indicated lack of success (Cushion and Ebbets, 1990; Durkin *et al.*, 1991; Sloand *et al.*, 1993). A report of continuous growth of *P. carinii* in culture has appeared, but this method has not been widely established, although several laboratories have attempted to use it (Merali *et al.*, 1999). At any rate, all culture methods reported so far require that the culture be seeded with a million or more *P. carinii*. Therefore, there is still no method for deriving clonal populations in culture. In addition, the number of *P. carinii* in the culture usually does not increase more than 10-fold, making culture a poor source if one needs larger numbers of *P. carinii*.

By contrast with culture, laboratory rats have been a reliable source of *P. carinii* in good numbers. Table 10.1 summarizes the three rat model systems that are used to obtain *P. carinii*. Most of the data described in this chapter have come from the natural transmission system because it is the best-established and most

Table 10.1 Rat systems that produce *P. carinii*

	System		
	Natural transmission	Inoculation	Latent *P. carinii*
Method	Rats that may or may not have latent infections are immunosuppressed and kept in open-air cages in a room with other rats that have PCP	Rats that do not have latent infections are immunosuppressed for 10 days then given *P. carinii* by inserting a blunt needle into the trachea via the mouth. The rats are kept in isolator cages to prevent infection by airborne organisms	Rats that have been presumably infected as neonates are immunosuppressed and kept in isolator cages to prevent infection by airborne organisms
Possible sources of *P. carinii*	Latent and airborne	Inoculum	Latent
Advantages	Reliable	Affords control over number and kind of *P. carinii* and time of infection	Ease
Disadvantages	Affords little control over the infection process	*P. carinii*-free rats have been rarely available	Yields are unpredictable and can be too low to be useful

commonly used model. However, this model has its limitations (see Table 10.1). Fortunately, the other two models are becoming more tractable with time, so we can expect continued progress in the investigation of antigenic variation in this species.

P. CARINII HAS A GENE FAMILY THAT CONFERS THE POTENTIAL FOR ANTIGENIC VARIATION

The MSG gene family was discovered via expression-cloning experiments targeted to a protein identified as a major immunoblot band with apparent size between 116 000 and 120 000 Da. This immunoblot band has been given many names, but most commonly it is known as either major surface glycoprotein (MSG) or glycoprotein A (gpA) (Gigliotti *et al.*, 1988; Pesanti and Shanley, 1988; Linke and Walzer, 1989; Linke *et al.*, 1989, 1994b; Nakamura *et al.*, 1989, 1991; Radding *et al.*, 1989a,b; Tanabe *et al.*, 1989; Lundgren *et al.*, 1991; Paulsrud *et al.*, 1991; Pottratz *et al.*, 1991; Andrews *et al.*, 1994; Gigliotti and McCool, 1996; Vasquez *et al.*, 1996). For the sake of brevity, we use MSG to refer to this entity.

As its name implies, MSG is a very abundant surface glycoprotein. Its abundance and location led to concerted efforts to clone the gene encoding it. Success in this endeavour was achieved by using antibodies that recognized MSG (on an immunoblot and by immunofluorescence) to screen expression libraries. Several cDNAs that encoded related proteins in either ferret-derived *Pneumocystis* or *P. carinii* were identified by such antibodies (Haidaris *et al.*, 1992; Kovacs *et al.*, 1993; Wada *et al.*, 1993). The relationship between MSG and the proteins encoded by the cDNAs was confirmed for *P. carinii* by showing that peptide sequences from MSG matched those encoded by one or more of the cDNAs (Kovacs *et al.*, 1993).

Isolation of cDNAs encoding MSG was a crucial breakthrough in the study of this antigen, which has been very difficult to study directly. Even though a large fraction of the protein in lysates prepared from *P. carinii* migrates at approximately 120 000 Da in a denaturing polyacrylamide gel, the amount of this protein that can be obtained is limited by the lack of a culture system, which limits the number of organisms available. Organisms obtained from lung tissue are a workable source of protein, but using *P. carinii* directly from rats introduces the possible complication of obtaining adventitious host material along with MSG. Added to these problems is the potential (explained below) for a multitude of different isoforms of MSG to be present among the population of *P. carinii* organisms, even when the population is from a single rat. It is not surprising, then, that MSG is not completely defined at the molecular level. In recognition of the difficulties faced in attempting to characterize MSG as a protein, we currently define MSG as the set of proteins produced by MSG genes.

The cDNAs described above provided the first access to the MSG gene family. DNA hybridization experiments showed that the cDNAs isolated using the anti-MSG antibody are related to the 3–1 family of repeated DNA sequences, which had previously been shown to have approximately 100 members, which occur in clusters distributed throughout the genome (Stringer *et al.*, 1991). Sequence analysis of a 3–1 family clone revealed the presence of two open reading frames (ORFs), which encoded two related 120 000 Da peptides (Sunkin *et al.*, 1994). We refer to these and other ORFs encoding MSG proteins as 'MSG genes'. This terminology is convenient but it must be noted that MSG ORFs should be regarded as nominal genes because they are not expressed unless associated with a specific locus called the UCS (see below). MSG genes have also been identified in other species of *Pneumocystis* including that from humans (Stringer *et al.*, 1993; Garbe and Stringer, 1994; Wright *et al.*, 1995a; Haidaris *et al.*, 1998; Mei *et al.*, 1998).

DNA hybridization showed that all of the 15 chromosome-sized bands resolved by pulsed field gel electrophoresis (PFGE) hybridized to an MSG DNA probe (Kovacs *et al.*, 1993; Wada *et al.*, 1993, 1995; Kitada *et al.*, 1994; Sunkin *et al.*, 1994). The number of *P. carinii* chromosomes is not known exactly. Clearly there are at least 15 but most likely there are 18, because three PFGE bands stain more intensely than others with DNA stains, suggesting that these bands contain two chromosomes each (Yoganathan *et al.*, 1989; Lundgren *et al.*, 1990; Cushion *et al.*, 1993). These data are consistent with the presence of MSG genes on all chromosomes.

As far as is known, all members of the MSG gene family reside at the ends of chromosomes. MSG genes were mapped to the ends of *P. carinii* chromosomes by showing that they are sensitive to digestion with an exonuclease called BAL 31 (Sunkin and Stringer, 1996). Supporting evidence for telomeric MSG genes came from cloning of *P. carinii* chromosome ends. The first such cloned DNA fragment began with the 3' end of an MSG gene and ended with a sequence known to be at sub-telomeric regions of the *P. carinii* genome (Underwood *et al.*, 1996). More recently, three full arrays of MSG genes have been cloned as cosmids and completely sequenced (Figure 10.1). Each of these three arrays begins in a unique sequence that resides at the end of a chromosome, and ends in either a telomere or a sub-telomere. In the arrays cloned so far, there are two or three MSG genes per array. The repeat structure in the cosmids is consistent with earlier data showing tandemly repeated MSG genes in lambda clones and in the genomes of *P. carinii*, *P. jiroveci* and ferret-derived *Pneumocystis* (Stringer *et al.*, 1991; Wada *et al.*, 1993; Garbe and Stringer, 1994; Sunkin *et al.*, 1994; Wada

Chromosome 15

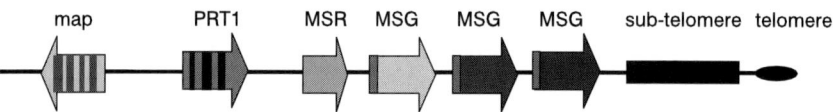

Chromosome 4

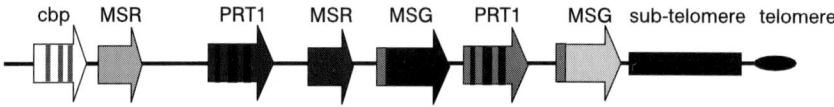

Chromosome 3

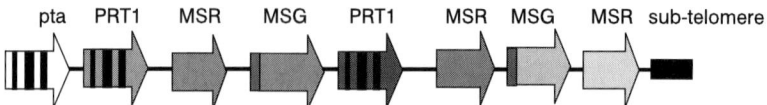

Figure 10.1 Structures of three telomeric gene arrays. Each array begins with a gene (*map*, *cbp* and *pta*) that maps to a single chromosome. There are then clusters of repeated genes, all pointed in the same direction, followed by sequences found at telomeres and sub-telomeres of *P. carinii*. Each repeated gene array contains at least one member of each of three gene families, MSG, MSR and PRT1. Only PRT1 genes have multiple introns, represented by the dark vertical bars. MSR genes have one intron at their 5' end. MSG genes have no introns.

and Nakamura, 1994; Wright *et al.*, 1995b). MSG gene clusters contain members of two other gene families, PRT1 and MSR. These genes will be described below, in the context of gene expression and evolution. The three arrays differ with respect to gene order, but all three have a PRT1 gene followed by an MSR gene followed by an MSG gene. This gene threesome may constitute a core repeat common to many or all arrays.

If there were to be two or three MSG genes at the ends of each of the 18 chromosomes in the genome, then the number of genes per genome would be between 72 and 108. These numbers agree well with the initial estimate of 100 copies of the 3–1 family per genome, obtained by DNA hybridization (Sunkin *et al.*, 1994). Sequencing data also indicate that the MSG gene family is large, and that many of these genes can be expressed in a single-organism population. In one population, we observed 26 different expressed MSG genes (Sunkin and Stringer, 1997). More recently, we observed 35 different expressed MSG genes in a single rat (Keely and Stringer, unpublished). These two studies entailed sequencing 50 or more randomly selected cloned copies of MSG. Many other populations of *P. carinii* have been subjected to less extensive analysis. All told, more than 100 different MSG sequences have been observed so far (Keely and Stringer, unpublished). Again, these data are consistent with there being approximately 100 MSG genes in the genome.

Analysis of sequences of MSG gene family members predicts that members of the MSG protein family have a similar molecular mass (approximately 120 000 Da) and are rich in cysteine residues, which are located at the same positions in different isoforms (Stringer, 1997). The ORFs of all MSG genes begin with a sequence encoding the amino acids, MARPVKRQ. Other amino acid motifs are conserved as well. However, the predicted protein isoforms of MSG vary in sequence by as much as 35%. While the DNA sequencing results are consistent with the presence of multiple MSG genes, the functionality of most of the open reading frames encoding MSG isoforms remains unproven because few of the predicted individual protein isoforms have been characterized biochemically in *P. carinii* organisms. Nevertheless, in the genomics era it is standard practice to assume that long open reading frames are functional genes, and, lacking evidence to the contrary, it seems reasonable to assume that all MSG open reading frames are able to produce an isoform of the MSG protein in *P. carinii*. In addition, when MSG has been prepared from *P. carinii* and from other members of the genus and this material has been subjected to biochemical and immunochemical analysis (Gigliotti *et al.*, 1988; Pesanti and Shanley, 1988; Linke and Walzer, 1989; Linke *et al.*, 1989, 1994b; Nakamura *et al.*,1989, 1991; Radding *et al.*, 1989a,b; Tanabe *et al.*, 1989; Lundgren *et al.*, 1991; Paulsrud *et al.*, 1991; Pottratz *et al.*, 1991; Andrews *et al.*, 1994; Gigliotti and McCool, 1996; Vasquez *et al.*, 1996), the results have been consistent with the expression of multiple isoforms of MSG in populations of the microbe. Moreover, immunochemical analysis of an epitope recognized by a monoclonal antibody showed that this epitope is encoded by only three MSG genes, and is detectable in some but not all *P. carinii* organisms (see below).

To summarize, the *P. carinii* genome carries numerous copies of ORFs (MSG genes) encoding different versions of a protein (MSG) known to be abundant and on the surface of the microbe. Hypothetically, this family of ORFs could confer the capacity to vary the surface, provided different family members are expressed in different microbes. Switching the antigen on the surface could be accomplished by switching off expression of one ORF and switching on another. There is a substantial amount of evidence supporting this hypothesis, including (1) the identification of a unique locus that is associated with expressed MSG genes, (2) association of diverse MSG genes with this expression locus, (3) correlation of expression of a surface epitope with the presence of the gene encoding it at the expression site. This evidence will be reviewed in the following sections. For the sake of continuity, consideration of the possibility that the MSG gene family is not involved in antigenic variation will be reserved for the end of this chapter.

THE MSG GENE FAMILY IS EXPRESSED IN A WAY THAT COULD CAUSE ANTIGENIC VARIATION

In other organisms, restricted expression of a gene family that confers antigenic variation has been detected by isolating several clonally derived populations and determining the number and types of antigen genes that are expressed. This approach to studying MSG has been precluded by the lack of a culture system capable of producing clonally derived *P. carinii*. Without a culture system, the only way to study MSG expression is to analyse *P. carinii* obtained from rats that have PCP. The first data from diseased rats seemed to argue against restriction of expression to a single MSG per organism because the original expression-cloning procedure yielded several different cDNAs (Kovacs *et al.*, 1993). However, interpretation of these cDNA cloning data was complicated by the fact that the organisms used to prepare the library were from rats infected by an unknown number of strains. Hence, the multiplicity of MSG cDNAs in the library could have been due to expression of one MSG gene in each of several different *P. carinii* strains. Subsequent analysis showed that pooling *P. carinii* from different rats was not required to obtain multiple different MSG cDNAs. In this study, a cDNA library, made from *P. carinii* from a single infected rat was hybridized to a single cloned MSG cDNA. Seven positive clones were isolated and sequenced. All seven clones encoded an MSG but none of the clones had the same sequence (Linke *et al.*, 1994a). Other studies had indicated that this rat was infected by a single strain of *P. carinii* (Cushion, 1998).

The studies described above showed that a population comprised of a single strain of *P. carinii* can express multiple MSG mRNAs. Nevertheless, the number of MSG genes expressed at any one time in an individual organism remained unclear. Expression of multiple MSG genes within a population of *P. carinii* could have been caused by different genes being expressed in different organisms. This situation could arise by either of two routes. The population might diversify as it expands within a rat lung. Alternatively, each rat might get infected by many organisms, which are expressing different MSG genes. The latter possibility exists because these studies were performed using the natural transmission model, where rats acquire *P. carinii* in a manner that is not controlled by the investigator (Table 10.1).

At this stage of the investigation, it was unclear whether it would be possible to distinguish between the two alternative situations consistent with the cDNA data: (1) expression of the same MSG genes in all organisms; (2) expression of different MSG genes in different organisms. Fortunately, basic studies on the structure of messenger RNAs encoding MSGs revealed a feature that provided a way to study regulation of MSG genes in populations of *P. carinii* obtained from naturally infected rats. These studies characterized the 5' ends MSG mRNAs by amplifying reverse-transcripts. This amplification of reverse-transcripts produced products of two sizes, 800 bp and 500 bp, but the larger product was 20 times more abundant. The 5' ends of 21 of the 800 bp products were sequenced. Surprisingly, the first 429 base pairs were the same in all 21, after which the cDNAs diverged to encode different isoforms of MSG (Wada *et al.*, 1995). Only the last 24 bp of these 429 bp was present in known MSG genes. These 24 bp are known as the CRJE and encode the peptide sequence MARPVKRQ, which may be the site of protease cleavage suspected to remove the amino terminus of the protein (see below and Figure 10.2). The presence of the 429 bp sequence on all MSG mRNAs was confirmed by another group using a similar approach (Edman *et al.*, 1996).

The first 405 bp of the sequence at the beginning of MSG mRNAs was named the type I Upstream Conserved Sequence (UCS). The sequence at the 5' ends of the 500 bp PCR product was named the type II UCS. Subsequent data showed that only the type I UCS is associated with MSG genes, and that the type II UCS is part of what we now call MSR genes (see below). Therefore, we use the term 'UCS' to refer to what was originally called the type I UCS. The presence of the UCS on MSG mRNAs was tested by hybridizing a UCS probe to northern blots, which showed that a single-size species of RNA was detected with the UCS probe. This RNA comigrated with RNA detected by an MSG probe (Wada *et al.*, 1995; Edman *et al.*, 1996). Thus, no MSG mRNAs lacking the UCS were detected by either 5' end analysis or northern blot analysis.

Finding the same 429 bp at the beginning of mRNAs encoding different MSGs was unexpected because the first 405 bp of this sequence was not present in the several MSG gene sequences that were known at the time. To determine whether this discrepancy might have been due to having looked at too few MSG genes, the UCS was used as a hybridization probe to map this sequence in the genome. The UCS hybridized to only one of the 15 chromosomes resolved by PFGE, whereas MSG gene probes hybridized to all 15 chromosomes (Wada *et al.*, 1995; Edman *et al.*, 1996). Therefore, the UCS is not present at 95% of the loci carrying an MSG sequence.

The localization of the UCS to a single PFGE band suggested that it might be unique in the genome. Results of quantitative hybridization experiments supported this hypothesis (Sunkin and Stringer, 1996). To determine the structure and copy number of the DNA encoding the UCS, genomic clones were isolated. Several lambda clones carrying the UCS were obtained by screening genomic libraries from *P. carinii* with a UCS DNA probe (Wada *et al.*, 1995; Edman *et al.*, 1996; Sunkin and Stringer, 1996; Wada and Nakamura, 1996). All of these clones contained an MSG gene attached directly to the UCS, with a CRJE in between. However, each of the cloned UCS loci contained a different MSG gene attached to the UCS. The UCS-linked MSG gene was followed by other MSG family members, showing that the clustering of MSG genes previously observed also occurs at the UCS locus. It was expected that the UCS would be telomeric because it is linked to MSG genes, which are telomeric. Experiments with exonuclease-digested chromosomes showed this to be the case (Sunkin and Stringer, 1996).

The observation that each cloned UCS was upstream of a different cluster of MSG genes suggested that the UCS locus in a given population of *P. carinii* might be heterogeneous in structure at points downstream of the UCS, and that the nature of the heterogeneity might differ between populations. Mapping of restriction enzyme cleavage sites downstream of the UCS produced results consistent with this hypothesis (Sunkin and Stringer, 1996). Further analysis of the UCS–MSG junction was performed by amplifying this region of the genome using the PCR. The junction was found to be heterogeneous in each of 37 rats analyzed (Sunkin and Stringer, 1997). Furthermore, the UCS–MSG junction appeared to be different in each of the 37 *P. carinii* populations. The large variety of MSG genes adjacent to the UCS locus was consistent with the previously observed presence of mRNAs encoding different MSG isoforms, yet each starting with a copy of the UCS.

By contrast with the heterogeneity of the region downstream of the UCS locus, the region upstream of this locus did not show evidence of polymorphism when subjected to restriction enzyme cleavage, which was capable of mapping the cleavage sites up to 10 kilobases (kb) upstream of the UCS (Sunkin and Stringer, 1996). Recently, three cosmid clones that each carry approximately 20 kb upstream of the UCS were isolated. Again, this region appeared to be the same in all three clones (Keely and Stringer, unpublished). Further evidence for a single UCS locus came from analysis of *P. carinii* obtained from rats that had been inoculated with a low dose of the microbe. In these studies, which are described in the following paragraph, a single UCS–MSG junction was present. Hence, all of the evidence indicates there is only one UCS locus. Nevertheless, this issue will not be fully resolved until both ends of the chromosome carrying the UCS are characterized. This goal should be reached soon because the *P. carinii* genome is being sequenced.

Most of the results described above were obtained by analysing *P. carinii* obtained from rats that develop PCP spontaneously upon immunosupression while kept in a room with animals that already have PCP (the natural transmission system described in Table 10.1). Therefore, the high degree of heterogeneity at the UCS could have been due to rats being infected by multiple organisms that have different MSG genes at the UCS. This hypothesis was tested by experiments in which rats were infected by inoculation with either a large or a small number of *P. carinii* from a population in which the UCS locus was very heterogeneous. In rats that received the high dose (10^7 *P. carinii*), the UCS locus in the *P. carinii* recovered from the inoculated rats had the same complex structure as in the population from which the inocula were drawn. By contrast, in rats that received the low dose (10 *P. carinii*), the UCS locus in the *P. carinii* recovered from the inoculated rats was drastically simplified. In some rats, a single MSG was attached to the UCS in 90% of the organisms in the population (Cushion *et al.*, 1999; Keely *et al.*, 1999). These data showed that heterogeneity at the UCS locus can be reduced, and suggested that the reduction was due to starting the infection with a single organism. Similar results were obtained with some rats that were carrying latent *P. carinii*, which were presumably acquired shortly after birth (Keely and Stringer, unpublished). These observation of less heterogeneity at the UCS locus in inoculated and latently infected rats suggests that the high degree of heterogeneity at the UCS in rats infected by being constantly exposed to rats with PCP (natural transmission) was due to infection by multiple organisms that have different MSG genes at the UCS. The time at which these different MSG genes became attached to the UCS is not known because the history of the organisms colonizing the rats in the natural transmission system colony is not known. Nevertheless, it is clear that the UCS locus is distinctly variable compared to the rest of the *P. carinii* genome. Populations of *P. carinii* have been compared at loci other than the UCS, but sequence differences either have not been observed, or have been limited to single nucleotide positions (Keely *et al.*, 1996).

The presence of the UCS at a single locus in the genome and on mRNAs encoding different MSG suggested that MSG genes must be attached to the UCS locus to be transcribed. Finding different MSG genes linked to the UCS locus supported this hypothesis. However, direct evidence linking expression of MSG mRNAs to the structure of the UCS locus was lacking and other possible mechanisms of MSG mRNA synthesis had not been excluded. The principal alternative mechanism that could have generated diverse mRNAs that all start with the same 5' end is *trans*-splicing, such as occurs in the kinetoplastida, where all mRNAs begin with the same sequence. In these organisms the 5' leader is transcribed from a separate locus, and then added to pre-mRNAs by splicing (Walder *et al.*, 1986). To determine whether *trans*-splicing was involved in putting the UCS on the 5' end of each MSG mRNA, the structure of the UCS–MSG junctions in mRNAs was compared to the structure of the UCS–MSG junctions in the genome. Six populations of *P. carinii*, each from a single rat, were studied. Each of the six populations expressed a different subset of the MSG gene family. All of the populations studied had the same UCS–MSG junctions in mRNAs as in the genome (Sunkin and Stringer, 1997). These data are not consistent with the *trans*-splicing hypothesis because splicing would be expected to produce mRNAs for which there is no corresponding UCS-linked MSG gene.

In summary, mRNAs encoding MSG proteins start with a common sequence (the UCS), which is encoded at one locus in the genome. In a population of *P. carinii* obtained from an individual rat, the UCS locus can be adjoined to a large number of different MSG genes and these genes correspond to the MSG mRNAs found in the same population of *P. carinii*. The simplest explanation for these findings is that MSG genes must be attached to the UCS to be transcribed and that different MSG genes can occupy the UCS locus in different organisms (see Figure 10.2). Such a system would work both to restrict expression to a single MSG per organism and to allow an organism to switch the MSG it expresses. It is easy to imagine how the switches take place because recombination is known to be capable of moving DNA sequences in the genome. Movement into the UCS locus would simultaneously switch on the newly arrived gene and switch off its predecessor (see Figure 10.2). Therefore, this system is an example of what the editors of this volume term 'simple' switching.

TRANSLATION AND TRANSPORT OF MSG PROBABLY INVOLVES THE UCS

The hypothesis that transcription of MSG genes is restricted to only UCS-linked MSG genes is supported by analysis of MSG proteins and the factors that influence their production and transport. The UCS has a potential translational start codon near its 5' end (Wada *et al.*, 1995; Edman *et al.*, 1996). This start codon is followed by a sequence encoding a putative signal peptide that could serve to direct the MSG into the endoplasmic reticulum (ER), from whence it would presumably be sent to the cell surface via the Golgi apparatus. Immunoblotting and immunoprecipitation experiments showed that UCS epitopes are present on a large protein (160 000 Da) containing both UCS and MSG epitopes (Sunkin *et al.*, 1998). Hence, the peptide encoded by the UCS could be part of an MSG protein precursor. The UCS epitopes were not present on the mature MSG found on the cell surface, suggesting that the UCS peptide is removed from the UCS–MSG precursor by a protease. The fate of the UCS after cleavage is not known. If it is removed after the precursor reaches the surface, which is suggested by the presence of the PRT1 protease on the surface (see below), then it may survive as a free unit. While there is no evidence either for or against the persistence of free UCS peptides, such a situation would explain the high degree of polymorphism in the genomic copy of the UCS (Sunkin and Stringer, 1996).

Although it has not been possible to test the functionality of the UCS peptide in *P. carinii*, this peptide appeared to be needed for trafficking MSG into the ER in insect cells (Sunkin *et al.*, 1998). The trafficking function was shown by

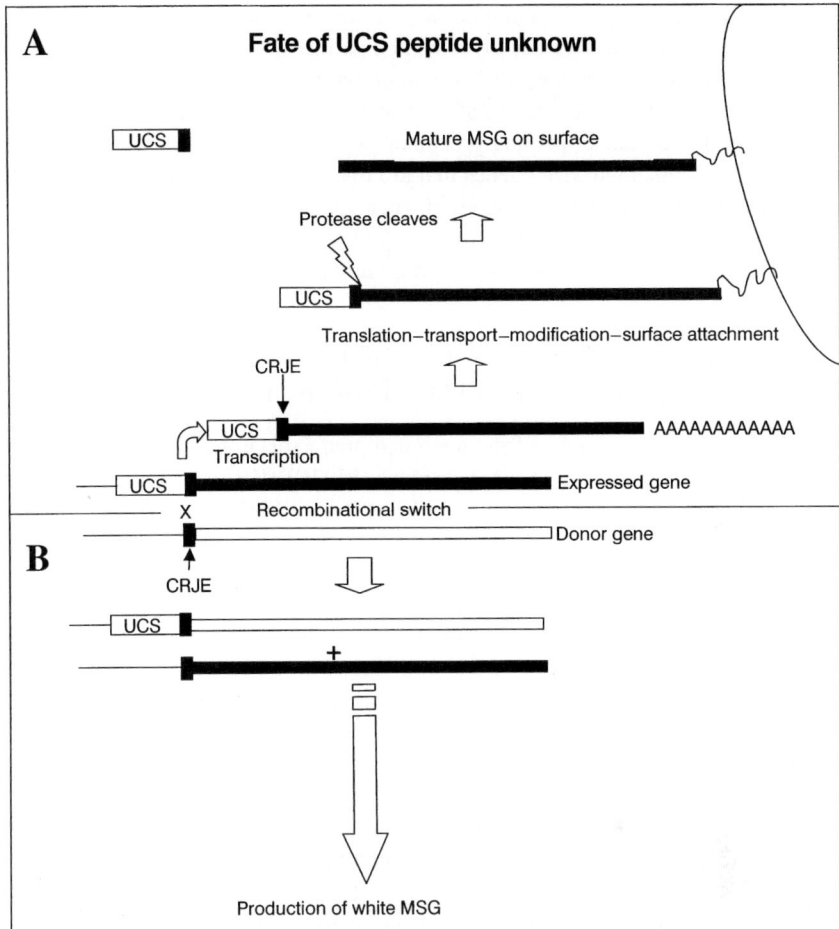

Figure 10.2 A model for expression of MSG genes. In panel (A) the black MSG gene is transcribed from the promoter upstream of the UCS. Translation starts in the UCS. The UCS peptide guides the protein into the secretory pathway, which is how it gets to the surface. As the UCS–MSG polypeptide is transported, a linker, such as GPI, is added to the carboxyl end, which serves to attach the protein to the surface of the cell. En route to the surface, or shortly thereafter, a PRT1 protease removes the UCS peptide, which may be discarded. By contrast, the white MSG gene is not transcribed because it is not installed at the UCS locus. Panel (B) shows one way in which the white gene could be installed at the UCS. A reciprocal exchange at the place marked with an **X** would replace the black gene with the white gene. Hence, the black gene would be turned off, and the white gene turned on.

expressing a gene encoding a UCS–MSG protein in insect cells in the presence or absence of tunicamycin, an inhibitor of glycosylation. When made in cells in the absence of tunicamycin, the UCS–MSG protein migrated slower than it did when made in cells incubated with tunicamycin. This behavior is generally taken to mean that the nascent polypeptide is glycosylated, which implies that it goes into the ER and Golgi. By contrast, a gene that encoded only an MSG ORF expressed protein that was not glycosylated in insect cells. These data suggest that the UCS is necessary and sufficient to direct the nascent polypeptide into the compartments known to be traversed by proteins bound for the cell surface. If the behavior of MSG transcripts and precursor peptides is the same in *P. carinii* as it is in insect cells, then MSG genes that are not linked to the UCS would not be expressed as surface proteins, even if such genes were to be transcribed to produce RNAs lacking a 5' UCS. Hence, the UCS appears to exert control over MSG transcription, translation, processing and transport (see Figure 10.2).

A PROTEASE FAMILY APPEARS TO HAVE CO-EVOLVED WITH THE MSG FAMILY

The UCS and MSG portions of the predicted precursor protein are divided by an invariant amino acid string of eight residues (MARPVKRQ) that contains a site (KR) that could be cut by a subtilisin-like protease. Molecular genetic analysis has identified a family of genes in *P. carinii* (the PRT1 family) that encodes proteases that might serve this purpose (Lugli *et al.*, 1997, 1999; Russian *et al.*, 1999; Wada and Nakamura, 1999a). At least some PRT1 family members appear to be surface proteins, which would explain why they vary in sequence (Lugli *et al.*, 1999; Russian *et al.*, 1999; Wada and Nakamura, 1999a).

As can be seen in Figure 10.1, PRT1 genes are found linked to MSG genes. This linkage suggests that the two gene families may be expressed in a coordinated way, thereby achieving surface variation with respect to both proteins. Whether they are coordinately expressed or not, the linkage of MSG and PRT1 genes in the same orientation suggests that they may have evolved from a common ancestral gene via gene amplification caused by unequal reciprocal crossing over.

The PRT1 family is notable for reasons separate from consideration of a possible role in MSG processing. Other fungal genomes encode related proteases – such as the kexins of *S. cerevisiae* (Julius *et al.*, 1984) – but *P. carinii* appears to encode many more different kexin-like proteases than other fungi (Lugli *et al.*, 1997, 1999; Russian *et al.*, 1999; Wada and Nakamura, 1999a). The reason for the large number of proteases in *P. carinii* is not known, but may be related to surface variation. However, it should be noted that mouse *Pneumocystis* organism (*P. muris*) has been reported to contain a single protease gene, which was called *Kex1* (Lee *et al.*, 2000). This difference underscores how different the species in this genus can be.

A THIRD GENE FAMILY CLOSELY RELATED TO MSG

A third gene family, MSR, for MSG-Related, seems to have evolved along with the MSG and PRT1 families (see Figure 10.1). MSR genes are very similar to

MSGs except they lack the sequence encoding MARPVKRQ, which is present in all MSG genes and is always present at the UCS–MSG boundary (Huang *et al.*, 1999; Schaffzin *et al.*, 1999a; Wada and Nakamura, 1999b,c). MSR genes have been called by other names, including 'variant MSG' and 'type II MSG' (Huang *et al.*, 1999; Wada and Nakamura, 1999c). However, our view is that MSR genes should not be called MSG genes, because MSG genes are defined by their association with the UCS. MSR genes do not attach directly to the UCS and do not have the CRJE. Nevertheless, MSG genes are very closely related to MSG genes. It can be difficult to distinguish an MSR sequence from an MSG sequence unless the 5' end sequence is available, in which case the absence of the CRJE is diagnostic of an MSR gene. The high similarity suggests that the two families have recently evolved from a common predecessor via unequal crossing over followed by divergence. MSR genes are similar enough to MSG genes to suggest that they may be able to contribute to MSG diversity by donating DNA into MSG genes via homologous recombination. The number of MSR genes in the genome is not known, but they appear to be located on at least 13 chromosomes (Schaffzin *et al.*, 1999b). All three of the cosmids in Figure 10.1 have one or more MSR genes, suggesting that there could be as many as 50 MSR genes in the genome. Whether MSR genes are regulated or not has not been determined, but it is clear that more than one can be expressed in a single population of *P. carinii* (Huang *et al.*, 1999; Schaffzin *et al.*, 1999b; Wada and Nakamura, 1999c).

EXPRESSION OF A SPECIFIC MSG IS CORRELATED WITH THE PRESENCE OF ITS COGNATE GENE AT THE UCS LOCUS

The singleness of the UCS locus and the role of this locus in transcription, translation and transport of MSG all suggest that only one MSG isoform is expressed per organism (see Figure 10.2). Studies on the surface protein itself have been limited by technical problems, but the available evidence supports this view. Indirect immunofluorescence studies showed that not all organisms within a population could be labelled with an antibody directed against a subset of MSG isoforms (Angus *et al.*, 1996; Sunkin and Stringer, 1996). In addition, the fraction of organisms labelled by such an antibody varied among populations. Additional evidence of limited MSG expression was provided by western blotting, which showed that some *P. carinii* populations contained a particular MSG epitope in abundance, and other populations did not (Vasquez *et al.*, 1996).

There is also some evidence that the UCS locus exerts control over the MSG that is on the cell surface (Schaffzin and Stringer, 1999). These studies utilized a monoclonal antibody called RA-C11, which recognizes a very small subset of the MSG isoforms encoded in the genome (Linke *et al.*, 1994b). In these studies, the C11 epitope and the nucleotide sequence encoding it (C11-epitope encoding sequence, or EES) were determined. Then, three populations of *P. carinii* were identified that varied over a range of 10-fold with respect to the fraction of cells with the C11 EES at the UCS locus. The same populations were analysed by immunofluorescence to determine the fraction of cells with detectable C11 epitope on their surface. There was a strong correlation between the proportion of C11 reactive organisms and the proportion of organisms with the C11 EES attached to the expression site.

It would appear that the MSG protein on the surface can be switched by changing the MSG gene at the expression site. How this is accomplished is not known. Studies on the C11 EES suggested that this sequence can be incorporated into an MSG gene that is at the UCS, perhaps by a gene-conversion event (Schaffzin and Stringer, unpublished). If such gene-conversion events were to occur, the number of different MSG isoforms that could be formed at the UCS locus would be virtually unlimited. On the other hand, analysis of the UCS locus has shown that changes can occur in the DNA more than 3 kb downstream of the UCS, suggesting that an entire array of genes can move from the end of a donor chromosome to the UCS via a crossover event, such as that depicted in Figure 10.2 (Wada *et al.*, 1995; Sunkin and Stringer, 1996). Both gene conversion and crossing over may be used to alter the MSG gene at the UCS locus.

MSG AND ANTIGENIC VARIATION

In other microbes that dwell in mammals, gene families serve to generate antigenic variation. Whether antigenic variation is a function of the MSG gene family has not been proved, but it is supported by several lines of evidence (see Table 10.2). These features suggest the following model (see Figure 10.2). A constitutively active promoter resides upstream of the UCS locus, and any MSG gene that becomes attached to it is transcribed. Because the UCS is unique, each organism expresses a single MSG isoform by transcribing the MSG gene that is attached to the UCS. The other MSG genes in the genome are not transcribed

Table 10.2 MSG gene family features that suggest antigenic variation

---

The genome contains approximately 100 MSG genes, which encode different proteins
Because the UCS locus is unique, only one MSG gene can be attached to it in a given cell
Many MSG family members can be linked to the UCS in one organism or another. Therefore, MSG genes must be able to move to the UCS, and do so at a rate high enough to generate an extraordinary degree of heterogeneity at this site
MSG genes attached to the UCS are transcribed to produce mRNAs that begin with the UCS and end with an ORF encoding an MSG isoform
The UCS sequence in the mRNA is translated to produce a putative UCS–MSG precursor protein
In insect cells, the UCS part of this protein is needed to send it into the ER and Golgi and ultimately to the cell surface
Populations of organisms in which a high fraction of cells exhibit a specific MSG-associated epitope have a similarly high fraction of cells that have DNA encoding that epitope at the UCS locus

because they lack promoters. The MSG transcripts are translated starting in the UCS and a UCS–MSG protein is made and sent into the ER, where it is glycosylated and perhaps attached to a membrane-anchor moiety such as glycosylphosphatidylinositol. The MSG then emerges on the cell surface. At some point, either prior to being placed on the surface or immediately thereafter, the UCS peptide is removed by a member of the PRT1 family. The MSG isoform that is on the surface can be altered by changing the MSG gene at the UCS. Such a change can take place either by a reciprocal exchange (crossover) between the MSG gene at the UCS and an MSG gene at some other telomeric locus, or by copying part or all of a silent MSG gene (or, perhaps an MSR gene) into the UCS-linked MSG gene (gene conversion). Switches at the UCS occur spontaneously, thereby continually generating organisms that vary with respect to surface MSG.

The phenotypic impact of a change at the UCS locus would depend on the fate of the old MSG and on the proportion of the surface that is covered by MSG. Once the UCS-linked gene is changed, the transcripts and protein made from the previous UCS resident might remain, but these molecules would decline in abundance as the cells propagate. To illustrate, after five rounds of cell division, only 3% of the mRNA and protein would be from the passé MSG gene. Of course the outmoded MSG mRNAs and proteins could be removed more rapidly through degradative mechanisms. A change at the UCS would have the greatest impact if the MSG were the only protein on the surface. The constitution of the surface is not known, but at least one other surface protein, PRT1, has been identified. The proteins made from the MSR gene family are probably also on the surface. In addition to these major constituents, there must be transporters, receptors and other proteins on the surface. It is not known if any of these proteins can vary at high frequency. The relative amounts of each surface protein are not known precisely, but when surface proteins were labelled and then analysed, the majority of the labelled protein migrated on a gel at the rate expected for MSG and MSR (Radding *et al.*, 1989a,b).

Preliminary studies have suggested that switches at the UCS occur no more often than 1 in every 100 organisms, but the actual rate of switching is not known and could be much lower (Keely *et al.*, 1999). However, even if comparatively few organisms were to switch, such as 1 in 10 000, there would be thousands of variants present by the time the population size reached 1 million. If the majority surface antigen were to be recognized by the host immune system, one or more rare variants has the opportunity to emerge as the predominant type. This is the pattern seen in several other pathogens, including African trypanosomes, and bacteria in the genus *Borellia*. These microbes use antigenic variation to sustain infection in the face of a strong immune response. However, *P. carinii* is unable to cause PCP in immunocompetent rats. Disease is seen only when the immune system is debilitated. The same is true in humans. Therefore, it seems that MSG-mediated antigenic variation is not able to overcome the immune response, at least not to an extent that would cause PCP. Nevertheless, the existence of the

MSG gene family and the sophisticated mechanism devoted to regulating its expression demand an explanation, and antigenic variation is the leading candidate.

Perhaps MSG variation is involved in allowing the microbe to colonize its host. All indications are that *Pneumocystis* organisms are an extremely widespread and enduring associate of mammals (Cushion, 1994; Mazars *et al.*, 1997; Nielsen *et al.*, 1998; Weisbroth *et al.*, 1999; Denis *et al.*, 2000; Palmer *et al.*, 2000; Demanche *et al.*, 2001; Icenhour *et al.*, 2001). At least one member of the *Pneumocystis* genus has been found in every mammal that has been examined, including a large number of animals captured in the wild. It is commonly reported that there are thousands of *Pneumocystis* organisms in the lungs of apparently healthy wild animals (Bishop *et al.*, 1997; Mazars *et al.*, 1997; Laakkonen, 1998; Laakkonen *et al.*, 1999; Demanche *et al.*, 2001). Laboratory animals are also frequently colonized, in spite of extensive efforts to eradicate *Pneumocystis* from colonies (Weisbroth *et al.*, 1999; Icenhour *et al.*, 2001).

Frequent colonization also fits with the observation that each host species has its own kind of *Pneumocystis*, suggesting that the microbe and host have co-evolved, and that the microbe depends on its specific host for survival (Wakefield *et al.*, 1998). This suggestion is supported by the fastidiousness of these microbes. They do not thrive in culture, and seem to be able to propagate continuously only in the lungs of the host in which they are found. Therefore, it is reasonable to speculate that the MSG system serves to foster survival of *P. carinii* in the lungs of immunocompetent rats, and that this survival is necessary to assure transmission to other rats, which are the only environment satisfying the needs of the microbe.

ON THE POSSIBILITY THAT THE MSG GENE FAMILY DOES NOT CAUSE ANTIGENIC VARIATION

Whereas the evidence in favour of the antigenic-variation model is strong, there are still many aspects of the system that remain uncharacterized (see Table 10.3). Therefore, it is appropriate to consider the possibility that the MSG gene family does not cause antigenic variation. This would be the case if every organism were to express the same set of MSG proteins. Immunodetection of MSG epitopes on *P. carinii* would appear to have demonstrated non-uniform expression of these epitopes, but such experiments cannot exclude the possibility that the epitope is present but not detectable via the immunofluorescence assay. Still, uniform transcription of the family is clearly not the case for MSG genes that are expressed from the UCS. Each of the six populations of *P. carinii* examined at this level differed with respect to the set of UCS-containing MSG mRNAs they had. However, uniform transcription would have been missed if MSG transcripts were to be made in a manner that is independent of the UCS. Although the UCS

The MSG gene family and antigenic variation in the fungus *P. carinii*   **219**

Table 10.3 Unknowns concerning MSG

The number of MSG genes that can become attached to the UCS is high, but it is not clear that all MSG genes contribute to the diversity seen at the UCS

The frequency at which MSG genes move to the UCS is not known

The mechanism by which MSG genes move to the UCS is not known

MSR genes may recombine with MSG genes that are at the UCS. Hence, MSR genes could contribute to MSG diversity, even though MSR genes do attach themselves to the UCS

It is possible that MSG genes that are not attached to the UCS may be transcribed as mRNAs that lack the UCS. These mRNAs would lack the ATG codon used by UCS-containing mRNAs, but might be translated from some other start codon. They would also lack the leader peptide provided by the UCS, but may be able to move into the ER anyway

The putative UCS–MSG precursor protein has not been shown to function as such

The number of MSG protein isoforms expressed per organism has not been established

---

appears to be needed for translation and transport to the surface, it is possible that these processes could occur without the UCS. (In entertaining the possibility of UCS-independent expression of MSGs, it is important to distinguish MSGs from MSRs, which are known to be expressed without being linked to the UCS.)

A major impediment to proving that any particular MSG gene is not expressed in a particular organism is the fact that the populations of organisms obtained from naturally infected rats tend to express many MSGs. This tendency means that lack of expression of a particular gene in some organisms in the population would be masked by expression of that gene in other organisms in the population. To ultimately either prove or disprove the antigenic-variation hypothesis, it will be necessary to produce clonally derived populations. The prediction of the antigenic-variation model is that such a population will express a single MSG isoform, which will match that encoded by the UCS-linked gene. Initial experiments with rats inoculated with few *P. carinii* have encouraged the hope that *P. carinii* can be cloned (Cushion *et al.*, 1999; Keely *et al.*, 1999). There was a single MSG gene at the UCS locus in 90% of the organisms in the inoculated rats. Unfortunately, the yield of microbes was insufficient to allow analysis of MSG protein or mRNA, but further progress in this direction should be possible.

CONCLUSION

Many microbial pathogens use high rates of surface variation to survive attacks by the immune system. One mechanism for achieving antigenic variation is to express a subset of a family of genes encoding various forms of a surface protein. *P. carinii* definitely has at least one such gene family, MSG. Therefore, the mere existence of this family of genes implies that it is used to produce a large variety of surface phenotypes. While it is conceivable that the same MSG genes might be expressed in all *P. carinii*, there is no evidence for this. Instead, all indications are that expression of this family occurs in an unusually complex fashion whereby transcription of a given gene, translation of the transcript and transport of the protein all require that the gene be moved to a particular site in the genome. This system would appear to allow only one MSG isoform to be made at a time in a given organism. Analysis of surface epitopes is in its infancy, but up to this point, the evidence indicates that surface variation occurs. Hence, it would be surprising in the extreme if it turned out that *P. carinii* do not differ with respect to the MSG on their surface.

Some of the best-known cases of such antigenic variation came to light because the host suffers from prolonged high-level infections made obvious by the attendant symptoms of disease. In these cases, the immune system is functional, but still unsuccessful in preventing episodes in which the pathogen burden becomes very heavy. By contrast, a robust immune system seems to be quite effective in preventing PCP. Therefore, it seems that the MSG system did not evolve to promote high levels of microbial population growth in a host that is deploying the full force of its defences. Instead, MSG variation may be used to help *P. carinii* maintain a colony that is small enough to not evoke a full immune response, yet large enough to ensure survival and transmission to new hosts.

ACKNOWLEDGMENTS

I gratefully acknowledge the late Ann Wakefield, for sharing her thoughts and unpublished data on the PRT1 gene family.

REFERENCES

Andrews, R. P., Theus, S. A., Cushion, M. T. and Walzer, P. D. (1994) *J. Eukaryot. Microbiol.* 41, 72S.

Angus, C. W., Tu, A., Vogel, P., Qin, M. and Kovacs, J. A. (1996) *J. Exp. Med.* 183, 1229–1234.

Bishop, R., Gurnell, J., Laakkonen, J., Whitwell, K. and Peters, S. (1997) *J. Eukaryot. Microbiol.* 44, 57S.

Cushion, M. T. (1994) In *Pneumocystis carinii Pneumonia* (ed. P. D. Walzer), Marcel Dekker, New York, pp. 123–140.

Cushion, M. T. (1998) *FEMS Immunol. Med. Microbiol.* 22, 51–58.

Cushion, M. T. and Ebbets, D. (1990) *J. Clin. Microbiol.* 28, 1385–1394.

Cushion, M. T., Kaselis, M., Stringer, S. L. and Stringer, J. R. (1993) *Infect. Immun.* 61, 4801–4813.

Cushion, M. T., Linke, M. J., Collins, M., Keely, S. P. and Stringer, J. R. (1999) *J. Eukaryot. Microbiol.* 46, 111S.

Demanche, C., Berthelemy, M., Petit, T., Polack, B., Wakefield, A. E., Dei-Cas, E. and Guillot, J. (2001) *J. Clin. Microbiol.* 39, 2126–2133.

Denis, C. M., Mazars, E., Guyot, K., Odberg-Ferragut, C., Viscogliosi, E., Dei-Cas, E. and Wakefield, A. E. (2000) *Med. Mycol.* 38, 289–300.

Durkin, M. M., Shaw, M. M., Bartlett, M. S. and Smith, J. W. (1991) *J. Protozool.* 38, 210S–212S.
Edman, J. C., Hatton, T. W., Nam, M., Turner, R., Mei, Q., Angus, C. W. and Kovacs, J. A. (1996) *DNA Cell Biol.* 15, 989–999.
Frenkel, J. K. (1999) *J. Eukaryot. Microbiol.* 46, 89S–92S.
Garbe, T. R. and Stringer, J. R. (1994) *Infect. Immun.* 62, 3092–3101.
Gigliotti, F. and McCool, T. (1996) *Parasitol. Res.* 82, 90–91.
Gigliotti, F., Ballou, L. R., Hughes, W. T. and Mosley, B. D. (1988) *J. Infect. Dis.* 158, 848–854.
Haidaris, C. G., Medzihradsky, O. F., Gigliotti, F. and Simpson-Haidaris, P. J. (1998) *DNA Res.* 5, 77–85.
Haidaris, P. J., Wright, T. W., Gigliotti, F. and Haidaris, C. G. (1992) *J. Infect. Dis.* 166, 1113–1123.
Huang, S. N., Angus, C. W., Turner, R. E., Sorial, V. and Kovacs, J. A. (1999) *J. Infect. Dis.* 179, 192–200.
Icenhour, C. R., Rebholz, S. L., Collins, M. S. and Cushion, M. T. (2001) *J. Clin. Microbiol.* 39, 3437–3441.
Julius, D., Brake, A., Blair, L., Kunisawa, R. and Thorner, J. (1984) *Cell* 37, 1075–1089.
Keely, S. P., Cushion, M. T. and Stringer, J. R. (1996) *J. Eukaryot. Microbiol.* 43, 49S.
Keely, S. P., Cushion, M. T. and Stringer, J. R. (1999) *J. Eukaryot. Microbiol.* 46, 128S.
Kitada, K., Wada, M. and Nakamura, Y. (1994) *DNA Res.* 1, 57–66.
Kovacs, J. A., Powell, F., Edman, J. C., Lundgren, B., Martinez, A., Drew, B. and Angus, C. W. (1993) *J. Biol. Chem.* 268, 6034–6040.
Laakkonen, J. (1998) *Int. J. Parasitol.* 28, 241–252.
Laakkonen, J., Henttonen, H., Niemimaa, J. and Soveri, T. (1999) *Parasitology* 118, 1–5.
Lee, L. H., Gigliotti, F., Wright, T. W., Simpson-Haidaris, P. J., Weinberg, G. A. and Haidaris, C. G. (2000) *Gene* 242(1–2), 141–150.
Linke, M. J. and Walzer, P. D. (1989) *J. Protozool.* 36, 60S–61S.
Linke, M. J., Cushion, M. T. and Walzer, P. D. (1989) *Infect. Immun.* 57, 1547–1555.
Linke, M. J., Smulian, A. G., Stringer, J. R. and Walzer, P. D. (1994a) *Parasitol. Res.* 80, 478–486.
Linke, M. J., Smulian, A. G., Yoshihara, P. and Walzer, P. D. (1994b) *J. Eukaryot. Microbiol.* 41, 99S–100S.
Lugli, E. B., Allen, A. G. and Wakefield, A. E. (1997) *Microbiology* 143, 2223–2236.
Lugli, E. B., Bampton, E. T., Ferguson, D. J. and Wakefield, A. E. (1999) *Mol. Microbiol.* 31, 1723–1733.
Lundgren, B., Cotton, R., Lundgren, J. D., Edman, J. C. and Kovacs, J. A. (1990) *Infect. Immun.* 58, 1705–1710.
Lundgren, B., Lipschik, G. Y. and Kovacs, J. A. (1991) *J. Clin. Invest.* 87, 163–170.
Mazars, E., Guyot, K., Fourmaintraux, S., Renaud, F., Petavy, F., Camus, D. and Dei-Cas, E. (1997) *J. Eukaryot. Microbiol.* 44, 39S.
Mei, Q., Turner, R. E., Sorial, V., Klivington, D., Angus, C. W. and Kovacs, J. A. (1998) *Infect. Immun.* 66, 4268–4273.
Merali, S., Frevert, U., Williams, J. H., Chin, K., Bryan, R. and Clarkson, A. B., Jr (1999) *Proc. Natl Acad. Sci. USA* 96, 2402–2407.

Nakamura, Y., Kitada, K., Wada, M. and Saito, M. (1991) *J. Protozool.* 38, 3S–4S.
Nakamura, Y., Tanabe, K. and Egawa, K. (1989) *J. Protozool.* 36, 58S–60S.
Nielsen, M. H., Settnes, O. P., Aliouat, E. M., Cailliez, J. C. and Dei-Cas, E. (1998) *APMIS* 106, 771–779.
Palmer, R. J., Settnes, O. P., Lodal, J. and Wakefield, A. E. (2000) *Appl. Environ. Microbiol.* 66, 4954–4961.
Paulsrud, J. R., Queener, S. F., Bartlett, M. S. and Smith, J. W. (1991) *J. Protozool.* 38, 10S–11S.
Pesanti, E. L. and Shanley, J. D. (1988) *J. Infect. Dis.* 158, 1353–1359.
Pottratz, S. T., Paulsrud, J., Smith, J. S. and Martin, W. J. II (1991) *J. Clin. Invest.* 88, 403–407.
Radding, J. A., Armstrong, M. Y., Bogucki, M. S. and Richards, F. F. (1989a) *J. Protozool.* 36, 61S–62S.
Radding, J. A., Armstrong, M. Y., Ullu, E. and Richards, F. F. (1989b) *Infect. Immun.* 57, 2149–2157.
Russian, D. A., Andrawis-Sorial, V., Goheen, M. P., Edman, J. C., Vogel, P., Turner, R. E., Klivington, D. L., Angus, C. W. and Kovacs, J. A. (1999) *Proc. Assoc. Am. Phys.* 111, 347–356.
Schaffzin, J. K. and Stringer, J. R. (1999) *J. Eukaryot. Microbiol.* 46, 127S.
Schaffzin, J. K., Garbe, T. R. and Stringer, J. R. (1999a) *Fungal. Genet. Biol.* 28, 214–226.
Schaffzin, J. K., Sunkin, S. M. and Stringer, J. R. (1999b) *Curr. Genet.* 35, 134–143.
Sloand, E., Laughon, B., Armstrong, M., Bartlett, M. S., Blumenfeld, W., Cushion, M., Kalica, A., Kovacs, J. A., Martin, W., Pitt, E. *et al.* (1993) *J. Eukaryot. Microbiol.* 40, 188–195.
Smulian, A. G. (2001) *Fungal. Genet. Biol.* 34, 145–154.
Stringer, S. L. (1997) In *Fungal Disease: Biology, Immunology and Diagnosis* (ed. P. H. Jacobs), L. Nall, New York. pp. 423–449.
Stringer, S. L., Garbe, T., Sunkin, S. M. and Stringer, J. R. (1993) *J. Eukaryot. Microbiol.* 40, 821–826.
Stringer, S. L., Hong, S. T., Giuntoli, D. and Stringer, J. R. (1991) *J. Clin. Microbiol.* 29, 1194–1201.
Sunkin, S. M. and Stringer, J. R. (1996) *Mol. Microbiol.* 19, 283–295.
Sunkin, S. M. and Stringer, J. R. (1997) *Mol. Microbiol.* 25, 147–160.
Sunkin, S. M., Linke, M. J., McCormack, F. X., Walzer, P. D. and Stringer, J. R. (1998) *Infect. Immun.* 66, 741–746.
Sunkin, S. M., Stringer, S. L. and Stringer, J. R. (1994) *J. Eukaryot. Microbiol.* 41, 292–300.
Tanabe, K., Takasaki, S., Watanabe, J., Kobata, A., Egawa, K. and Nakamura, Y. (1989) *Infect. Immun.* 57, 1363–1368.
Underwood, A. P., Louis, E. J., Borts, R. H., Stringer, J. R. and Wakefield, A. E. (1996) *Mol. Microbiol.* 19, 273–281.
Vasquez, J., Smulian, A. G., Linke, M. J. and Cushion, M. T. (1996) *Infect. Immun.* 64, 290–297.
Wada, M. and Nakamura, Y. (1994) *DNA Res.* 1, 163–168.
Wada, M. and Nakamura, Y. (1996) *DNA Res.* 3, 55–64.
Wada, M. and Nakamura, Y. (1999a) *J. Eukaryot. Microbiol.* 46, 151S–152S.
Wada, M. and Nakamura, Y. (1999b) *J. Eukaryot. Microbiol.* 46, 125S–126S.

Wada, M. and Nakamura, Y. (1999c) *DNA Res.* 6, 211–217.
Wada, M., Kitada, K., Saito, M., Egawa, K. and Nakamura, Y. (1993) *J. Infect. Dis.* 168, 979–985.
Wada, M., Sunkin, S. M., Stringer, J. R. and Nakamura, Y. (1995) *J. Infect. Dis.* 171, 1563–1568.
Wakefield, A. E., Stringer, J. R., Tamburrini, E. and Dei-Cas, E. (1998) *Med. Mycol.* 36 Suppl 1, 183–193.
Walder, J. A., Eder, P. S., Engman, D. M., Brentano, S. T., Walder, R. Y., Knutzon, D. S., Dorfman, D. M. and Donelson, J. E. (1986) *Science* 233, 569–571.
Weisbroth, S. H., Geistfeld, J., Weisbroth, S. P., Williams, B., Feldman, S. H., Linke, M. J., Orr, S. and Cushion, M. T. (1999) *J. Clin. Microbiol.* 37, 1441–1446.
Wright, T. W., Bissoondial, T. Y., Haidaris, C. G., Gigliotti, F. and Haidaris, P. J. (1995b) *DNA Res.* 2, 77–88.
Wright, T. W., Gigliotti, F., Haidaris, C. G. and Simpson-Haidaris, P. J. (1995a) *Gene* 167, 185–189.
Yoganathan, T., Lin, H. and Buck, G. A. (1989) *Mol. Microbiol.* 3, 1473–1480.

11

TRYPANOSOME ANTIGENIC VARIATION – A HEAVY INVESTMENT IN THE EVASION OF IMMUNITY

J. DAVID BARRY AND RICHARD McCULLOCH

Antigenic variation is of enormous importance to the African trypanosome. Whereas many eukaryotic parasites have evolved an intracellular habitat that helps conceal them from the host immune system, trypanosomes have obstinately remained extracellular throughout their entire life cycle. As a consequence, their cell surface is the primary target of vigorous, trypanocidal immune responses, and in turn presents the main line of defence against these responses. Defence like this is not a straightforward task, as a balance has had to be achieved between the accessibility of the parasite's surface to positively acting host molecules, such as nutrients (Borst and Fairlamb, 1998), and the protective nature of this interface with the host. Perhaps the most effective means of achieving the balance, because we can see that it has evolved independently in several other extracellular micropathogens living systemically in mammalian hosts, is antigenic variation. The trypanosome has invested very heavily in its antigenic-variation system, leading to further evolution in several directions, greater effectiveness and, notably, the appearance of some quite novel cellular mechanisms. Here we discuss this huge investment and the attendant novelties in the trypanosome.

BIOLOGY OF ANTIGENIC VARIATION

A range of relapsing diseases of humans and other animals are caused by *Trypanosoma brucei*, including human sleeping sickness of the chronic (*T. b.*

gambiense) and acute (*T. b. rhodesiense*) forms. Similar diseases are caused by *Trypanosoma congolense* and *Trypanosoma vivax*, most notably in livestock. All these species are transmitted by tsetse, although some have spread outside the tsetse belt through the use of simpler transmission modes. Both the relapsing nature and the chronicity of the trypanosomiases are a direct consequence of antigenic variation. *T. brucei* establishes infection in each host type (or anatomical niche) through multiplicative stages, including the long slender form in the mammalian host and the procyclic and epimastigote forms in the tsetse (Figure

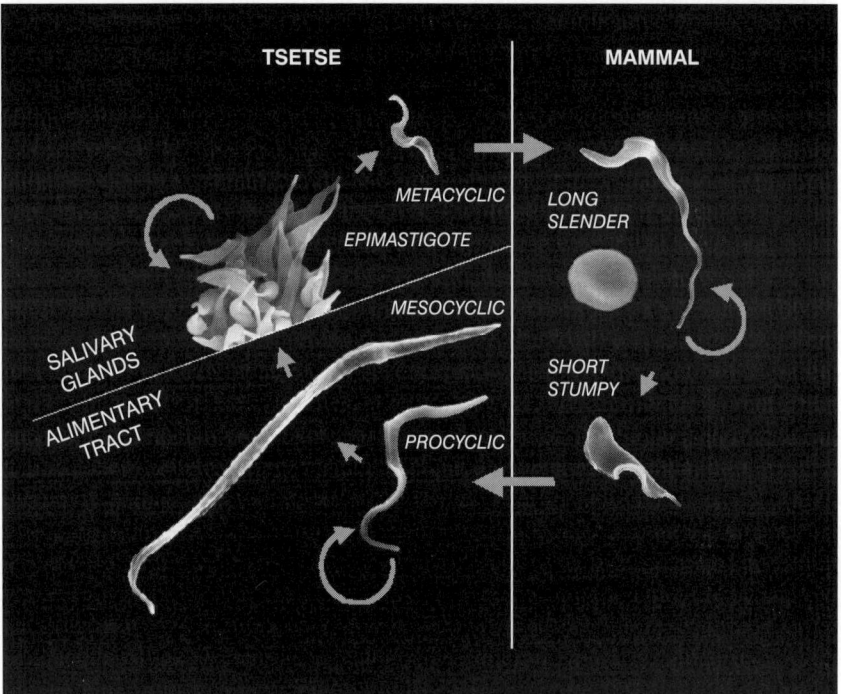

Figure 11.1 The life cycle of *Trypanosoma brucei*. Circular arrows depict replicating forms. The VSG coat is present on the metacyclic, long slender and short stumpy stages. (Reproduced with permission from Barry and McCulloch (2001) *Adv. Parasitol.* 49, 1–70.)

11.1). Transition between these forms is effected by non-multiplicative stages, including the stumpy form, the mesocyclic form – which differentiates to the epimastigote stage via a form that yields only one viable daughter cell on division (Van Den Abbeele *et al.*, 1999) – and the metacyclic form, which is the mammal-infective stage in the tsetse. All stages that exist in mammalian hosts have an essential surface coat composed of a single species of the variant surface glycoprotein (VSG) (Cross, 1975). The VSG is a glycosylphosphatidylinositol-anchored protein of about 550 amino acids, which has structural similarity with receptor molecules on the trypanosome, indicating from where it may have evolved, and exists as densely packed dimers on the cell surface (Carrington and Boothroyd, 1996; Salmon *et al.*, 1997). Different VSGs differ substantially over their amino terminal four-fifths, which is disposed away from the plasma membrane and, unlike some antigenically varying proteins such as PfEMP1 in *Plasmodium* (see Chapter 14), are thought to have no physiological or biochemical function. Structurally simpler VSGs are present on *T. congolense* and *T. vivax*, but they too display sequence hypervariability (Gardiner *et al.*, 1996; Urakawa *et al.*, 1997). The surface coat shields necessarily invariant surface molecules from immune responses and is thought to inhibit innate immune mechanisms. The weakness of this shielding strategy is that the VSG itself is a potent immunogen and that is why antigenic variation, the replacement of one VSG coat by another, is required to keep the trypanosome population extant in the host.

As with other contingency gene systems (Moxon *et al.*, 1994), trypanosome antigenic variation is rapid, occurring at about 10^{-2} switches/cell/generation (Turner and Barry, 1989). Its study generally has focused on short-term infections, yielding insight into the most predominant events, but we do not have a comprehensive view of chronic infection. The following account, which is illustrated in Figure 11.2 by the syringe-induced infection of a large host animal with a clonal population of *T. vivax* composed of predominantly a single variable antigen type (VAT) (Barry, 1986), is partly speculative. As this initial VAT expands by trypanosome division, IgM antibodies against its VSG remove most of the population, leaving switched parasites to survive and grow. Sometimes the remaining proportion of trypanosomes of new VAT is very high, resulting in an indistinct division between successive bouts of growth, such as occurred in the first 30 days, and even more strikingly between days 10 and 20 (Figure 11.2). Within a population, switching is always divergent, driving towards a complex mixture of different VATs, and it is unusual to find one dominant VAT (Van Meirvenne *et al.*, 1975; Miller and Turner, 1981; Barry, 1986).

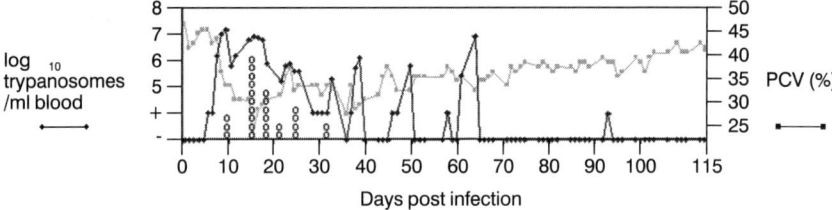

Figure 11.2 Parasitaemia in a cow infected by syringe with a clonal population of bloodstream *T. vivax*. The open circles represent new antibody responses against individual variable antigen types. The packed cell volume (PCV) is a measure of red blood cells and reveals anaemia, a symptom of the disease. As the parasitaemia decreases gradually, the PCV recovers. Note the high complexity of early peaks, which encompass many different VATs, and the relative simplicity of later peaks. An animal this large, maintained with unlimited food and water, can recover from infection, and it is likely that the trypanosome has expressed most of its *VSG* repertoire; it may even have exhausted its repertoire. (Data from Barry, 1986).

In *T. brucei*, antigenic variation basically is driven by differential expression of *VSG* genes; each VAT expresses a distinct *VSG*. The chronic nature of infection is thought to depend on a partially ordered, differential appearance of VATs, based on differences in the probability of activation for different *VSG* genes. Those VATs with the highest probability will appear early in infection and indeed subsequently are likely to appear repeatedly throughout infection, only to be removed by existing antibody (Van Meirvenne *et al.*, 1975; Capbern *et al.*, 1977; Barry, 1986). As the infection proceeds and the immune system experiences more VSGs, the proportion of switchers removed in this way increases progressively. Consequently, although the overall rate of switching remains constant throughout the infection, the *effective* rate decreases. Study of the time of appearance of different VATs expressing well-characterized *VSG* genes has suggested a possible genetic basis for differential *VSG*-activation probabilities, based on the type of genetic locus occupied by the silent gene that is becoming activated. Telomeric genes have a high probability of activation, chromosome internal genes are lower on the scale, and incomplete genes are lowest in the hierarchy; this we describe in more detail later on. Although the evidence favours this differential gene-activation system, rather than specific inter-VAT competition (discussed in Barry and McCulloch, 2001), as a basis for chronic infection and hierarchical VAT expression, there is separate evidence for a general interference between trypanosomes. Super-infection experiments show this most clearly (Turner *et al.*, 1996), but nothing else is known about the phenomenon. In considering the influence of antigenic variation on the core biology of the trypanosome, we can see that differential use of the *VSG* repertoire dominates the pattern of growth and the length of infection (and hence the probability of uptake by tsetse). In addition, the fact that the parasitaemia is maintained, or even reduces, during infection presumably reduces pathogenesis, rendering these diseases chronic in the onset of symptoms. The main symptoms of sleeping sickness include bouts of fever, anaemia (see the packed cell volume parameter in Figure 11.2) and eventual lethargy and coma. The fever bouts correlate with peaks of parasitaemia and destruction of trypanosomes; the anaemia is haemolytic and probably is due to the activity of trypanosome by-products; and the lethargy and coma arise from trypanosome infiltration of the brain. Although these diseases are chronic, there is a well-known exception, acute human sleeping sickness, which is a recently evolved host–parasite association (MacLeod *et al.*, 2000) has not yet developed to this chronic balance. *T. b. rhodesiense* grows rapidly in its human host, advancing substantially the onset of symptoms that usually are late.

Antigenic variation has evolved a subsystem that is a refinement of the main system. As trypanosomes begin to reacquire mammal infectivity in the tsetse, during differentiation to the metacyclic stage (Figure 11.1), they dispense with the procyclin surface coat, which is expressed by stages throughout the fly, and begin synthesis of the VSG coat. Each trypanosome activates, apparently at random, one of a set of genes known as *MVSGs* (metacyclic *VSGs*). Up to 27 different MVATs (metacyclic VATs) are expressed in the metacyclic population, each at a fixed proportion. Expression of *MVSG*s continues for several days following mammal infection, even after differentiation to the proliferating long slender bloodstream stage, and then the distinct system for expression and, crucially, switching of bloodstream *VSG*s comes into play (Le Ray *et al.*, 1977; Hajduk *et al.*, 1981; Barry and Emery, 1984). All evidence is that each MVAT switches to the full array of early bloodstream VATs, rather than switching to a distinct subset (Hajduk and Vickerman, 1981; Barry *et al.*, 1985). It is believed that, besides the synthesis of the coat being a necessary preadaptation for life in the mammalian host, there is a demand for differential *VSG* expression within the metacyclic population (Barry *et al.*, 1998). By introducing a diverse population to the new host when the fly bites, the parasite will have a greater chance of overcoming anti-VSG antibodies persisting from old infections. Molecular study has revealed that the trypanosome has had to go to some lengths in creating a trustworthy mechanism for generation of diversity in the metacyclic population. In essence, a group of *VSG* genes appears to have been partitioned for this purpose (Lenardo *et al.*, 1984; Cornelissen *et al.*, 1985; Graham *et al.*, 1999; Bringaud *et al.*, 2001), necessitating the development of a novel type of gene locus and a novel transcriptional control system. The trypanosome antigenic-variation system, therefore, is a highly evolved contingency gene system, as it now has developed to this dual system, both components of which contribute greatly to the transmission of the parasite between hosts, be it by prolonging infection or by ensuring diversity.

VSG GENES AND THE GENOME

A basic requirement for antigenic variation is a large pool of silent information, in the form of *VSG* genes. In bacterial pathogens, silent information used in contingency systems usually occurs as gene segments that can be used to alter the variable region of the expressed gene (Zhang and Norris, 1998; Barbet *et al.*, 2000), but silent whole genes are more common in protozoal contingency systems (Svard *et al.*, 1998; Allred *et al.*, 2000; Kyes *et al.*, 2001; Stringer and Keely, 2001). This may be because the single, circular chromosomal nature of bacterial genomes demands efficiency (Brayton *et al.*, 2001), whereas the fragmentation of eukaryotic genomes into many linear chromosomes relieves such a need; it is interesting that the prokaryotic genus *Borrelia* has evolved a genome of many linear DNA molecules and encodes its silent contingency information as a mixture of gene segments and whole genes (Barbour, 1993). Indirect methods have been used to estimate that there could be as many as 1000 silent *VSG* genes contained within the chromosomes of *T. brucei* (Van der Ploeg *et al.*, 1982). Many of these are arranged as tandem arrays and are bordered upstream by a set of repeats about 70 bp long (Liu *et al.*, 1983; Aline *et al.*, 1985; Shah *et al.*, 1987). Although most of the coding sequence is hypervariable, providing the variable epitopes important for antigenic variation (Carrington *et al.*, 1991), more conserved sequences are present towards the 3' end of the coding sequence, and extending further downstream (Liu *et al.*, 1983). *VSG* genes have often been considered as occupying cassettes, defined by their common upstream limit (the 70 bp repeats) and their common downstream limit.

The pool of silent genes has become augmented by the creation of an enormous set of minichromosomes (Weiden *et al.*, 1991; Alsford *et al.*, 2001). These have a remarkable structure, most being composed mainly of a 177 bp DNA sequence, repeated tandemly, and containing also a *VSG* cassette at the telomere. The repetitive region probably serves to make these molecules long enough to form stable chromosomes, but its presence, rather than coding sequences, underlines the likelihood that the minichromosomes exist merely to create a pool of telomeric *VSG*s. Such a large investment – creating as many as 100 new chromosomes just to carry less than 200 genes – not only must influence the generation time of the trypanosome, but also apparently overloads the nuclear-division machinery. Indeed, the trypanosome uses a novel and distinct segregation mechanism for them, which runs slightly out of synchrony with segregation of the main chromosomes (Ersfeld and Gull, 1997).

Individual trypanosomes express only one VSG, and this requires a very tight control mechanism, which is achieved by the use of special transcription units, located at telomeres. These bloodstream expression sites (BESs) (Kooter *et al.*, 1987; Pays *et al.*, 1989) are polycistronic transcription units composed of a strong promoter, several genes known as *ESAG*s (expression site associated genes), which encode proteins not known to have any direct involvement in antigenic variation (Figure 11.3), a very long array of 70 bp repeats and then the *VSG*. Immediately downstream of the BES (within several hundred bp) lies the telomere tract that constitutes the end of the chromosome. To become expressed, silent *VSG*s must be moved into the BES. One complication of the system is that there are an estimated 20–30 BESs in each trypanosome (Borst and Ulbert, 2001; Pays *et al.*, 2001), requiring further control to ensure only one is active. There are two possible reasons for this multiplicity. Silent BESs can allow the stepwise assembly of intact *VSG*s from pseudogenes, thereby expanding the silent repertoire to include incomplete and damaged genes (Roth *et al.*, 1986; Barbet and Kamper, 1993; Barry and McCulloch, 2001). In addition, the different BESs encode different isoforms of the transferrin receptor (in *ESAG6* and *ESAG7*), which differ in their affinity for transferrin from different potential host species, thereby possibly expanding host range (Bitter *et al.*, 1998; Gerrits *et al.*, 2002). The profusion of BESs means that many of the telomeres of the conventional trypanosome chromosomes have been colonized by the bloodstream antigenic-variation system. The rest of the telomeres of these chromosomes are thought to be occupied by the metacyclic expression sites (MESs), which appear to be degenerate BESs (Graham *et al.*, 1999) and are used for transcription of *MVSG*s (Graham and Barry, 1995; Nagoshi *et al.*, 1995). *T. brucei* thus has the unusual adaptation of keeping an essential set of genes at its telomeres.

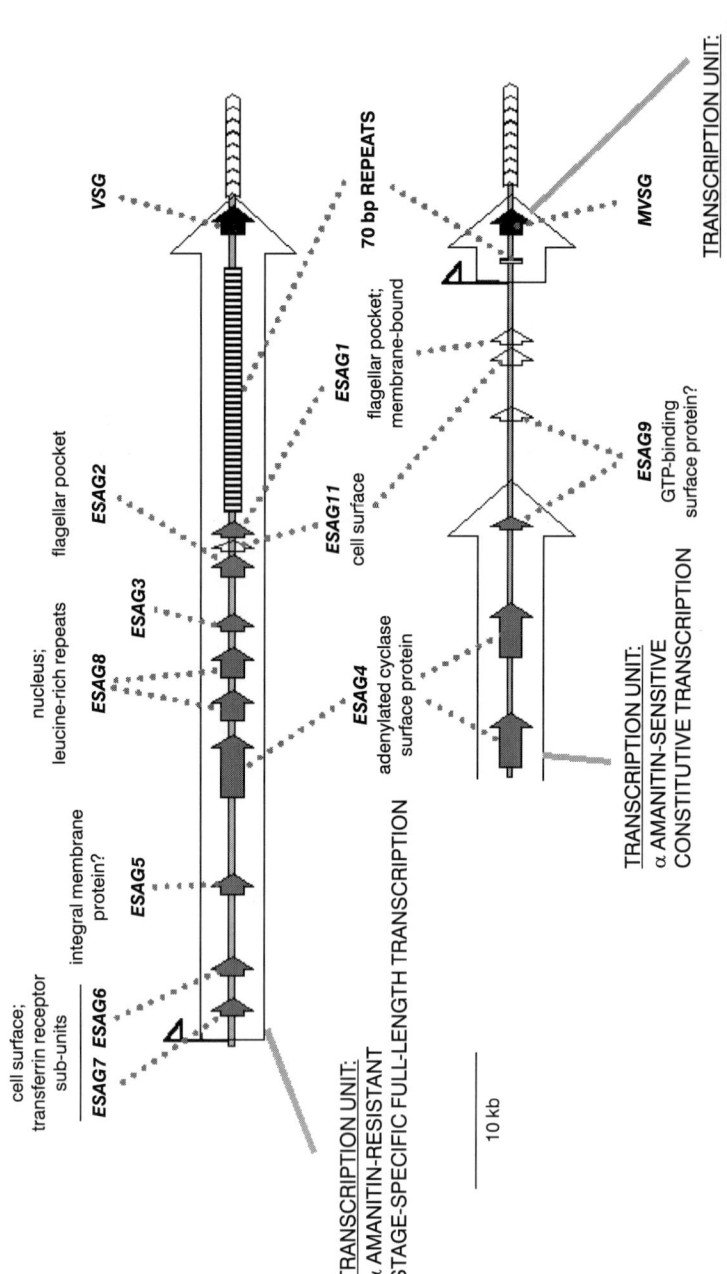

Figure 11.3 The bloodstream and metacyclic *VSG* expression sites. The sites shown are the prototypical AnTat 1.3 BES (Pays *et al.*, 1989) and the ILTat 1.61 MES (Graham *et al.*, 1999). The expression site associated genes (*ESAGs*) of the bloodstream site are identified, along with what is known about their products (reviewed in Pays *et al.*, 2001) and Graham *et al.* (1999). There is some variation in structure of these sites, as reviewed in Pays *et al.* (2001) and Graham *et al.* (1999). The 70 bp repeats are shown as a vertically striped box; intact genes are solid arrows; pseudogenes are open arrows; and telomere tract repeats are shown as block-arrows. *VSG* genes are shown in black.

TRANSCRIPTION OF *VSGS*

An abnormality of the transcription of *VSG*s is the use of RNA polymerase I, which is the classical transcribing enzyme for ribosomal RNA genes, but not for protein-coding genes. Judging mainly from the inhibitory effect of α amanitin (Kooter and Borst, 1984; Lee and Van der Ploeg, 1997), but also from some elegant *in vitro* studies (Laufer *et al.*, 1999), this activity operates on both the long BES and the short MES, as well as the transcription units for the procyclin genes. A likely reason for its use is that this enzyme can provide a high rate of transcription. Normally it cannot be used for transcription of protein-coding genes because only RNA polymerase II is capable of capping mRNA molecules, but trypanosomes and their relatives cap mRNA differently, through a *trans*-splicing process (Nilsen, 1995).

One longstanding question was how the trypanosome achieves expression of only one BES at a time. Once again, the answer involves a novel mechanism and structure in the trypanosome. Despite singular *VSG* transcription, all BES promoters are now thought to be active at any time, but those in inactive BESs are at a reduced level and transcribe over only a limited distance (Rudenko *et al.*, 1994; Vanhamme *et al.*, 2000). Nevertheless, the low level of transcription may allow production of some mature transcripts from the most promoter-proximal *ESAG*s (Ansorge *et al.*, 1999). The picture is that all the promoters of inactive sites are poised for activity but are missing a unique factor. The factor comes in the shape of a novel nuclear structure, the ESB (expression site body), which contains RNA polymerase I and the active BES, but not an inactive BES (Navarro and Gull, 2001). The ESB is not present in the procyclic stage, where no VSG is synthesized. Trypanosomes can change their VSG coat during antigenic variation by switching off the actively transcribed BES, and fully activating one of the silent BESs. It is not known how one BES can displace another from the ESB during such *in situ* switches, but a degree of communication between the two sites is probably involved (Chaves *et al.*, 1999). The constituent components of this body remain to be determined, and whether or not it is related to the nucleolus is an interesting question. It seems likely that trypanosomes have evolved the ESB for the purposes of exclusive *VSG* expression, and hence antigenic variation. A number of other questions remain: is this the site of all *VSG* switching, recombinational and transcriptional; does an ESB exist for MES expression; and does the structure contain all the factors required for initiation, elongation and processing of *VSG* transcription?

DNA RECOMBINATION AS THE MAIN DRIVER OF ANTIGENIC VARIATION

Over the years, a number of distinct mechanisms have been described for *VSG* switching. These fall into three categories. Transcriptional switching, described above, may have significance for host range, but does not occur frequently. The generation of multiple point mutations during *VSG* duplication has also been proposed (Lu *et al.*, 1993), but is not generally supported from experimental data (Graham and Barry, 1996). In a recent analysis of an analogous antigenic-variation system in *Borrelia hermsii*, it was found that generation of the diverse family of silent *vsp* contingency genes within a single genome had arisen not by gradual accumulation of point mutations, but rather by intragenic recombination (Rich *et al.*, 2001). Thus, whether during a switch or in the evolution of the silent repertoire, point mutation appears to be inadequate to produce the quantum differences required for change of all relevant epitopes on these contingency proteins. This leaves the third category of *VSG* switching, DNA recombination. Owing to the use of exclusive sites of *VSG* expression, allied to the dispersal of silent *VSG* sequences throughout the genome, the great majority of the *VSG* repertoire, as much as 98%, can be activated only by recombination into the BES. A specific machinery dedicated to such *VSG* recombination may seem like a sensible evolutionary investment. Ironically, given the number of antigenic-variation-related processes and structures that have evolved in this organism, the available evidence suggests trypanosome *VSG* switching does not involve a unique recombination pathway, but instead uses the general machinery of recombination, conserved throughout evolution, that is otherwise involved in repair of DNA damage.

Differential use of the different classes of *VSG* locus in recombination reactions can account, at least in part, for the hierarchical expression of VATs during chronic infection. Silent telomeric genes are strongly favoured, activating with high frequency and therefore being most apparent early in infection (Liu *et al.*, 1985; Robinson *et al.*, 1999). As most of these are on minichromosomes, they predominate, and the benefit of this recent adaptation – minichromosomes are present in *T. congolense* (Garside *et al.*, 1994) but are very rare in the more primitive *T. vivax* (Dickin and Gibson, 1989) – is evident. Chromosome-internal genes become activated later in infection, although it is not yet known whether they include a subhierarchy or are activated at random. Late in infection, *VSG* fragments or pseudogenes, which encode epitopes not yet expressed, can be assembled into novel chimaeric or mosaic genes, prolonging the infection perhaps beyond the achievement of immunity against the products of all intact *VSGs*. Accurate assembly will require many recombination attempts, explaining the recorded infrequency of appearance of such VATs. Nevertheless, considerable potential exists for temporary expansions of the repertoire by reactions of this type. In fact, some pathogens use mosaic gene formation as the sole, or main, means of achieving switches (Zhang and Norris, 1998; Barbet *et al.*, 2000). It is interesting that, so far in the sequencing of the trypanosome genome, there are very few intact *VSG* genes, but many pseudogenes. This raises the distinct possibility that mosaic formation is, in fact, a predominant route to VSG switching. If so, one explanation for hierarchical expression could be that an active gene is replaced by a segment of, or even completely by, a related gene encoding distinct epitopes, as proposed by Pays (Pays *et al.*, 1985).

That *VSG* switching utilizes homologous recombination is supported by several lines of evidence. Most events are gene conversions (also known as duplicative transpositions), where the *VSG* sequence in the BES is deleted and replaced by a copy generated from a silent *VSG*. The extent of sequence replaced normally encompasses the entire *VSG* open reading frame, but is flexible with regards to the extent of homologous neighbouring sequence used (Liu *et al.*, 1983; Michels *et al.*, 1983), which resembles homologous-recombination, rather than site-specific, reactions. Moreover, genetic inactivation of the major enzyme of eukaryotic homologous recombination, Rad51, substantially impairs antigenic variation (McCulloch and Barry, 1999). Whilst most of this analysis has been conducted in lab-adapted trypanosome strains that switch *VSG*s at rates comparable with background recombination (10^{-7} to 10^{-6} events/cell/generation), it is clear that gene-conversion reactions also predominate in high-switching (10^{-5} to 10^{-2} events/cell/generation) strains and that the same sequence flexibility in use of homologous flank sequences is found (Robinson *et al.*, 1999; P. Burton and J. D. Barry, unpublished).

Antigenic variation and homologous recombination impose contradictory requirements on the *VSG* system. In order to be switched, two *VSG*s must have sufficient sequence similarity that the homologous-recombination machinery can bring them together. On the other hand, to evade the immune response all exposed epitopes must be changed from one VSG to the next. To satisfy the requirement for homology, related sequences surround the *VSG*s. Most prominent are the 70 bp repeats. Small numbers of these are positioned upstream of most *VSG*s, and take the form of enormous arrays in the BES. It seems likely that these repeats are used specifically for *VSG* switching, since they have a nearly exclusive association with *VSG*s rather than being dispersed randomly throughout the genome. They have an unusual sequence composition, including imperfect TTA. TAA triplet repeats interrupted by more conserved domains, and appear to make the DNA helix predisposed to melting (Ohshima *et al.*, 1996). These characteristics may suit their functions as sites for homologous pairing during *VSG* switching. The short blocks of identity in the C terminal coding region and 3' untranslated sequences of *VSG*s are not nearly as striking as the 70 bp repeats.

Just why so many *VSG*s, including those in BESs, MESs and on minichromosomes, are telomeric has not yet received a satisfactory explanation. It has been speculated that some form of reversible telomere-mediated silencing may explain the inactivation of the majority of BESs and MESs (Horn and Cross, 1995; Rudenko *et al.*, 1995; Barry *et al.*, 1998) in the bloodstream stage. However, genetic inactivation of the *T. brucei* homologues of a major regulator of this process in *Saccharomyces cerevisiae*, the Ku70/80 heterodimer, does not result in any detectable effect on *VSG* switching or expression (Conway *et al.*, 2002). In these mutants, the telomere tract was also considerably shortened, suggesting that telomere length is not involved in a major way either. Moreover, the fact that the telomeric *VSG*s on minichromosomes are always silent in their own loci argues against the telomere *per se* being a major regulatory factor.

Perhaps the function of telomere location is to enhance *VSG* recombination reactions. In high-switching strains, the predominant *VSG*s activated early in infection are telomeric, and the majority reside on minichromosomes (Robinson *et al.*, 1999). Telomeric *VSG*s are also predominantly recombined in lab-adapted *T. brucei*, but here silent BESs, with greater overall levels of sequence homology, appear to be favoured (Robinson *et al.*, 1999). In the same strains, a single *VSG* has been shown to be more frequently duplicated when present in a telomeric BES compared with a chromosome-internal locus (Laurent *et al.*, 1984).

There could be a number of explanations for the preferred use of telomeric silent *VSG*s for recombination into the BES. The simplest may be that the presence of extensive telomeric repeats downstream of both *VSG*s simply provides a greater degree of sequence homology. Alternatively, telomeric and sub-telomeric sequences may be less constrained than other loci in their search for recombination partners, leading generally to higher rates of exchange. In fact, in yeast it appears that sequences around telomeres recombine freely with each other, but may be insulated from recombining with non-telomeric sequences (Haber, 2000). This may also underpin the frequent exchanges observed, at least during meiosis, in surface antigen genes found near *Plasmodium* telomeres (Freitas-Junior *et al.*, 2000). For trypanosomes, the use of telomeres could have the added benefit of avoiding potentially deleterious genome-wide rearrangements arising from *VSG* switching. Clearly, however, this putative constraint on recombination cannot be absolute, as chromosome-internal *VSG*s do gene convert into the BES. A final explanation may lie in allowing the trypanosomes flexibility in the molecular pathway of the *VSG* recombination reaction. Multiple recombination pathways appear to exist in eukaryotes, and telomeric *VSG* gene conversions could potentially exploit at least two of these (Figure 11.4). Gene-conversion reactions, including chromosome-internal *VSG* duplications, often occur over relatively short distances and have defined boundaries around the altered sequence. However, one class of homologous-recombination reaction, so-called break-induced replication (Paques and Haber, 1999), can be initiated at a single site and resolved by replicating the copied sequence until the chromosome end, in some cases over very long distances. Perhaps trypanosomes employ telomeric *VSG*s to allow them to exploit multiple reaction pathways, which have different catalytic and controlling factors.

The question arises whether trypanosomes have adapted general recombination reactions or pathways in ways specific for antigenic variation. Little can be said as yet. The high rate of *VSG* switching inherent to trypanosomes is greatly in excess of the background rate of homologous recombination, and is in line with the rate in contingency gene systems. Switching is not programmed and occurs spontaneously. The activating event in *VSG* switching is likely to be what dictates the rate, because homologous-recombination pathways will activate whenever needed. In the past, the use of a dedicated endonuclease has been invoked for this step (Michiels *et al.*, 1983; Pays, 1985; Pays *et al.*, 1994; Robinson *et al.*, 1999), in which the BES array of 70 bp repeats would be

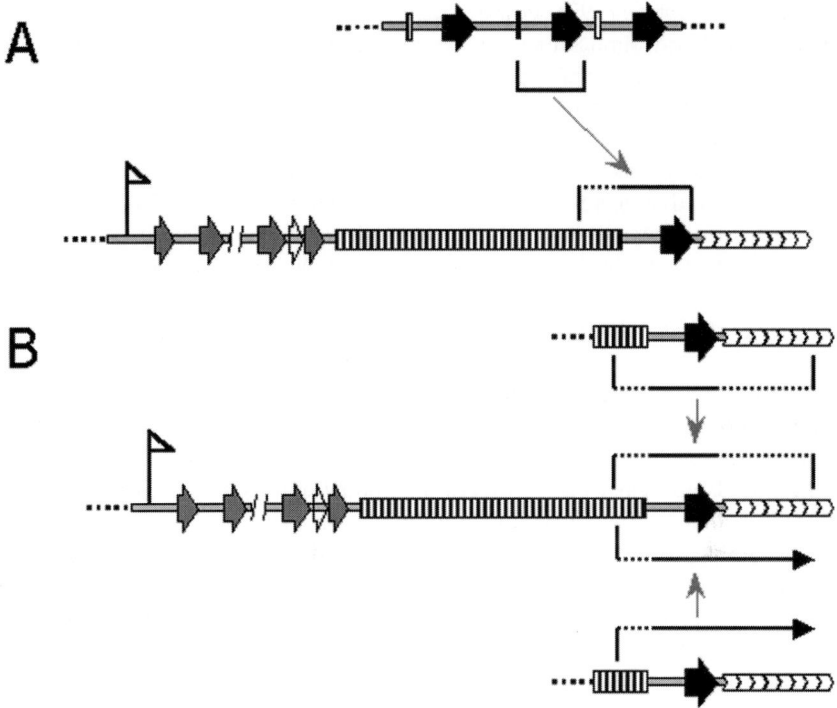

Figure 11.4 Possible recombination pathways involved in duplication of *VSG*s. (A) Gene conversion is proposed for duplication of chromosome-internal genes. (B) Break-induced replication is proposed as a means of duplicating telomeric genes. Symbols are as in Figure 11.3. The boundaries of the duplication reactions are shown, and a dashed line indicates where there is variability in, or uncertainty about, the precise boundary. Break-induced replication is depicted as copying silent sequence all the way to the end of the chromosome.

cleaved, leading to creation of a gap across the *VSG* that is repaired using a silent *VSG* as template. However, a simpler mechanism could be responsible. Commonly in bacteria, contingency switches operate through the activation or inactivation of the gene in question through DNA replication accidents. These involve DNA polymerase slippage over a homopolymeric nucleotide tract, or tetrameric or pentameric oligonucleotide repeats, in or upstream of the gene, leading to its inactivation (Deitsch *et al.*, 1997). A subsequent error can revert the accident, providing a true switch. Strong selection has fixed the presence of these impending accidents in the genome. In the trypanosome, the unstable nature of the 70 bp repeats could be conducive to formation of breaks during replication, with particularly high probability in the very long arrays within BESs. Study in the poorly switching monomorphic lines shows that the 70 bp repeats are not required for *VSG* switching (McCulloch *et al.*, 1997) and that a substantial proportion of switches use other homologous sequences further upstream (reviewed in Barry, 1997). In more rapidly switching trypanosomes, however, the few analysed duplications always map to the repeats (reviewed in Barry, 1997). If there is no 70 bp repeat-specific mechanism for initiating recombination, the difference in switch rates must be explained by a more general process, such as cell-cycle-related replication and DNA damage sensing functions. Again, there would be the irony of the trypanosome rarely not having evolved a novel system to aid antigenic variation! This would not preclude other trypanosome-specific adaptations of the recombination machinery(s), which might be required to cope with a few possible problems downstream of the activating event. *VSG* recombination probably involves genome-wide searching for homologous substrates rather than interaction between sister chromatids following replication, which would lead to fruitless self-switching. Although searching elsewhere may be helped by many *VSGs* being located at recombinationally efficient telomeres, the majority of silent *VSG* substrates are in the interior of chromosomes, truly ectopic sites compared with the BES. Finally, the sequences that the system uses as the basis of most shared exchanges are relatively short and degenerate (the 70 bp repeats), and might appear to be rather poor substrates. Our understanding of the molecular genetics of trypanosome antigenic variation is in its infancy, and a comprehensive discussion of this is not possible yet, but some recent functional observations may signal a beginning of understanding these issues.

Rad51 mutant lab-adapted trypanosomes are impaired in *VSG* switching, but gene conversion-like *VSG* switches can still be detected, suggesting that another recombination pathway operates in the background (McCulloch and Barry, 1999). Recent work in our laboratories has shown that this Rad51-independent reaction is capable of performing recombination reactions with as much precision as Rad51-dependent reactions, but may normally use shorter DNA substrates. We have currently no information on the factors that control this process. Nevertheless, such a pathway could have significant roles in switching between *VSG* genes with very limited flanking homologies, or even between distinct *VSGs* during an infection (mosaic gene formation, described above) or in evolution. The protein Mre11 forms a trimeric complex with two other proteins, Rad50 and Xrs2/NbsI, that is at the heart of much eukaryotic recombination (Hopfner *et al.*, 2001). This complex has nuclease activities that may be important in processing DNA breaks for strand exchange. It also appears to signal DNA damage and coordinate homologous and non-homologous repair, and to have roles in telomere maintenance. Mutants in the trypanosome homologue of the Mre11 protein have subtly different phenotypes when compared with other eukaryotes, perhaps pointing directly to a different use of DNA repair pathways in the parasite (Robinson *et al.*, 2002). When compared with other organisms, the trypanosome mutant appears to have a repair phenotype more specific to double strand breaks and has a different chromosome instability phenotype, which indicates that the parasite regulates rearrangements differently, separating telomere and chromosome-internal events.

TRANSCRIPTIONAL CONTROL OF METACYCLIC *VSGS*

Although homologous recombination can service the needs of antigenic variation in the bloodstream phase of the trypanosome's life cycle, transcriptional regulation controls differential expression of *MVSG*s in the metacyclic stage. During the final differentiation to the metacyclic stage, each trypanosome activates one of the *MVSG* promoters, ending its strict silencing, which has probably been in place all through the life cycle, since not long after its last transmission from a tsetse. This is distinct from the regulation of the BES promoters which, as far as is known, are subject only to up- and downregulation, remaining active in all tested life cycle stages. It is different also from what happens with any known promoter in trypanosomes, where constitutive gene expression is the norm. Again, we see a novelty in the trypanosome, in the form of a control mechanism developed for the purpose of antigenic variation. It would be interesting to know how this organism was able to (re)invent transcriptional control just for this relatively recently evolved subset of expression sites. Not much idea is gained from the sequence of the promoter itself. After some controversy over the identity of the MES promoter, we have recently confirmed that the element proposed by Donelson and colleagues (Kim and Donelson, 1997), from study in procyclic and bloodstream stages, is also the element used in trypanosomes recovered from the salivary glands of tsetse. This element appears to be derived from the BES promoter, and both are 65–70 bp long and transcribed by RNA polymerase I. However, one main difference is that the MES promoters are diverse in sequence, whereas the BES promoters are highly conserved, both within and between strains (Vanhamme *et al.*, 1995). Comparing the different *MVSG* promoter elements sequenced, and in some cases functionally tested, shows that they are related to each other in the range 29–94% identity, and are related to the BES promoters in the range 11–54% (M. L. Ginger and J. D. Barry, unpublished). Two short elements that are functionally important in the BES promoter are degenerate in the MES promoter, showing a fundamental difference in control. The diversity in the MES promoters may reflect that its important motifs are of relaxed sequence.

Functional study of MES promoters, in which they were tested in front of reporter genes in plasmids or in another locus, has revealed, first, that their silencing throughout the life cycle is alleviated when they are outside their own locus (Nagoshi *et al.*, 1995; Graham *et al.*, 1998) and, secondly, that *trans*-acting factors sufficient for their activation are present in bloodstream and procyclic stages. Under conditions of strong selection, it is even possible to obtain bloodstream trypanosomes expressing the VSG coat from an MES, using the MES promoter, confirming the presence of *trans*-acting factors (Pedram and Donelson, 1999). These expressors have possibly arisen through accidental chromatin alteration. There are several candidates for the basis of locus-associated repression. Most attractive is the telomere tract, because of the studies in *S. cerevisiae* demonstrating that proteins associated with the tract reversibly silence promoters experimentally inserted nearby, in what is known as the telomere position effect (TPE) (Gottschling *et al.*, 1990). Could it be that the trypanosome has achieved transcriptional repression by locating the MES promoter close to the telomere? This could also explain the monocistronic nature of the MES – the presence of other genes would push the promoter further away from the telomere tract. Unfortunately, deletion of the genes encoding the Ku dimer, which is necessary for silencing in yeast, does not lift the repression in the procyclic stage (Conway *et al.*, 2002), ruling out any role for this protein and also indicating that a greatly shortened telomere tract, one of the Ku mutant phenotypes, does not derepress either. Hence, *MVSG*s may not be subject to TPE, which is also the case for all bar one naturally telomeric genes in *S. cerevisiae* (VegaPalas *et al.*, 1997) and an experimentally inserted gene in *Saccharomyces pombe* (Manolis *et al.*, 2001). Perhaps other elements of the locus are responsible for silencing, and it remains a mystery what the trypanosome has invented to achieve life cycle promoter control. Whether it uses a system related to the tight silencing of chromosome-internal *VSG*s is worth considering.

SECONDARY PHENOTYPES ASSOCIATED WITH ANTIGENIC VARIATION

There are two known secondary phenotypes associated with the VSG and BES system (reviewed by Pays *et al.*, 2001). The siting of the transferrin receptor genes in BESs allows their cotranscription with the surface coat gene at a high level and enables, via BES switching, a major shift in expression between the receptor isoforms (Bitter *et al.*, 1998). This may allow better survival in different host species. Another gene associated with single BESs in *T. b. rhodesiense* is the partial *VSG* known as the serum resistance associated (*SRA*) gene (Xong *et al.*, 1998). The product of this gene is capable of conferring resistance to the lytic effect of human serum, and presumably causing infectivity to humans. When a trypanosome switches to transcription of that BES, the resistance phenotype appears. It is unclear whether presence in a BES confers any immediate benefit to the *SRA*.

CONCLUSIONS

Antigenic variation in trypanosomes probably began with the reassignment of a surface receptor to a more passive, coating role. As the coat became in direct conflict with the immune system, antigenic switching at a high rate became necessary, and we believe this was most readily achieved by the opportune use of an unstable DNA sequence, allowing the gene-duplication system to operate. The initial, presumably few, variant genes expanded into an enormous repertoire, and such is the success of antigenic variation, the trypanosome subsequently evolved some more elaborate structures and systems. We now see a highly evolved system that has necessitated an enormous investment by the parasite, and has insinuated itself into multiple core cellular and genetic processes.

REFERENCES

Aline, R., Macdonald, G., Brown, E., Allison, J., Myler, P., Rothwell, V. and Stuart, K. (1985) *Nucleic Acids Res.* 13, 3161–3177.

Allred, D. R., Carlton, J. M. R., Satcher, R. L., Long, J. A., Brown, W. C., Patterson, P. E., O'Connor, R. M. and Stroup, S. E. (2000) *Mol. Cell* 5, 153–162.

Alsford, S., Wickstead, B., Ersfeld, K. and Gull, K. (2001) *Mol. Biochem. Parasitol.* 113, 79–88.

Ansorge, I., Steverding, D., Melville, S., Hartmann, C. and Clayton, C. (1999) *Mol. Biochem. Parasitol.* 101, 81–94.

Barbet, A. F. and Kamper, S. M. (1993) *Parasitol. Today* 9, 63–66.

Barbet, A. F., Lundgren, A., Yi, J., Rurangirwa, F. R. and Palmer, G. H. (2000) *Infect. Immun.* 68, 6133–6138.

Barbour, A. G. (1993) *Trends Microbiol.* 1, 236–239.

Barry, J. D. (1986) *Parasitology* 92, 51–65.

Barry, J. D. (1997) *Parasitol. Today* 13, 212–218.

Barry, J. D. and Emery, D. L. (1984) *Parasitology* 88, 67–84.

Barry, J. D. and McCulloch, R. (2001) *Adv. Parasitol.* 49, 1–70.

Barry, J. D., Crowe, J. S. and Vickerman, K. (1985) *Parasitology* 90, 79–88.

Barry, J. D., Graham, S. V., Fotheringham, M., Graham, V. S., Kobryn, K. and Wymer, B. (1998) *Mol. Biochem. Parasitol.* 91, 93–105.

Bitter, W., Gerrits, H., Kieft, R. and Borst, P. (1998) *Nature* 391, 499–502.

Borst, P. and Fairlamb, A. H. (1998) *Annu. Rev. Microbiol.* 52, 745–778.

Borst, P. and Ulbert, S. (2001) *Mol. Biochem. Parasitol.* 114, 17–27.

Brayton, K. A., Knowles, D. P., McGuire, T. C. and Palmer, G. H. (2001) *Proc. Natl Acad. Sci. USA* 98, 4130–4135.

Bringaud, F., Biteau, N., Donelson, J. E. and Baltz, T. (2001) *Mol. Biochem. Parasitol.* 113, 67–78.

Capbern, A., Giroud, C., Baltz, T. and Mattern, P. (1977) *Exp. Parasitol.* 42, 6–13.

Carrington, M. and Boothroyd, J. (1996) *Mol. Biochem. Parasitol.* 81, 119–126.

Carrington, M., Miller, N., Blum, M., Roditi, I., Wiley, D. and Turner, M. (1991) *J. Mol. Biol.* 221, 823–835.

Chaves, I., Rudenko, G., Dirks-Mulder, A., Cross, M. and Borst, P. (1999) *EMBO J.* 18, 4846–4855.

Conway, C., McCulloch, R., Ginger, M. L., Robinson, N. P., Browitt, A. and Barry, J. D. (2002) *J. Biol. Chem.* 277, 21269–21277.

Cornelissen, A. W., Bakkeren, G. A., Barry, J. D., Michels, P. A. and Borst, P. (1985) *Nucleic Acids Res.* 13, 4661–4676.

Cross, G. A. M. (1975) *Parasitology* 71, 393–417.

Deitsch, K. W., Moxon, E. R. and Wellems, T. E. (1997) *Microbiol. Mol. Biol. Rev.* 61, 281–294.

Dickin, S. K. and Gibson, W. C. (1989) *Mol. Biochem. Parasitol.* 33, 135–142.
Ersfeld, K. and Gull, K. (1997) *Eur. J. Cell Biol.* 72, 240–240.
Freitas-Junior, L. H., Bottius, E., Pirrit, L. A., Deitsch, K. W., Scheidig, C., Guinet, F., Nehrbass, U., Wellems, T. E. and Scherf, A. (2000) *Nature* 407, 1018–1022.
Gardiner, P. R., Nene, V., Barry, M. M., Thatthi, R., Burleigh, B. and Clarke, M. W. (1996) *Mol. Biochem. Parasitol.* 82, 1–11.
Garside, L., Bailey, M. and Gibson, W. (1994) *Acta Trop.* 57, 21–28.
Gerrits, H., Mussmann, R., Bitter, W., Kieft, R. and Borst, P. (2002) *Mol. Biochem. Parasitol.* 119, 237–247.
Gottschling, D. E., Aparicio, O. M., Billington, B. L. and Zakian, V. A. (1990) *Cell* 63, 751–762.
Graham, S. V. and Barry, J. D. (1995) *Mol. Cell. Biol.* 15, 5945–5956.
Graham, S. V. and Barry, J. D. (1996) *Mol. Biochem. Parasitol.* 79, 35–45.
Graham, S. V., Terry, S. and Barry, J. D. (1999) *Mol. Biochem. Parasitol.* 103, 141–154.
Graham, S. V., Wymer, B. and Barry, J. D. (1998) *Mol. Cell. Biol.* 18, 1137–1146.
Haber, J. E. (2000) *Trends Genet.* 16, 259–264.
Hajduk, S. L. and Vickerman, K. (1981) *Parasitology* 83, 609–621.
Hajduk, S. L., Cameron, C. R., Barry, J. D. and Vickerman, K. (1981) *Parasitology* 83, 595–607.
Hopfner, K. P., Karcher, A., Craig, L., Woo, T. T., Carney, J. P. and Tainer, J. A. (2001) *Cell* 105, 473–485.
Horn, D. and Cross, G. A. M. (1995) *Cell* 83, 555–561.
Kim, K. S. and Donelson, J. E. (1997) *J. Biol. Chem.* 272, 24637–24645.
Kooter, J. M. and Borst, P. (1984) *Nucleic Acids Res.* 12, 9457–9472.
Kooter, J. M., Van der Spek, H. J., Wagter, R., d'Oliveira, C. E., Van der Hoeven, F., Johnson, P. J. and Borst, P. (1987) *Cell* 51, 261–272.
Kyes, S., Horrocks, P. and Newbold, C. (2001) *Annu. Rev. Microbiol.* 55, 673–707.
Laufer, G., Schaaf, G., Bollgonn, S. and Gunzl, A. (1999) *Mol. Cell. Biol.* 19, 5466–5473.
Laurent, M., Pays, E., Van der Werf, A., Aerts, D., Magnus, E., Van Meirvenne, N. and Steinert, M. (1984) *Nucleic Acids Res.* 12, 8319–8328.
Le Ray, D., Barry, J. D., Easton, C. and Vickerman, K. (1977) *Ann. Soc. Belg. Med. Trop.* 57, 369–381.
Lee, Mg S. and Van der Ploeg, L. H. T. (1997) *Annu. Rev. Microbiol.* 51, 463–489.
Lenardo, M. J., Rice-Ficht, A. C., Kelly, G., Esser, K. M. and Donelson, J. E. (1984) *Proc. Natl Acad. Sci. USA* 81, 6642–6646.
Liu, A. Y. C., Michels, P. A. M., Bernards, A. and Borst, P. (1985) *J. Mol. Biol.* 182, 383–396.
Liu, A. Y. C., Van der Ploeg, L. H. T., Rijsewijk, F. A. M. and Borst, P. (1983) *J. Mol. Biol.* 167, 57–75.
Lu, Y., Hall, T., Gay, L. S. and Donelson, J. E. (1993) *Cell* 72, 397–406.
MacLeod, A., Tweedie, A., Welburn, S. C., Maudlin, I., Turner, C. M. R. and Tait, A. (2000) *Proc. Natl Acad. Sci. USA* 97, 13442–13447.
Manolis, K. G., Nimmo, E. R., Hartsuiker, E., Carr, A. M., Jeggo, P. A. and Allshire, R. C. (2001) *EMBO J.* 20, 210–221.
McCulloch, R. and Barry, J. D. (1999) *Genes Dev.* 13, 2875–2888.
McCulloch, R., Rudenko, G. and Borst, P. (1997) *Mol. Cell. Biol.* 17, 833–843.

Michels, P. A. M., Liu, A. Y. C., Bernards, A., Sloof, P., Vanderbijl, M. M. W., Schinkel, A. H., Menke, H. H., Borst, P., Veeneman, G. H., Tromp, M. C. and Vanboom, J. H. (1983) *J. Mol. Biol.* 166, 537–556.
Michiels, F., Matthyssens, G., Kronenberger, P., Pays, E., Dero, B., Van Assel, S., Darville, M., Cravador, A., Steinert, M. and Hamers, R. (1983) *EMBO J.* 2, 1185–1192.
Miller, E. N. and Turner, M. J. (1981) *Parasitology* 82, 63–80.
Moxon, E. R., Rainey, P. B., Nowak, M. A. and Lenski, R. E. (1994) *Curr. Biol.* 4, 24–33.
Nagoshi, Y. L., Alarcon, C. M. and Donelson, J. E. (1995) *Mol. Biochem. Parasitol.* 72, 33–45.
Navarro, M. and Gull, K. (2001) *Nature* 414, 759–763.
Nilsen, T. W. (1995) *Mol. Biochem. Parasitol.* 73, 1–6.
Ohshima, K., Kang, S., Larson, J. E. and Wells, R. D. (1996) *J. Biol. Chem.* 271, 16784–16791.
Paques, F. and Haber, J. E. (1999) *Microbiol. Mol. Biol. Rev.* 63, 349.
Pays, E. (1985) *Prog. Nucleic Acid Res. Mol. Biol.* 32, 1–26.
Pays, E., Houard, S., Pays, A., Van Assel, S., Dupont, F., Aerts, D., Huetduvillier, G., Gomes, V., Richet, C., Degand, P., Van Meirvenne, N. and Steinert, M. (1985) *Cell* 42, 821–829.
Pays, E., Lips, S., Nolan, D., Vanhamme, L. and Perez-Morga, D. (2001) *Mol. Biochem. Parasitol.* 114–116.
Pays, E., Tebabi, P., Pays, A., Coquelet, H., Revelard, P., Salmon, D. and Steinert, M. (1989) *Cell* 57, 835–845.
Pays, E., Vanhamme, L. and Berberof, M. (1994) *Annu. Rev. Microbiol.* 48, 25–52.
Pedram, M. and Donelson, J. E. (1999) *J. Biol. Chem.* 274, 16876–16883.
Rich, S. M., Sawyer, S. A. and Barbour, A. G. (2001) *Proc. Natl Acad. Sci. USA* 98, 15038–15043.
Robinson, N. P., Burman, N., Melville, S. E. and Barry, J. D. (1999) *Mol. Cell. Biol.* 19, 5839–5846.
Robinson, N. P., McCulloch, R., Conway, C., Browitt, A. and Barry, J. D. (2002) *J. Biol. Chem.* 277, 26185–26193.
Roth, C. W., Longacre, S., Raibaud, A., Baltz, T. and Eisen, H. (1986) *EMBO J.* 5, 1065–1070.
Rudenko, G., Blundell, P. A., Dirksmulder, A., Kieft, R. and Borst, P. (1995) *Cell* 83, 547–553.
Rudenko, G., Blundell, P. A., Taylor, M. C., Kieft, R. and Borst, P. (1994) *EMBO J.* 13, 5470–5482.
Salmon, D., Hanocq-Quertier, J., Paturiaux-Hanocq, F., Pays, A., Tebabi, P., Nolan, D. P., Michel, A. and Pays, E. (1997) *EMBO J.* 16, 7272–7278.
Shah, J. S., Young, J. R., Kimmel, B. E., Iams, K. P. and Williams, R. O. (1987) *Mol. Biochem. Parasitol.* 24, 163–174.
Stringer, J. R. and Keely, S. P. (2001) *Infect. Immun.* 69, 627–639.
Svard, S. G., Meng, T. C., Hetsko, M. L., McCaffery, J. M. and Gillin, F. D. (1998) *Mol. Microbiol.* 30, 979–989.
Turner, C. M. R. and Barry, J. D. (1989) *Parasitology* 99, 67–75.
Turner, C. M. R., Aslam, N. and Angus, S. D. (1996) *Parasitol. Res.* 82, 61–66.
Urakawa, T., Eshita, Y. and Majiwa, P. A. O. (1997) *Exp. Parasitol.* 85, 215–224.
Van Den Abbeele, J., Claes, Y., Van Bockstaele, D., Le Ray, D. and Coosemans, M. (1999) *Parasitology* 118, 469–478.

Van der Ploeg, L. H. T., Bernards, A., Rijsewijk, F. A. M. and Borst, P. (1982) *Nucleic Acids Res.* 10, 593–609.

Van Meirvenne, N., Janssens, P. G. and Magnus, E. (1975) *Ann. Soc. Belg. Med. Trop.* 55, 1–23.

Vanhamme, L., Pays, A., Tebabi, P., Alexandre, S. and Pays, E. (1995) *Mol. Cell. Biol.* 15, 5598–5606.

Vanhamme, L., Poelvoorde, P., Pays, A., Tebabi, P., Xong, H. V. and Pays, E. (2000) *Mol. Microbiol.* 36, 328–340.

VegaPalas, M. A., Venditti, S. and DiMauro, E. (1997) *Nat. Genet.* 15, 232–233.

Weiden, M., Osheim, Y. N., Beyer, A. L. and Van der Ploeg, L. H. T. (1991) *Mol. Cell. Biol.* 11, 3823–3834.

Xong, H. V., Vanhamme, L., Chamekh, M., Chimfwembe, C. E., Van Den Abbeele, J., Pays, A., Van Meirvenne, N., Hamers, R., Debaetselier, P. and Pays, E. (1998) *Cell* 95, 839–846.

Zhang, J. R. and Norris, S. J. (1998) *Infect. Immun.* 66, 3698–3704.

12

ANTIGENIC VARIATION IN *ANAPLASMA MARGINALE* AND *EHRLICHIA* (*COWDRIA*) *RUMINANTIUM*

SUMAN M. MAHAN

INTRODUCTION

This chapter will review the mechanisms of antigenic variation of the two economically important tick-borne pathogens of domestic ruminants, *Anaplasma marginale* and *Cowdria ruminantium*, recently renamed *Ehrlichia ruminantium*.

The genus *Anaplasma* contains four members, *Anaplasma marginale*, *Anaplasma centrale*, *Anaplasma caudatum* and *Anaplasma ovis*, that cause anaplasmosis in domestic ruminants. All members of this genus infect erythrocytes in their respective ruminant hosts (sheep, cattle and goats) and are transmitted by ixodid ticks (Theiler, 1910; Ristic and Kreier, 1984; Losos, 1986). *A. centrale* is a benign pathogen and is used to immunize cattle against *A. marginale* infections, because it reduces the severity of *A. marginale* infection (Kuttler, 1967). *A. ovis* affects sheep and goats and *A. caudatum* affects cattle and occurs in mixed infections with *A. marginale*. *A. marginale* is the most pathogenic and important of these four pathogens and causes severe bovine anaplasmosis. Owing to its recognized relative importance, *A. marginale* has been characterized in detail and will be the subject of this chapter. Similarities and differences will be highlighted with the other *Anaplasma* species wherever appropriate and possible.

Cowdria ruminantium is the sole member of the genus *Cowdria* (Ristic and Huxsoll, 1984). However, recently, the classification of *C. ruminantium*, *A. marginale* and other ehrlichial pathogens has been the subject of a discussion which has supported their re-classification. *C. ruminantium* has been subsequently renamed *Ehrlichia ruminantium* (see below).

A. marginale and *E. (C.) ruminantium* are tick-borne ehrlichial pathogens of significant economic importance because they cause illnesses in domestic ruminants which result in decreased livestock production (Uilenberg, 1983; Losos, 1986; Mukhebi *et al.*, 1999; Mahan *et al.*, 2001). *A. marginale* has a worldwide distribution and causes anaplasmosis or gall sickness in cattle. It is primarily transmitted by the biological vectors *Dermacenter* and *Boophilus* ticks, but can also be transmitted by biting flies and by using needles contaminated with infected blood. *C. ruminantium* causes an acute and fatal infection called cowdriosis or heartwater in sheep, goats, cattle and certain wildlife species and is widely distributed in sub-Saharan Africa and on three islands in the Caribbean region (Uilenberg, 1983; Walker and Olwage, 1987). The principal method of transmission of *C. ruminantium* is via *Amblyomma* tick species. Experimental infection of *C. ruminantium* can be initiated by using infected blood or cell culture material. *A. marginale* and *C. ruminantium* cause high morbidity and mortality in domestic ruminants and their control requires persistent prevention strategies that cost the farming community and veterinary authorities billions of dollars annually.

PHYLOGENETIC RE-CLASSIFICATION OF *A. MARGINALE* AND *C. RUMINANTIUM*

A. marginale and *C. ruminantium* are obligately intracellular pathogens that grow in vacuoles and were classified under the order Rickettsiales (Ristic and Huxsoll, 1984). Their classification under different families or tribes has been based on morphology and host cell tropism rather than on genetic phylogeny. *A. marginale* was classified under the family *Anaplasmataceae*, genus *Anaplasma* and *C. ruminantium* under the family *Rickettsiaceae*, tribe *Ehrlichieae*, genus *Cowdria* (Ristic and Huxsoll, 1984, Ristic and Krier, 1984). However, based on the 16s ribosomal DNA sequence homology, *A. marginale* and *C. ruminantium* have been shown to belong to two different genogroups of the ehrlichia complex (Figure 12.1; Dame *et al.*, 1992; van Vliet *et al.*, 1992; Mahan *et al.*, 1999), demonstrating that these and other ehrlichial pathogens are closely related and there is a need for re-classification of these pathogens. Phylogenetic analysis based on alignments of other conserved genes such as GroEL and citrate synthase further support the re-classification of these pathogens deduced from 16s ribosomal DNA analyses (Inokuma *et al.*, 2001; Yu *et al.*, 2001). Furthermore, molecular characterizations of genes encoding outer membrane proteins of these related pathogens supports reorganization of certain members of the order Rickettsiales, families *Anaplasmataceae* and *Rickettsiaceae*. Therefore, it has been proposed that the tribes *Rickettsiae*, *Ehrlichieae* and *Wolbachiaea* be abolished and all the members within these tribes be transferred to the family *Anaplasmataceae*. The family *Anaplasmataceae* has been broadened to include

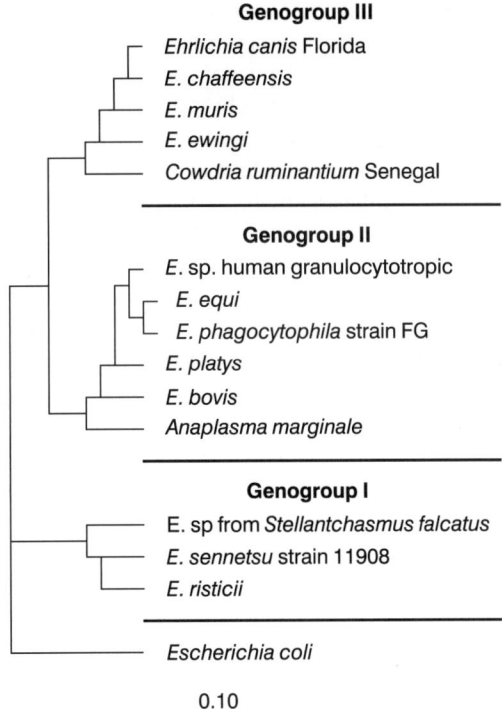

Figure 12.1 Maximum likelihood phylogenetic tree of some *Ehrlichiae*, based on 16S ribosomal RNA gene sequences, outgroup-rooted using *Escherichia coli*. The scale bar represents 10 nucleotide substitutions/100 bases. (Reproduced from *Parasitology Today*, vol. 15, Mahan, S. M., Allsopp, B., Kocan, K. M., Palmer, G. H. and Jongejan, F., pp. 290–294. Copyright ©1999, with permission from Elsevier.)

the genera *Anaplasma*, *Cowdria*, *Ehrlichia*, *Neorickettsia* and *Wolbachia*, which are all obligate intracellular bacteria that replicate within a vacuole derived from host cell membrane. *Cowdria*, *Ehrlichia*, *Neorickettsia* and *Wolbachia* were previously all members of the family *Rickettsiaceae* (Ristic and Huxsoll, 1984). These genera were reorganized and based on genetic similarities, *C. ruminantium* has been placed within the genus *Ehrlichia* and its name changed to *Ehrlichia ruminantium* and combined with existing species *Ehrlichia canis*, *Ehrlichia chaffeensis*, *Ehrlichia ewingii* and *Ehrlichia muris*, whereas previously it was the lone member of the genus *Cowdria* (Ristic and Huxsoll, 1984; Dumler et al., 2001). From here on *C. ruminantium* will be referred to as *Ehrlichia ruminantium* or *E. ruminantium*.

DISEASE PATHOGENESIS

A. marginale is only known to infect erythrocytes. In susceptible cattle, *A. marginale* can infect 10–90% of the erythrocytes (Richey, 1981), resulting in severe anaemia, marked weight loss, icterus, abortion and death of a proportion of affected animals (Losos, 1986; Palmer, 1989). Anaemia occurs as a result of removal of infected erythrocytes by phagocytosis without haemoglubinuria or haemoglobinaemia. These are the main clinical signs of *A. marginale* infection and they coincide with acute or primary infection. Animals that recover from acute infection become persistently infected and exhibit no clinical signs. *A. marginale* can persistently infect cattle for at least seven years and this infection may persist for the life of the animal (Eriks *et al.*, 1989, 1993). The persistence of infection in bovine hosts ensures that *A. marginale* is available to maintain an infected tick population, since transovarial passage of the pathogen within the tick does not occur (Stich *et al.*, 1989).

E. ruminantium infects nucleated cells, namely vascular endothelial cells, monocytes and neutrophils. Infected animals can suffer mortalities ranging from 20% to 90% based on breed, species, age of the hosts and virulence of *E. ruminantium* (Uilenberg, 1983; Mahan *et al.*, 2001). Infected animals become febrile, anorexic, depressed and suffer from respiratory and cardiovascular system distress. Sudden death usually follows these clinical signs. In some cases, nervous signs develop followed by death. Pathognomic autopsy signs include hydropericardium (hence the name of the disease, heartwater), hydrothorax and ascites with haemorrhagic lesions on serosal and mucosal surfaces. Animals that recover from *E. ruminantium* infection become resistant to re-challenge, are persistently infected and are a source of transmissible infection for ticks (Andrew and Norval, 1989).

Protective immune responses are directed at the outer membrane proteins (OMPs), also referred to as major surface proteins (MSPs) in *A. marginale* or major antigenic proteins (MAP) in *E. ruminantium* (Barbet *et al.*, 1994; Brown *et al.*, 1998a,b; Mwangi *et al.*, 1998; Nyika *et al.*, 1998, 2002; Totté *et al.*, 1998, 1999; Palmer *et al.*, 1999; Byrom *et al.*, 2000a,b; van Kleef *et al.*, 2000). Protective immunity mediated by high-titred IgG2 antibody responses against B cell epitopes of MSPs, in concert with CD 4^+ T lymphocyte-mediated macrophage activation for opsonization and killing, promotes clearance of *A. marginale* (Buening, 1976; Carson *et al.*, 1976; Brown *et al.*, 1998a,b; Palmer and McElwain, 1995). In addition, the role of interferon gamma is thought to be important as it enhances bovine IgG2 antibody levels (Estes *et al.*, 1994). In anaplasmosis, structural variation of these MSPs plays a primary role in establishing persistent infection by evading the host's immune response and is discussed in detail below.

Protective immunity against *E. ruminantium* is cell-mediated and is derived from activation of the T Helper 1 pathway. The activated cells produce IFN gamma, which is known to inhibit *E. ruminantium* growth (Mahan *et al.*, 1994, 1996; Mwangi *et al.*, 1998; Totté *et al.*, 1999; Byrom *et al.*, 2000a,b; van Kleef *et al.*, 2000). MAP-1 of *E. ruminantium* induces protective immunity (Nyika *et al.*, 1998, 2002; Mwangi *et al.*, 2002). Detailed studies on all OMPs of *E. ruminantium* are lacking at this time. However, owing to high sequence identity of the genes that encode the OMPs of *Ehrlichia*, common mechanisms of generating diversity and persistence of infection may be employed by these related pathogens (Palmer *et al.*, 1994; Reddy *et al.*, 1996, 1998; Mahan *et al.*, 1999; Sulsona *et al.*, 1999; Yu *et al.*, 2000; Ohashi *et al.*, 2001).

Information from molecular characterization of *A. marginale* and *E. ruminantium* is now accumulating and genome-sequencing projects for both pathogens are in progress. By pulse field gel electrophoresis, *A. marginale* has a genome size of between 1200 and 1260 kbp with a G+C content of 56% (Alleman *et al.*, 1993) and the genome size of *E. ruminantium* is predicted to be of approximately 1576 kbp (de Villers *et al.*, 2000a). These pathogens have a small genome, which they utilize effectively for their survival.

ANAPLASMA MARGINALE

A. marginale are dense, homogenous, bluish-purple, round bodies measuring 0.3 to 1.0 micron in diameter (Ristic, 1968). *A. marginale* was first described by Theiler (1910). *A. marginale* was discovered as small, punctiform inclusion bodies inside bovine erythrocytes with a peripheral location. The name *Anaplasma marginale* was coined because the organisms were without cytoplasm and because they were located on the periphery of erythrocytes. Each body contains several subunits called initial bodies and these structures reproduce by binary fission. The *A. marginale* inclusion bodies appear as ring or comet forms. The inclusion bodies are attached to a structure called the inclusion appendage ('tail' or 'band') (Kocan *et al.*, 1978a,b). The inclusion appendage consists of tightly packed interconnected laminae and assumes a loop, dumbbell or comet configuration. The inclusion appendage and the outer membrane of the *A. marginale* inclusion body have been shown to have antigenic determinants as they react with antisera from infected cattle.

Antigenic variation in *A. marginale*

Antigenic variation is the primary mechanism by which *A. marginale* persists in a fully immunocompetent host following recovery from acute infection (Palmer *et al.*, 1999). This phenomenon is facilitated by the periodic appearance of phenotypically distinct organisms bearing unique OMPs during persistent *A. marginale* infection of bovine hosts. Persistence of infection otherwise seems not to have any adverse effect on the well-being or productivity of the host (Kieser *et al.*, 1990).

The role of outer membrane proteins of *A. marginale* in antigenic variation

There are six OMPs of *A. marginale* that are highly immunogenic (Table 12.1; Palmer and McGuire, 1984; Vidotto *et al.*, 1994) and provoke potent B and T cell

Table 12.1 Outer membrane proteins (OMPs) of *A. marginale*

OMPs	Molecular size	Function	Gene organization	Role in antigenic variation
MSP 1				
MSP 1a	70–105 kDa	Adhesin	Polymorphic single copy	Unlikely
MSP 1b	97–100 kDa	Adhesin	Polymorphic multigene family	Unknown
MSP 2	36 kDa	Adhesin	Polymorphic multigene family	Yes
MSP 3	86 kDa	Unknown	Polymorphic multigene family	Yes
MSP 4	31 kDa	Unknown	Single copy conserved	Unlikely
MSP 5	19 kDa	Unknown	Single copy conserved	Unlikely
MSP ?	61 kDa	Unknown	Not characterized	Unknown

responses (Tebele *et al.*, 1991; Brown *et al.*, 1998a,b; French *et al.*, 1998, 1999; Rurangirwa *et al.*, 1999). With the exception of the 61 kDa OMP, the genes encoding the other OMPs have been cloned, sequenced, expressed and some demonstrate polymorphism between strains and are encoded by multigene families. Some of these OMPs are targets of protective immune responses and therefore candidates for the development of a recombinant vaccine against anaplasmosis (Barbet *et al.*, 1987; Palmer *et al.*, 1986, 1988, 1989; A. F. Barbet unpublished observations on MSP 4).

MSP 1, MSP 2, MSP 3, MSP 4 and MSP 5 of *A. marginale* occur as an aggregation of cross-linked proteins (Vidotto *et al.*, 1994). MSP 1 occurs as a complex and is non-covalently associated with the other MSPs. Outer membranes of *A. marginale* consisting of the MSPs mentioned above (and also listed in Table 12.1) are able to protect cattle against clinical disease (Tebele *et al.*, 1991). Homologues of the major MSP of *A. marginale* are found in *A. centrale* and *A. ovis*. These are MSP 1, 2, 3, 4 and 5 (Shkap *et al.*, 1991, 2002a,b; Visser *et al.*, 1992; Palmer and McElwain, 1995; Palmer *et al.*, 1998). Based on accumulating data, some of these OMPs have been shown to be involved in antigenic and strain diversity in *A. marginale*. The role of each MSP in generation of antigenic variation is discussed below.

Major surface protein (MSP) 1

MSP 1 (previously known as Am105; Palmer *et al.*, 1986) is a heterodimer consisting of two unrelated proteins MSP 1a and 1b (also known as Am105U and Am 105L; Barbet *et al.*, 1987; Oberle *et al.*, 1988; Palmer *et al.*, 1989), which are encoded by the *msp 1α* and *1β* genes (Barbet *et al.*, 1987; Allred *et al.*, 1990; Barbet and Allred, 1991). Both sub-units of MSP 1 are involved in adhesion to cells and facilitate invasion (McGarey and Allred, 1994; McGarey *et al.*, 1994). MSP 1a is an adhesin for both bovine erythrocytes and tick cells, whereas MSP

1b is an adhesin only for bovine erythrocytes (de la Fuente *et al.*, 2001a). MSP 1a plays a key role in mediating attachment to tick gut cells to facilitate infection in salivary glands (de la Fuente *et al.*, 2001b). Both polypeptides are recognized by the bovine immune response and have surface-exposed regions. The MSP 1 complex is conserved in intra-erythrocytic *A. marginale* and in the tick salivary glands (Kocan *et al.*, 2000), elicits a protective immune response and is a candidate for development of recombinant vaccine for anaplasmosis (Palmer *et al.*, 1986, 1989).

MSP 1a demonstrates size polymorphism between isolates of *A. marginale* (ranging from 70 to 100 kDa) and is encoded by a single copy *msp 1α* gene (Oberle *et al.*, 1988; Allred *et al.*, 1990). The size polymorphism of MSP 1a is as a result of the varying number of 28 or 29 amino acid tandem repeats present in its structure. These repeats bear a highly conserved antibody-neutralization epitope, which is present in all strains of *A. marginale* (Palmer *et al.*, 1987; Allred *et al.*, 1990). In contrast, MSP 1b is encoded by a polymorphic multigene family of five *msp 1β* genes and has minimal size polymorphism between isolates (Barbet *et al.*, 1987; Viseshakul *et al.*, 2000; Bowie *et al.*, 2002). Two complete genes *msp 1β$_1$* and *msp 1β$_2$* and three partial genes *msp 1β$_1$ pg*, *msp 1β$_2$ pg* and *msp 1β$_3$ pg* have been analysed from the Florida *A. marginale* strain. The complete genes contain initiation codons and transcription terminators and are translated into polypeptide products (Viseshakul *et al.*, 2000). By RT-PCR, cDNA of *msp 1β$_1$* and *msp 1β$_2$* was detected, suggesting that these genes were being expressed. The *msp 1β$_1$* and *msp 1β$_2$* genes share a 85% homology. Between the partial genes, there were rearrangements of coding segments of the two complete genes. Both partial genes, *msp 1β$_1$ pg* and *msp 1β$_2$ pg* begin with sequences homologous to the 3' sequence of *msp 1β$_1$* and then change to sequence homologous to the 5' regions of *msp 1β$_1$*. The third partial gene *msp 1β$_1$ pg3* does not contain any 3' to 5' switching, but seems to contain gene segments from the *msp 1β$_1$* and *msp 1β$_2$*. This observation suggests that intragenic recombination events could occur between the *msp 1β$_1$* gene family members, which potentially can generate genetic and antigenic diversity. By RT-PCR, cDNA of only the *msp 1β$_1$ pg* was detected. In another study on the Florida strain, the *msp1β$_2$* gene was not detected, but three other full-length expressed *msp1β* genes were detected (Camacho-Nuez *et al.*, 2000), which were not detected by Viseshakul *et al.* (2000). These contrasting data may reflect minor polymorphic sequences not detected previously, but indicate that the functional role of *msp 1β* genes in generation of antigenic diversity is not fully known. However, it is known that the *msp 1β$_1$* gene copies are highly polymorphic between *A. marginale* strains but conserved within the strains. In subsequent studies on the sequence diversity of *msp 1α* and *msp1β*, little evidence was found for change in either MSP 1a or MSP 1b during different cyclic transmissions between ticks and cattle (Palmer *et al.*, 2001; Bowie *et al.*, 2002). The role of these proteins in antigenic variation is, therefore, unclear at this time. Genome-sequencing projects in progress will clarify the role of the genes encoding these related polypeptides.

MSP 2 and MSP 3 play a crucial role in antigenic variation of *A. marginale*

MSP 2 and MSP 3 are polymorphic immunodominant OMPs of *A. marginale* (Palmer *et al.*, 1994; Alleman *et al.*, 1997). MSP 2 is an adhesin and induces protective immunity against anaplasmosis (Palmer *et al.*, 1988). MSP 2 and MSP 3 are encoded by the *msp 2* and *msp 3* multigene families and each family accounts for >1% of *A. marginale* genome (Palmer *et al.*, 1994; Alleman *et al.*, 1997). These multigene families are related to each other structurally and in their arrangement. In *A. ovis*, MSP 2 and MSP 3 are also encoded by *msp 2 and 3* multigene families but their role in pathogenesis and antigenic variation is unknown at present (Palmer *et al.*, 1998). MSP 2 and MSP 3 have been identified in *A. centrale* and it has been suggested that the cross-protection between *A. centrale* and *A. marginale* is due to shared protective epitopes between these proteins (Shkap *et al.*, 1991, 2002a).

Intra-erythrocytic *A. marginale* express antigenically variant MSP 2 molecules

A. marginale expressing a progeny of antigenically distinct MSP 2 are found in the blood of cattle. Evidence for this statement is from data that show that unique, gene-copy-specific, B cell epitope-expressing MSP 2 variants can be detected in the blood of infected cattle. These MSP 2 variants are antigenically distinct because they bind to some but not all the monoclonal antibodies that bind to native MSP 2 (Palmer *et al.*, 1994; Eid *et al.*, 1996). The MSP 2 variants are encoded by the *msp 2* multigene family, which consists of an estimated ten or more *msp 2* gene-related copies, which are found widely distributed throughout its genome (Figures 12.2 and 12.3; Palmer *et al.*, 1994). Analysis of these gene copies revealed a surface-exposed central hypervariable region of ~100 amino acids. Genomic and hence antigenic polymorphism is greatest in this central hypervariable region, whereas the 5' and 3' termini of *msp 2* gene are more conserved (Palmer *et al.*, 1994; Barbet *et al.*, 2000; Meeus and Barbet, 2001; Figure 12.4).

This polymorphism is responsible for generation of phenotypically distinct MSP 2 variant types, which are expressed within strains and are co-expressed by individual *A. marginale* during acute infection of cattle. Sequence identities between the *msp 2* gene copies range from 47% to 95% as a result of multiple sequence additions, deletions and substitutions in the central hypervariable region (Eid *et al.*, 1996). These sequence changes also influence MSP 2 size variations and are responsible for simultaneous emergence of a highly heterogeneous population of *A. marginale* during infection in cattle. When *A. marginale* is transmitted by ticks, the rickettsaemia at the onset of acute infection in naive cattle contain the MSP 2 variant types that were expressed by *A.*

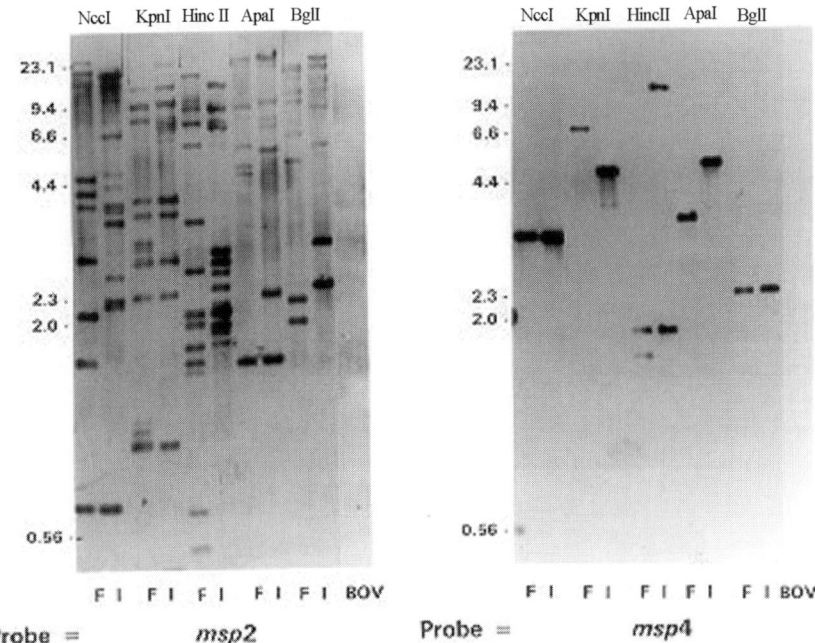

Figure 12.2 Presence of multiple *msp 2* gene copies in the *A. marginale* genome. *A. marginale* DNAs from either the Florida (F) or the South Idaho (I) strain were digested with restriction enzymes indicated at the top of each lane and Southern blotted. Undigested bovine DNA was used as a negative control (BOV). Blots were hybridized with either whole *msp 2* (1,216bp; nt10–1226) or whole *msp 4* gene, the single copy gene, used to control for multiple bands due to partial restriction enzyme digestion; there is a single *Hin*dIII site in *msp* 4. Molecular size markers in kilobases are indicated on the left margin. (Reproduced from Palmer *et al.* (1994) *Infect. Immun.* 62, 3808–3816, with permission of the authors and the American Society of Microbiologists.)

marginale in the salivary glands of ticks (Rurangirwa *et al.*, 1999, 2000; Barbet *et al.*, 2001a). Animals that survive acute infection, become persistently infected as a result of *A. marginale* being able to repeatedly generate progenies of organisms bearing antigenically variant MSP 2, which are distinct from those found during acute infection (French *et al.*, 1998, 1999).

Persistent infection in cattle by *A. marginale* can be demonstrated by detection of low-level cyclical rickettsaemia, utilizing nucleic acid probes and PCR techniques (Eriks *et al.*, 1989; Keiser *et al.*, 1990; French *et al.*, 1999; Palmer *et al.*, 2000). The rickettsial peaks are detected at 4–8-week intervals and represent emergence of *A. marginale* organisms bearing new MSP 2 antigenic variants (French *et al.*, 1998; Palmer *et al.*, 2000). In each rickettsaemic cycle, there are at least four *msp 2* genetic variants, which were identified by analysis of the cDNA of *A. marginale* (French *et al.*, 1998). By antibody binding assays, at least three of these four genetic variants were antigenically distinct. Primary antibody responses appear after the variant type's emergence but not before, demonstrating that each

252 Antigenic Variation

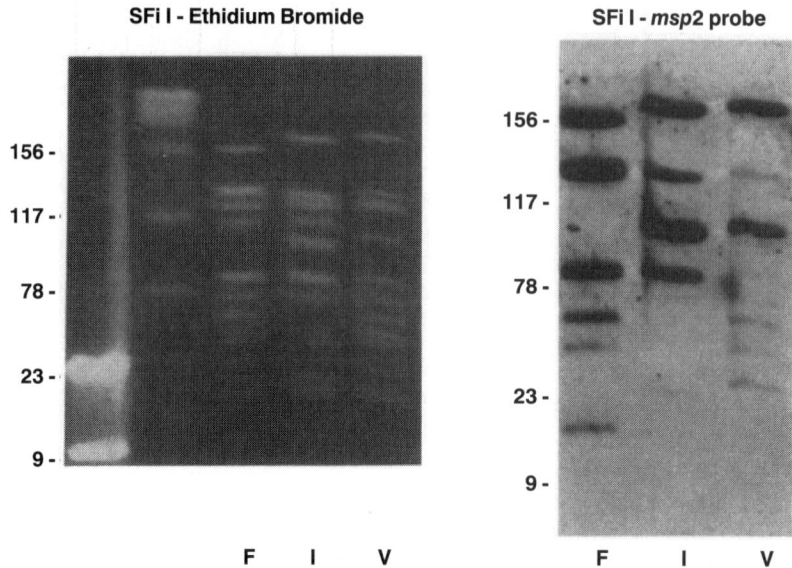

Figure 12.3 Distribution of *msp 2* copies in the *A. marginale* chromosome. *Sfi*-I digested *A. marginale* genomic DNA (F, Florida strain; I, South Idaho strain; V, Virginia strain) separated by clamped homogeneous electric field gel electrophoresis and stained with ethidium bromide. The first two lanes contain lambda DNA-*Hind*III fragments and Promega Delta 39 markers, respectively, as size markers. The fragments were Southern blotted and hybridized with whole *msp 2*. Molecular size markers in kilobases are indicated at the left of each panel. (Reproduced from Palmer *et al.* (1994) *Infect. Immun.* 62, 3808–3816, with permission of the authors and the American Society of Microbiologists.)

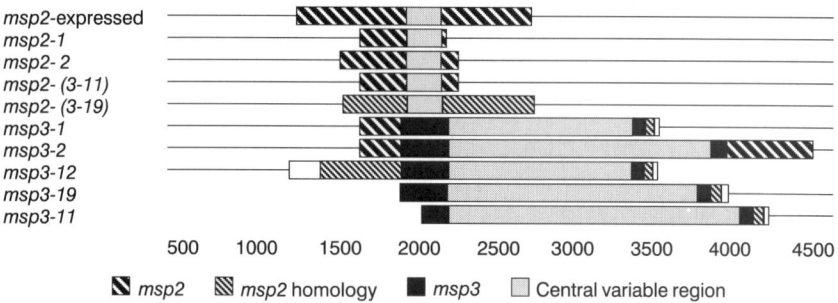

Figure 12.4 Conserved sequence blocks in *msp 2* and *msp 3* genes. The nucleotide sequences of *msp 2* and *msp3* genes were aligned using information from database searches and computer algorithms for divergent sequences containing blocks of moderate to strong homology (MACAW and GLASS). These methods identified the conserved sequence blocks at the 5' and 3' ends of the genes despite long intervening variable regions of different lengths. Regions of nearly identical sequence are depicted as boxes with thick diagonal stripes (*msp 2*) or black (*msp 3*). Regions that are similar to *msp 2* but less homologous are shown as boxes with thin diagonal stripes. The central variable region of the genes is grey and the remaining sections are white or colourless. A single black line represents unrelated flanking sequence. *msp 3–12* is truncated at the 3' end by a junction with a plasmid vector; *msp 3–11* and *msp 3–19* are truncated at the 5' end. GenBank accession numbers for these sequences are AF290593, AF305077, AMU60778, AMU60780 and AMU60779. (Reproduced from *Trends in Microbiology*, vol. 9, Meeus, P. F. M. and Barbet, A. F., pp. 353–355. Copyright ©2001, with permission from Elsevier.)

variant type contains unique B cell epitopes (French *et al.*, 1998, 1999). MSP 2 variants were generated by multiple substitutions, additions and deletions in the central hypervariable region of *msp 2* gene. Within and between cycles of rickettsaemia, amino acid identity of between 75% and 100% can be detected in *msp 2* variant genes. The mechanisms by which clearance of rickettsaemic peaks occurs during acute and persistent infection is similar and is mediated by antibody responses to the MSP 2 variant types present in the respective peak. Variant-specific antibody responses to one variant type appeared after the appearance of the variant type but not before and did not recognize other variants, which is consistent with the phenomenon of antigenic variation (French *et al.*, 1999).

Molecular mechanisms of MSP 2 antigenic variation in *A. marginale*

A. marginale utilizes an ingenious molecular mechanism to generate MSP 2 antigenic diversity. MSP 2 protein of *A. marginale* is encoded on a polycistronic mRNA transcript in erythrocytes (Barbet *et al.*, 2000). This transcript could be amplified by RT-PCR and contained an operon with the *msp 2* gene and three other paralogues or open reading frames (*orfs*) *2*, *3* and *4*, which are located upstream of *msp 2*. *msp 2* and *orf 2* are separated by only 12bp and *orf 2* and *3* by 23bp, which is an arrangement characteristic of prokaryotic polycistronic mRNAs. All three *orfs* are predicted to encode OMPs. The *orf 2* encodes a protein of 13 219 kDa and the *orfs 3* and *4* encode 31 and 29 kDa proteins, respectively. There was no sequence in the database that resembled *orf 2*, but *orf 3* and *orf 4* had significant sequence homologies with OMP1b (p28) of *E. chaffeensis*. Whereas several copies of *msp 2* were detected throughout the *A. marginale* genome, only single copies of *orf 2*, *orf 3* and *orf 4* are detected (Figure 12.5; Barbet *et al.*, 2000). The sequence of *orf 2* and *orf 3* is conserved between different *A. marginale* strains and stages of infection. *orf 4* is the exception to this rule and demonstrates some amino acid substitutions between strains (Barbet *et al.*, 2000, Barbet, Yi *et al.*, 2001). The *msp 2* gene of the *msp 2* operon demonstrated the most variability between strains of *A. marginale*. This variability was concentrated in the central hypervariable region due to insertions, deletions and substitutions and occurred regardless of the strain of *A. marginale* and timing of analyses during the acute infection (Barbet *et al.*, 2000). Although there are multiple copies of *msp 2* detected in the *A. marginale* genome, there is only one full-length *msp 2* gene, which is found in the expression site of the operon (Barbet *et al.*, 2000; Brayton *et al.*, 2001). The multiple *msp 2* copies that exist in the genome outside the operon are partial genes, or pseudogenes (Palmer *et al.*, 1994; Brayton *et al.*, 2001). Therefore, how does *A. marginale* generate diversity in the central hypervariable region of the *msp 2* gene? It has been proposed that the generation of new hypervariable sequences occurs by segmental gene conversion of the *msp 2* expression site (Barbet *et al.*, 2000; Brayton *et al.*, 2001). This can be explained by the following observations.

1 Whereas numerous *msp 2*-related sequences are detected in the *A. marginale* genome, only one *orf 2* gene is found and although *msp 2* is associated with *orf 2* in the expression site, multiple *msp 2* sequences found in the genome are not associated with *orf 2*.
2 Several pseudogenes (10–20 in number) bearing only the hypervariable regions but lacking the conserved 5' and 3' sequences can be found in the genome (Figures 12.5 and 12.6; Brayton *et al.*, 2001; Meeus and Barbet, 2001). These pseudogenes are not transcriptionally functional units as they lack promoter sequences within 350 bp of the first methionine and are expressed by insertion into the *msp 2* expression site (Brayton *et al.*, 2001).
3 Intragenic recombination of genetic material also aids in generating MSP 2 diversity. Many *msp 2* sequence mosaics can be detected, whereby sequences of different variants are detected within a single *msp 2* gene expression site (Barbet *et al.*, 2000).

Hence, in persistently infected cattle, numerous *msp 2* variants are generated, which ensures the long-term survival of the pathogen against a hostile host environment. Given that each rickettsaemic peak in a persistently infected animal occurs every 4–8 weeks and each cycle contains at least four new variants, annually, between 24 and 48 new variants are likely to be expressed in such an

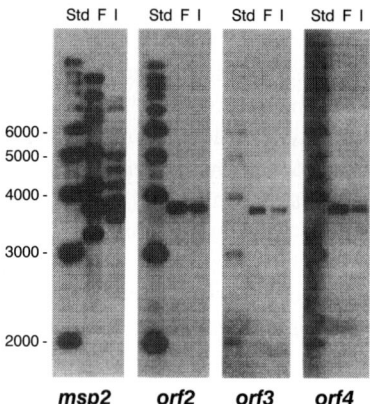

Figure 12.5 Structure of *msp 2* and *orf 2* to *orf 4* in genomic DNA of Florida and South Idaho strains of *A. marginale*. Southern blots of Florida (F) and South Idaho (I) genomic DNA digested with the restriction enzyme *Fsp* I and hybridized with probes specific for *msp 2*, *orf 2*, *orf 3* or *orf 4* are shown. *Fsp* I cleaves 41 nucleotides 5' to *orf 4* and 268 nucleotides 3' to *msp 2* to release a fragment of 3.76 kbp containing the complete polycistronic *msp 2* expression-site sequence from both Florida and South Idaho genomic DNAs. Molecular size standards are shown in the left lane of each blot. Multiple *msp 2*-related sequences are detected in genomic DNA of both strains; only *msp 2* sequences located in the expression site are contiguous with *orf 2* to *orf 4*. Only single copies of *orf 2* to *orf 4* are detected in genomic DNA of the two *A. marginale* strains. (Reproduced from Barbet *et al.* (2000) *Infect. Immun.* 68, 6133–6138 with permission of the authors and the American Society of Microbiologists.)

animal. So far, the sequence of 50 *msp 2* variants has been reported, 24 by Barbet, Yi *et al.* (2001) and 26 are in the Genbank (Brayton *et al.*, 2002). It has been further argued that insertion of pseudogenes alone cannot sustain a life long persistence of *A. marginale* infection (Brayton *et al.*, 2002). Brayton *et al.* (2002) have shown that *A. marginale* has a repertoire to generate an unlimited number of *msp 2* variants, which is facilitated by gene conversion, whereby small segments (four have been detected so far) of pseudogenes recombine in the hypervariable region of expressed new variants. This mechanism facilitates prolonged antigenic diversity to succeed. From the South Idaho *A. marginale* strain alone, nine pseudogene sequences have been reported. The potential to generate new *msp 2* variants is hence quite impressive. If there are four segmental conversions from nine pseudogenes, a minimum of 6561 (9^4) different combinations of pseudogenes can be inserted into the *msp 2* expression site. If as predicted there are up to 20 pseudogenes in an *A. marginale* genome, the potential for generation of additional variants is greater.

MSP 2 expression in cyclic transmission of *A. marginale* between ticks and cattle

MSP 2 is a protective antigen of *A. marginale*. However, its inclusion into a recombinant vaccine against *A. marginale* infection would have a major drawback, because *A. marginale* can generate and express an unlimited number of MSP 2 variants. Hence, it was hypothesized that the diversity of *A. marginale* may be restricted in the vector host, because the bovine immune response is absent and a vaccine against tick stages of *A. marginale* may be feasible. Initial observations proved this assumption to be correct. The *msp 2* variants expressed by the South Idaho strain of *A. marginale* in the tick salivary glands were restricted regardless of the heterogeneity of variant types expressed in the blood at the time of acquisition feeding by the ticks. These variants were designated salivary gland variant 1 (SGV1) and SGV2 (Rurangirwa *et al.*, 1999). Tick transmission of these variants to cattle resulted in their expression in the blood during the acute infection rickettsaemia. Feeding of ticks at different stages of the infection (acute or persistent) with the South Idaho *A. marginale* strain, on different cattle, resulted in reversion of a heterogenous population of variants to the same restricted expression in the tick salivary glands, namely SGV1 and SGV2 (Rurangirwa *et al.*, 1999). To examine whether reversion to a restricted variant type in ticks is universal, similar studies were performed with other *A. marginale* strains. This was found not to be true when Idaho, Virginia and Oklahoma strains of *A. marginale* were tested (Rurangirwa *et al.*, 2000; Barbet *et al.*, 2001a). The same heterogeneity of *msp 2* variants was expressed in the salivary gland, which was present in the blood of cattle at the time when infection was acquired by ticks, and reversion to any dominant variant types did not occur. In some *msp 2* variants, the heterogeneity was quite extreme (Rurangirwa *et al.*,

2000). It was later demonstrated that *msp 2* variants could be generated in tick salivary glands independent of the host's immune response (de la Fluente and Kocan, 2001a). Brayton *et al.* (2002) also examined the *msp 2* sequences of *A. marginale* in the tick salivary glands reported by de la Fluente and colleagues (2001a). They found that segmental changes were identified in the hypervariable region of the *msp 2* variants and suggest that *A. marginale* uses the same mechanisms for generation of antigenic diversity in the ticks as it does in the bovine host, a process driven by different environmental pressures.

Expression of *msp 2* in the tick salivary glands is also polycistronic, from one expression site, as described for blood stages of *A. marginale* and contains an operon of four genes, *msp 2*, *orf 2*, *orf 3* and *orf 4*. Multiple copies of *msp 2* genes were detected in *A. marginale* derived from the tick salivary glands, but only single copies of *orf 2*, *orf 3 and orf 4* were detected (Barbet *et al.*, 2000). The generation of diversity of MSP 2 in ticks was as a result of the polymorphism in the central hypervariable region of *msp 2*. This occurred by insertions, deletions, substitutions and formation of mosaics, demonstrating that intragenic recombination could occur between the various *msp 2* genes and the polycistronic site (Barbet *et al.*, 2001a). The diversity of *A. marginale msp 2* found in the salivary glands of ticks is preserved during cyclical transmission, from cattle to ticks and back into cattle. When infections in cattle were initiated by using *A. marginale*-infected tick cell cultures, the *msp 2* variants expressed in blood were those that were expressed in the cultures. The same MSP 2 variants that were expressed in cattle during the acute infection were expressed in the salivary glands of ticks that were fed on these cattle and were in turn found in blood of cattle when the infection was transmitted cyclically by the ticks. In conclusion, the overall structure of the polycistronic expression site for the *msp 2* gene in different strains and life cycle stages of *A. marginale* is conserved and maintains a capacity to generate antigenic diversity, which allows adaptation of and persistence in different host environments.

The role of MSP 3 in antigenic variation of *A. marginale*

MSP 3 also plays an important role in antigenic variation of *A. marginale*. MSP 3 is encoded by the *msp 3* multigene family and is closely related to MSP 2 (Alleman *et al.*, 1997). Multiple MSP 3-related polypeptides which react with distinct monoclonal antibodies are found within a given strain of *A. marginale*. Between seven and fifteen related *msp 3* gene copies (number varies with the strain) are found distributed throughout the genome of *A. marginale* (Alleman *et al.*, 1997). The *msp 3* and *msp 2* multigene families have structural similarities and contain homologous amino acid regions (Alleman *et al.*, 1997; Brayton *et al.*, 2001; Meeus and Barbet, 2001; see Figure 12.4). The *msp 3* genes contain conserved regions that flank a hypervariable region, but the hypervariable region of *msp 3* is longer than that of *msp 2* genes. The *msp 3* gene is also transcribed on a polycistronic mRNA, and several *orfs* were found upstream of the expressed *msp 3* gene (P. Meeus and A. Barbet, unpublished observations). Pseudogenes of *msp 3* have also been detected in the *A. marginale* genome (Alleman *et al.*, 1997; Brayton *et al.*, 2001; Meeus and Barbet, 2001). These pseudogenes have a 5' sequence, which is conserved in the pseudogenes of *msp 2* (Brayton *et al.*, 2001). Antigenic diversity in *msp 3* is generated by insertion, deletion and substitution of sequences within the hypervariable region, as well as insertion of complete or partial pseudogenes. In addition to the structural similarity between the *msp 2* and *msp 3* genes, their respective pseudogenes also appear together in a head-to-tail arrangement (Alleman *et al.*, 1997; Brayton *et al.*, 2001; Meeus and Barbet, 2001; Figure 12.6). The structural similarities of the *msp 2* and *msp 3* genes and the close proximity of their respective pseudogenes represents a coordinated control of expression of these two gene families, which probably includes recombination between these

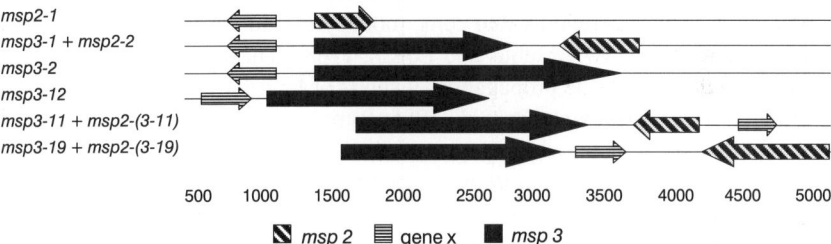

Figure 12.6 Genomic structure and orientation of *msp 2* and *msp 3* genes. The same genes as in Figure 12.4 are represented on the same scale to show their relationships in the *A. marginale* genome. Gene X is a previously undescribed gene present in the flanking regions of many *msp 2* and *msp 3* copies. The *msp 2* gene family is depicted by arrows with thick diagonal lines, the *msp 3* family by black arrows and gene X by short arrows with horizontal thin lines. The 5'-3' orientation of the genes is indicated by arrows. (Reprinted from *Trends in Microbiology*, vol. 9, Meeus, P. F. M. and Barbet, A. F., pp. 353–355. Copyright ©2001, with permission from the authors and Elsevier.)

gene families. Many *msp 2* and *msp 3* pseudogenes have 600 bp long conserved flanking regions, which are also thought to be involved in recombination events and generation of antigenic variation (Brayton *et al.*, 2001; Gene X in Figure 12.6; Meeus and Barbet, 2001). Collectively, *A. marginale* has developed an efficient and ingenious mechanism of antigenic variation through the use of *msp 2* and *3* multigene families, which facilitates persistent infection in cattle whilst engaging the host immune response.

MSP 4 and MSP 5

MSP 4 and MSP 5 are conserved OMPs of *A. marginale* and are encoded by single copy genes (Visser *et al.*, 1992; Oberle *et al.*, 1993; see Figure 12.2). MSP 4, a 31 kDa protein and MSP 5, a 19 kDa protein, are recognized by serum from cattle immunized with outer membrane proteins of *A. marginale* (Tebele *et al.*, 1991) and by neutralizing serum (Palmer and McGuire, 1984). The *msp 4* gene has significant homology to *msp 2* gene (60% similarity at the amino acid level; Palmer *et al.*, 1994). There is no data to endorse that MSP 4 and MSP 5 proteins are involved in antigenic variation of *A. marginale*. MSP 5 is significantly similar to the MAP 2 protein of *C. ruminantium* (55% similarity; Mahan *et al.*, 1994). Owing to the high level of conservation between *A. marginale* strains and the fact that they are both encoded by single copy genes, MSP 4 and MSP 5 are considered not to be involved in antigenic variation.

EHRLICHIA (COWDRIA) RUMINANTIUM

Ehrlichia ruminantium, previously known as *Cowdria ruminantium* (Dumler *et al.*, 2001), is an intracellular coccoid-shaped organism (size range 0.5–2.7 μm) with a predilection for endothelial cells, monocytes and neutrophils (Uilenberg, 1983; Logan *et al.*, 1987; Mebus and Logan, 1988). These cell types have been established *in vitro* for the propagation of short- or long-term infected cultures of *E. ruminantium* (Bezuidenhout *et al.*, 1985; Byrom and Yunker, 1990). Recently, *E. ruminantium* has also been propagated in cell cultures derived from ticks (Bell-Sakyi *et al.*, 2000). *E. ruminantium* are observed as intra-cytoplasmic vacuoles within host cells and multiply by binary fission resulting in large colonies.

E. ruminantium causes heartwater, a severe and fatal illness in domestic ruminants. Animals recovering from infection are usually resistant to challenge with the homologous strain and become persistently infected as shown by PCR and infectivity of such hosts for ticks (Andrew and Norval, 1989; Semu *et al.*, 2001). Animals that recover from infection may not be protected against all heterologous *E. ruminantium* strains. This lack of cross-protection is the biological evidence for antigenic disparity between *E. ruminantium* strains (Uilenberg *et al.*, 1983; Jongejan *et al.*, 1988, 1991; Mahan *et al.*, 2001). Other evidence of genetic diversity in *E. ruminantium* is obtained from macrorestriction DNA fragment profiles followed by pulse field gel electrophoresis (de Villers *et al.*, 2000b). Random amplified polymorphic DNA (RAPD) analyses also provides evidence of genetic diversity in *E. ruminantium* (Perez *et al.*, 1997). The correlation of these genetic differences with cross-protection between *E. ruminantium* strains is unclear, because exhaustive molecular evaluations of the genome of *E. ruminantium* have not been carried out. The profiles derived from these two molecular techniques are likely to provide DNA fingerprints for individual *E. ruminantium* strains and may be useful in understanding the molecular pathogenesis and epidemiology of heartwater. Molecular characterizations and comparisons of genomic sequences encoding OMPs of *E. ruminantium* are likely to reveal important information on cross-protection between strains, antigenic variation and its relevance to pathogenicity and persistence of this organism.

Antigenic variation in *E. ruminantium*

The phenomenon of antigenic variation in *E. ruminantium* has not been identified yet. Antigenic variation has been identified in pathogens that are closely related to *E. ruminantium*. The likelihood of antigenic variation occurring in *E. ruminantium* is high. This is due to sequence and structural similarities of the genes encoding some of the OMPs of *E. ruminantium* and those involved in antigenic variation in the related ehrlichial agents.

Several immunogenic proteins of *E. ruminantium* have been identified (Table 12.2) with the aim of evaluating their use in recombinant vaccines and diagnosis of heartwater (Jongejan and Theilemans, 1989; Rossouw *et al.*, 1990; Mahan *et al.*, 1993, 1994; Barbet *et al.*, 1994, 2001b; Lally *et al.*, 1995; Nyika *et al.*, 1998, 2002). Genes encoding some of these proteins have been cloned and their inter- and intra-strain differences characterized. The *map 1* and *map 2* genes, which encode two of the immunodominant proteins, MAP 1 (Major Antigenic Protein 1) and MAP 2, have been subjected to detailed characterization (Barbet *et al.*, 1994; Mahan *et al.*, 1994; van Vliet *et al.*, 1994, 1995; Reddy *et al.*, 1996, 1998; Bowie *et al.*, 1999; Sulsona *et al.*, 1999; Allsopp *et al.*, 2001). The *map 1* and *map 2* genes have orthologues in other ehrlichial pathogens (Mahan *et al.*, 1994; Reddy *et al.*, 1996, 1998; Bowie *et al.*, 1999; Sulsona *et al.*, 1999; Yu *et al.*, 2000; Ohashi *et al.*, 2001). Although data from sequence analyses of *E. ruminantium* DNA is accumulating, minimal information exists on intra-strain characterization of the genes that encode other immunogenic proteins (Barbet *et al.*, 2001b). Two reports have been published on sequencing large sections of *E. ruminantium* DNA. One presents a preliminary genetic map of the *E. ruminantium* DNA (de Villers *et al.*, 2000a) and the other, a more detailed evaluation of genes that may have value in vaccine and diagnostics development (Barbet *et al.*, 2001b). The latter authors have reported new gene sequences of *E. ruminantium* which

Table 12.2 OMPs and immunogenic proteins of *E. ruminantium*

Proteins	Molecular size	Function	Gene organization	Role in antigenic variation
MAP 1	28–30 kDa	Unknown	Multigene family	Possibly
MAP 2	21 kDa	Unknown	Single gene copy	Unlikely
1HW	36 kDa	Unknown	Unknown	Unknown
4HW	19 kDa	Unknown	Unknown	Unknown
18HW	28 kDa	Unknown	Unknown	Unknown
27HW	28 kDa	Unknown	Unknown	Unknown
Heat-shock protein	58 kDa	Antigen presentation	Single copy gene	Unlikely
3GD	16 kDa	Unknown	Single gene copy	Unknown
pCS20	Two ORFs (1 and 2)	ORF2 is a RNA polymerase	Single gene copy	Unknown

encompass about 5% of the organism's genome (of the Highway strain of Zimbabwe) and report on homologies of the *E. ruminantium*, sequences with those available in the gene data banks. In addition, unedited sequence reads are available from the *E. ruminantium* genome sequencing project – http://www.sanger.ac.uk/Projects/C_ruminantium – of the Welgevonden strain of South Africa.

Major Antigenic Protein 1 (MAP 1)

The MAP 1 protein of *E. ruminantium* (originally called CR32) is an immunodominant OMP of approximately 30–32 kDa molecular size (Jongejan and Thielemans, 1989), which induces protective immune responses in mammalian hosts (Mwangi *et al.*, 1998, 2002; Nyika *et al.*, 1998, 2002; Byrom *et al.*, 2000a) and is encoded by a multigene family (Sulsona *et al.*, 1999; Bekker *et al.*, 2002; van Heerden *et al.*, 2002). This protein can vary in molecular size between *E. ruminantium* strains (Barbet *et al.*, 1994), due to gene sequence polymorphisms (Reddy *et al.*, 1996), and has immunogenic B and T cell epitopes (Figure 12.7) (van Vliet *et al.*, 1995; Mwangi *et al.*, 1998, 2002; Nyika *et al.*, 1998, 2002; Reddy *et al.*, 1998; Totté *et al.*, 1998; Semu *et al.*, 2001). The MAP 1 protein of *E. ruminantium* is closely related to the MSP 2 protein of *A. marginale* and p28 and p30 OMPs of *E. chaffeensis* and *E. canis* respectively. Unlike MSP 2 and MSP 3 proteins of *A. marginale*, variants of MAP 1 protein within an infecting strain and variant specific antibodies have not been discovered so far. Serum antibodies from sheep, cattle and goats, infected with a diverse range of *E. ruminantium* field strains, recognize the MAP 1 protein from various *E. ruminantium* strains, demonstrating that the expressed MAP 1 protein in all strains contains highly cross-reactive B cell epitopes (Mahan *et al.*, 1993; Barbet *et al.*, 1994; van Vliet *et al.*, 1995; Reddy *et al.*, 1998; Semu *et al.*, 2001).

Molecular characterization of *map 1* gene and its probable role in antigenic variation

The *map 1* gene encodes a 28 kDa MAP 1 protein based on the sequence data and was first cloned and characterized from the Senegal strain (van Vliet *et al.*, 1994). This gene has been under detailed investigation by various research groups and, at present, there are over 30 *map 1* gene sequences available in the gene databases. *map 1* is a member of a multigene family. Its related genes or *orfs* are arranged in tandem in the *E. ruminantium* genome and not widely distributed (Sulsona *et al.*, 1999). All known *map 1* genes have areas of high sequence homology (especially the 5' end), but in the whole gene, several amino acid substitutions are observed (Reddy *et al.*, 1996, 1998). The conserved regions are interspersed with three hypervariable regions, referred to as hypervariable region

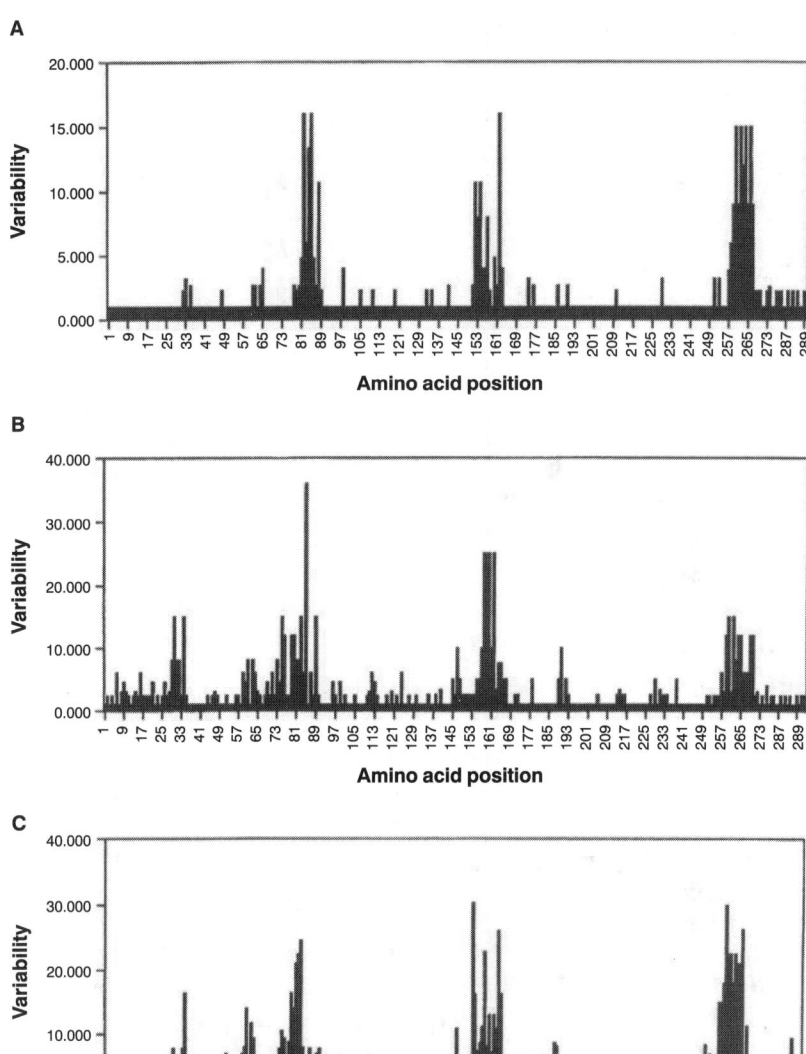

Figure 12.7 Variability in the protein-coding sequence of MAP 1 and its orthologues. (A) The variability profiles of the *E. ruminantium* MAP 1 alone (MAP 1 sequences used here were reported in Reddy *et al.* (1996); (B) *E. chaffeensis* and *E. canis* 28/30 kDA MAP 1 orthologue gene-coding sequences; the semi-variable region is shown before the first hypervariable region and (C) sequences encoding MAP 1 orthologues of all three ehrlichial pathogens are presented together to demonstrate that the immunogenic regions are mapped to the hypervariable regions of the three OMPs. The underlined regions depict the areas consisting of B cell epitopes and correspond to the regions shown in Figure 12.8. (Reproduced from Reddy *et al.* (1998) *Biochem. Biophys. Res. Commun.* 247, 636–643 with permission from Academic Press.)

I, II and III (Figure 12.7). These hypervariable regions are generated by deletions or insertions. Hypervariable region I is hydrophilic and contains dominant B cell epitopes (Figure 12.8; Reddy *et al.*, 1998). The overall similarity between the *map 1* genes of *E. ruminantium* strains at the nucleotide level is 86–99.4% and translates to 0.8–10% variation in the predicted protein sequences. Three additional *orfs* or paralogues of *map 1* were identified, namely *orf 2*, *orf 3* and *orf 4* (Sulsona *et al.*, 1999; Bekker *et al.*, 2002; van Heerden *et al.*, 2002). *orf 2*, *orf 3* and *orf 4* are also referred to as *map 1–1*, *map 1–2* and *map 1+1*, respectively (Bekker *et al.*, 2002; van Heerden *et al.*, 2002). *orf 2* and *orf 3* are located upstream and *orf 4* (*map 1+1*) is located downstream of the expressed *map 1* gene. *orf 4* is located upstream of gene sequences homologous to *SecA* genes of other organisms (van Heerden *et al.*, 2002). *SecA* is involved in transport of OMPs. *orfs 2, 3* and *4* have a similar overall arrangement as *map 1*, containing conserved and variable regions, but are more homologous to the OMP-encoding multigene families of *E. chaffeensis* and *E. canis* than *map 1* (Sulsona *et al.*, 1999; van Heerden *et al.*, 2002). The *map 1* gene has an 89–100% homology and *orf 2* has a 100% homology between *E. ruminantium* strains (Sulsona *et al.*, 1999). However, the sequence identity between *orf 2* and *map 1* is only 43–47% and between *orf 2* and OMP 1B (also called *p28*) of *E. chaffeensis* is 73–74%. The mechanism by which *map 1* generates sequence polymorphisms (and if it is involved in antigenic variation), or influences the pathogenesis and facilitates persistence of *E. ruminantium* infection is unknown. In addition, the interaction of *map 1* with its paralogues *orfs 2, 3* and *4* in generating diversity is also unknown at present. It is anticipated, however, that *map 1* is a member of a large multigene family, as is observed for its orthologues in *E. chaffeensis and E. canis* (Yu *et al.*, 2000; Ohashi *et al.*, 2001). It is also likely that the *map 1* multigene family is transcribed by the same mechanism as the *p28* and *p30* multigene families of *E. chaffeensis* and *E. canis*. This assumption is being made firstly, because the *p28* and *p30* multigene families are also tandemly arranged, as are the *map 1* family genes, and secondly, because there is significant structural and sequence identity between these gene families. Some preliminary data on transcriptional analysis of the *map 1* gene family has revealed that *map 1* is transcribed in both virulent and attenuated *E. ruminantium* strains (Senegal, Gardel and Sankat 430) *in vitro* or in ticks (Bekker *et al.*, 2002). However, *orf 2* or *map 1–1* is only transcribed in attenuated cultures of Senegal and Gardel strains of *E. ruminantium*. The *map 1–1* transcript was also detected in *A. variegatum* ticks infected with the virulent Senegal strain, suggesting that differential expression was observed. Transcripts of *map 1–2* were not detected in these studies. The authors suggest that the expression of the *map 1* gene family is polycistronic (Bekker *et al.*, 2002). Further, extended studies will reveal the mechanism of expression of this multigene family and how this gene contributes towards genetic diversity and pathogenesis of infection.

One piece of biological data suggests that *map 1* has a role in determining antigenic differences between strains which can be mapped to its hypervariable

Antigenic variation in *A. marginale* and *E. ruminantium*   263

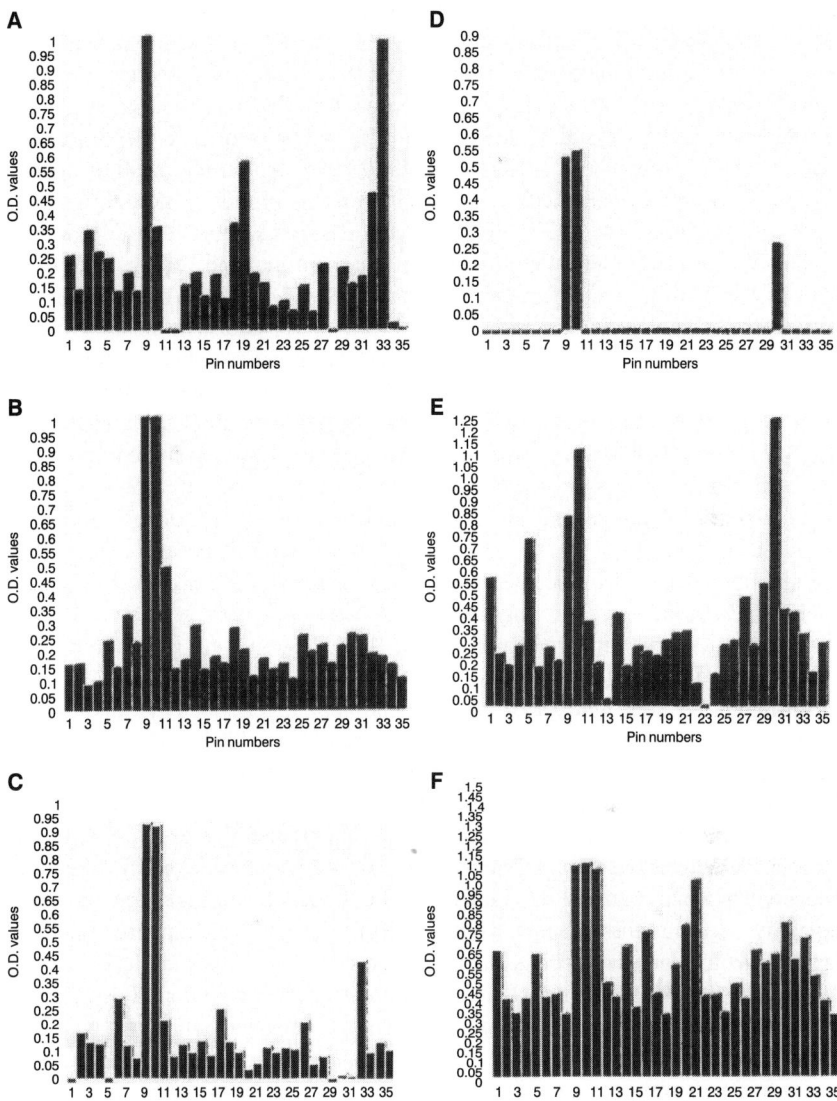

Figure 12.8 Recognition of MAP 1 regions by serum from animals infected with diverse strains of *E. ruminantium*. Peptide pins were coated with overlapping 15 amino acid sequences derived from the MAP 1 of Senegal (van Vliet *et al.*, 1994) and ELISAs were performed with various strain-specific serum: (A) Senegal; (B) Welgevonden; (C) Nigeria; (D) Gardel; (E) Crystal Springs; and (F) dog serum specific for *E. canis* to demonstrate cross-reaction between MAP 1 and antigens of *E. canis*. The immunogenic regions of the MAP 1 protein map to the hypervariable regions and are also shown in Figure 12.7C as underlined areas. (Reproduced from Reddy *et al.* (1998) *Biochem. Biophys. Res. Commun.* 247, 636–643 with permission from Academic Press.)

regions. Data supporting this statement are derived from *map 1* DNA vaccine studies done in a mouse model of infection (Nyika *et al.*, 1998, 2002). A *map 1* DNA vaccine protects mice against homologous Crystal Springs strain challenge. However, mice immunized with the *map 1* DNA vaccine constructed from the Mbizi strain of *E. ruminantium* were not protected against heterologous challenge from the Crystal Springs strain. The overall sequence identity of the *map 1* genes from these two strains is 96.2% and amino acid sequence alignments reveal a difference of thirteen amino acids. Eight of these differences are in the hypervariable region I and two are in hypervariable region III. Based on this data, one may infer that the differences in amino acid sequences in these regions represent protective epitopes and are likely to influence cross-protection between *E. ruminantium* strains (Nyika *et al.*, 2002) and also aid in immune evasion. In contrast, other researchers suggest that positive selection pressure is not responsible for the *map 1* sequence heterogeneity, based on statistical evaluations of thirty *map 1* sequences, and that this gene or its protein product may not be involved in immune evasion (Allsopp *et al.*, 2001). These researchers report that the polymorphisms detected in *map 1* sequences are not specific to any geographical region and therefore there is no clustering of the sequences based on these regions. Clustering is detected, but seems unrelated to origin of the strains and has too much variation in the sequence to be distinguished by cluster-specific probes. The polymorphisms detected in this gene were mainly located in the hypervariable regions I and III and these were considered to be non-silent sequence changes (Allsopp *et al.*, 2001).

Although the total number of genes in the *map 1* gene locus is unknown at present, it is presumed that *map 1* paralogues or *orfs* are likely to be involved in generating diversity of *E. ruminantium*. How is diversity of *map 1* generated and what is its relevance to the survival of the pathogen? There are two schools of thought on this: the first mechanism proposes that recombinatorial events occur between the members of *map 1* gene family, thus generating diversity. The more conserved *orfs* such as *orf 2* provide a repository of genetic information to generate diversity and immune evasion, while conserving regions to support functional properties of the gene, which may include adhesion and/or invasion (Sulsona *et al.*, 1999). The second or alternative mechanism is that the generation of diversity in these genes may have resulted from random mutations over a very long period of time. A study on the generation of MAP 1 variants during the course of infection (acute and persistent) using a cloned *E. ruminantium* strain will provide valuable insight into generation of its genetic and hence antigenic diversity.

Implications of MAP 1 homology with OMPs of other *Ehrlichia*

The major implication of the shared homology between MAP 1 and the OMPs of other *Ehrlichia* is that they may be involved in similar functions in their respective pathogen. Hence common control strategies may be applied to the complex of diseases caused by these related pathogens (Mahan *et al.*, 1999; Palmer *et al.*, 1999) The *map 1* gene shares significant homology with the *msp 2* and *msp 4* genes of *A. marginale*, despite some structural differences between them. Considering their overall similarity, MAP 1 may have similar functions to MSP 2 and MSP 4. MSP 2 is an adhesin (McGarey and Allred, 1994) and both MSP 2 and MSP 4 induce protective immune responses against *A. marginale* challenge. MAP 1 induces protective immune responses in cattle and protects mice against lethal challenge when used as a DNA vaccine (Nyika *et al.*, 1998, 2002; Totté *et al.*, 1998; Mwangi *et al.*, 2002).

map 1 and its paralogues are more related or homologous to the multigene families encoding the 28–30 kDa OMPs of *E. chaffeensis* (*p28*) and *E. canis* (*p30*) than to the *msp 2* gene family of *A. marginale* (Sulsona *et al.*, 1999; Yu *et al.*, 2000; Ohashi *et al.*, 2001; van Heerden *et al.*, 2002). The *map 1*, *p28* and *p30* genes also have significant homology with the *p28* gene of *E. ewingii* (Gusa *et al.*, 2001). The *p28* and *p30* genes of *E. chaffeensis* and *E. canis* are arranged in tandem and contain three hypervariable regions interspersed with conserved regions, like in *map 1*. The only difference is that the *p28* and *p30* genes have a semi-variable region upstream of the first hypervariable region, which is not present in *map 1* (Reddy *et al.*, 1998). The hypervariable region I of the *p28* and *p30* genes is predicted to be surface exposed, as in *map 1* of *E. ruminantium* (Yu *et al.*, 2000; Ohashi *et al.*, 2001). Hence, molecular characterization of the *p28* and *p30* gene families may allow us to predict the functional role and factors controlling expression of *map 1*. A detailed evaluation of the *p28* and *p30* multigene families of *E. chaffeensis* and *E. canis* revealed the presence of a cluster of 22 paralogues in one gene locus (Yu *et al.*, 2000; Ohashi *et al.*, 2001). A sequence identity of 20–83% and of 19–72% was detected between the paralogues of *E. chaffeensis* and *E. canis*, respectively. The identity of the entire DNA sequences between the *p28* and *p30* genes was 64.3%. The *p28* and *p30* gene paralogues are transcriptionally active in monocyte cultures (Yu *et al.*, 2000; Ohashi *et al.*, 2001; Long *et al.*, 2002) and universal start codons were found in all paralogues of these two gene families. In addition, putative signal peptides of 25–31 amino acids were also found in the N terminal of each gene. Yu *et al.* (2000) reached the conclusion that the *p28* gene family cluster of *E. chaffeensis* is transcribed monocistronically. However, Ohashi *et al.* (2001) proposed that the transcription of the 5' end half of the *p28* gene family of *E. chaffeensis* and the *p30* gene cluster of *E. canis* is primarily polycistronic, whereas the genes located in the 3' end half are monocistronic. The reasoning for this conclusion is from the fact that the intergenic spaces on the 5' end are only 8–27 bp long, whereas the intergenic spaces on the 3' end range from 213 to 623 bp. Despite the contrasting opinions on mode of transcription, the *p28* of *E. chaffeensis* and the *p30* genes of *E. canis* are active genes and encode polymorphic forms of the OMPs. Presence of multiple active genes and the fact that these OMPs are divergent (based on sequence data and some monoclonal antibody-binding data) between strains demonstrates another unique (compared to the *msp 2* gene family) method of generation of antigenic diversity in these pathogens (Long *et al.*, 2002; Unver *et al.*, 2002). Long *et al.* (2002) came to the conclusion that the *p28* paralogues of *E. chaffeensis* were expressed differentially in monocyte cultures. Unver *et al.* (2002) observed that in dogs infected with *E. chaffeensis*, the paralogues of *p28* genes are transcribed differentially. Sixteen of the 22 paralogues were actively transcribed in dog monocytes during a 56-day infection period. However, only one of these paralogues was transcribed in the *A. americanum* tick stages before or after transmission of infection to dogs. Currently, however, there seems to be no direct evidence of gene-recombination events occurring in these *p28* and *p30* gene family clusters. It is highly likely that an equally large *map 1* multigene family also exists in *E. ruminantium* and it employs a similar mechanism for generation of diversity as in the multigene families of *E. chaffeensis* and *E. canis*. The intergenic space between the known *orfs* or paralogues of *map 1* are large and therefore the expression of the members of this gene family is likely to be monocistronic (Sulsona *et al.*, 1999). In contrast to this suggestion, Bekker *et al.* (2002) suggest a polycsitronic expression of the *map 1* genes. Further evaluation of the three gene families is likely to reveal their function in immune evasion, pathogenesis, and how and if they influence the pathogen's survival in the host.

MAP 2 and newly cloned *E. ruminantium* genes

The *map 2* gene encodes an immunogenic 21 kDa OMP of *E. ruminantium* and has orthologues in *A. marginale*, *E. chaffeensis* and *E. canis* (Mahan *et al.*, 1994; Bowie *et al.*, 1999), but no paralogues have been detected. Inter-strain comparisons have revealed that this gene is highly conserved (Bowie *et al.*, 1999). When comparing its sequence between five *E. ruminantium* strains, only 10 nucleotide differences were detected, which translated to three amino acid changes in the whole protein. Owing to its high sequence conservation, *map 2* is targeted for use in vaccine and diagnostic tests development for heartwater. The *map 2* orthologues of *A. marginale*, *E. chaffeensis* and *E. canis* also encode a polypeptide of approximately 19 kDa. A 55.5% sequence identity exists between the *map 2* gene of *E. ruminantium* and *msp 5* gene of *A. marginale* (Mahan *et al.*, 1994); and a 84.4% and 83.4% identity exists between *map 2* and its orthologues in *E. chaffeensis* and *E. canis*, respectively (Bowie *et al.*, 1999). This high sequence identity between the *map 2* and its orthologues in *E. chaffeensis* and *E. canis* complicates its use in species-specific serodiagnosis of heartwater or monocytic ehrlichiosis in humans and dogs. *map 2* is highly conserved and hence is unlikely to be involved in antigenic variation.

Several new genes have been isolated and characterized from *E. ruminantium* genomic DNA libraries (Table 12.2; Barbet *et al.*, 2001b). Their protein products are recognized by immune sera and peripheral blood mononuclear cells from domestic ruminants that are immune to heartwater. At present, data on their function or contribution to antigenic variation and pathogenesis is lacking.

However, several of these genes have orthologues in rickettsial and bacterial agents that are important to human and animal health. Some of the new *E. ruminantium* genes have significant similarity to genes that encode outer membrane proteins and lipoproteins of *Rickettsia prowazekii*. In addition, homologues of genes encoding an OMP of *Brucella abortus* and *Coxiella burnetii* have been isolated (Barbet et al., 2001b).

DISCUSSION AND CONCLUSIONS

Persistence is a hallmark of the tick-borne *Ehrlichia* that cause infections in animals and humans. Persistence of infection can be life long and plays a critical role in the epidemiology of these diseases. Otherwise chronic infections caused by *Ehrlichiae* seem not to be detrimental to the host. Antigenic variation is a common feature of persistent infection by various pathogens, but the mechanisms used differ. In *A. marginale* infections, persistence is a result of an ingenious manipulation of genomic coding sequences of the *msp 2* and *msp 3* multigene families. The pathogen utilizes these gene families to generate antigenic variants, which precede the immune responses that eventually clear them. This phenomenon is a true form of antigenic variation and resembles the antigenic variation seen in trypanosomes (Roth et al., 1989; Barbet and Kamper, 1993). Trypanosomes ensure persistent infection by generating new antigenic variants by gene conversion of a polycistronic expression site by pseudogenes encoding immunogenic surface glycoproteins (Roth et al., 1989). Recombination of genes encoding surface proteins occurs in *Borrelia burgdorferi*. Persistent infection by *Borrelia* is characterized by appearance of new variant types by recombination of the *vmp/vls* genes encoding a surface-exposed lipoprotein VlsE. In *Borrelia*, antigenic variation involves a plasmid containing a tandem array of pseudogenes (Zhang et al., 1997) and in *A. marginale* or *E. ruminantium* no plasmids have been located. Antigenic variation in *Neisseria gonorrhoeae* and *Mycoplamsa* species results from use of pseudogenes which are distributed throughout the genome (Kenri et al., 1999; Noormohammadi et al., 2000). The ehrlichial agent that causes human granulocytic ehrlichiosis (HGE agent) has a multigene family (*p44*) that is structurally similar to the *msp 2* multigene family (Zhi et al., 1999). Pseudogenes of *p44* have also been found and are distributed throughout the genome of the HGE agent. Although the HGE agent infects different host cells compared with *A. marginale*, they probably use the same mechanisms to generate antigenic variants as those used by the *msp 2* gene (Brayton et al., 2001). Recently the HGE agent has been renamed *Anaplasma phagocytophila* and by molecular characterizations cannot be differentiated from *Ehrlichia phagocytophila* and *Ehrlichia equi* (Dumler et al., 2001), justifying the grouping of these pathogens together with *A. marginale* (see Figure 12.1).

Antigenic variation has not been demonstrated in *E. ruminantium* yet, but there is every likelihood that this pathogen generates diversity by the same mechanism used by *E. chaffeensis* and *E. canis*. This is anticipated because there is a high sequence homology and similar structural arrangement of the *map 1*, *p28* and *p30* genes. The *p28* and *p30* orthologues of *map 1* in *E. chaffeensis* and *E. canis* are expressed as a multigene family of 22 paralogues. Each gene is thought to be transcriptionally active in monocyte cultures. *In vivo*, 11 of 14 paralogues of *E. canis p30* genes were transcribed in experimentally infected dogs as detected by RT-PCR. In ticks, however, only one of the same 14 paralogues was transcribed (Unver *et al.*, 2001). Presence of antibody responses seemed not to influence the expression of the *p30* paralogues. The expression of the *p28* multigene family of *E. chaffeensis* has also been studied in dogs (Unver *et al.*, 2002). Sixteen of 22 *p28* paralogues were transcribed differentially in dog monocytes over a 56-day period of infection. In contrast, only one of the paralogues was transcribed in the three tick stages of *A. americanum* ticks before and after transmission of the infection to dogs. This data suggests that the higher number of transcribed paralogues *in vivo* may have resulted from the immune response stimulus. These data demonstrate that although there are differences in the structure of the *map1* gene orthologues compared with the *msp 2* gene, the respective pathogens utilize these multigene families to generate diversity, albeit by different mechanisms.

REFERENCES

Alleman, A. R., Kamper, S. M., Viseshakul, N. and Barbet, A. F. (1993) *J. Gen. Microbiol.* 139, 2439–2444.

Alleman, A. R., Palmer, G. H., McGuire, T. C., McElwain, T. F., Perryman, L. E. and Barbet, A. F. (1997) *Infect. Immun.* 65, 156–163.

Allred, D. R., McGuire, T. C., Palmer, G. H., Leib, S. R., Harkins, T. M., McElwain, T. F. and Barbet, A. F. (1990) *Proc. Natl Acad. Sci. USA* 87, 3220–3224.

Allsopp, M. T. E. P., Dorfling, C. M., Maillard, J. C., Bensiad, A., Haydon, D. T., van Heerden, H. and Allsopp, B. A. (2001) *J. Clin. Microbiol.* 39, 4200–4203.

Andrew, H. R. and Norval, R. A. I. (1989) *Vet. Parasitol.* 34, 261–266.

Barbet, A. F. and Allred, D. R. (1991) *Infect. Immun.* 59, 971–976.

Barbet, A. F. and Kamper, S. M. (1993) *Parasitol. Today* 9, 63–66.

Barbet, A. F., Lundgren, A., Yi, J., Rurangirwa, F. R. and Palmer, G. H. (2000) *Infect. Immun.* 68, 6133–6138.

Barbet, A. F., Palmer, G. H., Myler, P. J. and McGuire, T. C. (1987) *Infect. Immun.* 55, 2428–2435.

Barbet, A. F, Semu, S. M., Chigagure, N., Kelly, P. J., Jongejan, F. and Mahan, S. M. (1994) *Clin. Diagn. Lab. Immunol.* 1, 744–746.

Barbet, A. F., Whitmire, W. M., Kamper, S. M., Simbi, B. H., Reddy, G. R., Moreland, A. L., Mwangi, D. M., Mcguire, T. C. and Mahan, S. M. (2001b) *Gene* 275, 287–298.

Barbet, A. F., Yi, J., Lundgren, A., McEwen, B. R., Blouin, E. F. and Kocan, K. M. (2001a) *Infect. Immun.* 69, 3057–3066.

Bekker, C. P. J., Bell-Sakyi, L., Paxton, E. A., Martinez, D., Bensaid, A. and Jongejan, F. (2002) *Gene* 285, 193–201.

Bell-Sakyi, L., Paxton, E. A., Munderloh, U. G. and Sumption, K. J. (2000) *J. Clin. Microbiol.* 38, 1238–1240.

Bezuidenhout, J. D., Paterson, C. L. and Barnard, B. H. D. (1985) *Onderstepoort J. Vet. Res.* 5, 113–120.
Bowie, M. V., de la Fuente, J., Kocan, K. M., Blouin, E. F. and Barbet, A. F. (2002) *Gene* 282, 95–102.
Bowie, M. V., Reddy, G. R., Semu, S. M., Mahan, S. M. and Barbet, A. F. (1999) *Clin. Diagn. Lab. Immunol.* 6, 209–215.
Brayton, K. A., Knowles, D. P., McGuire, T. C. and Palmer, G. H. (2001) *Proc. Natl Acad. Sci. USA* 98, 4130–4135.
Brayton, K. A., Palmer, G. H., Lundgren A., Yi, J. and Barbet, A. F. (2002) *Mol. Microbiol.* 43, 1151–1159.
Brown, W. C., Shkap, V., Zhu, D., McGuire, T. C., Tuo, W., McElwain, T. F. and Palmer, G. H. (1998a) *Infect. Immun.* 66, 5406–5413.
Brown, W. C., Zhu, D., Shkap, V., McGuire, T. C., Blouin, E. F., Kocan, K. M. and Palmer, G. H. (1998b) *Infect. Immun.* 66, 5414–5422.
Buening, G. M. (1976) *Am. J. Vet. Res.* 34, 1215–1218.
Byrom, B. and Yunker, C. E. (1990) *Cytotechnology* 4, 285–290.
Byrom, B., Obwolo, M., Barbet, A. F. and Mahan, S. M. (2000a) *J. Parsitol.* 86, 983–992.
Byrom, B., Obwolo, M., Barbet, A. F. and Mahan, S. M. (2000b) *Vet. Parasitol.* 93, 159–172.
Camacho-Nuez, M., De Lourdes Muñoz, M., Suarez, C. E., McGuire, T. C., Brown, W. C. and Palmer, G. H. (2000) *Infect. Immun.* 68, 1946–1952.
Carson, C. A., Sells, D. M. and Ristic, M. (1976) *Vet. Parasitol.* 2, 75–81.
Dame, J. B., Mahan, S. M. and Yowell, C. A. (1992) *Int. J. Syst. Bacteriol.* 42, 270–274.
De la Fuente, J., Garcia-Garcia, J. C., Blouin, E. F. and Kocan, K. M. (2001a) *Int. J. Parasitol.* 31, 145–153.
De la Fuente, J., Garcia-Garcia, J. C., Blouin, E. F., McEwen, B. R., Clawson, D. and Kocan, K. M. (2001b) *Int. J. Parasitol.* 31, 1705–1714.
De Villiers, E. P., Brayton, K. A., Zweygarth, E. and Allsopp, B. A. (2000a) *Microbiology* 146, 2627–2634.
De Villiers, E. P., Brayton, K. A., Zweygarth, E. and Allsopp, B. A. (2000b) *J. Clin. Microbiol.* 38, 1967–1970.
Dumler, J. S., Barbet, A. F., Bekker, C., Dasch, G. A., Jongejan, F., Palmer, G. H., Ray, S. C., Rikihisa, Y. and Rurangirwa, F. (2001) *Int. J. Syst. Evol. Bacteriol.* 51, 2145–2165.
Eid, G., French, D. M., Lundgren, A. M., Barbet, A. F., McElwain, T. F. and Palmer, G. H. (1996) *Infect. Immun.* 64, 836–841.
Eriks, I. S., Palmer, G. H., McGuire, T. C., Allred, D. R. and Barbet, A. F. (1989) *J. Clin. Microbiol.* 27, 279–284.
Eriks, I. S., Stiller, D. and Palmer, G. H. (1993) *J. Clin. Microbiol.* 31, 2091–2096.
Estes, D. M., Closser, N. M. and Allen, G. K. (1994) *Cell Immunol.* 154, 287–295.
French, D. M., Brown, W. C. and Palmer, G. H. (1999) *Infect. Immun.* 67, 5834–5840.
French, D. M., McElwain, T. F., McGuire, T. C. and Palmer, G. H. (1998) [published erratum appears in *Infect. Immun.*, 1998 May; 66, 2400]. *Infect. Immun.* 66, 1200–1207.
Gusa, A. A., Buller, R. S., Storch, G. A., Huycke, M. M., Machado, L. J., Slater, L. N., Stockham, S. L. and Massung, R. F. (2001) *J. Clin. Microbiol.* 39, 3871–3876.

Inokuma, H., Brouqui, P., Drancourt, M. and Raoult D. (2001) *J. Clin. Microbiol.* 39, 3031–3039.

Jongejan, F. and Thielemans, M. J. C. (1989) *Infect. Immun.* 57, 3243–3246.

Jongejan, F., Thielemans, M. J. C., Brière, C. and Uilenberg, G. (1991) *Res. Vet. Sci.* 51, 24–28.

Jongejan, F., Uilenberg, G., Franssen, F. F. J., Gueye, A. and Nieuwenhuijs, J. (1988) *Res. Vet. Sci.* 44, 186–189.

Kenri, T., Taniguchi, R., Sasaki, Y., Okazaki, N., Narita, M., Izumikawa, K., Umetsu, M. and Sasaki, T. (1999) *Infect. Immun.* 67, 4557–4562.

Kieser, S. C., Eriks, I. S. and Palmer, G. H. (1990) *Infect. Immun.* 58, 1117–1119.

Kocan, K. M., Blouin, E. F. and Barbet, A. F. (2000) *Ann. NY Acad. Sci.* 916, 501–509.

Kocan, K. M., Venable, J. H. and Brock, W. E. (1978a) *Am. J. Vet. Res.* 39, 1123–1130.

Kocan, K. M., Venable, J. H., Hsu, K. C. and Brock, W. E. (1978b) *Am. J. Vet. Res.* 39, 1131–1135.

Kuttler, K. L. (1967) *Res. Vet. Sci.* 8, 467–471.

Lally, N. C., Nicoll, S., Paxton, E. A., Cary, C. M. and Sumption, K. J. (1995) *Microbiology* 141, 2091–2100.

Logan, L. L., Whyhard, T. C., Quintero, J. C. and Mebus, C. A. (1987) *Onderstepoort J. Vet. Res.* 54, 197–204.

Long, S. W., Zhang, X. F., Qi, H., Standaert, S., Walker, D. H. and Yu, X. J. (2002) *Infect. Immun.* 70, 1824–1831.

Losos, G. J. (1986) In *Infectious Tropical Diseases of Domestic Animals* (ed. G. J. Losos), Longman Press, London, pp. 742–795.

Mahan, S. M., Allsopp, B., Kocan, K. M., Palmer, G. H. and Jongejan, F. (1999) *Parasitol. Today* 15, 290–294.

Mahan, S. M., Barbet, A. F., Tebele, N., Nyathi, C., Wassink, L. A., Semu, S., Peter, T. and Kelly, P. J. (1993) *J. Clin. Microbiol.* 31, 2729–2737.

Mahan, S. M., McGuire, T., Jongejan, F. and Barbet, A. F. (1994) *Microbiology* 140, 2135–2142.

Mahan, S. M., Sileghem, M., Smith, G. E. and Byrom, B. (1996) *Parasite Immunol.* 18:6, 317–324.

Mahan, S. M., Smith, G. E., Kumbula, D., Burridge, M. J. and Barbet, A. F. (2001) *Vet. Parasitol.* 97, 295–308.

McGarey, D. J. and Allred, D. R. (1994) *Infect. Immun.* 62, 4587–4593.

McGarey, D. J., Barbet, A. F., Palmer, G. H., McGuire, T. C. and Allred, D. R. (1994) *Infect. Immun.* 62, 4594–4601.

Mebus, C. A. and Logan, L. L. (1988) *JAVMA* 192, 950–952.

Meeus, P. F. M. and Barbet, A. F. (2001) *Trends Microbiol.* 9, 353–355.

Mukhebi, A. W., Chamboko, T., O'Callaghan, C. J., Peter, T. F., Kruska, R. L., Medley, G. F., Mahan, S. M. and Perry, B. D. (1999) *Prev. Vet. Med.* 39, 173–189.

Mwangi, D. M., Mahan, S. M., Nyanjui, J. K., Taracha, E. L. and McKeever, D. J. (1998) *Infect. Immun.* 66, 855–860.

Mwangi, D. M., McKeever, D. J., Nyanjui, J. K., Barbet, A. F. and Mahan, S. M. (2002) *Vet. Immunol. Immunopathol.* 85, 23–32.

Noormohammadi, A. H., Markham, P. F., Kanci, A., Whithear, K. G. and Browning, G. F. (2000) *Molec. Microbiol.* 35, 911–923.

Nyika, A., Barbet, A. F., Burridge, M. J. and Mahan, S. M. (2002) *Vaccine* 20, 1215–1225.
Nyika, A., Mahan, S. M., Burridge, M. J., McGuire, T. C., Rurangirwa, F. and Barbet, A. F. (1998) *Parasite Immunol.* 20, 111–119.
Oberle, S. M., Palmer, G. H. and Barbet, A. F. (1993) *Infect. Immun.* 61, 5245–5251.
Oberle, S. M., Palmer, G. H., Barbet, A. F. and McGuire, T. C. (1988) *Infect. Immun.* 56, 1567–1573.
Ohashi, N., Rikihisa, Y. and Unver, A. (2001) *Infect. Immun.* 69, 2083–2091.
Palmer, G. H. (1989) In *Veterinary Protozoan and Hemoparasite Vaccines* (ed. I. G. Wright), CRC Press, Boca Raton, FL, pp. 1–29.
Palmer, G. H. and McElwain, T. F. (1995) *Vet. Parasitol.* 57, 233–253.
Palmer, G. H. and McGuire, T. C. (1984) *J. Immunol.* 133, 1010–1015.
Palmer, G. H., Abbott, J. R., French, D. M. and McElwain, T. F. (1998) *Infect. Immun.* 66, 6035–6039.
Palmer, G. H., Barbet, A. F., Cantor, G. H. and McGuire, T. C. (1989) *Infect. Immun.* 57, 3666–3669.
Palmer, G. H., Barbet, A. F., Davis, W. C. and McGuire, T. C. (1986) *Science* 231, 1299–1302.
Palmer G. H., Brown, W. C. and Rurangirwa, F. R. (2000) *Microbes Infect.* 2, 167–176.
Palmer, G. H., Eid, G., Barbet, A. F., McGuire, T. C. and McElwain, T. F. (1994) *Infect. Immun.* 62, 3808–3816.
Palmer, G. H., Oberle, S. M., Barbet, A. F., Davis, W. C. and McGuire, T. C. (1988) *Infect. Immun.* 56, 1526–1531.
Palmer, G. H., Rurangirwa, F. R., Kocan, K. M. and Brown, W. C. (1999) *Parasitol. Today* 15, 281–286.
Palmer, G. H., Rurangirwa, F. R. and McElwain, T. F. (2001) *J. Clin. Microbiol.* 39, 631–635.
Palmer, G. H., Waghela, S. D., Barbet, A. F., Davis, W. C. and McGuire, T. C. (1987) *Int. J. Parasitol.* 17, 1279–1285.
Perez, J. M., Martinez, D., Debus, A., Sheikboudou, C. and Bensaid, A. (1997) *FEMS Microbiol. Lett.* 154, 73–79.
Reddy, G. R., Sulsona, C. R., Barbet, A. F., Mahan, S. M., Burridge, M. J. and Alleman, A. R. (1998) *Biochem. Biophys. Res. Commun.* 247, 636–643.
Reddy, G. R., Sulsona, C. R., Harrison, R. H., Mahan, S. M., Burridge, M. J. and Barbet, A. F. (1996) *Clin. Diagn. Lab. Immunol.* 3, 417–422.
Richey, E. J. (1981) In *Current Veterinary Therapy: Food Animal Practice* (ed. R. J. Howard), W. B. Saunders Co., Philadelphia, pp. 767–772.
Ristic, M. (1968) In *Infectious Diseases of Man and Animals*, Vol. II (eds D. Weinman and M Ristic), Academic Press, New York, pp. 473–542.
Ristic, M. and Huxsoll, D. L. (1984) Tribe II. Ehrlichieae Philip, 1959, 948 AL. In *Bergey's Manual of Systematic Bacteriology* (eds N. R. Krieg and J. G. Holt), Williams & Wilkins, Baltimore, MD, pp. 704–709.
Ristic, M. and Kreier, J. P. (1984) Family III. Anaplasmataceae Philip, 1957, 980 AL. In *Bergey's Manual of Systematic Bacteriology* (eds N. R. Krieg and J. G. Holt), Williams & Wilkins, Baltimore, MD, pp. 719–723.
Rossouw, M., Neitz, A. W. H., De Waal, D. T., Du Plessis, J. L., Van Gas, L. and Brett, S. (1990) *Onderstepoort J. Vet. Res.* 57, 215–221.
Roth, C. F., Bringaud, R., Layden, R., Baltz, T. and Eisen, H. (1989) *Proc. Natl Acad. Sci. USA* 96, 3171–3176.

Rurangirwa, F. R., Stiller, D., French, D. M. and Palmer, G. H. (1999) *Proc. Natl Acad. Sci. USA* 96, 3171–3176.

Rurangirwa, F. R., Stiller, D. and Palmer, G. H. (2000) *Infect. Immun.* 68, 3023–3027.

Semu, S. M., Peter, T. F., Mukwedeya, D., Barbet, A. F., Jongejan, F. and Mahan S. M. (2001) *Clin. Diagn. Immunol. Lab.* 8, 388–396.

Shkap V., Molad, T., Brayton, K. A., Brown, W. C. and Palmer, G. H. (2002a) *Infect. Immun.* 70, 642–648.

Shkap, V., Molad, T., Fish, L. and Palmer, G. H. (2002b) *Parasitol. Res.* 88, 546–552.

Shkap, V., Pipano, E., McGuire, T. C. and Palmer, G. H. (1991) *Vet. Immunol. Immunopathol.* 29, 31–40.

Stich, R. W., Kocan, K. M., Palmer, G. H., Ewing, S. A., Hair, J. A. and Barron, S. J. (1989) *Am. J. Vet. Res.* 50, 1377–1380.

Sulsona, C. R., Mahan, S. M. and Barbet, A. F. (1999) *Biochem. Biophys. Res. Commun.* 257, 300–305.

Tebele, N., McGuire, T. C. and Palmer, G. H. (1991) *Infect. Immun.* 59, 3199–3204.

Theiler, A. (1910) *Transvaal S. Afr. Rep. Vet. Bacteriol. Dept Agr.* 1908–9, 7–64.

Totté, P., Bensaid, A., Mahan, S. M., Martinez, D. and McKeever, D. J. (1999) *Parasitol. Today* 15, 286–290.

Totté, P., McKeever, D., Jongejan, F., Barbet, A. F., Mahan, S. M., Mwangi, D. and Bensaid, A. (1998) *Ann. NY Acad. Sci.* 849, 155–160.

Uilenberg, G. (1983) *Adv. Vet. Sci. Comp. Med.* 27, 427–480.

Uilenberg, G., Zivkovic, D., Dwinger, R. H., Ter Huurne, A. A. H. M. and Perie, N. M. (1983) *Res. Vet. Sci.* 35, 200–205.

Unver, A., Ohashi, N., Tajima, T., Stich, R. W., Grover, D. and Rikihisa, Y. (2001) *Infect. Immun.* 69, 6172–6178.

Unver, A., Rikihisa, Y., Stich, R. W., Ohashi, N. and Felek. S. (2002) *Infect. Immun.* 70, 4701–4704.

Van Kleef., Gunter, N. J., Macmillan, H., Allsopp, B. A., Shkap, V. and Brown, W. C. (2000) *Infect. Immun.* 68, 603–614.

Van Heerden, H., Collins, N. E., Allsopp, M. T. E. P. and Allsopp, B. A. (2002) *Proc. NY Acad. Sci.* 969, 131–134.

Van Vliet, A. H., Jongejan, F., van Kleef, M. and van der Zeijst, B. A. M. (1994) *Infect. Immun.* 62, 1451–1456.

Van Vliet, A. H., Jongejan, F. and van der Zeijst, B. A. (1992) *Int. J. Syst. Bacteriol.* 42, 494–498.

Van Vliet, A. H., van der Zeijst, B. A., Camus, E., Mahan, S. M., Martinez, D. and Jongejan, F. (1995) *J. Clin. Microbiol.* 33, 405–410.

Vidotto, M. C., McGuire, T. C., McElwain, T. F., Palmer, G. H. and Knowles D. P., Jr (1994) *Infect. Immun.* 62, 2940–2946.

Viseshakul, N., Kamper, S., Bowie, M. V. and Barbet, A. F. (2000) *Gene* 253, 4553.

Visser, E. S., McGuire, T. C., Palmer, G. H., Davis, W. C., Shkap, V., Pipano, E. and Knowles, D. P., Jr (1992) *Infect. Immun.* 60, 5139–5144.

Walker, J. and Olwage, A. (1987) *Onderstepoort J. Vet. Res.* 54, 353–379.

Yu, X., McBride, J. W., Zhang, X. and Walker, D. H. (2000) *Gene* 248, 59–68.

Yu X., Zhang X., McBride, J. W., Zhang, Y. and Walker, D. H. (2001) *Int. J. Sys. Evolut. Microbiol.* 51, 1143–1146.

Zhang, J. R., Hardham, J. M., Barbour, A. G. and Norris, S. J. (1997) *Cell* 89, 275–285.

Zhi, N., Ohashi, N. and Rikihisa, Y. (1999) *J. Biol. Chem.* 274, 17828–17836.

13

ANTIGENIC VARIATION AND ITS SIGNIFICANCE TO BABESIA

David R. Allred, Basima Al-Khedery and Roberta M. O'Connor

INTRODUCTION

The first protozoal parasites for which transmission by an arthropod vector (*Boophilus* spp. ticks) was established were the agents of Texas redwater fever, *Babesia bigemina* and *Babesia bovis* (Smith and Kilborne, 1893). In one of the major successes of improved animal husbandry practices, the disease was eradicated in the United States over a period of approximately 45 years (1907–50). Using what may have been a prophetic approach, this was not accomplished by targeting the parasite. Rather, intensive dipping of cattle in arsenical acaricides was used to eliminate the tick vectors that determined the parasites' distributions (Graham and Hourrigan, 1977). Ironically, even had an efficacious drug treatment been available at the time, this may still have been the only effective approach. The reason for this is that cattle that seem to be free of the parasite can harbour low-level, persistent infections. These animals serve as carriers of the disease, infecting ticks, which then transmit the disease to immunologically naive animals, re-establishing and spreading foci of infection. Widespread drug treatment of seemingly 'uninfected' animals is unlikely to have been practised, and it was therefore only with elimination of the tick vector that the disease was controlled.

It was not appreciated until relatively recently that, once infected with *B. bovis*, animals nearly uniformly remain infected for long periods. In 1973, it was demonstrated that 21 of 22 animals receiving *B. bovis* once remained infected, without re-challenge, for 4 years or more (Mahoney *et al.*, 1973). In contrast, in the same study, only 2 of 22 animals remained infected with *B. bigemina* for such an extended period, most having lost their infection before two years had elapsed. This behaviour of establishing persistent infections is very common among babesial parasites. Although all the 'tricks' used by these parasites are not known, it is known that antigenic variation occurs in at least two species of babesial parasites: *Babesia rodhaini* and *B. bovis*. Other mechanisms may also contribute to persistence, but go beyond the scope of this chapter. Here we will attempt to describe what is currently known about the mechanisms by which these parasites accomplish antigenic variation, and how it may contribute to parasite survival.

THE PHENOTYPE OF ANTIGENIC VARIATION IN *BABESIA BOVIS*

Antigenic variation is a term that seems to mean different things to different individuals. We will define this term here as *the rapid structural and antigenic alterations of **specific** parasite components*. *In vivo* antigenic variation is observed during the course of an ongoing infection within an individual animal. It is observed as waves of antigenically distinct parasite populations (i.e. variants) selected by the host immune system, and by slightly delayed induction of immunological responses to the progressively varying parasite populations. Although the genetic and structural changes of antigenic variation also occur *in vitro*, the selection for variant populations is generally absent. As a result, variation may be less obvious and perhaps overlooked. Indeed, it was not until the careful studies by Phillips of the immunogenicity of lethally irradiated *B. rodhaini* (Phillips, 1971) that the occurrence of this phenomenon in babesial parasites was established. In these experiments, it was demonstrated that rats could be immunized against disease by infection with lethally irradiated *B. rodhaini*-infected blood. When re-challenged with the same 'reference parasite' population, these rats were immune to disease. On the other hand, if parasites were collected at later dates from the blood of these same rats, and used to challenge a second group that had been immunized only by infection with the original, lethally irradiated reference parasite population, the recipients suffered serious disease symptoms. These results strongly supported rapid changes in the antigenicity of the parasite populations present in the blood of immune rats, a classic expectation in antigenic variation. Tested initially with uncloned parasite populations, these observations were subsequently confirmed in clonal parasite populations (Roberts and Tracey-Patte, 1975). Despite these pioneering studies, the *B. rodhaini* components involved in rapid antigenic variation still have not been described, although their compartmentalization to the infected host erythrocyte membrane has been suggested (Thoongsuwan and Cox, 1973). On the other hand, multiple merozoite surface components have been observed, which share sequence homology among themselves, and even with host components (Snary and Smith, 1986, 1988). As components of antigenic variation are generally encoded by members of multigene families sharing at least some common sequence elements, it is possible that these antigens are involved in this process. However, this possibility remains unresolved.

B. bovis, a parasite of cattle and buffalo, is the only other babesial parasite for which antigenic variation *per se* has actually been demonstrated. In early studies on the antigenicity of *B. bovis*-infected bovine erythrocytes, Curnow demonstrated that the host erythrocyte surface becomes altered antigenically (Curnow, 1968), a result that was independently confirmed by alternative, less subjective procedures (Allred and Ahrens, 1993). Such antigenic changes appeared to be restricted to parasitized cells, implying biochemical modification(s) of the infected-erythrocyte membrane. This observation was significant, for *B. bovis*-infected erythrocytes cytoadhere to the endothelium and sequester in the deep vasculature through knob-like structural alterations (Wright, 1972, 1973; Aikawa *et al.*, 1992). This behaviour is thought to contribute to the circulatory disturbances and other pathologic sequelae caused by this parasite (Wright *et al.*, 1988; Allred, 1995). It was felt that if the IRBC membrane were altered biochemically, and such modifications contributed to pathology, then the altered components might be targets of a protective immune response. This reasoning led to the targeting of infected-erythrocyte components for the development of a partially purified, killed-parasite anti-babesial vaccine (Goodger *et al.*, 1980; Mahoney *et al.*, 1981; Wright *et al.*, 1983). While the IRBC represents a logical target, this approach met with only limited success, particularly when attempted against heterologous isolates (Goodger *et al.*, 1984, 1985), a result reminiscent of the *B. rodhaini* immunization experiments.

The probable reason for the lack of significant protection against heterologous *B. bovis* isolates and breakthrough infections with homologous isolates was exposed much earlier. Curnow demonstrated that by infecting a calf with *B. bovis*, then testing the agglutinability of infected erythrocytes recovered on different dates post-infection with similarly collected sera, infected-erythrocyte agglutination occurred primarily with sera collected later in the infection. Serum collected earlier in infection, or on the same date as the 'antigen', was ineffective at agglutinating the infected cells (Curnow, 1973). Thus, the antigenicity of the IRBC surface appeared to vary dynamically, a classic characteristic of antigenic variation. However, there were caveats to these observations. First, the infection was initiated with a parasite isolate taken from an infected animal in an endemic area. Because of this, the genetic uniformity of the parasites present in the blood was suspect, and the possibility even existed of a mixed infection with *B. bigemina*. Further, the use of agglutination, while a decidedly surface-specific phenomenon, is subject to many technical vagaries, including the titres of antibodies at the time of collection, competing antibody specificities and potential cross-reactivities that might interfere with agglutination, levels of parasitaemia and subjectivity in interpretation. It also does not identify the host or parasite origins of the antigens being recognized. Whereas all these factors might compromise a strict interpretation, it was clear there were dynamic changes occurring in the agglutinability of the parasite population with progression of the infection. These early observations thus represent the first demonstration of population antigenic variation in *B. bovis*.

To better understand the nature of antigenic variation in *B. bovis*, it was necessary to study isogenic populations of at least initially uniform antigenicity. This was facilitated by the development in 1983 of methods for *in vitro* cloning of this parasite (Rodriguez *et al.*, 1983). Using this methodology, the haploid asexual stage of this parasite was cloned *in vitro* and used to initiate infection by a single injection of clonal parasites into an immunologically naive calf. Sera and parasitized erythrocytes, collected at various dates post-infection, were assayed for reactivity with the IRBC surface by a live-cell immunofluorescence assay in a manner analogous to the agglutination studies of Curnow. The results, shown in Figure 13.1, clearly demonstrated two points: (1) progressive changes were occurring in IRBC surface antigenicity over time in an isogenic population of parasites; and (2) the changes in antigenicity were accompanied by the

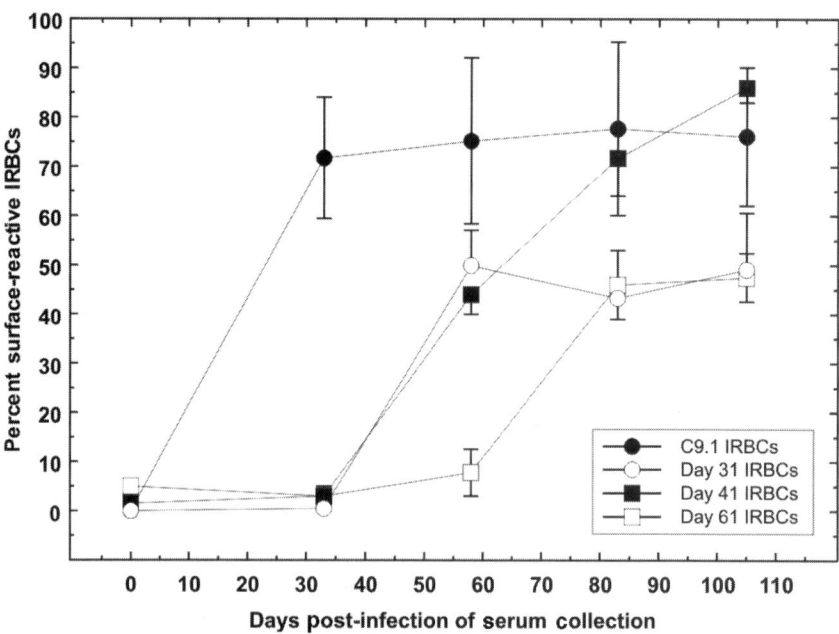

Figure 13.1 Antigenic variation on the infected-erythrocyte surface in *B. bovis*. A calf was infected once with the C9.1 clonal *B. bovis* line, then blood was collected at different dates post-infection for recovery of antisera and parasites. This plot demonstrates the relationship and kinetics between parasite variation and the abilities of the ensuing immune responses to recognize the surface of the infected erythrocytes in a live-cell immunofluorescence assay. (Reproduced by permission of the American Society for Microbiology from Allred *et al.* (1994) *Infect. Immun.* 62, 91–98.)

concomitant development of serum antibodies recognizing parasite populations present earlier but not later in infection (Allred et al., 1994). These results are diagnostic of clonal antigenic variation, in this case at the surface of the infected erythrocytes. As antigenic variation is a phenomenon of parasite-derived products and is driven by parasite genetics, it is important to identify the parasite components involved in this process and the genetic mechanisms responsible for their varied generation.

Identification of the variant antigens of B. bovis

In early studies of the B. bovis-infected erythrocyte membrane, Howard and co-workers observed unique patterns of radioiodinated proteins associated with infected-erythrocyte membranes (Howard et al., 1980). However, it was unclear whether these apparently novel proteins represented parasite-mediated modifications of host components or were novel parasite-derived components. More recently, a surface-specific immunoprecipitation method was devised to detect parasite-derived products on the infected-erythrocyte surface (described in Allred, 1997). Using this methodology, a size-polymorphic, isolate-specific doublet comprised of bands of approximately 105–115 kDa and 120–135 kDa was observed on the IRBC surface (Allred et al., 1993). When this approach was applied to parasite populations acquired from peripheral blood on successive dates post-infection, using similarly collected sera, the results demonstrated that these antigens varied over time in both antigenicity and size, with a recognition pattern similar to that observed by live-cell immunofluorescence (Allred et al., 1994). Mouse monoclonal antibodies, 3F7.1H11 and 4D9.1G1, which on western blots recognize only the larger of the two polypeptides, confirmed both the highly variant nature of this polypeptide and its identity with the larger of the two polypeptides captured by bovine sera. These monoclonal antibodies, which were raised to the B. bovis C9.1 clonal line, were found to recognize in live-cell immunofluorescence assays the surface of erythrocytes infected only with the C9.1 line, and not that of either progenitor or progeny populations (O'Connor et al., 1997).

A comparison of partial proteolytic digestion products of the larger and smaller polypeptides, now referred to as the Variant Erythrocyte Surface Antigen sub-units 1a and 1b (i.e. VESA1a and 1b), suggested that the two are probably products of different genes (O'Connor et al., 1997). However, as some proteolytic fragments are apparently shared between the two polypeptides it remains possible that the two are alternative forms of the same polypeptide, and alternative explanations may hold for the disparities in size and antigenic reactivity. Importantly, upon immunoprecipitation of the VESA1 protein from B. bovis IRBCs by surface-specific immunoprecipitation with bovine infection sera, or by surface-specific or conventional immunoprecipitation with mouse monoclonal antibodies recognizing the VESA1a subunit, both sub-units are always recovered (Allred *et al.*, 1993; O'Connor *et al.*, 1997). This behaviour strongly suggests a stable association between the two polypeptides, probably giving rise to a VESA1 holoprotein. The VESA1 holoprotein thus would appear to be comprised of a polymorphic heterodimer of the 1a and 1b sub-units. At this time, the exact stoichiometric relationship of the two within the hypothetical holoprotein has not been established. However, gaining a full understanding of antigenic variation will ultimately require having detailed knowledge about the genes encoding both sub-units, their individual conformations, and sub-unit stoichiometries and arrangements. This will be crucial, as the bovine humoral response to the VESA1 protein appears to be directed almost entirely to conformational epitopes found on the native protein (unpublished observations).

Identification of the *B. bovis ves* multigene family

The availability of monoclonal antibodies specific to a component of rapid antigenic variation provided tools to recover the $ves1\alpha$ gene encoding the 1a sub-unit of the VESA1 holoprotein, and began the molecular dissection of antigenic variation in *B. bovis*. This was initiated by immunoscreening a cDNA expression library created from poly-A^+ RNA isolated from the *B. bovis* C9.1 clonal line with a pool of the monoclonal antibodies 3F7.1H11 and 4D9.1G1. Based upon restriction endonuclease cleavage sites the cDNAs recovered could be grouped into two clearly related but distinct types (imaginatively called 'type A' and 'type B') (Allred *et al.*, 2000). These cDNAs were validated by a number of criteria to encode the C9.1 VESA1a polypeptide. First, they were identified by the fact that they encoded polypeptide products that are immunoreactive with monoclonal antibodies 3F7.1H11 and/or 4D9.1G1. Second, p9.6.2, one of the recovered type B cDNAs, contained a complete open reading frame of 3966 nt, long enough to encode a polypeptide of approximately 125–130 kDa. Third, when expressed as a recombinant protein in *Escherichia coli*, the p9.6.2-encoded polypeptide was immunoreactive with monoclonal antibody 3F7.1H11 and exactly co-migrated with the VESA1a polypeptide from C9.1 line extracts. Fourth, calves immunized with crude *E. coli* lysates containing the recombinant polypeptide produced antisera that recognized an antigen on the surface of the *B. bovis* IRBC. In a live-cell immunofluorescence assay, this antigen was observed in the same finely punctate pattern as that detected with the monoclonal antibodies. Fifth, when used as a probe on southern blots of restriction-digested *B. bovis* gDNAs, the p9.6.2 cDNA hybridized under stringent conditions with a large number of bands, indicative of probable membership in a multigene family. Further, this multigene family (now called *ves*, for Variant Erythrocyte Surface antigen genes) is distributed on at least 3 of the 4 *B. bovis* chromosomes (Allred *et al.*, 2000). Finally, the recovery of distinct but related cDNA classes is itself a result consistent with a highly and rapidly variant antigen.

The recovery of two cDNA classes suggested that either all parasites were making multiple, similar transcripts, or alterations were occurring in the structure of the expressed gene(s) within individual parasites and then expanded within the parasite population. In either scenario, the population could contain more than one major transcript form for the expressed gene. Currently available data do not adequately discriminate between these major possibilities. Attempts to clarify this situation were made by re-cloning the C9.1 line by limiting dilution, then using RT-PCR amplification of parasite RNAs to determine the structures of the transcripts present. This experiment was first performed using a reverse primer known to be common to both transcript types along with a forward primer specific to one type or the other. By this approach, most subcloned lines appeared to transcribe either type A or type B transcripts, and three were not amplified with these primers. Only a few appeared to transcribe both cDNA types, usually with one present in trace amounts. Because this approach would bias the results toward detection of already known transcripts, this experiment was repeated on the same cDNA preparations but using conserved primers flanking a highly variable region of the *ves*1α gene. Under these conditions, individual subclones yielded multiple related but distinct transcripts. To make matters more confusing, multiple transcripts from individual subclones often shared closer homology to one another than to those derived from other subclones. Also, significant patches of sequence were often shared, which were not present in transcripts from other subclones. Thus, these results failed to distinguish among at least three major possibilities: (1) simultaneous transcription by individual parasites of multiple similar but unique gene copies (i.e. no or incomplete allelic exclusion); (2) transcription by individual parasites of a single gene copy that is rapidly altered, for example through gene conversion; or (3) rapid *in situ* switching of transcription between very similar gene copies by individual parasites. However, northern blot analysis of RNAs from the C9.1 line and multiple isogenic, antigenically variant lines, using the p9.6.2 cDNA as probe, yielded significant hybridization only to C9.1 (Allred *et al.*, 2000). This result would suggest that a lack of allelic exclusion is unlikely to exist. To answer this question unambiguously will probably require looking at transcription in individual parasitized cells, for example by single-cell RT-PCR with universal *ves*1α gene primers. A directly analogous approach was helpful in clarifying *var* gene transcription in *Plasmodium falciparum* (Chen *et al.*, 1998).

Structure of the *ves*1α gene and the VESA1a polypeptide

The *ves*1α gene hybridizes under stringent conditions with a large number of bands on southern blots of *B. bovis* gDNA, indicative of the existence of a large number of closely related sequences. This is not surprising, as most variant antigens are encoded by genes that are members of multigene families, and *ves*1α appears very conventional in this respect. These restriction fragments are overall highly stable in size and hybridization signal intensity among various isogenic clonal lines of *B. bovis* (Allred *et al.*, 2000), suggesting corresponding strong sequence stability and conservation of most gene copies over time.

The intron/exon structure of this gene was defined by comparing genomic sequences encoding the cDNA with the cDNA itself. Genomic sequences were cloned using PCR amplification of gDNA with primers representing the very 5' and 3' ends of the p9.6.2 orf. Colony PCR with internal primers identified a recombinant that matched the transcribed gene well. Comparison of the sequence of the cloned segment, pgC3c, with that of the p9.6.2 cDNA revealed two introns, both situated near the 5' end of the orf. In similarity to other described *B. bovis* introns (Silins *et al.*, 1996; Suarez *et al.*, 1998), both are short (111 and 72 bp) and possess the canonical /GU...AG/ splice junctions (Allred *et al.*, 2000). It is possible that additional introns are present in the 5' or 3' UTR regions, but genomic sequences flanking the orf of the transcribed gene copy have not yet been characterized and no information is currently available.

In silico translation of the p9.6.2 cDNA open reading frame revealed a polypeptide predicted to be 1322 amino acids in length, with a molecular mass of 145 kDa and a pI of 8.21. This length is consistent with the 123 kDa apparent mass of the C9.1 line VESA1a sub-unit when observed by SDS-polyacrylamide gel electrophoresis (Allred *et al.*, 1994). Interestingly, among seven cloned *ves*1α genes (two cDNA and five gDNA sequences), the predicted isoelectric points of the encoded polypeptides range only from pI = 7.58 to 8.21 for the predicted translational products. This 0.6 pH unit range is in stark contrast to the situation in the VSG protein of African trypanosomes, where, despite only modest size polymorphism, the pI can range over several pH units (Rovis *et al.*, 1978). In the absence of post-translational modifications, this narrow, slightly alkaline pI range suggests that most copies of the VESA1a polypeptide will be positively charged at physiological pH, perhaps to facilitate another function.

The VESA1a polypeptide possesses several unique features, each of which may play a role in its probable function. The first is a cysteine- and lysine-rich domain in the N-terminal third of the polypeptide, called the 'CKRD' (using the single-letter amino acid code). In the p9.6.2 cDNA-encoded polypeptide there are a remarkable 22 cysteines and 17 lysines interspersed between amino acids 127–252! Overall content within the polypeptide is 58 cysteines and 116 lysines. This domain also contains a sequence motif, $CX_8CX_3CX_{3-4}CXC$, that is found in the CIDR domain of CD36-binding forms of PfEMP1, a *P. falciparum*-infected erythrocyte variant surface antigen (Baruch *et al.*, 1997). However, this motif is not always found, and its detection in the p9.6.2-encoded polypeptide may be merely a serendipitous byproduct of the exaggerated cysteine content.

A second significant structural element is the VDCS domain (for 'variant domain conserved sequences') found in the central region of the polypeptide. The VDCS domain is comprised of two subdomains (VDCS1 and VDCS2) of invariant length and which contain very highly conserved sequence motifs. These are separated by an extremely variant interdomain region of variable length. The VDCS1 contains the highly conserved motif, $KX_2IX_3IX_6L$. Like the CKRD,

VDCS2 is relatively cysteine-rich and is defined by the large and apparently invariant motif, $AX_6IX_5CX_2CX_2HX_2KCGX_7CX_2CXQX_4G$. The VDCS2 sub-domain also contains the epitope recognized by monoclonal antibody 3F7.1H11 (D. R. Allred, unpublished data), the significance of which will be discussed later. These may be unique structural motifs, as neither identifies any non-$ves1\alpha$ potential gene homologues in the Genbank non-redundant or EST databases (as of 09 April 2003).

A predicted helical transmembrane domain near the C-terminal end of the polypeptide probably serves as an anchoring domain for the polypeptide on the IRBC surface. VESA1 is situated over the tips of the knob-like protrusions that are formed by maturing parasites (O'Connor and Allred, 2000). Only a short domain at the C-terminus of the polypeptide, approximately 42 amino acids in length, is predicted to be oriented in the cytoplasmic compartment of the IRBC. Presumably, this cytoplasmic tail is responsible for any necessary associations with erythrocyte cytoskeletal components needed for organization over the knob structures. Such a role would probably involve conserved sequences for proper targeting and assembly, and may be present in the short, highly conserved sequence, DW(I/T/A)(H/Y)MRSHWLRGG, found immediately following the transmembrane segment.

A less obvious feature of the VESA1a polypeptide sequence, and one that was initially overlooked, is a pair of closely spaced, predicted coiled-coil segments. In the p9.6.2-encoded polypeptide these segments involve amino acids 694–736 and 781–808, with the more C-terminal segment encompassing the VDCS1 subdomain. Such segments are predicted, in the same region, in all variants of the polypeptide available to date, although the predictive probability values vary. Coiled-coil segments are generally involved in protein–protein interactions or assembly of apoprotein sub-units. In VESA1a they could mediate association with the VESA1b sub-unit or perhaps binding to endothelial components during cytoadhesion, a probable function of the VESA1 protein (O'Connor and Allred, 2000). This remains to be determined.

PROBABLE MECHANISMS OF ANTIGENIC VARIATION IN *B. BOVIS*

The study of the molecular basis for antigenic variation in *B. bovis* is in its infancy. However, a few key observations suggest mechanisms that may be at work and provide logical directions for future research. Identification of the *ves* multigene family encoding at least one component of antigenic variation, and a close comparison of multiple members of the two cDNA types that were recovered reveals interesting patterns of similarity and difference. Throughout most of the lengths of both cDNA types the sequence is identical, consistent with a probable common origin. On the other hand, short patches of informative sequence exist, where all type A cDNAs match and all type B cDNAs match, but the two types differ, suggesting segmental modification of a common ancestral form of the gene (Figure 13.2, underlined regions). On the other hand, a patch of sequence exists where there is a match among all type B cDNAs and the type A cDNA, p8.4.1, but all differ from the sequence present in the remaining type A cDNAs (Figure 13.2, boxed region). This observation suggests that such changes are ongoing processes and that not all members of the population express identical polypeptides.

Figure 13.2 Alignment of *B. bovis* C9.1 line *ves*1α cDNAs. These five cDNAs were cloned using a pool of monoclonal antibodies 3F7.1H11 and 4D9.1G1 that are specific for the VESA1a polypeptide of only the clonal *B. bovis* C9.1 line. The cDNAs p9.6.2 and p9.7.6 are both type B, whereas p8.4.2, p8.4.4 and p8.4.1 are type A cDNAs. Underlined segments indicate where the two cDNA types differ consistently, based on their type. Asterisks indicate where single amino acid changes, due to non-silent single base changes, fail to match others of the same type. The boxed region shows where p8.4.1, a type A cDNA, contains an extended sequence segment matching type B cDNAs. Numbering of sequences is based on that of the p9.6.2 predicted polypeptide. (Reproduced with permission from Allred (2001) *Microbes Infect.* 3, 481–491, Figure 4.)

A second significant observation relative to probable mechanisms was made when gDNAs of isogenic but antigenically variant lines were compared, using p9.6.2 cDNA as a probe. Nearly all bands were identical among the lines in both sizes and relative intensities. However, two unique bands were observed in gDNA from the C9.1 line, and these matched internal fragments generated from the p9.6.2 cDNA. This result indicates that unique sequences are present in the genome, which are associated with the transcribed copy of the gene but which are absent from the progenitor and progeny of the C9.1 line. In other words, a unique copy of the gene was present in C9.1, which is absent from isogenic antigenic variants, and this unique copy is associated with transcription of an expressed polypeptide. This result again is consistent with the creation of a novel gene copy by ongoing sequence modifications, and thus the creation and/or loss of restriction sites. Since the remainder of the hybridization pattern was seemingly stable, it also suggested that there may be a duplication of sequences involved in such modifications.

The concept of sequence duplication was confirmed by hybridization analysis using an end-labelled oligonucleotide probe, DA32, representing one of the central variant segments differing between the type A and type B cDNAs of C9.1. Using such a defined probe, it was apparent that there were two copies of the DA32 sequence present in C9.1 gDNA, but only one copy in the variant progeny gDNAs (Allred *et al.*, 2000). This observation has since been confirmed with additional oligonucleotide probes to sequences at the 5' and 3' ends of the open reading frame, and in the 5' UTR, the CKRD and VDCS2 regions (unpublished observations). In each case, there is a repertoire of from one to four bands that are constant in their sizes and relative hybridization intensities among the isogenic clonal variants, and one additional band in C9.1 that corresponds to the transcribed gene sequence. With certain combinations of probe and parasite lines, an additional size-polymorphic band was sometimes present in one or more of the other lines as well. These results are consistent with, and strongly suggestive of, progressive modification of a transcribed gene copy that undergoes persistent and frequent modification, as shown schematically in Figure 13.3. Duplication and movement of an entire donor gene copy to a site of transcription has not yet been observed, but this possibility cannot be ruled out.

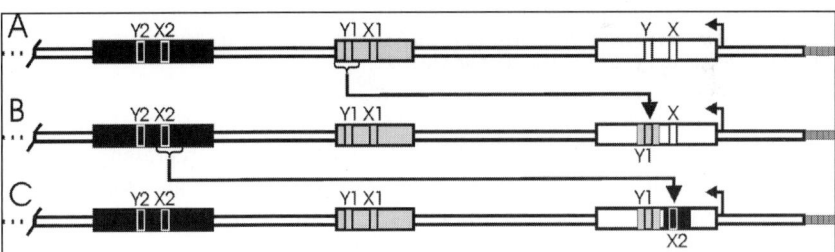

Figure 13.3 Hypothetical model of sequence movements in the *ves* multigene family. Preliminary data indicate that expressed VESA1a polypeptides are associated with transcribed genes containing sequences duplicated and moved from elsewhere within the genome (Allred *et al.*, 2000; additional unpublished observations). Sequences, indicated here as X, Y, etc., are shown being duplicated into a site of active *ves*1α transcription. The movement of these sequences from hypothetical donor gene copies is shown as occurring in two steps. One of the donors is shown as a full-length gene and one as a short pseudogene, although all *ves*1α genes recovered to date are apparently full length (unpublished results). The end result of these sequence movements is transcription of mosaic *ves*1α genes by parasite populations 'B' and 'C', without alteration of the donor loci. The right-angle arrow indicates the promoter driving *ves*1α transcription; centromeric sequences are off-frame to the left, and reiterated telomeric sequences (short, grey blocks) to the right. The orientation and positioning of the gene copies are hypothetical only. (Modified with permission from Allred (2001) *Microbes Infect.* 3, 481–491, Figure 7.)

Gene conversion is thought to occur through strand invasion, extension of the 3' ends of the invading strands and resolution of the heteromolecular complex. The way in which resolution of the intermediate heteroduplex DNA occurs affects whether there is reciprocal or unidirectional sequence exchange (Cromie and Leach, 2000; Allers and Lichten, 2001). A particularly intriguing possibility for efficient partial gene conversion and mosaic gene formation in *B. bovis* was uncovered while pursuing the molecular cloning of genomic fragments providing duplicated sequences. Using the oligonucleotide probe, DA32, to screen a cosmid library made from *B. bovis* C9.1 line gDNA, several cosmids were recovered that contained 4.5 kbp KpnI fragments, which co-migrated with the seemingly invariant gene copy present in C9.1 and each of the isogenic clonal lines. These cosmids were found to contain two *ves*1α gene copies in a very tightly juxtaposed, head-to-head orientation (Figure 13.4). As these gene copies are 79% identical and

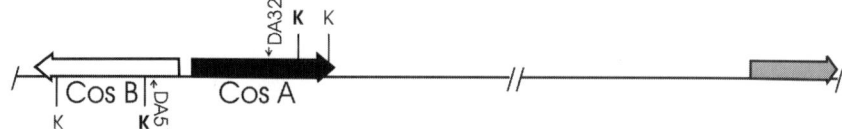

Figure 13.4 Close juxtaposition and head-to-head orientation of two *ves*1α genes. Two complete *ves*1α genes, designated 'Cos A' (black-filled arrow) and 'Cos B' (white-filled arrow), were identified within a single cosmid fragment of approximately 35 kbp. No apparent promoter structure was identified in the 5' UTR of the two genes. Additionally, a third, partial gene copy (grey-filled arrow) was found at the opposite end of the cosmid, which had been truncated at its 3' end during the cloning process. Within the Cos A and Cos B genes were identified the 'DA32' and 'DA51' oligonucleotide sequences, respectively, that are present in the p9.6.2 cDNA. These sequences are present within a size-invariant 4.5 kbp KpnI fragment found in variant lines isogenic to C9.1, as well as in a duplicated form within the actively transcribed gene in the C9.1 line (Allred *et al.*, 2000). The duplicated sequences are absent in most other lines. K, KpnI sites in the region of the juxtaposed genes; DA32 and DA51, locations of each oligonucleotide sequence.

contain numerous segments of identical sequence, they represent a quasi-palindrome. It may be possible for the quasi-palindrome to assume a topologically (nearly) equivalent quasi-cruciform structure. If strand invasion into the quasi-cruciform structure by the transcriptionally active *ves*1α gene were to occur it may be possible for the invading gene to simultaneously acquire sequences from both donor gene copies. If this were to occur, this would contrast with the step-wise acquisition of such sequences, as suggested in Figure 13.3. A comparison of the p9.6.2 cDNA sequence with those of the two gene copies in this cosmid reveals the presence of numerous sequence segments from and otherwise unique to each 'stable' donor gene copy (unpublished observations). While preliminary, these results suggest an intriguing potential mechanism for the rapid creation of mosaic *ves*1α genes containing many unique sequences.

It would seem clear from the available evidence that one mechanism used by *B. bovis* to generate antigenic variants is through gene conversion. The seeming stability of most sequences in the *ves* multigene family and the obvious polymorphism in transcribed *ves*1α gene(s) suggest that only the transcribed *ves*1α gene is modified and that sequence movements are unidirectional. If so, the transcribed *ves*1α gene may reside within a specialized expression site, and become modified while residing there. Such an expression site may be analogous to, although probably very different from, the polycistronic expression sites used by African trypanosomes (reviewed in Vanhamme *et al.*, 2001) and *Anaplasma marginale* (Barbet *et al.*, 2000).

Should gene conversion – either complete or segmental – prove to be a major or perhaps even the primary mechanism of antigenic variation in *B. bovis*, this obviously does not preclude the possibility that other mechanisms are also important for expression of novel antigen forms. For example, during the intraerythrocytic part of its life cycle *B. bovis* replicates asexually and is haploid in its genetic complement. As an integral part of its normal life cycle *B. bovis* undergoes sexual reproduction with reductive meiotic divisions in the tick host (Mackenstedt *et al.*, 1995). The parasite returns to haploidy prior to re-establishment of infection in the bovine host. During the meiotic divisions, interchromosomal meiotic recombination undoubtedly occurs, and could help to generate structural diversity in *ves*1α genes, although this has not been demonstrated. Other possibilities include *in situ* switching of actively transcribed gene copies through epigenetic regulation, similar to that observed in *P. falciparum* (Scherf *et al.*, 1998; Deitsch *et al.*, 1999), switching and allelic exclusion among specialized expression sites as occurs in the African trypanosomes (reviewed in Borst *et al.*, 1998; Borst and Ulbert, 2001; Vanhamme *et al.*, 2001), and sequence inversions, frame shifts and other possibilities like those observed in prokaryotic organisms (reviewed in Barbour and Restrepo, 2000; Brayton *et al.*, 2001; Meeus and Barbet, 2001). At present there is simply no direct evidence either for or against such possibilities in babesial parasites, with one possible exception: *in situ* switching. One prediction of *in situ* switching events is that by most techniques the genome should appear stable; that is, there should be no evidence of rearrangements associated with antigenic variation. For example, there would be no apparent restriction fragment length polymorphisms of the variant multigene family during southern blot analyses of related variants. With the exception of sequences associated with a known transcribed copy (represented by the p9.6.2 cDNA), this result has been consistently observed in *B. bovis*. In addition, of four *ves*1α gene copies recovered from gDNA by PCR or cosmid cloning, all contain complete open reading frames with introns at appropriate sites, suggesting at least theoretical competence for *in situ* transcription. Thus, *in situ* switching remains a distinct possibility, which may be clarified once evidence becomes available regarding the actual number and make-up (relative to genomic sequences) of *ves*1α genes transcribed by individual parasites, and the consistency with which gene-conversion events are observed in different antigenic variants.

ASSOCIATION OF ANTIGENIC VARIATION WITH PATHOLOGY

In similarity to the well-known example of falciparum malaria, *B. bovis* cytoadheres to the capillary and post-capillary venous endothelium, resulting in sequestration of mature parasites in the deep microvasculature. Presumably, sequestration provides a means for the parasite to avoid 'non-specific' splenic clearance of infected erythrocytes (reviewed in Allred, 1995). Sequestration is thought to play a significant role in the pathology caused by *B. bovis*, including cerebral complications (Callow and McGavin, 1963; Wright, 1972; Wright *et al.*, 1988; Aikawa *et al.*, 1992). However, in an important experiment it was observed that immunologically naive calves could be protected against *B. bovis* disease pathology by the passive administration of bovine infection serum from immune cattle. Although the basis for this protection was not firmly established, the authors deduced from the kinetics of parasite removal that immunity was manifested against both the extraerythrocytic parasites and the infected erythrocyte (Mahoney, 1967; Mahoney *et al.*, 1979). Similarly, opsonic phagocytosis has been demonstrated *in vitro* (Jacobson *et al.*, 1993) and is probably more effective *in vivo*.

An *in vitro* assay has been established for the dissection of *B. bovis* cytoadhesion to bovine brain endothelial cells (BBEC) (O'Connor *et al.*, 1999). Significant to the validation of this assay and to our understanding of cytoadhesion were two observations: (1) cytoadhesion was dependent upon nascent parasite protein synthesis and transport, directly implicating parasite-derived components in this process; and (2) adhesion to BBEC occurred through the knob-like protrusions on the IRBC surface (O'Connor *et al.*, 1999), the same site as the *in vivo* adhesive interaction (Wright, 1972, 1973). Using the *in vitro* cytoadhesion assay, it was demonstrated that the adhesion of IRBC to BBEC could be blocked and reversed with immune sera that recognized the IRBC surface in live-cell immunofluorescence assays (O'Connor and Allred, 2000), an observation that helps to explain the ability of immune sera to passively protect naive animals. Adhesion-blocking activity was resident in purified IgG from such sera, although the active subclass(es) were not identified. When adhesion-blocking sera or purified IgG were used to identify parasite components in a surface-specific immunoprecipitation assay, the VESA1 heterodimer was captured. Importantly, by applying selection for changes in either the adhesive or the antigenic phenotype (in the *in vitro* assay) of IRBCs, co-selection for changes in the second, presumably unrelated phenotype also occurred. This finding strongly implies a direct connection between the VESA1 protein and cytoadhesion, perhaps even in a role as the adhesive ligand. It might seem intuitively inconsistent that a highly variant component should serve a key survival function such as cytoadhesion. However, further support for this conjecture was obtained when monoclonal antibodies 3F7.1H11 and 4D9.1G1, specific for the VESA1a sub-unit, were shown to both effectively block and efficiently reverse *in vitro* cytoadhesion (O'Connor and Allred, 2000). In addition, individual adhesive clonal parasite lines demonstrate specificity for different endothelial cell lines (unpublished data), consistent with the potential to express many different adhesive ligands. Even structural characteristics of the VESA1a polypeptide are consistent with such a role. For example, the presence of the invariant sequence motif present in the VDCS2 domain and two coiled-coil protein assembly/interaction domains in the same exposed region of all VESA1a polypeptides, just preceding and encompassing the VDCS1 subdomain, are consistent with requirements for specific interactions between different polypeptide segments. This could be during folding and/or assembly of the apoprotein into the holoprotein, or may specifically mediate interactions with endothelial components during cytoadhesion. In this regard it may be significant that the epitope recognized by monoclonal antibody 3F7.1H11 maps to the beginning of the VDCS2 subdomain (unpublished data).

If the VESA1 antigen were to serve only as an antigenically variant target for the immune system, strong conservation of sequence and structural elements would not be expected to occur. On the other hand, if the VESA1 protein were to mediate cytoadhesion, rapid variation through segmental gene conversion could provide the potential to display an enormous variety of antigenically distinct, alternative VESA1a structures. This, in turn, could provide a large variety of potentially functional ligands that could circumvent allelic polymorphism among host receptors. Although the results of the cytoadhesion studies were derived *in vitro* and are not definitive, when considered as a whole they strongly suggest that a direct and perhaps logical connection exists between the phenomena of antigenic variation and cytoadhesion. Therefore, a close association of antigenic variation with *in vivo* pathology and virulence, such as that already firmly established for falciparum malaria, is to be expected.

FUTURE DIRECTIONS

There are a great many significant challenges to be surmounted before we will understand, at any meaningful level, the mechanisms by which antigenic variation occurs and its true significance to the parasite–host interaction. Among these are included such basic questions as: how rapidly does antigenic variation occur; are mechanisms of antigenic variation affected by the host immune response; does antigenic variation have any effect on the immune response itself; what mechanisms are involved in variation besides gene conversion-like mechanisms, and which predominate; which enzymatic machinery are responsible for rapid, targeted gene conversion; what is the organization of the *ves* multigene family in the *B. bovis* genome; what is the developmental regulation of *ves*1α gene transcription; how are the VESA1a and 1b polypeptides targeted and transported to the erythrocyte surface, and when and where do they assemble into VESA1 holoprotein; and is VESA1 really the cytoadhesion receptor for the

IRBC, and if so are there consistent binding domains formed that mediate this process?. These questions are only the beginning. Obviously, what remains unknown ominously dwarfs what little is known.

Considering that many babesial parasites establish persistent infections, often of long duration, there must be many examples of antigenic variation in this genus, besides *B. bovis* and *B. rodhaini*, simply waiting to be discovered. In addition, it is already known that other distinctly different mechanisms of immune evasion may also come into play. For example, *B. bigemina*, another bovine parasite, also modifies the IRBC surface antigenically, including the deposition of parasite-derived polypeptides on the IRBC surface (Shompole *et al.*, 1994, 1995). Unlike those on the *B. bovis* IRBC surface, the *B. bigemina* antigens carry readily accessible isolate-common epitopes and do not appear to vary rapidly. On the other hand, the *B. bigemina* IRBC surface also binds the μ heavy chain of host plasma IgM in an immunologically non-specific fashion (Echaide *et al.*, 1998), presumably as a form of 'antigen masking'. Whether host or parasite-derived components are responsible for this activity is unknown, but clearly these parasites have devised the ways and means to survive in the very hostile environment provided by the host immune system.

ACKNOWLEDGMENTS

The authors would like to thank Julie Crabtree, Jennifer Long, Laura Stockman and Kristi Warren for technical assistance. Supported by grants from the United States Department of Agriculture (2001–35204–10144) and the American Heart Association (0051422B).

REFERENCES

Aikawa, M., Pongponratn, E., Tegoshi, T., Nakamura, K. I., Nagatake, T., Cochrane, A. and Ozaki, L. S. (1992) *Mem. Inst. Oswaldo Cruz* 87 (Suppl. III), 297–301.
Allers, T. and Lichten, M. (2001) *Cell* 106, 47–57.
Allred, D. R. (1995) *Parasitol. Today* 11, 100–105.
Allred, D. R. (1997) *Methods* 13, 177–189.
Allred, D. R. (2001) *Microbes Infect.* 3, 481–491.
Allred, D. R. and Ahrens, K. P. (1993) *J. Parasitol.* 79, 274–277.
Allred, D. R., Carlton, J. M. R., Satcher, R. L., Long, J. A., Brown, W. C., Patterson, P. E., O'Connor, R. M. and Stroup, S. E. (2000) *Mol. Cell* 5, 153–162.
Allred, D. R., Cinque, R. M., Lane, T. J. and Ahrens, K. P. (1994) *Infect. Immun.* 62, 91–98.
Allred, D. R., Hines, S. A. and Ahrens, K. P. (1993) *Mol. Biochem. Parasitol.* 60, 121–132.
Barbet, A. F., Lundgren, A., Yi, J., Rurangirwa, F. R. and Palmer, G. H. (2000) *Infect. Immun.* 68, 6133–6138.

Barbour, A. G. and Restrepo, B. I. (2000) *Emerg. Infect. Dis.* 6, 449–457.
Baruch, D. I., Ma, X. C., Singh, H. B., Bi, X., Pasloske, B. L. and Howard, R. J. (1997) *Blood* 90, 3766–3775.
Borst, P. and Ulbert, S. (2001) *Mol. Biochem. Parasitol.* 114, 17–27.
Borst, P., Bitter, W., Blundell, P. A., Chaves, I., Cross, M., Gerrits, H., Van Leeuwen, F., McCulloch, R., Taylor, M. and Rudenko, G. (1998) *Mol. Biochem. Parasitol.* 91, 67–76.
Brayton, K. A., Knowles, D. P., McGuire, T. C. and Palmer, G. H. (2001) *Proc. Natl Acad. Sci. USA* 98, 4130–4135.
Callow, L. L. and McGavin, M. D. (1963) *Aust. Vet. J.* 39, 15–21.
Chen, Q., Fernandez, V., Sundström, A., Schlichtherle, M., Datta, S., Hagblom, P. and Wahlgren, M. (1998) *Nature* 394, 392–395.
Cromie, G. A. and Leach, D. R. F. (2000) *Mol. Cell* 6, 815–826.
Curnow, J. A. (1968) *Nature* 217, 267–268.
Curnow, J. A. (1973) *Aust. Vet. J.* 49, 279–283.
Deitsch, K. W., del Pinal, A. and Wellems, T. E. (1999) *Mol. Biochem. Parasitol.* 101, 107–116.
Echaide, I. E., Hines, S. A., McElwain, T. F., Suarez, C. E., McGuire, T. C. and Palmer, G. H. (1998) *Infect. Immun.* 66, 2922–2927.
Goodger, B. V., Commins, M. A., Wright, I. G. and Mirre, G. B. (1984) *Z. Parasitenk.* 70, 321–329.
Goodger, B. V., Wright, I. G., Mahoney, D. F. and McKenna, R. V. (1980) *Int. J. Parasitol.* 10, 33–36.
Goodger, B. V., Wright, I. G., Waltisbuhl, D. J. and Mirre, G. B. (1985) *Int. J. Parasitol.* 15, 175–179.
Graham, O. H. and Hourrigan, J. L. (1977) *J. Med. Entomol.* 13, 629.
Howard, R. J., Rodwell, B. J., Smith, P. M., Callow, L. L. and Mitchell, G. F. (1980) *J. Protozool.* 27, 241–247.
Jacobson, R. H., Parrodi, F., Wright, I. G., Fitzgerald, C. J. and Dobson, C. (1993) *Parasitol. Res.* 79, 221–226.
Mackenstedt, U., Gauer, M., Fuchs, P., Zapf, F., Schein, E. and Mehlhorn, H. (1995) *Parasitol. Res.* 81, 595–604.
Mahoney, D. F. (1967) *Exp. Parasitol.* 20, 119–124.
Mahoney, D. F., Kerr, J. D., Goodger, B. V. and Wright, I. G. (1979) *Int. J. Parasitol.* 9, 297–306.
Mahoney, D. F., Wright, I. G. and Goodger, B. V. (1981) *Vet. Immunol. Immunopathol.* 2, 145–156.
Mahoney, D. F., Wright, I. G. and Mirre, G. B. (1973) *Ann. Trop. Med. Parasitol.* 67, 197–203.
Meeus, P. F. M. and Barbet, A. F. (2001) *Trends Microbiol.* 9, 353–355.
O'Connor, R. M. and Allred, D. R. (2000) *J. Immunol.* 164, 2037–2045.
O'Connor, R. M., Lane, T. J., Stroup, S. E. and Allred, D. R. (1997) *Mol. Biochem. Parasitol.* 89, 259–270.
O'Connor, R. M., Long, J. A. and Allred, D. R. (1999) *Infect. Immun.* 67, 3921–3928.
Phillips, R. S. (1971) *Parasitology* 63, 315–322.
Roberts, J. A. and Tracey-Patte, P. (1975) *Int. J. Parasitol.* 5, 573–576.
Rodriguez, S. D., Buening, G. M., Green, T. J. and Carson, C. A. (1983) *Infect. Immun.* 42, 15–18.

Rovis, L., Barbet, A. F. and Williams, R. O. (1978) *Nature* 271, 654–656.

Scherf, A., Hernandez-Rivas, R., Buffet, P., Bottius, E., Benatar, C., Pouvelle, B., Gysin, J. and Lanzer, M. (1998) *EMBO J.* 17, 5418–5426.

Shompole, S., McElwain, T. F., Jasmer, D. P., Hines, S. A., Katende, J., Musoke, A. J., Rurangirwa, F. R. and McGuire, T. C. (1994) *Parasite Immunol.* 16, 119–127.

Shompole, S., Perryman, L. E., Rurangirwa, F. R., McElwain, T. F., Jasmer, D. P., Musoke, A. J., Wells, C. W. and McGuire, T. C. (1995) *Infect. Immun.* 63, 3507–3513.

Silins, G. U., Blakeley, R. L. and Riddles, P. W. (1996) *Mol. Biochem. Parasitol.* 76, 231–244.

Smith, T. and Kilborne, F. L. (1893) *Bull. Bur. Anim. Ind. US Dept Agric.* 1, 1.

Snary, D. and Smith, M. A. (1986) *Mol. Biochem. Parasitol.* 20, 101–109.

Snary, D. and Smith, M. A. (1988) *Mol. Biochem. Parasitol.* 27, 303–312.

Suarez, C. E., Palmer, G. H., Hötzel, I. and McElwain, T. F. (1998) *Mol. Biochem. Parasitol.* 93, 215–224.

Thoongsuwan, S. and Cox, H. W. (1973) *Ann. Trop. Med. Parasitol.* 67, 373–385.

Vanhamme, L., Pays, E., McCulloch, R. and Barry, J. D. (2001) *Trends Parasitol.* 17, 338–343.

Wright, I. G. (1972) *Int. J. Parasitol.* 2, 209–215.

Wright, I. G. (1973) *J. Parasitol.* 59, 735–736.

Wright, I. G., Goodger, B. V. and Clark, I. A. (1988) *Parasitol. Today* 4, 214–218.

Wright, I. G., White, M., Tracey-Patte, P. D., Donaldson, R. A., Goodger, B. V., Waltisbuhl, D. J. and Mahoney, D. F. (1983) *Infect. Immun.* 41, 244–250.

14

ANTIGENIC VARIATION IN *PLASMODIUM FALCIPARUM* AND OTHER *PLASMODIUM* SPECIES

Mallika Kaviratne, Victor Fernandez, William Jarra, Deirdre Cunningham, Mary R. Galinski, Mats Wahlgren and Peter R. Preiser

INTRODUCTION

Pathogens constrained to survive within a mammalian host are under evolutionary pressure to acquire mechanisms that favour chronic infection. The main obstacle to establishing a chronic infection is the host immune response. In consequence, the evolved mechanisms all incorporate some element of immune evasion, for example, hiding within cells, the induction of immunosuppression and antigenic variation. The latter is a feature of many microbial pathogens. In the case of *Plasmodium falciparum*, which depends on an insect vector for transmission and completion of its life cycle, the demands on the parasite are determined by the climatic and epidemiological setting. In areas of highly seasonal transmission, the episodes of clinical disease are concentrated in a short period of the year following the rainy season. In order to endure the dry season, the parasites must establish long or chronic infections to survive for many months in the human host, waiting for the following transmission season. In areas where malaria is holoendemic and transmission is stable over most of the year, the challenge to the parasite is conceivably different. Here the organism must find ways to overcome the high degree of acquired immunity prevalent in the host population. Manifestations of the success with which *P. falciparum* evades the immune control are the occurrence of continued superinfections with new parasites, the paucity of sterile immunity, the establishment of chronic asymptomatic parasitaemia, or the recrudescence of parasites. Instrumental in this evasion function is the expression of parasite-derived proteins on the surface of the infected erythrocyte (IE) with the ability to undergo clonal antigenic variation. Thus, in the natural vertebrate host infections tend towards chronicity with low-grade but relapsing parasitaemia lasting up to 3 years in *P. falciparum* (see Figure 14.1) and more than 30 years in *Plasmodium malariae*. Only in *Plasmodium vivax* and *Plasmodium ovale* infections do clinical episodes due to dormant liver stages extend beyond the first 12 months after the initial infection. There are a number of criteria that must be met by the parasite to ensure a successful infection. During each erythrocytic cycle a sufficient number of infected red blood cells (rbc) must evade the host immune response and clearance in the spleen. Furthermore, merozoites released after schizont rupture must survive the brief period spent in the midst of the host bloodstream, and the infection will only continue if the merozoites are able to find a suitable rbc for invasion. If these criteria are not met the infection will falter. On the other hand, it is also important that the efficiency at each step is not so high as to overwhelm the host, yet allows sufficient time to proliferate in the host in order to generate sexual forms so that the parasite may be further transmitted. While it is not yet completely understood how the parasite achieves all this despite facing a vigorous immune response by the host, antigenic variation seems to play an important role.

ANTIGENIC VARIATION OF INFECTED RED BLOOD CELLS

The first experimental observations suggesting phenotypic variation among malaria parasite species were made early in the twentieth century. With the use of malaria infections for the treatment of syphilis, knowledge was acquired on how different isolates showed different but reproducible virulence patterns (James *et al.*, 1932). Further experiments on neurosyphilitic patients demonstrated that immunity to malaria is parasite species- and strain-specific (Jeffery, 1966) (Figure 14.1). Antigenic variation was first observed in animal models before it was described in *P. falciparum*. One of the first suggestions of antigenic variability during the course of infection in the blood was in *Plasmodium berghei* in laboratory mice (Cox, 1962). From a very different experimental set up, however, an indication of antigenic variation was first obtained as early as 1938. It was then shown that serum from a monkey infected with *Plasmodium knowlesi* could agglutinate infected, but not uninfected, cells, in what since then has been known as the schizont infected cell agglutination (SICA) test (Eaton, 1938). The SICA test also showed that this reaction only occurred with the maturer parasite stages, implying the presence of parasite antigens on the surface. Many years later, using the SICA test with *P. knowlesi* in the rhesus monkey, it was demonstrated that

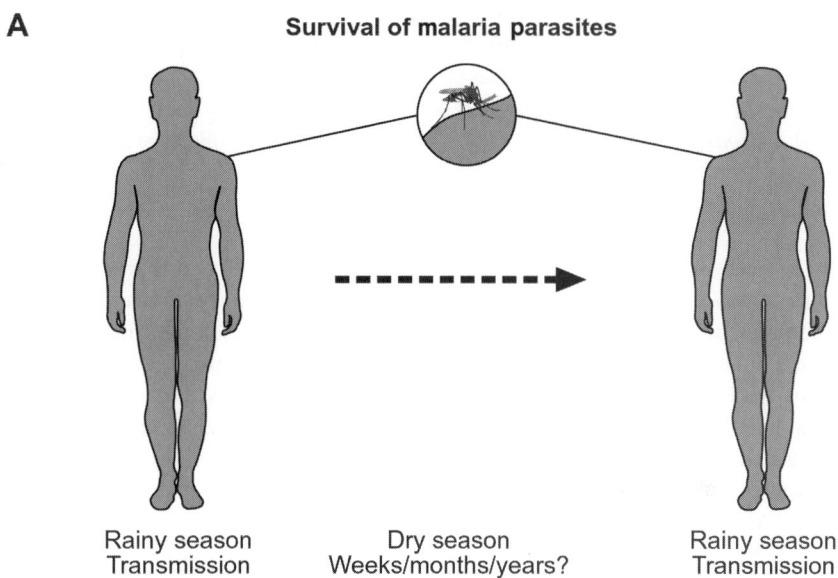

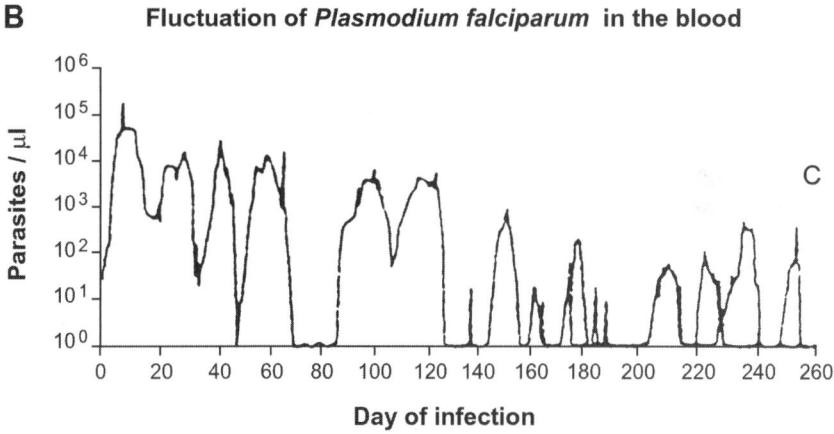

C = Curative dose of chloroquine

Figure 14.1 Chronic infection of *Plasmodium* in humans. (A) To ensure survival during periods of low or no transmission the parasite needs to establish a chronic long-lasting infection in its host. (B) Chronic infection of *P. falciparum* in blood after a single infectious mosquito bite. Changes in parasitaemia are shown for a period of 260 days. Parasitaemia was finally cleared by a curative dose of chloroquine (based on Miller *et al.*, 1994).

chronic infections were maintained by the serial expression of different antigenic types, and that immunity to infection was associated with the development of variant-specific antibodies (Brown and Brown, 1965; Brown, 1971). Subsequently, clonal antigenic variation was formally proven in *Plasmodium knowlesi* (Barnwell *et al.*, 1983a), in the rodent *Plasmodium chabaudi* (McLean *et al.*, 1986) and the simian *Plasmodium fragile* (Handunnetti *et al.*, 1987) parasites, as well as in *P. falciparum* (Hommel *et al.*, 1983, Biggs *et al.*, 1991, Roberts *et al.*, 1992).

Switching of variant antigens and role of the spleen

It has been proposed that selection and expansion of pre-existing minor variant populations explains antigenic variation within a *P. falciparum* infection. An alternative hypothesis suggests that binding of homologous anti-variant antibody to the corresponding antigen on the IE surface triggers antigenic switching by some undefined signalling system (Brown, 1973). Similarly, induced switching was later postulated as a mechanism underlying the appearance of new *P. knowlesi* variant types (Barnwell *et al.*, 1983b). Among several reasons for the appeal of this model, one is the economic use of the variant repertoire, as the switching would not occur until after the appearance of anti-variant antibody. However, no experimental evidence supporting this model has been produced that unambiguously rules out the effect of anti-variant antibody-mediated selection from a pool of small incipient variant stocks.

Several investigations have shown the important effect that the spleen has over the maintenance of variant antigen expression. Studies have been performed with splenectomized human and animal hosts infected either with *P. falciparum* or with other malaria species such as *P. fragile*, *P. knowlesi* or *P. chabaudi*. Repeated parasite passage in such hosts results in the reproducible appearance of parasites that do not sequester, that no longer express the variant antigen or express a different antigenic type (Barnwell *et al.*, 1983b; Hommel *et al.*, 1983; Handunnetti *et al.*, 1987; Gilks *et al.*, 1990). It has been argued that these phenomena are not the consequence of selection of minor variants in the parasite population since passage back into intact animals results in the gradual re-emergence of variant antigen expression and, so far, there is no published evidence that non-sequestering parasites have any growth advantage. One purported conclusion from some of these studies is that the spleen is required for antigenic switching, which would depend on a host signal reliant on the presence of an intact organ. In this respect, the experimental evidence from different animal models is contradictory. Furthermore, solid experimental data show that clonal switching occurs *in vitro*, that is, in the absence of spleen effect.

Antigenic variation and role of antibody in protection

Acquired immunity to malaria relies on antibody as an important effector, and at least a part of the protective effect depends on opsonization and phagocytosis of the infected erythrocyte. A number of studies in animal models show that passive transfer of immune antibodies efficiently reduces or eliminates parasitaemia, and highlight the species-, strain-, stage- and variant-specific condition of the humoral immune response in the animal host. In the human the situation is not that clear. In general, adult sera recognize the PRBC surface of a high proportion of isolates, whereas sera from young children have a more restricted specificity range (Marsh and Howard, 1986; Barragan et al., 1998). Studies based on agglutination assays have shown that large numbers of *P. falciparum* parasite variants are circulating, even in small communities, and that antibodies to variant surface antigens (VSA) of the IE, one of them the *P. falciparum* erythrocyte membrane protein 1 (PfEMP1), appear to develop readily and to agglutinate IE in a predominantly strain-specific manner (Forsyth et al., 1989; Bull et al., 1998). Cross-reaction between parasites from widely separated geographical regions is not rare (Aguiar et al., 1992; Barragan et al., 1998), suggesting that some variants share identical or very similar epitopes. With regard to protection, in two studies involving Gambian children it was found that the capacity to disrupt rosettes (see below) was more frequent in children with mild malaria as compared with children with severe disease (Carlson et al., 1990a; Treutiger et al., 1992). Further, it has been found that Kenyan children who show the presence of antibody to a particular VSA have a significantly reduced chance of becoming clinically ill after infection with that variant (Bull et al., 1998, 1999). Thus, while the functional protective capabilities of anti-PfEMP1 antibodies remain unclear, high levels of antibodies to a range of VSA in patients' sera is the best predictor of protection (Marsh et al., 1989). The data therefore support the idea that pre-existing anti-variant antibodies can select the variants expressed in a new infection and suggest the existence of a dominant subset of surface antigen variants. An interpretation of these results is that in humans, as in animal models, anti-variant antibody plays a crucial protective role in the immune response.

P. falciparum

Pathology and antigenic variation in Plasmodium falciparum

In addition to acting as a means of immune evasion, PfEMP1 is also an important virulence factor because it mediates adhesion to a variety of host cell types. An important adhesive interaction of IE is the phenomenon of cytoadherence, in which pRBC adhere to endothelial cells lining the microvasculature protecting IE from entrapment and loss in the spleen. As a result of cytoadherence, cells infected by mature *P. falciparum* parasites are absent from the peripheral circulation, a feature known as sequestration. Although the property of cytoadherence is universal, under some conditions the parasites appear to accumulate in particular organs possibly due to the co-expression of rosetting, another adhesive property of *P. falciparum* IE. Rosetting, the ability of IE to adhere to uninfected rbc also mediated by PfEMP1, is a parasite phenotype associated with the occurrence of severe malaria (Carlson et al., 1990a; Treutiger et al., 1992). Sequestration results in the obstruction of blood vessels, compromising blood flow to the affected organs. This is considered the primary cause of many forms of severe malaria. Since sequestration occurs in *P. falciparum* malaria but not with the other *Plasmodium* species infecting humans, it is widely believed that it is the sequestered parasites that cause complicated pathology, being a key determinant for the particular virulence of *P. falciparum* and the outcome of severe disease. *P. falciparum* is also a major cause of maternal death, abortion, stillbirth, premature delivery and low birth-weight in endemic areas. The placenta is a preferred site for sequestration of IE, resulting in increased susceptibility of pregnant women to malaria. Thus, the pathology associated with *P. falciparum* infections is largely believed to be due to the adhesive properties of the infected erythrocytes. It is important to mention that anaemia also contributes significantly to the severe disease associated with malaria.

Sequestration may serve the parasite twofold, as residence in the hypoxic postcapillary vessels allows optimal growth and prevents the IE from circulating and being destroyed in the spleen. Adhesion to endothelial cells and to uninfected rbc, as well as the aggregation of IE, directly (autoagglutination), are adhesive traits of *P. falciparum*-infected erythrocytes which are studied *in vitro* and that are thought to reflect the events bringing about sequestration and enhanced accumulation of parasites in vascular beds.

Modifications of the host red cell surface induced by P. falciparum

During its intraerythrocytic life, the parasite is located within the parasitophorous vacuole. Any molecular exchange with the host cell requires transport across the membrane of this compartment, which is formed during the course of invasion by the invagination of the red cell membrane. In the first hours of development there is no evidence of noticeable changes outside the parasitophorous vacuole or in the erythrocyte membrane. As the parasite grows, alterations in the host cell can be observed first at 12–14 hours post-invasion and thereafter. The main modifications include a visible change in the shape and a reduction in the deformability of the cell, together with an increase in the permeability and transport across the erythrocyte membrane. Characteristic of the gradual development of the more mature intraerythrocytic stages are the appearance of electron-dense protrusions known as knobs (K) in the outer face of the plasma membrane and the expression of parasite-encoded variant antigens on the IE surface, including PfEMP1 mediating cytoadherence and rosetting (Figure 14.3, see page 298).

PfEMP1 encoded for by the var *multigene family*

Molecules with biochemical properties similar to the SICA proteins of *P. knowlesi* (Howard *et al.*, 1983; Howard and Barnwell, 1984) were identified on the surface of *P. falciparum*-infected erythrocytes after radioiodination of intact IE and electrophoretic analysis of the Triton X-100 insoluble/SDS soluble polypeptides (Leech *et al.*, 1984). These large polypeptides (>200 kDa) were not detected after mild trypsinization of intact IE, and they could be labelled

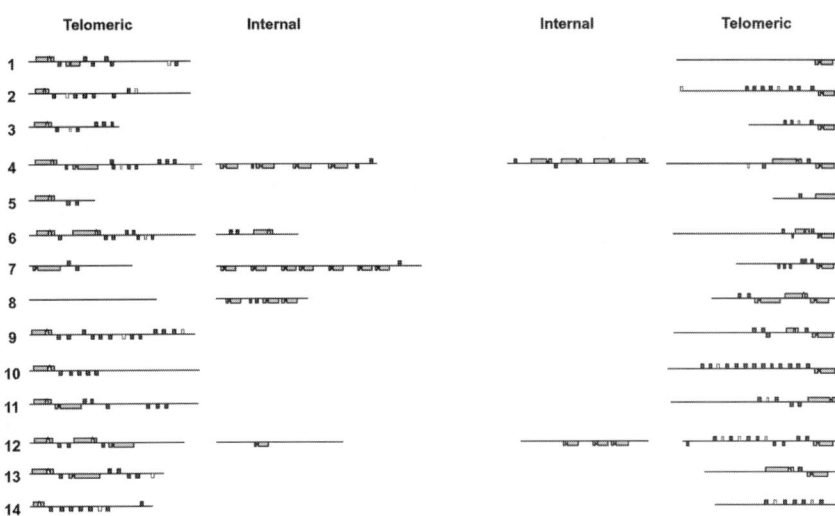

Figure 14.2 Chromosomal arrangement of variant antigens thought to be involved in antigenic variation in *P. falciparum*. Telomeric ends of each chromosome are shown as well as the internal regions that contain variant antigens. Relative positions of *var* (light grey), *rif* (dark grey) and *stevor* (white) genes on all the chromosomes are shown (based on Gardner *et al.*, 2002). Nearly all the genes involved in antigenic variation are located close to the telomeric ends of the chromosomes.

metabolically. The size of the labelled polypeptides varied among different strains and they were immunoprecipitated by rabbit immune serum in a strain-specific fashion. In this way, PfEMP1 was operationally defined and established as a parasite-derived variant antigen expressed on the IE surface, probably interacting with the erythrocyte cytoskeleton. Initially, PfEMP1 was detected in strains with K^+ IE that could cytoadhere to C32 melanoma cells, but not in knobless, non-cytoadherent strains (Aley *et al.*, 1986). Trypsin treatment of IE that cleaved PfEMP1 simultaneously ablated cytoadherent capacity (David *et al.*, 1983; Leech *et al.*, 1984). Moreover, immune sera immunoprecipitated PfEMP1 and inhibited cytoadherence with the same strain specificity (Udeinya *et al.*, 1983; Leech *et al.*, 1984). These observations shaped the idea that the variant antigen was the cytoadherent ligand-mediating sequestration of pRBC. Experiments in which different parasites were selected by 'panning' on melanoma and endothelial cells showed that the cytoadherent capacity of IE closely correlated with the intensity of expression of a family of polypeptides with identical biochemical and immunochemical properties as the variant antigen (Magowan *et al.*, 1988; Biggs *et al.*, 1992). The genes encoding PfEMP1 proteins were identified as the *var* multigene family (Baruch *et al.*, 1995; Su *et al.*, 1995). Approximately 50 *var* genes (per haploid genome) seem to be localized in all chromosomes (1–14) of *P. falciparum*, although the presence of full-size *var* gene(s) in chromosome 14 is still subject to confirmation (Rubio *et al.*, 1996; Thompson *et al.*, 1997). The number and position of *var* genes in some chromosomes almost certainly varies between parasite strains (Thompson *et al.*, 1997). Members of the *var* family are mostly found in the sub-telomeric regions, where they usually constitute the first open reading frame centromeric to the telomeric repeats. Clusters of *var* genes are also found in central regions of chromosomes 4, 7, 8 and 12 (Figure 14.2) (Su *et al.*, 1995; Fischer *et al.*, 1997; Hernandez-Rivas *et al.*, 1997; Thompson *et al.*, 1997; Gardner *et al.*, 1998; Bowman *et al.*, 1999; http://www.sanger.ac.uk/Projects/P_falciparum/).

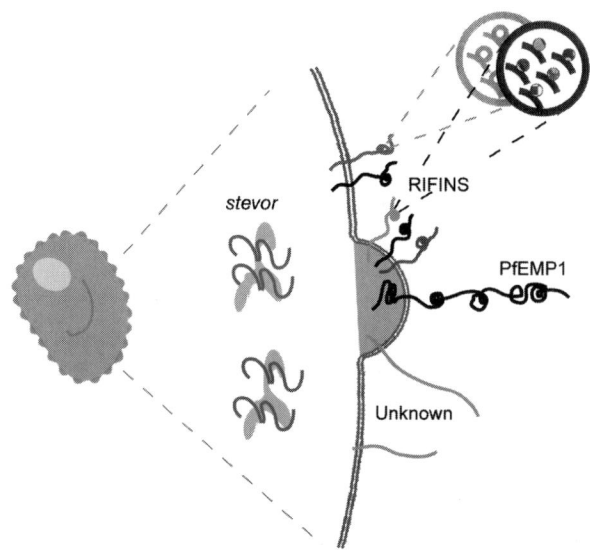

Figure 14.3 Antigenic variation at the infected-erythrocyte surface. PfEMP1 and RIFIN polypeptides encoded by the multigene families *var* and *rif*, respectively, are the major source of antigenic variation but today unknown antigens may also play a role. The polypeptides encoded by the *stevor* family of genes are not transported to the erythrocyte surface but are associated with the Maurer's clefts.

With a two-exon structure, *var* genes comprise a large, highly variable 5' exon of 4–10 kb, and with the exception of one 0.23 kb intron (Buffet *et al.*, 1999), a relatively conserved 0.8–1.2 kb intron, and a highly conserved 3' exon ~1.5 kb long. The base composition of the *var* exons is similar to that typical in coding regions of *P. falciparum*, i.e. ~75% (A+T). Intervening sequences of *var* genes have a higher overall A+T content (≥80%). The 5' exon (exon 1) of *var* encodes the extracellular segment and the putative transmembrane region of PfEMP1. Prominent in the 5' exon are a variable number (2–7) of motifs 300–400 amino acids long that have sequence similarity to the Duffy Binding Ligand in *P. vivax* and have therefore been termed DBL domains (Peterson *et al.*, 1995). The extracellular segment of PfEMP1 comprises an N-terminal head structure always composed by the N-terminal segment (NTS) starting at the initiating methionine, a semiconserved DBL1 domain, and a less conserved cysteine-rich interdomain region 1 (CIDR1) with ~300 amino acids. This is followed by a variable number of polymorphic DBL domains interspersed with highly diverse sequences and the domains C2 and CIDR2, not present in every variant. Five sequence homology classes of DBL domains (α, β, γ, δ, ϵ), and three of CIDR domains (α, β, γ), have been identified so far (Smith *et al.*, 2000a). The 3' end of exon 1 encodes a conserved consensus membrane-spanning sequence 25–30 amino acids long. The C-terminal portion of PfEMP1 proteins, which by all accounts resides intracellularly, is encoded by the 3' exon (exon 2). It comprises 450–500 amino acids and features >75% conservation at the protein level between variants. Because of the relative high content of acidic residues (D, E) in this domain, it has been dubbed the acidic terminal segment or ATS (Su *et al.*, 1995). The insolubility of PfEMP1 in non-ionic detergents suggests an interaction with erythrocyte cytoskeletal proteins or with parasite polypeptides forming the knob structure where the variant antigen is clustered on the surface. Accordingly, recombinant ATS proteins were found to bind with relatively high affinity to erythrocytic spectrin and actin (Oh *et al.*, 2000), as well as to two regions of the knob-associated, histidine-rich protein (KAHRP), which is a major component of the knob (Waller *et al.*, 1999). Knobs seem to constitute the 'normal' parasite-induced superstructure on the IE outer membrane where PfEMP1 is targeted and anchored to mediate binding events under the flow stress of circulation (Crabb *et al.*, 1997). Nevertheless, PfEMP1 polypeptides are also expressed on the surface of some knobless IE, where they can mediate cytoadhesion, as assayed under static conditions (Udomsangpetch *et al.*, 1989).

To date, more than 10 complete *var* from cDNA of different *P. falciparum* strains and more than 50 genes from the 3D7 strain genome have been sequenced (Smith *et al.*, 2000b; http://www.sanger.ac.uk/Projects/P_falciparum/). This data, together with a number of functional studies performed with cloned domains and fragments from several PfEMP1 variants, support beyond ambiguity the roles in adhesion and as a scaffold for antigenic variation ascribed early on to this family of molecules. Mapping of ligands within PfEMP1 indicates that the ability of IE from different parasite strains to express various combinations of binding phenotypes is a reflection of the number and class of adhesion domains present in the expressed *var* gene. Hypothetically, the overall binding profile of any PfEMP1 can be additionally modulated by sequence polymorphisms determining changes in strength of binding and/or in conformational arrangements exposing or burying adhesive motifs. The combined data from functional mapping studies show that the distal (relative to the IE membrane) head structure of PfEMP1 includes a DBL1α domain, which harbours specific sites for all the major rosetting receptors described to date, i.e. complement receptor 1 (CR1), heparan sulphate-like GAGs, and ABO blood group antigens (Rowe *et al.*, 1997; Barragan *et al.*, 2000; Chen *et al.*, 2000). Rosetting ligands have not been described anywhere else in the PfEMP1 molecule. Another important region in the head structure is the CIDR domain, almost always of the CIDRα class, containing the consensus motif $CX_{(7-12)}CX_{(3-5)}CX_3CX_{1-2}CX_3WX_{7-8}W$ and ligand specificities for CD36 and, as recently determined, for PECAM-1 (Baruch *et al.*, 1997; Smith *et al.*, 1998; Chen *et al.*, 2000). The binding of immunoglobulin molecules at region(s) different from the antigen-combining site is another adhesive feature of PfEMP1. In the case of another member of the Ig superfamily, the endothelial receptor intracellular adhesion molecule (ICAM)-1, an extended region composed of a DBL of the β class and the contiguous C2 domain has been reported to form the binding pocket (Smith *et al.*, 2000b). Binding sites in two distinct domain classes have been defined for PECAM-1. In addition to the CIDRα, a DBLδ domain was shown to support binding to this receptor (Chen *et al.*, 2000). At least two different DBLγ domains have been implicated in binding to chondroitin sulphate A (CSA) (Buffet *et al.*, 1999; Reeder *et al.*, 1999) and immunization against this domain induces a transcending and adhesion-blocking antibody response against CSA binding to pRBC (Lekana Douki *et al.*, 2002). Similarly, CIDRα recombinant proteins or fragments thereof have also been shown to bind or interfere with binding to CSA, albeit, at least in one case, IE of the strain used to clone the gene do not adhere to this receptor (Degen *et al.*, 2000; Reeder *et al.*, 2000). These data perhaps illustrate the limitations of functional analysis strategies, in which fragments of PfEMP1 more or less arbitrarily defined are assayed outside their molecular context.

Another form of immunoescape, besides sequestration, in which PfEMP1 may play a role is through the manipulation of cells instrumental for the adaptive immune response in the host. Hindrance of dendritic cell maturation *in vitro* upon interaction with erythrocytes infected with mature parasites was described by Urban *et al.* (1999). It is not yet clear whether the PfEMP1 mediates this effect. However, the CIDRα domain of a PfEMP1 polypeptide with IgM binding activity was found to trigger, in a superantigen-like fashion, the polyclonal activation of B cells expressing IgM on the surface. (Zhang *et al*, submitted manuscript). Although preliminary, these first pieces of experimental data point to a potential immunomodulatory function of the variant antigen.

In an original proposition that turns the line of reasoning by 180 degrees, it has been suggested that PfEMP1 antigens are deliberately displayed outside the IE in order to provide a target for the immune system. In this way the parasite would regulate its own growth, sparing the life of its host and securing transmission (Saul, 1999).

Switching and transcriptional control of **var** *genes*

While rapid switching has been shown to occur in the erythrocyte surface antigens of *P. falciparum* at 2.4% per generation, the bulk of parasites, 97.6%, still express the same *var* gene in a clonal population (Roberts *et al.*, 1992; Chen *et al.*, 1998; Fernandez *et al.*, 1998). Thus the parasites must be pre-determined to enter a default pathway that might be controlled at the genomic level by genomic imprinting and *cis*-acting regulatory elements. This could suggest that neither recombinations nor DNA modifications are major factors involved in *P. falciparum var* gene regulation. Moreover, *var* gene expression apparently does not occur from a unique (transcriptionally active) site, as expressed *var* genes have been mapped to several different chromosomal loci and all *var* genes can be transcribed irrespective of their chromosomal location (Su *et al.*, 1995; Fischer *et al.*, 1997; Hernandez-Rivas *et al.*, 1997). Investigations on the nature of *var* transcription were carried out in parasite clones serially panned over cells expressing different adhesion molecules. Parasite populations with defined adhesive phenotypes for ICAM-1, CSA or CD36 receptors expressed a single and distinct *var* variant (Scherf *et al.*, 1998). The chromosomal location of each *var* variant remained unchanged irrespective of its transcriptional status and appeared to occur *in situ* with no requirement for the transposition of the *var* gene to a transcriptionally active site. Finally, regulation of *var* expression appears to be mediated at the transcriptional level as demonstrated by nuclear run on analyses, suggesting that while one promoter is activated the others are silenced (Scherf *et al.*, 1998).

It has previously been reported that many *var* genes are simultaneously transcribed in *in vitro* cultures of clonal parasites early during parasite intraerythrocytic development, although limited to a few genes in each parasite (Fernandez *et al.*, 2002), followed by the selection of a single transcript when the parasite develops into a more mature parasite, i.e. trophozoite (Chen *et al.*, 1998; Scherf *et al.*, 1998). In a recent study of 28 *var* genes in single ring-infected erythrocytes, unique patterns (type and number) of *var* transcripts were found in each individual IE with 1–15 mRNA species detected per cell 2–4 hours post-invasion. Other investigators have suggested the presence of only one full-length transcript (Taylor *et al.*, 2000a), while Noviyanti *et al.* (2001) and Duffy *et al.* (2002) argue that several full-length *var* transcripts can be detected at trophozoite stages but only one dominant transcript is expressed. Data on the *P. knowlesi SICAvar* gene family (Al-Khedery *et al.*, 1999) is in general agreement with the latter conclusions (Corredor and Galinski, submitted). In any case, current data suggest that the *var* genes are not duplicated or moved, but rather that there is a local activation of each *var* gene at its own chromosomal location. The crux of this issue is why only one transcript survives to become translated; does this depend on the stability of the RNA due to 5' or 3' alterations? Or, is it merely due to an earlier onset of transcription with subsequent higher concentration of the relevant RNA? Other possibilities are that the transport of the relevant RNA represents the selection point, or that a silencing mechanism acts during the development of the parasite. In *P. falciparum* it was found that genes that are normally only transcribed in mosquitoe-stage sporozoites were also activated in the asexual stages of the parasite, suggesting the occurrence of a more general, but transient activation of the genome (Chen *et al.*, 1998). If the DNA modifications (which force the parasite to enter a default pathway) are unstable it could be envisaged that the initial ubiquitous activation process is relevant in both the switching manoeuvre and the exclusion of all alleles but one. Taken together, it may be that an epigenetic mechanism is involved in the early activation of the genome, when chromatin-mediated transcriptional silencing may be reversed by modifications of histones.

Deletions, transpositions or recombinations are infrequent during asexual growth of the parasite, although *var* gene recombination events do happen regularly during meiosis in the mosquito. This is very different from the VSG genes of trypanosomes, which are dependent on a single active sub-telomeric expression site, which is only used after duplicative translocation of a previously silent VSG gene from a pool of approximately 1000 genes. The data therefore suggest that *Plasmodium* uses a different mechanism of switching from the VSG genes and argues that each *var* gene has its own transcriptional regulation unit. The modification of nucleotides can play a significant role in the control of gene expression as demonstrated in African trypanosomes (Gommers-Ampt *et al.*, 1993) and there is evidence for DNA methylation in *P. falciparum* (Pollack *et al.*, 1991). However, experiments performed to investigate this suggest that changes in the methylation pattern or nucleosomal phasing do not play a role in *var* gene switching (Scherf *et al.*, 1998; Horrocks *et al.*, 2002). The role of *trans*-acting factors in controlling *var* transcription has been examined through the use of transient transfection experiments by Deitsch and colleagues (Deitsch *et al.*, 2001). They showed that silent *var* promoters become transcriptionally active when removed from their chromosomal context, suggesting that switching is not mediated by *trans*-acting factors alone. This group has demonstrated in *in vitro* studies that the conserved intron sequence (which separates the two *var* exons) is responsible for the silencing of its promoter, and that this silencing mechanism is specific and occurs by cooperative action with the upstream *var* 5′-flanking region. This silencing was S-phase-dependent, and suggests that cooperative DNA-binding complexes and changes in chromatin structure could be involved in the control of transcription (Darkin-Rattray *et al.*, 1996; Wolffe and Guschin, 2000). Further, a constitutively transcribed *var* gene that lacks an intron has been identified in the 3D7 genome (3D7var5.2; var_{common}) (Vazquez-Macias *et al.*, 2002) arguing that the *in vitro* studies performed by Deitsch *et al.* are also relevant in the chromosomal context (Winter *et al.*, 2003). The role of the var_{common} transcript in the biology of the parasite is not known. Preliminary evidence indicates that transmission through the insect stage might reset the *var* repertoire to express a particular limited set of genes (Peters *et al.*, 2002), as has been suggested for trypanosomes.

There are many questions concerning transcription of *P. falciparum* genes in general, and *var* genes in particular, that remain unanswered. The nature and significance of the ring-stage *var* transcripts is one of them. Another group of transcripts, 1.8–2.4 kb mRNA mapping to the second exon of *var* and running into the intron, are abundantly transcribed in erythrocytic stages (Su *et al.*, 1995). It has been hypothesized that these RNAs may participate in the regulation of *var* expression. Alternatively, they could represent a by-product from the molecular mechanism of *var* silencing, in which intron sequences participate. Serial analysis of gene expression in *P. falciparum* has recently revealed that antisense RNA transcription is widespread and it occurs in asexual and sexual stages (Patankar *et al.*, 2001). It is not known whether antisense *var* RNA is transcribed.

Besides the potential role of antisense transcripts in post-transcriptional control of protein expression, it has been suggested that they may reflect mechanisms of transcriptional initiation and termination.

The *var* gene repertoire cannot be limited to 50 different genes in an area where malaria is endemic, as DNA hybridization analysis has revealed that common groups of *var* genes are not shared between different parasites (Su *et al.*, 1995; Freitas-Junior *et al.*, 2000). It may be so that the *var* gene repertoire is restructured or expanded depending on the endemicity of the parasite in the location studied. If the transmission rate is high it is expected, for example, that the *var* gene repertoire is significantly enlarged by frequent fusions of different haploid genomes in the mosquito, while the opposite might be true in pockets of restricted transmission. Yet self-fertilization has been found to produce genetically distinct parasites as well (Scherf *et al.*, 1998).

The close proximity of the multigene families to each other and their location at the sub-telomeric ends of the chromosomes may be advantageous to the parasite (see Figure 14.2). Genes located at telomeric ends are often part of specialized higher-order nuclear architecture, such as the chromosome-end cluster at the periphery of nuclei as demonstrated by the clustering of *P. falciparum* chromosome ends (Freitas-Junior *et al.*, 2000). This may be a significant element in promoting ectopic exchange of sequences in the sub-telomeric regions of *P. falciparum*. Telomere-mediated enhancement of recombination would perhaps augment the effect of the relatively small *var* gene complement in individual parasites during chronic asexual blood stage infection. It has been calculated that recombination between telomeric *var* genes is some eightfold higher than that measured in the genome as a whole (Taylor *et al.*, 2000b).

Other gene families of **P. falciparum** *associated with antigenic variation*

The first observations of low molecular weight polypeptides appearing *de novo* on the surface of IE are essentially contemporaneous with the original identification of PfEMP1. A comparison of surface radioiodinated polypeptides of uninfected and *P. falciparum*-infected erythrocytes from malaria-ill children led to the identification of a major labelled band with Mr ~30 000 (Aley *et al.*, 1986). Almost simultaneously, surface-labelled polypeptides with Mr values 20 000, 35 000, ~40 000, 45 000 and 55 000 were detected in similar experiments performed with parasites adapted to laboratory culture (Stanley and Reese, 1986). Several years later, the first investigations attempting the identification of the parasite ligand(s) mediating rosetting binding were conducted. Using a monoclonal antibody to the KAHRP1 protein that at low dilutions disrupts rosettes of some isolates and strains, a polypeptide with Mr ~28 000 was detected after immunoprecipitation of [^{3}H]histidine metabolically (although not from surface-) labelled parasites of a rosetting strain (Carlson *et al.*, 1990b). Surface iodination of rosetting and non-rosetting parasites helped to identify another labelled band with Mr 22 000, which was present in the Triton X-100-insoluble extracts of two rosette-forming parasites (Helmby *et al.*, 1993). A range of genes potentially coding for these proteins has now been identified.

rif/RIFIN and *stevor*

The *rif* family of repetitive interspersed DNA was identified based on the isolation of a λgt11 clone from an expression library that gave a complex and polymorphic pattern when hybridized to restriction fragments of genomic DNA from different parasites. Probing of chromosomal blots suggested that the sequence localized to most of the chromosomes. The sequence of the clone (RIF-1) contained a 1 kb ORF, and a positive northern blot showed that this sequence was transcribed in blood stages (Weber, 1988). The apparent lack of a translation start site meant that these transcribed sequences were not studied further. However, sequence analysis arising from the genome project revealed a spliced exon, ~300 bp upstream of each *rif*, that encoded a translational start site and a putative secretory signal sequence (Cheng *et al.*, 1998). Further analysis has revealed 13 *rif* as well as four closely related *stevor* genes on chromosome 2 (Gardner *et al.*, 1998) and, by extrapolation of sequence data from chromosomes 2 and 3, it has been suggested that there are in excess of 200 *rif* copies per haploid genome (Kyes *et al.*, 1999). Like *var*, the *rif* and *stevor* genes are located in close association to each other in clusters within 50 kb of the telomeres (see Figure 14.2) and have subsequently been located on all chromosomes (Cheng *et al.*, 1998). All *rif* genes are composed of two exons. A short exon 1 encodes the 19 amino acid-long signal peptide followed by an intron of 140–270 bp, and a second exon of variable length (~1 kb) (Cheng *et al.*, 1998). The products of *rif* genes (RIFINs) consist of a highly conserved N-terminal sequence (exon 1) with some characteristics of a signal peptide. In addition, two predicted putative transmembrane domains flank the highly polymorphic region of exon 2. Protein structure predictions strongly indicate that RIFINs are membrane-bound proteins with the C-terminal end in the cytoplasm and the polymorphic region exposed on the outside (Cheng *et al.*, 1998). Comparison of aligned *rif* genes shows that the N-terminal half of the molecule contains multiple short polymorphic and semiconserved blocks and is rich in cysteines. The other half of the sequence is highly polymorphic except for the last 40 amino acids, the most conserved region in this family, which comprises a strongly predicted transmembrane region and a highly positively charged C-terminal domain. Transcription of *rif* genes has been confirmed in blood stages of many, but not all, laboratory-adapted parasites (Cheng *et al.*, 1998; Gardner *et al.*, 1998; Kyes *et al.*, 1999, 2000) and occurs primarily at the late ring/early trophozoite stage (18–23 hours). RIFINs are immunogenic in natural infections as individual strains are specifically recognized when products of surface-labelled IE are immunoprecipitated by human immune sera (Fernandez *et al.*, 1999; Kyes *et al.*, 1999). Although no definitive function has been ascribed to members of this family, there is some experimental evidence that RIFINs are thought to play an accessory role in rosetting and CD31 binding (Fernandez *et al.*, 1999; Kyes *et al.*, 1999).

The prototype member of the *stevor* (sub-telomeric variable open reading frames) group of genes was 7H8/6, a clone selected from a *P. falciparum* genomic expression library, which hybridized to DNA isolated from all tested strains giving a distinctive multiple binding pattern (Limpaiboon *et al.*, 1991). There are 30–40 *stevor* genes per haploid genome, located adjacent to *rif* genes within 50 kb of the telomere of most, if not all chromosomes. In one parasite, the Dd2 strain, 34 *stevor* copies were identified (Cheng *et al.*, 1998). All *stevor* genes and their deduced proteins share the basic structural features of *rif*, including a first short exon coding for a putative signal peptide or anchor, which may or not be cleaved, and a second exon encoding products of approximately 30 kDa. Because of the extent of structural and amino acid conservation, *rif* and *stevor* are considered to form a gene superfamily. Both groups of genes are predicted to encode membrane-associated products. A detailed examination of *stevor* suggests that an early *rif* element has evolved and further expanded divergently. Moreover, the finding of *stevor* (but not *rif*) pseudogenes may suggest a distinct function and/or cellular location for *stevor*, perhaps free of direct immune selective pressure.

Recent work has shown that s*tevor* is transcribed after both *var* and *rif* during a very tight window of the parasite life cycle (Kaviratne *et al.*, 2002). Individual micromanipulated parasites transcribe more than one *stevor* gene but it has not yet been established whether they are all expressed (Kaviratne *et al.*, 2002). Unlike PfEMP1 and RIFIN, the STEVOR protein does not seem to migrate to the rbc surface but is located within discrete sites (Maurer's clefts) of the infected red blood cell in late trophozoites and schizonts (Kaviratne *et al.*, 2002). Moreover, STEVOR is also expressed in gametocytes (Sutherland, 2001; Preiser, personal communication). These results suggest that STEVOR may have a distinct role in *P. falciparum* biology.

Clag

As discussed previously, in some *P. falciparum* parasites adapted to growth in the laboratory, continuous *in vitro* culture results in the irreversible loss of cytoadherent and/or gametocytogenesis capacity. At least in one case, karyotype analysis by pulse-field gel electrophoresis could show that expression of both phenotypes mapped to a 0.3 Mb sub-telomeric deletion of chromosome 9 (Day *et al.*, 1993). Upon fine mapping and sequencing of the deleted region, a gene believed to be necessary for cytoadherence was identified. The gene in question, the cytoadherence-linked asexual gene 9 (*clag9*), comprised nine exons and was found to be transcribed in asexual stages (Trenholme *et al.*, 2000). The product encoded by *clag9* was predicted as a multipass membrane protein. Targeted gene disruption of *clag9* ablated cytoadherence to C32 melanoma cells, a binding normally mediated by CD36. Consistently, *clag9* knockouts did not bind to purified CD36. Transfection of parasites with a *clag9* antisense RNA vector resulted in a significant lowering of cytoadherent capacity (Gardiner *et al.*, 2000). Experiments of hybridization to chromosomal blots and the ongoing sequencing of the parasite genome have led to the identification of at least nine genes related to *clag9*, indicating that an additional gene family is encoded in the sub-telomeric regions of several chromosomes, within 150 kb from the telomere. The location of *clag9* protein has not been defined and, until now, there is no evidence of its presence on the infected red cell surface. It has been suggested that products of this gene may exert their effect on binding indirectly, through some yet undefined interaction with PfEMP1 proteins.

P. vivax

The vir *gene family*

Fewer multigene families have been identified in *P. vivax*. While homologues of the *P. falciparum* SERA genes, exhibiting considerable sequence variation, which may have the potential for the generation of antigenic diversity, have been cloned from *P. vivax* (Kiefer *et al.*, 1996), v*ar* gene homologues have not been found. Recently the *vir* multigene family was described in *P. vivax* (del Portillo *et al.*, 2001). There are thought to be between 600 and 1000 *vir* copies per haploid genome and they are also located sub-telomerically. Based on the degree of homology between different members, the *vir* genes of *P. vivax* can be subdivided into six subfamilies. Each *vir* gene has three exons: a short first exon lacking a consensus signal peptide sequence; a longer second exon containing a predicted transmembrane domain at its 3' end; and a third exon with a highly conserved region. The *vir* multigene family encodes immunovariant proteins, which may function in antigenic variation and hence immune evasion. The VIR proteins have a size range of 30–37 kDa, which shares no homology to known proteins in BLAST analysis (del Portillo *et al.*, 2001). There are three major conserved blocks of amino acids, interspersed with more polymorphic regions. The proteins contain a (~5aa) amino terminal putative signal peptide and a putative highly conserved carboxy terminal transmembrane region. A number of cysteines are conserved, with four conserved cysteines occurring in the second exon.

P. knowlesi

Using the SICA test Brown and Brown (1965) first clearly demonstrated antigenic variation in *P. knowlesi* in chronically infected, semi-immune rhesus monkeys. Infections in these monkeys were characterized by repeated recrudescences of parasitaemia and the decline of each of these recrudescences (variants) correlated with the appearance of variant-specific SICA antibodies. As monkeys became increasingly more efficient at controlling their recrudescences, this correlated with the appearance/presence of high titres of variant-specific opsonizing, as well as variant-specific agglutinating antibodies. These antibodies enhanced the internalization and destruction of IE by macrophages. This raised the intriguing possibility that chronicity was maintained in this malaria model, in part, by a balance between the production of SICA antibodies, which induced antigenic variation, and opsonizing antibodies, which helped prevent parasitaemia overwhelming the host (Brown, 1971; Brown and Brown, 1965; Brown and Hills, 1974).

The SICAvar gene family

The long-sought, strain-specific determinants of antigenic variation in *P. knowlesi*, or SICA antigens, were biochemically defined on the IE surface as parasite-encoded proteins of high molecular weight (Howard *et al.*, 1983). These antigens were found to be polymorphic in size and antigenicity, trypsin sensitive, accessible to surface radioiodination, and insoluble in nonionic detergents. The SICA antigens are produced by intracellular trophozoites and schizonts (dividing asexual parasites) and are detected on the IE surface throughout these latter stages of the parasite's 24-hour erythrocytic cycle as polypeptides of M_r 180 000–210 000 (Howard *et al.*, 1983). The SICA antigens, like PfEMP1, are transported through the erythrocyte cytoplasm to the erythrocyte surface; that they are only soluble in ionic-detergents suggests their stable insertion into the host cytoskeleton. As Al-Khedery (Al-Khedery *et al.*, 1999) and previous investigators have indicated, these and the orthologous PfEMP1-variant antigens play a role in maintaining chronic infections and may also have yet unidentified essential functions at this membrane location.

The SICA assay has revealed an extensive antigenic variability of the *P.knowlesi* IE surface. Two SICA polypeptides are present on the IE surface yet the size difference between them is not large: not exceeding $\approx M_r$ 20 000. It was previously noted that the smaller of the two SICA polypeptides could perhaps be a processing or degradation product of the larger, but Al-Khedery *et al.* (1999) suggest that this doublet may in fact represent the expression of two *SICAvar* genes. Data to date shows that the two SICA polypeptides are antigenically distinct: antibodies raised to one of them do not react with the other. The variability of the IE surface may therefore be due to changes of either one or both of the two surface polypeptides. Both SICA polypeptides are in any case involved in the deception of the immune system.

The linear amino acid sequence of the SICA antigens is very different from that of PfEMP1, as would be expected given the evolutionary distance of *P. knowlesi* and *P. falciparum*. However, the SICA antigens share a predominantly external variable domain containing a series of distinct although related cysteine-rich domains, a transmembrane region and a cytoplasmic domain. These multiple cysteine-rich domains in the SICA antigens do not bear any resemblance to the cysteine-rich domains of PfEMP1. Nor are they related to the Duffy-binding antigen families involved in merozoite invasion of *P. falciparum* or *P. vivax*. However, as *P. knowlesi* does not adhere in a manner analogous to that seen in *P. falciparum* this dissimilarity is not surprising.

Interestingly, the *P. knowlesi* SICA antigen gene reported to date (Al Khedery *et al.*, 1999) is encoded by a 10-exon/9-intron *var* gene structure that is unusual compared to *P. falciparum*'s more typical 2- or 3-exon structure, common for protein-encoding genes (Gardner *et al.*, 1998). Only a few other *Plasmodium* genes that are similarly organized have been found, in contrast to the multi-intron/exon gene structure that is widespread in multicellular eukaryotes. The interesting multi-exon/intron structure of the *P. knowlesi var* genes, particularly where each SICAvar coding region is interrupted by an intron, may allow recombination of the A/T-rich introns with a subsequent shuffling of the exons, as suggested by Al-Khedery *et al.* (1999). This could be a way to further generate diversity of the *SICAvar* gene-family repertoire, in a process unrelated to switching. Regardless of the unique exon/intron structure of the *P. knowlesi SICAvar* genes, data to date indicates that the *SICAvar* genes like the *P. falciparum var* genes are present on all chromosomes, are transcribed *in situ* and are abundant in the ring stages, but less so in the trophozoite stages (Al-Khedery *et al.* 1999; Corredor and Galinski, submitted). The data being generated from these studies further indicate that the majority of the *SICAvar* transcripts produced in the ring stages, although produced in their entirety, are immediately silenced (Corredor and Galinski, submitted). These investigators have proposed that the *SICAvar* 3' genomic DNA rearrangement they observed previously may function to 'tag' (i.e. protect) the transcripts that are destined to be translated and expressed at the surface of the infected erythrocyte and their recent data studying transcription in several SICA[+] and SICA[–] isogenic clones, derived one from the other in rhesus monkeys, remains consistent with this novel hypothesis.

Another fascinating property of malaria antigenic variation noted from the *P. knowlesi* and the SICA-antigen system is that switching may be induced by antibodies reactive with the SICA antigens at the schizont-infected erythrocyte surface (Brown and Brown, 1965, Barnwell *et al.*, 1983a). When rhesus monkeys with agglutinating antibody titres to a particular variant were reinfected with the homologous parasite, they switch SICA antigens while parasites of a heterologous challenge or similar parasite inoculations in naive animals do not switch (Barnwell *et al.*, 1983a). It is not trivial to be able to tickle the parasite from the outside as it is clearly intracellular, behind three membranes, in an erythrocyte that lacks the conventional cellular machinery. An induction signal for the *SICAvar* genes to switch must consequently reach the inside via a series of signals initiated by an antibody bound to an exterior antigenic moiety of the parasite. Hypothetically, a phosphorylation might occur at the cytoplasmic tail of the SICA antigen, which mediates a change transmitted through the erythrocyte cytoplasm, transversing the parasitophorous vacuolar membrane and space before reaching the parasite itself. One prediction from this model is that the switching occurs in a programmed way; whether there is a certain hierarchy in the switching between the genes of the *SICAvar* gene family is another question.

Plasmodium yoelii, P. chabaudi, P. berghei
Homologues of the **P. vivax vir** *gene family*
Homologues of the *P. vivax vir* family have not been found in *P. falciparum*; they do, however, exist in all three rodent malaria databases, where they have been termed *yir, cir* and *bir* for *P. yoelii, P. chabaudi* and *P. berghei* respectively (Table 14.1) (Janssen *et al.*, 2002). In *P. yoelii, yir* genes are located on approximately 75% of sub-telomeric *P. yoelii* contigs examined and >800 genes have been identified in the *P. yoelii* database (Carlton *et al.*, 2002). The genes are contiguous, with some contigs having up to five *vir* genes, situated between 1 and 8 kb apart (Carlton *et al.*, 2002). Phylogenetic analysis distinguishes between five subgroups of *yir*. Based on the current available sequence information for *P. berghei* and *P. chabaudi* a similar high copy number is expected for *bir* and *cir*. Each *yir* gene consists of a short first exon, a (700–900 bp) polymorphic second exon and a third (~90 bp), highly conserved exon. Intron size varies between 80 and 150 bp. The region spanning the splice junction of exons 2 and 3 is conserved both between the genes and between species. Sequences corresponding to this region are represented in both the *P. yoelii* and the *P. berghei* EST databases, indicating transcription as well as correct splicing. Expression of VIR in *P. vivax* has been demonstrated experimentally during intra-erythrocytic development in human (del Portillo *et al.*, 2001) as well as rodent infections (Cunningham *et al.*, manuscript in preparation). Immunolocalization of the YIR protein to the surface of *P. yoelii*-infected erythrocyte has recently been demonstrated (Cunningham *et al.*, manuscript in preparation). In addition, data mining of the *P. yoelii* expression databases indicates expression of YIR in sporozoites and gametocytes.

The discovery of direct homologues to a human malaria multigene family in rodent malaria will facilitate the study of the role of these proteins in immune evasion. The rodent model offers the potential for manipulation of the immune response through immune selection and the use of animals genetically modified (or knockout?) at immune response gene loci.

PHENOTYPIC/ANTIGENIC VARIATION OF MEROZOITES

After schizont rupture merozoites are directly exposed to the host immune system. Crucial for parasite survival is that merozoites avoid neutralization by the host defences as well as that they are able to identify and penetrate a suitable host cell. However, up to now there has been no clear evidence that merozoites undergo antigenic variation similar to that observed for the infected rbc. Most of the proteins identified on the surface of the merozoite seem to be coded for by single-copy genes and there is no indication for clonal variation of any of these proteins. The only evidence to date that merozoites are able to undergo clonal variation comes from work on the rodent parasite *P. yoelii*. In this parasite a

Table 14.1 *Plasmodium* multigene families thought to be important in antigenic variation. Basic characteristics for each gene family are indicated

Gene family	Number of genes	Parasite	Protein	Size (kDa)	Final destination (location) in parasite	Chromosomal location	Transcription
var	40–50 copies	*P. falciparum*	PfEMP1	200–350	Erythrocyte surface	Sub-telomeric Some internal	Early rings and trophozoites
rif	~200 copies	*P. falciparum*	RIFIN	35–44	Erythrocyte surface	Sub-telomeric	Late rings/early trophozoites Gametocytes
stevor	30–40 copies	*P. falciparum*	STEVOR	~30	Maurer's cleft	Sub-telomeric	Late trophozoites/early schizonts Gametocytes
vir	>600 copies	*P. vivax* *P. yoelii* *P. chabaudi* *P. berghei*	VIR YIR CIR BIR	~26–68 ~30–37	Reticulocyte membrane Erythrocyte membrane	Sub-telomeric	Rings, trophozoites and schizonts Gametocytes Sporozoites
SICAvar	>50 copies	*P. knowlesi*	SICA	180–225	Erythrocyte surface	Not known	Early rings and trophozoites
Py235	>14 copies	*P. yoelii* Homologues in *P. chabaudi* *P. berghei* *P. vivax* *P. falciparum*	PY235	235	Rhoptry	Sub-telomeric	Trophozoite, schizont, merozoite Sporozoite

235 kDa rhoptry protein (Py235) that has been implicated in host cell recognition is coded for by a multigene family. There are at least 14 distinct members of Py235 that can be identified in the *P. yoelii* genome database (Carlton *et al.*, 2002). *Py235* is distributed similarly throughout the genomes of both the YM and 17X lines of *P. yoelii* and like *var* appears to be located close to telomeres (Owen *et al.*, 1999). Unlike *var* though, *Py235* genes are not located on all chromosomes (Owen *et al.*, 1999). By contrast to *var*, a different method of transcriptional control appears to be utilized in the *Py235* multigene family. Single schizonts transcribe multiple members of *Py235* whereas an individual merozoite contains only a single transcript (Preiser *et al.*, 1999). Therefore, each progeny arising from a single schizont can express a different member of Py235. This form of clonal phenotypic variation confers two advantages to the parasite. First of all it gives the parasite an adaptive advantage, enabling it to adjust to changes in the host environment by ensuring that at least one merozoite is able to successfully invade a red blood cell. Secondly, this mechanism is an effective way of immune evasion, as immune responses to specific types of Py235 will only eliminate a proportion of the merozoites released from a schizont. A mechanism like this enhances the parasites' chances for survival and helps in establishing a chronic infection. It is speculated that Py235 may mediate merozoite antigenic and phenotypic variation by facilitating the ability of the merozoite to invade rbc of different ages as well as evade the host immune response. This hypothesis is supported by the demonstration that monoclonal antibodies against Py235, when passively transferred to mice, were able to protect mice from a subsequent infection with the virulent *P. yoelii* strain YM. During these experiments the invasion property of the virulent parasite was restricted to reticulocytes only, leading to an avirulent infection (Freeman *et al.*, 1980; Holder and Freeman, 1981; Oka *et al.*, 1984). The monoclonal anti-Py235 antibodies therefore only seem to prevent the invasion of mature rbc without influencing invasion of reticulocytes, implying that reticulocyte invasion is mediated by a different member of Py235 not affected by these antibodies.

An important question is whether the phenomenon described in *P. yoelii* is also applicable to other *Plasmodium* species. Homologues of Py235 have been identified in the genome of other rodent malaria parasites (Owen *et al.*, 1999; Preiser, unpublished observation) as well as *P. falciparum* (Rayner *et al.*, 2000, 2001; Taylor *et al.*, 2001; Triglia *et al.*, 2001), *P. vivax* (Galinski *et al.*, 1992, 2000; Keen *et al.*, 1994), and several simian malaria species (M.R.G., personal communication). As noted for the homologous *P. vivax* reticulocyte binding proteins -1 and -2 (Galinski *et al.*, 1992, 2000), the simian malaria species seem to have only two expressed homologues and, akin to at least one of the *P. falciparum* homologues (Rayner *et al.*, 2000), they are localized at the apical pole of the merozoite but not within in the rhoptries.

P. falciparum is able to utilize at least three different erythrocyte surface ligands to successfully invade erythrocytes (Dolan *et al.*, 1990, 1994; Sim *et al.*, 1994). While some *P. falciparum* parasites are only able to utilize one of these receptors, others can use more (Mitchell et al., 1986; Soubes et al., 1997). Most interestingly, some parasite clones have been shown to switch the invasion pathway they use when selective pressure is applied (Dolan et al., 1990) and field isolates also seem to adapt their invasion pathways (Okoyeh et al., 1999). It is thought that the switching is at least in part due to the activation of different members of the Erythrocyte Binding Like (EBL) gene family (Reed et al., 2000; Adams et al., 2001) and possibly the *Py235* homologues (Rayner et al., 2001). Expression of different members of these genes would lead to phenotypically distinct merozoites thereby enabling *P. falciparum* merozoites to adapt and possibly evade host immune pressure. There is currently no evidence that this involves the same mechanism as seen in *P. yoelii*.

ANTIGENIC DIVERSITY, ANOTHER WAY OF INCREASING THE REPERTOIRE OF GENES

Antigenic diversity of these multigene families is also another way of increasing the repertoire of genes. Antigenic diversity can be defined as the expression of antigenically different forms of parasite gene products by parasites of different genotype within any given isolate. This heterogeneity is maintained by genetic recombination, in addition to mutation (Walliker et al., 1975, 1987). The malaria parasite exists as a series of genetically and biologically distinct species and lines. Isolates collected in the wild from the natural host may represent complex mixtures of different lines and/or species of parasites (Thaithong et al., 1984). Analysis of cloned parasite lines, derived from such islolates, has allowed the recognition and definitive characterization of parasite biodiversity. As a result of intensive studies on the comparison of DNA and amino acid sequences from well-characterized parasite genes and gene products from different parasite isolates, antigenic diversity has been demonstrated not only amongst multigene families, but also in many different antigens from different parasite stages. These include the circumsporozoite protein (CSP) in mature sporozoites (Jongwutiwes et al., 1994), merozoite surface antigens MSP-1, MSP-2 and MSP-3 (Miller et al., 1993; Farnert et al., 1997; McColl and Anders, 1997; Rayner et al., 2002), the reticulocyte binding proteins (Rayner et al., submitted), rhoptry antigens (Howard et al., 1986) in the erythrocytic stage and gamete/gametocyte antigens and zygote/ookinete antigens in sexual stages (Foo et al., 1991). The diversification of antigens may confer on the parasite the ability to infect a host that has previously only been exposed to parasites of a different genotype, as the immunity directed at these parasites is species- and strain-specific.

CONCLUSIONS

The diversity of adhesion mechanisms used by intraerythrocytic stages of *P. falciparum* to ensure sequestration, and the plasticity of the evasion systems that the parasite has developed as a response to immune pressure in the host are great. Available data also show that the individual parasite tightly controls the exposure of its large, but not infinite, binding and antigenic repertoire. Examples of other complexities in the parasite and parasite–host biology are, among many others, the existence of redundant invasion pathways and the genomic plasticity to expand gene families in the face of immune pressure seemingly vital to the parasite.

REFERENCES

Adams, J. H., Blair, P. L., Kaneko, O. and Peterson, D. S. (2001) *Trends Parasitol.* 17, 297–299.
Aguiar, J. C., Albrecht, G. R., Cegielski, P., Greenwood, B. M., Jensen, J. B., Lallinger, G., Martinez, A., McGregor, I. A., Minjas, J. N., Neequaye, J. *et al.* (1992) *Am. J. Trop. Med. Hyg.* 47, 621–632.
Aley, S. B., Sherwood, J. A., Marsh, K., Eidelman, O. and Howard, R. J. (1986) *Parasitology* 92, 511–525.
Al-Khedery, B., Barnwell, J. W. and Galinski, M. R. (1999) *Mol. Cell* 3, 131–141.
Barnwell, J. W., Howard, R. J., Coon, H. G. and Miller, L. H. (1983a) *Infect. Immun.* 40, 985–994.
Barnwell, J. W., Howard, R. J. and Miller, L. H. (1983b) *Ciba Found. Symp.* 94, 117–136.
Barragan, A., Fernandez, V., Chen, Q., von Euler, A., Wahlgren, M. and Spillmann, D. (2000) *Blood* 95, 3594–3599.
Barragan, A., Kremsner, P. G., Weiss, W., Wahlgren, M. and Carlson, J. (1998) *Infect. Immun.* 66, 4783–4787.
Baruch, D. I., Ma, X. C., Singh, H. B., Bi, X., Pasloske, B. L. and Howard, R. J. (1997) *Blood* 90, 3766–3775.
Baruch, D. I., Pasloske, B. L., Singh, H. B., Bi, X., Ma, X. C., Feldman, M., Taraschi, T. F. and Howard, R. J. (1995) *Cell* 82, 77–87.
Biggs, B. A., Anders, R. F., Dillon, H. E., Davern, K. M., Martin, M., Petersen, C. and Brown, G. V. (1992) *J. Immunol.* 149, 2047–2054.
Biggs, B. A., Gooze, L., Wycherley, K., Wollish, W., Southwell, B., Leech, J. H. and Brown, G. V. (1991) *Proc. Natl Acad. Sci. USA* 88, 9171–9174.
Bowman, S., Lawson, D., Basham, D., Brown, D., Chillingworth, T., Churcher, C. M., Craig, A., Davies, R. M., Devlin, K., Feltwell, T., Gentles, S., Gwilliam, R., Hamlin, N., Harris, D., Holroyd, S., Hornsby, T., Horrocks, P., Jagels, K., Jassal, B., Kyes, S., McLean, J., Moule, S., Mungall, K., Murphy, L., Barrell, B. G. *et al.* (1999) *Nature* 400, 532–538.
Brown, K. N. (1971) *Trans. R. Soc. Trop. Med. Hyg.* 65, 6.
Brown, K. N. (1973) *Nature* 242, 49–50.
Brown, K. N. and Brown, I. N. (1965) *Nature* 208, 1286–1288.
Brown, K. N. and Hills, L. A. (1974) *Trans. R. Soc. Trop. Med. Hyg.* 68, 139–142.
Buffet, P. A., Gamain, B., Scheidig, C., Baruch, D., Smith, J. D., Hernandez-Rivas, R., Pouvelle, B., Oishi, S., Fujii, N., Fusai, T., Parzy, D., Miller, L. H., Gysin, J. and Scherf, A. (1999) *Proc. Natl Acad. Sci. USA* 96, 12743–12748.

Bull, P. C., Lowe, B. S., Kortok, M. and Marsh, K. (1999) *Infect. Immun.* 67, 733–739.

Bull, P. C., Lowe, B. S., Kortok, M., Molyneux, C. S., Newbold, C. I. and Marsh, K. (1998) *Nat. Med.* 4, 358–360.

Carlson, J., Helmby, H., Hill, A. V., Brewster, D., Greenwood, B. M. and Wahlgren, M. (1990a) *Lancet* 336, 1457–1460.

Carlson, J., Holmquist, G., Taylor, D. W., Perlmann, P. and Wahlgren, M. (1990b) *Proc. Natl Acad. Sci. USA* 87, 2511–2515.

Carlton, J. M. et al. (2002) *Nature* 419, 512–519.

Chen, Q., Fernandez, V., Sundstrom, A., Schlichtherle, M., Datta, S., Hagblom, P. and Wahlgren, M. (1998) *Nature* 394, 392–395.

Chen, Q., Heddini, A., Barragan, A., Fernandez, V., Pearce, S. F. and Wahlgren, M. (2000) *J. Exp. Med.* 192, 1–10.

Cheng, Q., Cloonan, N., Fischer, K., Thompson, J., Waine, G., Lanzer, M. and Saul, A. (1998) *Mol. Biochem. Parasitol.* 97, 161–176.

Cox, H. W. (1962) *J. Protozool.* 9, 114–118.

Crabb, B. S., Cooke, B. M., Reeder, J. C., Waller, R. F., Caruana, S. R., Davern, K. M., Wickham, M. E., Brown, G. V., Coppel, R. L. and Cowman, A. F. (1997) *Cell* 89, 287–296.

Darkin-Rattray, S. J., Gurnett, A. M., Myers, R. W., Dulski, P. M., Crumley, T. M., Allocco, J. J., Cannova, C., Meinke, P. T., Colletti, S. L., Bednarek, M. A., Singh, S. B., Goetz, M. A., Dombrowski, A. W., Polishook, J. D. and Schmatz, D. M. (1996) *Proc. Natl Acad. Sci. USA* 93, 13143–13147.

David, P. H., Hommel, M., Miller, L. H., Udeinya, I. J. and Oligino, L. D. (1983) *Proc. Natl Acad. Sci. USA* 80, 5075–5079.

Day, K. P., Karamalis, F., Thompson, J., Barnes, D. A., Peterson, C., Brown, H., Brown, G. V. and Kemp, D. J. (1993) *Proc. Natl Acad. Sci. USA* 90, 8292–8296.

Degen, R., Weiss, N. and Beck, H. P. (2000) *Exp. Parasitol.* 95, 113–121.

Deitsch, K. W., Calderwood, M. S. and Wellems, T. E. (2001) *Nature* 412, 875–876.

del Portillo, H. A., Fernandez-Becerra, C., Bowman, S., Oliver, K., Preuss, M., Sanchez, C. P., Schneider, N. K., Villalobos, J. M., Rajandream, M. A., Harris, D., Pereira da Silva, L. H., Barrell, B. and Lanzer, M. (2001) *Nature* 410, 839–842.

Dolan, S. A., Miller, L. H. and Wellems, T. E. (1990) *J. Clin. Invest.* 86, 618–624.

Dolan, S. A., Proctor, J. L., Alling, D. W., Okubo, Y., Wellems, T. E. and Miller, L. H. (1994) *Mol. Biochem. Parasitol.* 64, 55–63.

Duffy, M. F., Brown, G. V., Basuki, W., Krejany, E. O., Noviyanti, R., Cowman, A. F. and Reeder, J. C. (2002) *Mol. Microbiol.* 43, 1285–1293.

Eaton, M. D. (1938) *J. Exp. Med.* 67, 857–873.

Farnert, A., Snounou, G., Rooth, I. and Bjorkman, A. (1997) *Am. J. Trop. Med. Hyg.* 56, 538–547.

Fernandez, V., Chen, Q., Sundstrom, A., Scherf, A., Hagblom, P. and Wahlgren, M. (2002) *Mol. Biochem. Parasitol.* 121, 195–203.

Fernandez, V., Hommel, M., Chen, Q., Hagblom, P. and Wahlgren, M. (1999) *J. Exp. Med.* 190, 1393–1404.

Fernandez, V., Treutiger, C. J., Nash, G. B. and Wahlgren, M. (1998) *Infect. Immun.* 66, 2969–2975.

Fischer, K., Horrocks, P., Preuss, M., Wiesner, J., Wunsch, S., Camargo, A. A. and Lanzer, M. (1997) *Mol. Cell Biol.* 17, 3679–3686.

Foo, A., Carter, R., Lambros, C., Graves, P., Quakyi, I., Targett, G. A., Ponnudurai, T. and Lewis, G. E., Jr (1991) *Am. J. Trop. Med. Hyg.* 44, 623–631.
Forsyth, K. P., Philip, G., Smith, T., Kum, E., Southwell, B. and Brown, G. V. (1989) *Am. J. Trop. Med. Hyg.* 41, 259–265.
Freeman, R. R., Trejdosiewicz, A. J. and Cross, G. A. (1980) *Nature* 284, 366–368.
Freitas-Junior, L. H., Bottius, E., Pirrit, L. A., Deitsch, K. W., Scheidig, C., Guinet, F., Nehrbass, U., Wellems, T. E. and Scherf, A. (2000) *Nature* 407, 1018–1022.
Galinski, M. R., Medina, C. C., Ingravallo, P. and Barnwell, J. W. (1992) *Cell* 69, 1213–1226.
Galinski, M. R., Xu, M. and Barnwell, J. W. (2000) *Mol. Biochem. Parasitol.* 108, 257–262.
Gardiner, D. L., Holt, D. C., Thomas, E. A., Kemp, D. J. and Trenholme, K. R. (2000) *Mol. Biochem. Parasitol.* 110, 33–41.
Gardner, M. J. *et al.* (2002) *Nature* 419, 498–511.
Gardner, M. J., Tettelin, H., Carucci, D. J., Cummings, L. M., Aravind, L., Koonin, E. V., Shallom, S., Mason, T., Yu, K., Fujii, C., Pederson, J., Shen, K., Jing, J., Aston, C., Lai, Z., Schwartz, D. C., Pertea, M., Salzberg, S., Zhou, L., Sutton, G. G., Clayton, R., White, O., Smith, H. O., Fraser, C. M., Hoffman, S. L. *et al.* (1998) *Science* 282, 1126–1132.
Gilks, C. F., Walliker, D. and Newbold, C. I. (1990) *Parasite Immunol.* 12, 45–64.
Gommers-Ampt, J. H., Van Leeuwen, F., de Beer, A. L., Vliegenthart, J. F., Dizdaroglu, M., Kowalak, J. A., Crain, P. F. and Borst, P. (1993) *Cell* 75, 1129–1136.
Handunnetti, S. M., Mendis, K. N. and David, P. H. (1987) *J. Exp. Med.* 165, 1269–1283.
Helmby, H., Cavelier, L., Pettersson, U. and Wahlgren, M. (1993) *Infect. Immun.* 61, 284–288.
Hernandez-Rivas, R., Mattei, D., Sterkers, Y., Peterson, D. S., Wellems, T. E. and Scherf, A. (1997) *Mol. Cell Biol.* 17, 604–611.
Holder, A. A. and Freeman, R. R. (1981) *Nature* 294, 361–364.
Hommel, M., David, P. H. and Oligino, L. D. (1983) *J. Exp. Med.* 157, 1137–1148.
Horrocks, P., Pinches, R., Kriek, N. and Newbold, C. (2002) *Int. J. Parasitol.* 32, 1203.
Howard, R. J. and Barnwell, J. W. (1984) *Parasitology* 88, 27–36.
Howard, R. J., Barnwell, J. W. and Kao, V. (1983) *Proc. Natl Acad. Sci. USA* 80, 4129–4133.
Howard, R. J., McBride, J. S., Aley, S. B. and Marsh, K. (1986) *Parasite Immunol.* 8, 57–68.
James, S. P., Nicol, V. D. and Shute, P. G. (1932) *Proc. R. Soc. Med.* 25, 1153–1186.
Janssen, C. S., Barrett, M. P., Turner, C. M. and Phillips, R. S. (2002) *Proc. R . Soc. Lond. B Biol. Sci.* 269, 431–436.
Jeffery, G. M. (1966) *Bull. WHO* 35, 873–882.
Jongwutiwes, S., Tanabe, K., Hughes, M. K., Kanbara, H. and Hughes, A. L. (1994) *Am. J. Trop. Med. Hyg.* 51, 659–668.
Kaviratne, M., Khan, S., Jarra, W. and Preiser, P. R. (2002) *Eukaryotic Cell* 1, 926–935.
Keen, J. K., Sinha, K. A., Brown, K. N. and Holder, A. A. (1994) *Mol. Biochem. Parasitol.* 65, 171–177.
Kiefer, M. C., Crawford, K. A., Boley, L. J., Landsberg, K. E., Gibson, H. L., Kaslow, D. C. and Barr, P. J. (1996) *Mol. Biochem. Parasitol.* 78, 55–65.

Kyes, S., Pinches, R. and Newbold, C. (2000) *Mol. Biochem. Parasitol.* 105, 311–315.
Kyes, S. A., Rowe, J. A., Kriek, N. and Newbold, C. I. (1999) *Proc. Natl Acad. Sci. USA* 96, 9333–9338.
Leech, J. H., Barnwell, J. W., Miller, L. H. and Howard, R. J. (1984) *J. Exp. Med.* 159, 1567–1575.
Lekana Douki, J. B., Traore, B., Costa, F. T., Fusai, T., Pouvelle, B., Sterkers, Y., Scherf, A. and Gysin, J. (2002) *Blood* 100, 1478–1483.
Limpaiboon, T., Shirley, M. W., Kemp, D. J. and Saul, A. (1991) *Mol. Biochem. Parasitol.* 47, 197–206.
Magowan, C., Wollish, W., Anderson, L. and Leech, J. (1988) *J. Exp. Med.* 168, 1307–1320.
Marsh, K. and Howard, R. J. (1986) *Science* 231, 150–153.
Marsh, K., Otoo, L., Hayes, R. J., Carson, D. C. and Greenwood, B. M. (1989) *Trans. R. Soc. Trop. Med. Hyg.* 83, 293–303.
McColl, D. J. and Anders, R. F. (1997) *Mol. Biochem. Parasitol.* 90, 21–31.
McLean, S. A., Pearson, C. D. and Phillips, R. S. (1986) *Parasite Immunol.* 8, 415–424.
Miller, L. H., Good, M. F. and Milon, G. (1994) *Science* 264, 1878–1883.
Miller, L. H., Roberts, T., Shahabuddin, M. and McCutchan, T. F. (1993) *Mol. Biochem. Parasitol.* 59, 1–14.
Mitchell, G. H., Hadley, T. J., McGinniss, M. H., Klotz, F. W. and Miller, L. H. (1986) *Blood* 67, 1519–1521.
Noviyanti, R., Brown, G. V., Wickham, M. E., Duffy, M. F., Cowman, A. F. and Reeder, J. C. (2001) *Mol. Biochem. Parasitol.* 114, 227–237.
Oh, S. S., Voigt, S., Fisher, D., Yi, S. J., LeRoy, P. J., Derick, L. H., Liu, S. and Chishti, A. H. (2000) *Mol. Biochem. Parasitol.* 108, 237–247.
Oka, M., Aikawa, M., Freeman, R. R., Holder, A. A. and Fine, E. (1984) *Am. J. Trop. Med. Hyg.* 33, 342–346.
Okoyeh, J. N., Pillai, C. R. and Chitnis, C. E. (1999) *Infect. Immun.* 67, 5784–5791.
Owen, C. A., Sinha, K. A., Keen, J. K., Ogun, S. A. and Holder, A. A. (1999) *Mol. Biochem. Parasitol.* 99, 183–192.
Patankar, S., Munasinghe, A., Shoaibi, A., Cummings, L. M. and Wirth, D. F. (2001) *Mol. Biol. Cell* 12, 3114–3125.
Peters, J., Fowler, E., Gatton, M., Chen, N., Saul, A. and Cheng, Q. (2002) *Proc. Natl Acad. Sci. USA* 99, 10689–10694.
Peterson, D. S., Miller, L. H. and Wellems, T. E. (1995) *Proc. Natl Acad. Sci. USA* 92, 7100–7104.
Pollack, Y., Kogan, N. and Golenser, J. (1991) *Exp. Parasitol.* 72, 339–344.
Preiser, P. R., Jarra, W., Capiod, T. and Snounou, G. (1999) *Nature* 398, 618–622.
Rayner, J. C., Galinski, M. R., Ingravallo, P. and Barnwell, J. W. (2000) *Proc. Natl Acad. Sci. USA* 97, 9648–9653.
Rayner, J. C., Vargas-Serrato, E., Huber, C. S., Galinski, M. R. and Barnwell, J. W. (2001) *J. Exp. Med.* 194, 1571–1581.
Rayner, J. C., Corredor, U., Feldman, D., Ingravello, P., Iderabdullah, F., Galinski, M. R. and Barnwell, J. W. (2002) *Parasitology* 125, 393–405.
Reed, M. B., Caruana, S. R., Batchelor, A. H., Thompson, J. K., Crabb, B. S. and Cowman, A. F. (2000) *Proc. Natl Acad. Sci. USA* 97, 7509–7514.
Reeder, J. C., Cowman, A. F., Davern, K. M., Beeson, J. G., Thompson, J. K., Rogerson, S. J. and Brown, G. V. (1999) *Proc. Natl Acad. Sci. USA* 96, 5198–5202.

Reeder, J. C., Hodder, A. N., Beeson, J. G. and Brown, G. V. (2000) *Infect. Immun.* 68, 3923–3926.
Roberts, D. J., Craig, A. G., Berendt, A. R., Pinches, R., Nash, G., Marsh, K. and Newbold, C. I. (1992) *Nature* 357, 689–692.
Rowe, J. A., Moulds, J. M., Newbold, C. I. and Miller, L. H. (1997) *Nature* 388, 292–295.
Rubio, J. P., Thompson, J. K. and Cowman, A. F. (1996) *EMBO J.* 15, 4069–4077.
Saul, A. (1999) *Parasitol. Today* 15, 455–457.
Scherf, A., Hernandez-Rivas, R., Buffet, P., Bottius, E., Benatar, C., Pouvelle, B., Gysin, J. and Lanzer, M. (1998) *EMBO J.* 17, 5418–5426.
Sim, B. K., Chitnis, C. E., Wasniowska, K., Hadley, T. J. and Miller, L. H. (1994) *Science* 264, 1941–1944.
Smith, J. D., Craig, A. G., Kriek, N., Hudson-Taylor, D., Kyes, S., Fagen, T., Pinches, R., Baruch, D. I., Newbold, C. I. and Miller, L. H. (2000b) *Proc. Natl Acad. Sci. USA* 97, 1766–1771.
Smith, J. D., Kyes, S., Craig, A. G., Fagan, T., Hudson-Taylor, D., Miller, L. H., Baruch, D. I. and Newbold, C. I. (1998) *Mol. Biochem. Parasitol.* 97, 133–148.
Smith, J. D., Subramanian, G., Gamain, B., Baruch, D. I. and Miller, L. H. (2000a) *Mol. Biochem. Parasitol.* 110, 293–310.
Soubes, S. C., Wellems, T. E. and Miller, L. H. (1997) *Exp. Parasitol.* 86, 79–83.
Stanley, H. A. and Reese, R. T. (1986) *Proc. Natl Acad. Sci. USA* 83, 6093–6097.
Su, X. Z., Heatwole, V. M., Wertheimer, S. P., Guinet, F., Herrfeldt, J. A., Peterson, D. S., Ravetch, J. A. and Wellems, T. E. (1995) *Cell* 82, 89–100.
Sutherland, C. J. (2001) *Mol. Biochem. Parasitol.* 113, 331–335.
Taylor, H. M., Kyes, S. A., Harris, D., Kriek, N. and Newbold, C. I. (2000a) *Mol. Biochem. Parasitol.* 105, 13–23.
Taylor, H. M., Kyes, S. A. and Newbold, C. I. (2000b) *Mol. Biochem. Parasitol.* 110, 391–397.
Taylor, H. M., Triglia, T., Thompson, J., Sajid, M., Fowler, R., Wickham, M. E., Cowman, A. F. and Holder, A. A. (2001) *Infect. Immun.* 69, 3635–3645.
Thaithong, S., Beale, G. H., Fenton, B., McBride, J., Rosario, V., Walker, A. and Walliker, D. (1984) *Trans. R. Soc. Trop. Med. Hyg.* 78, 242–245.
Thompson, J. K., Rubio, J. P., Caruana, S., Brockman, A., Wickham, M. E. and Cowman, A. F. (1997) *Mol. Biochem. Parasitol.* 87, 49–60.
Trenholme, K. R., Gardiner, D. L., Holt, D. C., Thomas, E. A., Cowman, A. F. and Kemp, D. J. (2000) *Proc. Natl Acad. Sci. USA* 97, 4029–4033.
Treutiger, C. J., Hedlund, I., Helmby, H., Carlson, J., Jepson, A., Twumasi, P., Kwiatkowski, D., Greenwood, B. M. and Wahlgren, M. (1992) *Am. J. Trop. Med. Hyg.* 46, 503–510.
Triglia, T., Thompson, J., Caruana, S. R., Delorenzi, M., Speed, T. and Cowman, A. F. (2001) *Infect. Immun.* 69, 1084–1092.
Udeinya, I. J., Miller, L. H., McGregor, I. A. and Jensen, J. B. (1983) *Nature* 303, 429–431.
Udomsangpetch, R., Aikawa, M., Berzins, K., Wahlgren, M. and Perlmann, P. (1989) *Nature* 338, 763–765.
Urban, B. C., Ferguson, D. J., Pain, A., Willcox, N., Plebanski, M., Austyn, J. M. and Roberts, D. J. (1999) *Nature* 400, 73–77.
Vazquez-Macias, A., Martinez-Cruz, P., Castaneda-Patlan, M. C., Scheidig, C., Gysin, J., Scherf, A. and Hernandez-Rivas, R. (2002) *Mol. Microbiol.* 45, 155–167.

Wahlgren, M., Fernandez, V., Chen, Q., Svard, S. and Hagblom, P. (1999) *Cell* 96, 603–606.

Waller, K. L., Cooke, B. M., Nunomura, W., Mohandas, N. and Coppel, R. L. (1999) *J. Biol. Chem.* 274, 23808–23813.

Walliker, D., Carter, R. and Sanderson, A. (1975) *Parasitology* 70, 19–24.

Walliker, D., Quakyi, I. A., Wellems, T. E., McCutchan, T. F., Szarfman, A., London, W. T., Corcoran, L. M., Burkot, T. R. and Carter, R. (1987) *Science* 236, 1661–1666.

Weber, J. L. (1988) *Mol. Biochem. Parasitol.* 29, 117–124.

Winter, G., Chen, Q., Flick, K., Kremsner, P., Fernandez, V. and Wahlgren, M. (2003) *Mol. Biochem. Parasitol.* 127, 179–191.

Wolffe, A. P. and Guschin, D. (2000) *J. Struct. Biol.* 129, 102–122.

15

ANTIGENIC VARIATION IN *BORRELIA*: RELAPSING FEVER AND LYME BORRELIOSIS

Alan G. Barbour

INTRODUCTION

Relapsing fever (RF) was the first recognized example of antigenic variation (Felsenfeld, 1971). Clinical descriptions of relapsing fever date to ancient Greece; no other type of infection is characterized by three or more episodes of high fever separated by week-long periods of comparative well-being. The typical pattern of illness followed by recovery followed by a relapse attracted the attention of early immunologists like Ehrlich and Metchnikoff (Russell, 1936). A century ago they realized that the relapse phenomenon offered clues to the specificity of the vertebrates' adaptive immune system. Most of the key features of the biology of the infection were described in the first decades of the last century. With the technical advances of the second half of the twentieth century, understanding of antigenic variation and disease pathogenesis at the molecular level was possible.

Untreated RF usually lasts no more than a few weeks and leads to either death in about 1 in 10 patients or complete recovery in the survivors (Barbour, 1999a). The antibody response to the large numbers of spirochetes circulating in the blood during RF applies a powerful selection on the population of bacteria. These pathogens have responded to the host's immune system with the tactic of multiphasic antigenic variation, the consequence of which for the host are recurrences of fever. The large repertoire of polymorphic surface proteins occurs within individual lineages of RF *Borrelia* spp. The variation in surface proteins may also serve the pathogen for niche selection, which in this context means the differential localization of the bacteria in different tissues according to bacterial phenotype.

Lyme borreliosis (LB), sometimes called Lyme disease, may also have afflicted humans for centuries (Barbour, 1998). LB usually starts as a localized skin disorder instead of heavy bacteraemia. When bacteraemia occurs during LB the density of spirochetes is orders of magnitude lower than in relapsing fever. The course of LB infection is more indolent and plays out over months to years in the skin, the joints, the central nervous system and other organs. There is antigenic variation of at least one type of surface protein within clonal populations of LB agents, but the allele for another highly polymorphic surface protein exists in a single copy in the spirochete's genome. In this case, the heterogeneity is manifested by groups or communities of strains that share vectors and reservoirs within a particular environment.

In both RF and LB agents there are prominent changes in major surface proteins of the spirochetes during the spirochete's life cycle. These phenotypic variations arise through differential gene expression, not DNA rearrangements or gene conversion and they are mainly associated with specific adaptions for residence in different hosts or for transmission between hosts. An example is the shift from expression of the OspA protein within the unfed tick to expression of OspC in the tick after its blood meal (Schwan *et al.*, 1995). This type of variation in surface proteins of *Borrelia* spp. will not be reviewed in depth here. Instead, I focus on variation in surface proteins of RF and LB *Borrelia* spp. that occurs through DNA rearrangements or gene conversions in a clonal population and that serves for evasion of the immune response and for niche selection within individual vertebrate hosts.

Not surprisingly, there are also differences in antigenic composition between strains of *Borrelia* spp. Such antigenic differences between separate lineages are not considered as 'antigenic variation' by a criterion requiring clonality, but I propose that intra- and inter-strain antigenic differences are more comparable than has been assumed. If simultaneous or sequential infections of the same vertebrate host by different strains of the same pathogen occur and if the susceptibility to a second or third infection is a consequence of poor antigenic cross-reactivity between these strains, then this is effectively the same phenomenon as sequential expression of different antigens by an individual strain. When we shift the focus from a single strain with a large repertoire of polymorphic alleles to a group of organisms, each of which has only one version instead of multiple versions of the allele in its cell, then for the latter case we need to consider and characterize the community of strains in a given area of disease transmission.

ECOLOGY AND EPIDEMIOLOGY

RF occurs in two forms: louse-borne and tick-borne (Barbour, 1999b). Both forms of RF have had global distributions. Isolated cases and epidemics have occurred in equatorial, temperate and the sub-arctic regions in both hemispheres. The only agent of louse-borne RF is *Borrelia recurrentis*. *B. recurrentis* is transmitted by the body louse, *Pediculus humanus*, which only feeds on humans. Only *B. recurrentis* occurs in epidemics, often with louse-borne typhus. In the twentieth century there were millions of cases of louse-borne RF, mainly around the times of the world wars. The mortality of untreated louse-borne RF is about 10%. Because *B. recurrentis* has only been cultivated recently and does not readily infect other animals besides humans (Cutler *et al.*, 1994), there is comparatively less data on the antigenic variation by this organism.

There are several species that cause tick-borne RF (see below). These are transmitted by soft-bodied (argasid) ticks of the genus *Ornithodoros* (Figure 15.1). These types of ticks tend to remain close to the habitations of animals and humans (Felsenfeld, 1971). *Ornithodoros* ticks usually feed on just one type of animal, squirrels, for example, during their lifetimes; the blood feeding itself takes less than an hour and usually occurs at night. The vertebrate reservoirs of the RF borrelias include a variety of mammals, including humans, but most commonly they are rodents (Barbour and Hayes, 1986).

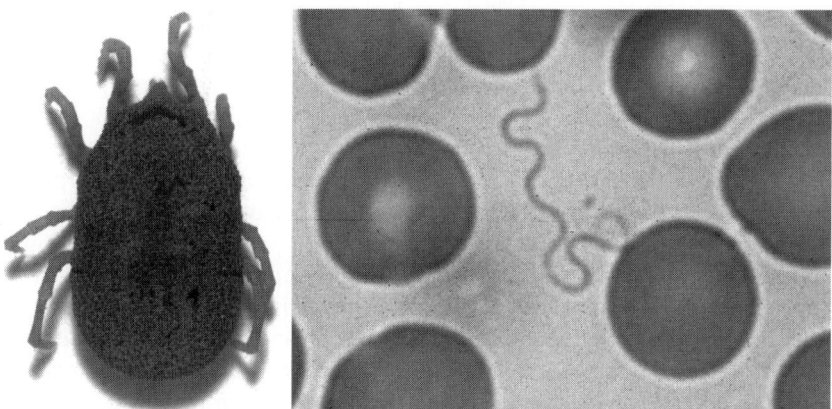

Figure 15.1 Adult *Ornithodoros hermsi* soft (left) and *Borrelia hermsii* spirochete in the Wright stained-blood smear of a patient with tick-borne relapsing fever (right).

Transmission of LB to humans occurs predominantly in temperate regions of North America and Eurasia (reviewed by Barbour and Fish, 1993). While LB is rarely, if ever, fatal, cumulative morbidity from untreated, persistent infection has been considerable. Species of *Borrelia* that cause LB have been found in hard (ixodid) ticks *Ixodes ricinus*, *Ixodes scapularis*, *Ixodes pacificus* and *Ixodes persulcatus*. A characteristic of these ticks is that they generally feed on a variety of hosts, including large and small mammals, birds and reptiles. The immature stages of larva and nymph feed on small mammals, ground-feeding birds and lizards, and the adults feed on large free-roaming animals like deer or cattle. The tick vectors of LB generally quest in grass or shrubs for passing animals, instead of residing in the burrow or nest of their host like the RF vectors. *I. scapularis* and other vectors may be distributed over the range of the host, including over large distances by migratory birds.

In enzootic areas in northeastern United States, the major reservoirs for *Borrelia burgdorferi*, such as the white-footed mouse, *Peromyscus leucopus*, become infected during the spring, when the nymphal ticks feed. The mice generally are bacteraemic for 2–3 weeks, but their tissues may remain infected for the rest of their lives. The mice are continually infested with nymphal ticks, which have a prevalence of infection with *B. burgdorferi* of about 25%. Even though a mouse had been infected previously, it can be infected with another strain. Mixed infections with two, three, or four different strains have been found in ticks and patients (Guttman *et al.*, 1996; Liveris *et al.*, 1999; Seinost *et al.*, 1999a).

CLINICAL MANIFESTATIONS AND PATHOLOGY

The clinical hallmark of RF is the sudden onset of two or more episodes of fever of 39 ° C or more, spaced by afebrile periods of 4–14 days (reviewed in Barbour, 1999a). The fever usually starts within a few days of exposure to argasid ticks; the actual tick bites are not usually noted by patients. Headache, neck stiffness, arthralgia, myalgia and other constitutional symptoms accompany the fevers. The first fever episode ends by crisis, which is characterized by rigors, hyperpyrexia and elevations of pulse and blood pressure over about 15 to 30 minutes. This phase is followed by a few to several hours of profuse diaphoresis, falling temperature and hypotension. During and shortly before the fever peak there are typically 10^5-10^7 extracellular spirochetes per millilitre of blood, which are easily detected by microscopic examination of smears of unconcentrated blood (Figure 15.1). The presence of spirochetes in the blood is called spirochetaemia. Between fevers spirochetes are not visually detectable in the blood. Untreated louse-borne RF and tick-borne RF have mortality rates of 10–70% and 4–10%, respectively.

LB, in contrast, is rarely fatal but can result in considerable morbidity from late manifestations; infections may last months to years (reviewed by Steere, 2001). Between 1 and 3 weeks after a tick bite the majority of infected individuals develop a flat, non-painful red rash at the site of the bite, often with central clearing or concentric ring pattern. This rash is called erythema migrans and may be accompanied by a low-grade fever, mild constitutional symptoms and lymphadenopathy. As the spirochetes move peripherally, the rash expands over several days to a few weeks to diameters of several centimetres. Spirochetes can be cultured from the blood or detected by polymerase chain reaction for a few days to a few weeks during the early phase of infection, but the spirochetes are too few to be directly observed in the blood.

Organs and tissues seeded with *B. burgdorferi* during the spirochetaemia of early infection include the heart, large joints, the peripheral and central nervous system and distant parts of the skin (Sigal, 1997). This stage of the disease is analogous to secondary syphilis. During early dissemination patients may have carditis, meningoencephalitis, meningopolyneuritis, polyarthritis and/or multiple erythema migrans lesions. Almost all patients have detectable antibodies to *B. burgdorferi* during this stage. As many as 50% of untreated patients with early LB in the United States subsequently have a remitting oligoarthritis that lasts several months to years (Steere, 1997). Untreated patients with Lyme arthritis have *B. burgdorferi* in the synovial tissue by polymerase chain reaction. The spirochetes are seldom seen by histologic examination or cultured at this stage, though.

Borrelias are primarily extracellular organisms. They move through or between endothelial cells as they leave the blood for tissues, but they do not appear to proliferate in these or phagocytic cells (Comstock and Thomas, 1989, 1991). The spirochetes are usually found in perivascular locations, sometimes as tangles of bacteria. In autopsy cases of RF and experimentally infected animals, spirochetes are commonly found in the spleen, liver, brain, eye and kidney (Southern and Sanford, 1969; Cadavid and Barbour, 1998). Unlike the LB agents, the RF *Borrelia* spp. do not localize in skin tissues. RF is primarily an infection of the blood. Involvement of other organs, such as the central nervous system, the eye and liver, is the consequence of large numbers of intravascular organisms.

MORPHOLOGY AND PHYSIOLOGY

Borrelias, like other spirochetes, are filamentous bacteria with more than three waves or helices. They have both inner and outer cell membranes (Figure 15.2). Members of the genus *Borrelia* have widths of approximately 0.2 μm and lengths of 10–30 μm; they can be visualized by dark field or phase contrast microscopy but not standard light microscopy without special stains. Several flagellas, which are inserted at each end, traverse the length of the cell through the periplasmic space of *Borrelia* spp. (Holt, 1978). Unlike the flagellas of other bacteria, spirochetal flagellas are not exposed to the host's antibodies or tissues. The major structural flagellar protein is FlaB (Barbour *et al.*, 1986).

Known species of *Borrelia* are host-associated microorganisms and are found either in an arthropod, usually a tick, or in a mammal or bird and not free-living in the environment. The complete genome of *B. burgdorferi*, an agent of LB, revealed a limited biosynthetic potential that is consistent with its parasitic lifestyle and complex nutritional requirements for cultivation in the laboratory (Fraser *et al.*, 1997). Some *Borrelia* species have yet to be cultivated in serial

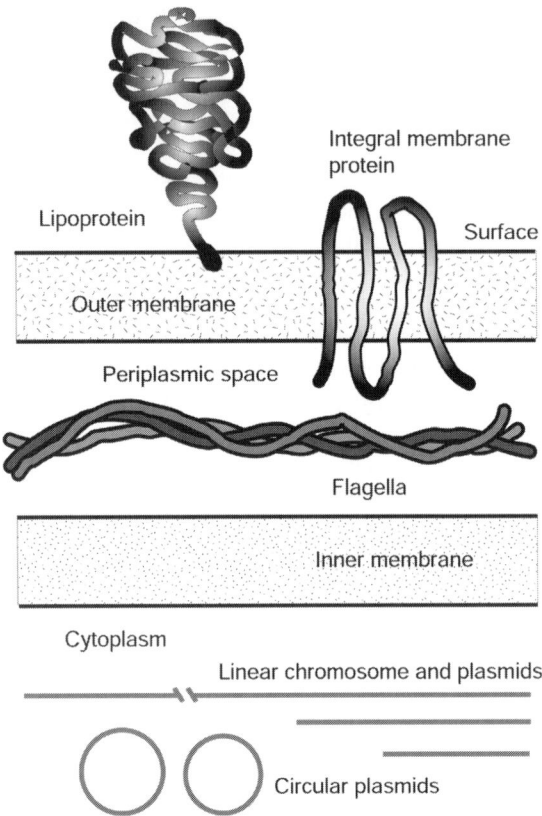

Figure 15.2 Cellular organization of *Borrelia* spirochetes.

passage outside of a natural or experimental host animal. Many *Borrelia* spp. lose their capacity to infect animals after a few passages in culture medium in the laboratory (Barbour and Hayes, 1986).

Borrelia spirochetes do not have lipopolysaccharides with lipid A (Takayama *et al.*, 1987) and there is little evidence they have other potent exotoxins. The outer membranes of *Borrelia* spp. have few integral membrane proteins in comparison to *Escherichia coli*, another bacterium with two membranes (Walker *et al.*, 1991; Radolf *et al.*, 1994). *Borrelia* cells have instead a large number of lipoproteins anchored in the fluid outer membrane by N-terminal fatty acids (Figure 15.2) (Bergstrom *et al.*, 1989; Brandt *et al.*, 1990; Carter *et al.*, 1994). *Borrelia* lipoproteins are distinct from the lipid-modified proteins of trypanosomes and other eukaryotes, which are anchored in the membrane at lipids at their C-termini instead. *Borrelia* lipoproteins can be aggregated in the membrane by a cross-linking ligand, such as antibodies, leading to the appearance of patches of antigens on the cells (Barbour *et al.*, 1983a).

PHYLOGENY AND GENETICS

The genus *Borrelia* represents a deep division among the bacterial phylum of spirochetes (Paster *et al.*, 1991; Fukunaga *et al.*, 1996; Ras *et al.*, 1996). On the basis of DNA sequences, members of the genus are more closely related to treponemes, such as *Treponema pallidum*, than they are to leptospires, a large group of animal and human pathogens. Characteristics of *Borrelia* spp. that together serve to distinguish the genus from other spirochetes are a dependence on arthropods for transmission, GC contents of about 30%, multiple flagella and a tolerance of oxygen.

Phylogenetic trees of *Borrelia* species that are based on DNA sequences have confirmed in many instances a taxonomy based mainly on biological characteristics, such as the associations of certain species with certain types of ticks (reviewed by Barbour, 2001) (Figure 15.3). DNA hybridization studies and

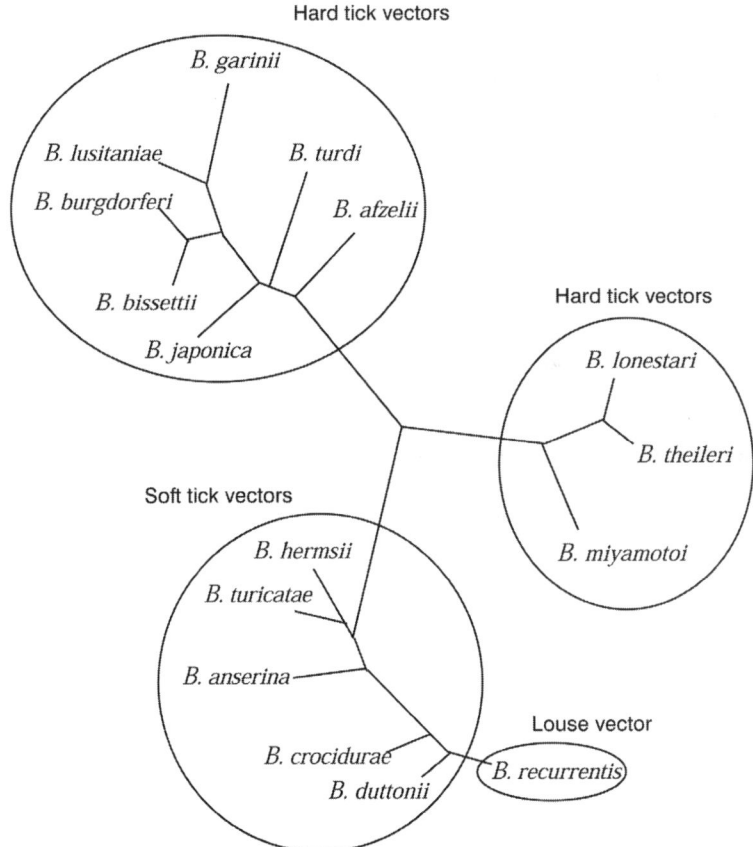

Figure 15.3 Unrooted neighbour-joining distance phylogram of the *flaB* sequences of selected *Borrelia* spp. and the associations with their arthropod vectors.

subsequent DNA sequence analysis of ribosomal RNA and other genes established that *Borrelia* spp. fall into three major groups: the RF group, which are primarily transmitted by soft-bodied or argasid ticks; the LB group, which are transmitted by hard-bodied or ixodid ticks; and a third group that has genetic similarity to RF species but is transmitted by ixodid ticks instead of argasid ticks.

The several tick-borne RF species are further grouped into Old World species, such as *Borrelia duttonii*, *Borrelia persica* and *Borrelia crocidurae* and New World species, such as *Borrelia hermsii*, *Borrelia turicatae* and *Borrelia venezuelensis* (Barbour, 2001). The sole louse-borne species, *B. recurrentis*, clusters with the Old World species of tick-borne species; the bird pathogen, *Borrelia anserina*, also groups with the RF species. The LB group includes *B. burgdorferi*, *Borrelia afzelii* and *Borrelia garinii*, which are known to cause LB, as well as several other species, such as *Borrelia bissettii* and *Borrelia turdi*, which have not been associated with human infection. *B. afzelii* and *B. garinii* have been found on the Eurasian continent but not in North America; *B. burgdorferi* occurs on both continents. The third group includes *Borrelia miyamotoi* and uncultivable organisms such as *Borrelia theileri* and *Borrelia lonestari* (Barbour, 2001).

The genomes of *Borrelia* species are largely linear (Figure 15.2) (Plasterk *et al.*, 1985; Baril *et al.*, 1989; Ferdows and Barbour, 1989; Casjens *et al.*, 2000). Each *Borrelia* genome consists of a linear chromosome of about 1000 kb and one copy each of different types of linear and circular plasmids. *Borrelia* cells are polyploid; there are 5–20 duplicate genomes distributed along the length of these long cells (Hinnebusch and Barbour, 1992; Kitten and Barbour, 1992). The linear plasmids range in size from about 10 kb to 180 kb and the circular plasmids are about 8–40 kb (Ferdows *et al.*, 1996). One group of circular plasmids of *B. burgdorferi* has been identified as bacteriophages (Eggers and Samuels, 1999) and it is likely that other *Borrelia* plasmids, both linear and circular, trace their origins to viruses.

The relatedness of linear and circular plasmids was demonstrated by a mutant with a circular version of a linear plasmid (Ferdows *et al.*, 1996) and by the finding of the same paralogous genes on both linear and circular plasmids (Dunn *et al.*, 1994; Barbour *et al.*, 1996; Zuckert and Meyer, 1996). The telomeres of the linear plasmids are sealed as hairpins (Hinnebusch and Barbour, 1991); denaturation of a linear duplex plasmid yields a single-stranded circle with a circumference that is twice the length of the linear plasmid (Barbour and Garon, 1987). The right and left ends of the linear plasmids are short inverted repeats (Hinnebusch *et al.*, 1990; Casjens *et al.*, 1997). Replication begins in the centre of the linear plasmids and linear chromosome (Picardeau *et al.*, 1999, 2000). Each type of plasmid has a selection of genes that are homologous to replication or partition genes of other bacteria, but most of the open reading frames encode polypeptides of unknown function (Barbour *et al.*, 1996; Fraser *et al.*, 1997; Casjens *et al.*, 2000).

Possible mediators of recombination reactions in *B. burgdorferi* are the RecA and RecBCD pathways (Fraser *et al.*, 1997). Unlike *T. pallidum*, *B. burgdorferi* does not have a RecF pathway. Other genes with possible recombination function in the *B. burgdorferi* genome are *ruvA*, *ruvB*, *mutL*, *mutS*, *recG* and *recJ*. Sequencing to date has revealed *recA*, *recBCD* and *recG* homologues in *B. hermsii* (Putteet-Driver *et al.*, unpublished findings). *B. burgdorferi* does not have genes that are discernibly homologous to *lexA*, *mutH*, or *ruvC*. Both *B. burgdorferi* and *B. hermsii* have plasmid-borne genes that are homologous to the transposase genes of *E. coli* and *Dichelobacter nodosus* and have frame-shift mutations that could be activated by slipped-strand mismatching. *B. hermsii* and some other *Borrelia* species have *dam*-type and *dcm*-type methylases (Hughes and Johnson, 1990; T. McDonald *et al.*, unpublished findings).

PATHOGENESIS

A *Borrelia* infection of humans almost always begins with contact with a tick or louse bearing the spirochetes. Only rarely is the disease acquired through contact of infected blood with mucous membranes or transplacental transmission from mother to fetus (Barbour, 1999a). Transmission to humans by aerosol, fomites, human saliva, urine, faeces, or sexual contact has not been documented.

Relapsing fever

At the time of the tick feeding, the spirochetes of RF *Borrelia* spp. are already present in the salivary glands and enter the host's skin and blood during the short feeding period of argasid ticks (Barbour, 1999b). (In louse-borne RF the spirochetes are in the deposited faeces of the insect and are introduced into the skin when the human host scratches the bite site.) Figure 15.4 shows infections in laboratory rats infected by tick bite or by needle injection; the patterns of spirochetaemia are similar for each type of transmission. A single spirochete is sufficient to initiate an RF infection (Schuhardt and Wilkerson, 1951; Stoenner *et al.*, 1982). Having gained access to the blood, the organisms multiply there at a rate of one cell division every 6–12 hours until they number 10^6–10^8 per millilitre of blood. From the blood the spirochetes may invade the central nervous system, the eye, the liver and other organs (Cadavid *et al.*, 1993, 1994, 2001; Cadavid and Barbour, 1998), but the blood is the main site of replication during the acute illness. The longer the spirochetes persist in their host's blood, the greater the chance that at least one of their number will be acquired by another tick or louse. Although spirochetes do not completely disappear from the blood between febrile periods, they are uncommon enough to be undetectable by microscopic examination of unconcentrated blood (Stoenner *et al.*, 1982).

Early last century investigators recognized that each new crop of spirochetes differed antigenically from the population it succeeded and the population that

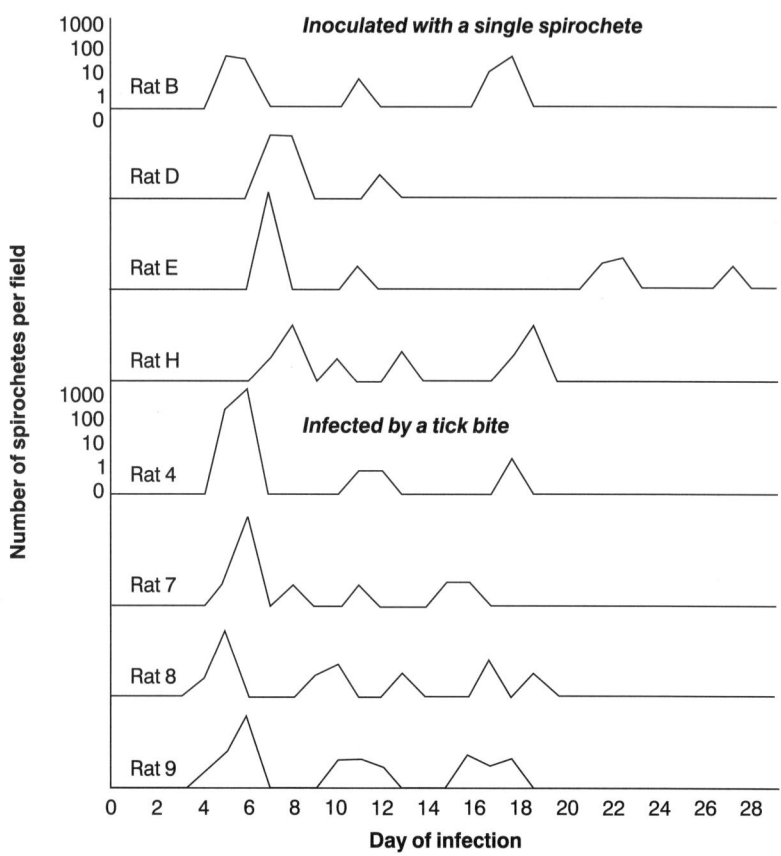

Figure 15.4 Antigenic variation in rats infected with *B. turicatae* by inoculation with single spirochetes or by bite of *Ornithodoros turicata*. The spirochetes in the blood were counted microscopically daily. (Data from Schuhardt and Wilkerson, 1951.)

will in turn succeed it (Meleney, 1928; Barbour, 1987). The phenomenon shown in Figure 15.4 indicated that the genetic information for producing relapses of spirochetemia was contained within a single cell. Thus, the repertoire for producing all the antigens is present within clonal populations. The antigenically different variants within the population are called serotypes. Serotyping was originally performed with polyclonal antisera, obtained from either infected patients or immunized animal (Coffey and Eveland, 1967a), but now is done with monoclonal antibodies (Barbour *et al.*, 1982; Barstad *et al.*, 1985).

The rats from Figure 15.4 that were inoculated with single spirochetes had an initial infection with a particular variant, for instance serotype 1, of *B. turicatae*. This serotype caused the first peak in spirochetes in the blood about 7 days after start of the infection. A few days later the rats had relapses. These could have been caused by one of several different serotypes: 2, 3, 4 and so forth. The appearance of a new serotype in the population occurs at a frequency of approximately 10^{-3} to 10^{-4} per cell generation (Stoenner et al., 1982). This occurs in culture medium and immunodeficient mice, so the host's immune system provides the selection, not the primary stimulus for variation. There may be a rough order to the appearance of serotypes (Coffey and Eveland, 1967b; Stoenner et al., 1982; Barbour and Stoenner, 1985), but any of the serotypes represented in a cell's repertoire can appear next (Stoenner et al., 1982). In *B. hermsii* 27 different serotypes have been derived from a population in a mouse infected with a single cell (Stoenner et al., 1982; Restrepo et al., 1992). Specific antisera to 24 of these serotypes accounted for at least 90% of the variants that appeared during relapses of infection in mice in prospective experiments (Stoenner et al., 1982).

The neurotropism of some species of *Borrelia*, such as *B. duttonii* in Africa and *B. turicatae* in North America, have long been noted (reviewed by Cadavid and Barbour, 1998). A newly described species of RF *Borrelia* in Spain prominently featured invasion of the central nervous system (Garcia-Monco et al., 1997). Cadavid et al. demonstrated that some serotypes within a given strain of *B. turicatae* were more neurotropic than others (Cadavid et al., 1994, 1997, 2000). What was originally called serotype A (and now renamed serotype 1) of the Oz1 strain of *B. turicatae* was present in significantly higher numbers in the leptomeninges of the brain of infected mice than was serotype B (now called serotype 2) (Cadavid et al., 1994; Pennington et al., 1999a). Serotype 2, on the other hand, achieved cell densities in the blood 10–20 times higher than that of serotype 1. This resulted in greater numbers of spirochetes in organs and tissues and in higher mortality of infant mice infected with serotype 2. If serotype 2 with its more marked spirochetaemia facilitated transmission by ticks to other rodents, then serotype 1 engendered persistence of the organism in the immunoprivileged brain during periods of infrequent tick exposures and low population densities of other possible hosts. This is a form of niche selection.

RF and LB spirochetes undoubtedly have other mechanisms of pathogenesis than multiphasic antigenic variation. One way in which *B. crocidurae*, an Old World RF species, seems to delay immune clearance is by aggregating in clumps or rosettes with erythrocytes of infected animals (Burman et al., 1998). In *B. crocidurae* infections antibodies appear and rise in the blood more slowly than is observed with non-aggregating *Borrelia* spp. (Burman et al., 1998). The rosettes of eythrocytes and spirochetes activate endothelial cells (Shamaei-Tousi et al., 2000) and cause microvascular thrombi and emboli that lead to pathologic changes in tissues (Shamaei-Tousi et al., 1999, 2001).

The rosetting phenomeon illustrates the importance of adhesion to host cells and tissues by the spirochetes. There have been several possible adhesin–ligand pairs in pathogen–host interactions. Two of the best-documented are the differential binding of *Borrelia* spp. to integrins, in particular integrins of platelets (Coburn et al., 1994, 1998; Leong et al., 1998; Magoun et al., 2000;

Alugupalli *et al.*, 2001), and the binding of LB spirochetes to decorin, a cell matrix protein (Guo *et al.*, 1995; Ulbrandt *et al.*, 2001).

Lyme borreliosis

The minimum infectious inoculum for a low-passage isolate of *B. burgdorferi* by syringe inoculation is between 10 and 100 organisms. However, cultured organisms probably differ in several respects from organisms in ticks and it is possible that less than 10 spirochetes are sufficient to initiate an infection under natural transmission conditions. At the time the tick first embeds in the skin and begins feeding, the spirochetes are not yet in the salivary glands and saliva, unlike the situation for RF spirochetes. At attachment, the spirochetes are still in the intestine and remain there for 24–48 hours (Piesman, 1993; Ohnishi *et al.*, 2001). Then, as the intestine fills with blood, the spirochetes multiply and migrate through the intestinal wall into the body fluid that bathes the internal organs. From there the spirochetes move to the salivary glands of the tick and enter with the saliva as it feeds.

Dissemination of *B. burgdorferi* through the skin or blood may occur a few days later (Shih *et al.*, 1992). It is possible that spirochetes also disseminate via lymph channels or peripheral nerves. Whatever the route, the bacteria can be found in experimental animals and in infected humans at sites distant from the tick bite within days to weeks of the start of the infection. Spirochetes have been few in number in whatever tissue they have been found (Duray, 1987; Berger, 1989). For instance, at the height of the blood infection in mice there are only about 10^3 spirochetes per millilitre of blood (Sadziene *et al.*, 1993a). After several days to a few weeks in the blood, the LB spirochetes disappear from the blood. Thereafter they are only found in tissues in mice and other experimental animals, in particular the skin, spleen, liver, joints and bladder, but, again, at low densities (Barbour, 1999a). Infection of the central nervous system is usually transient (Pachner *et al.*, 1995; Kazragis *et al.*, 1996), but other tissues may be infected for months (Barthold *et al.*, 1993).

Although LB is not characterized as having relapses as dramatic of those of RF, there are indications of antigenic variation during LB. Burgdorfer and colleagues showed that periods when xenodiagnostic ticks readily acquired *B. burgdorferi* from infected rabbits or cotton rats alternated with periods of lower infectiousness (Burgdorfer and Gage, 1987; Burgdorfer and Schwan, 1991). Steere has noted that Lyme arthritis has a remitting course (Steere, 1997) and the appearance of a new IgM response several months into infections has been observed (Craft *et al.*, 1986).

To reach other organs, the LB and RF spirochetes appear to pass from the blood through the endothelium directly and into the tissues (Comstock and Thomas, 1989; Szczepanski *et al.*, 1990). A less virulent mutant of *B. burgdorferi* had also lost ability to penetrate endothelium (Sadziene *et al.*, 1993a). Another

B. burgdorferi mutant, which lacked flagellas and was non-motile, had poor penetration of vascular endothelium in comparison to wild-type cells under experimental conditions (Sadziene *et al.*, 1991).

IMMUNITY

A mouse infected with *B. hermsii* and *B. turicatae* has 10^6-10^8 spirochetes per millilitre of blood about a week after injection with a single spirochete (Stoenner *et al.*, 1982; Pennington *et al.*, 1997). Within 24–48 hours after reaching this peak the spirochetes are undetectable by microscopic examination of the blood. Serum obtained from the mouse at that time passively protects other mice from infection with that serotype but not other serotypes (Barbour, 1987). Meanwhile, the new serotype or serotypes, which have been proliferating in the expanding population, reach their peak a few days after the initial serotype is eliminated (Stoenner *et al.*, 1982, Barbour, 1987).

Serotype-specific IgM antibodies alone are sufficient to limit or eliminate infection by an RF *Borrelia* sp. from the blood (Barbour and Bundoc, 2001; Connolly and Benach, 2001). Experimental animals with deficient or non-existent T cell function, such as nude mice, are as effective in clearing spirochetes from the blood as their immunocompetent counterparts (Newman and Johnson, 1984). IgM monoclonal antibodies to serotype-specific antigens clear spirochetes from the blood but not from the brain when intravenously infused into infected, immunodeficient mice (Cadavid *et al.*, 1993). Whether T cells play a more important role in long-term protection from infection or in control of the infection in the central nervous system is not known. Because specific antibodies, even Fab monomers, to outer membrane proteins can kill borrelias in the absence of complement or phagocytes (Escudero *et al.*, 1997; Sadziene *et al.*, 1993b; 1994), deficiencies in the latter immune effectors may not hinder clearance of borrelias from the blood. Immune deficiencies that do increase susceptibility to more serious manifestations of RF or result in prolonged disease are absence of the spleen and impaired B cell function (Stoenner *et al.*, 1982; Barbour, 1987).

Although there are reports of adoptive transfer of immunity against challenge with T cell lines alone (Rao and Frey, 1995; Pride *et al.*, 1998), most experimental evidence indicates that antibodies are necessary for protection against infectious challenge and for recovery from infection with *B. burgdorferi* (reviewed in Seiler and Weis, 1996). T cell-independent antibody responses are important in recovery from LB as well as from RF (McKisic and Barthold, 2000). Although OspA, the basis for a human vaccine against LB, is broadly cross-reactive among strains within a species, such as *B. burgdorferi,* this antigen is seldom if ever expressed in vertebrates (Schwan *et al.*, 1995; de Silva and Fikrig, 1997). The antibodies to OspA kill the spirochetes within the tick or shortly thereafter within the host, before infection is under way. In natural infections, when the mammalian host does not encounter the highly conserved OspA proteins, immunity appears to be more strain-specific (Hofmeister et al., 1999; Bunikis et al., 2003a).

VARIABLE ANTIGENS

Early investigators of RF recognized that the recurrence of spirochetes in blood once, twice, or more times after the initial spirochetaemia was cleared meant that the infecting population was staying one step ahead of the host's adaptive response (Meleney, 1928). This is an example of multiphasic antigenic variation: a clonal population of the organisms has a repertoire of three or more immunodominant antigens that they can display during the course of infection. (If there were only two antigenic states, such as occurs with the flagella of *Salmonella* species and some pilus appendages of *E. coli*, it would be called biphasic antigenic variation.) Other examples of pathogenic organisms that use the strategy of multiphasic antigenic variation for immune evasion are African trypanosomes and *Neisseria gonorrhoeae* (see Chapters 8 and 11).

In the RF *Borrelia* spp. serotype identity is determined by what we have called as a group, variable membrane proteins (Barbour et al., 1982). The variable membrane proteins are divided about equally between two different families: variable large proteins (Vlp) of about 36 kDa and variable small proteins (Vsp) of about 20 kDa (Barbour et al., 1983b; Burman et al., 1990; Restrepo et al., 1992). Figure 15.5 shows on the left a polyacrylamide gel of whole cell lysates of five serotypes of *B. hermsii* strain HS1: 33, 26, 21, 7 and 14. The two most abundant cellular proteins are FlaB, the major structural protein of the periplasmic flagellas, and the variable membrane proteins, Vsp and Vlp.

These abundant lipoproteins are anchored in the outer membrane by their lipid moieties (Carter et al., 1994) (Figure 15.2). The *vlp* and *vsp* genes use the same locus for expression and may have near-identical signal peptides, but the processed Vsp and Vlp proteins, i.e. after cleavage of the signal peptide, differ by as much as 60% in sequence (Restrepo et al., 1992). *B. recurrentis* (Vidal et al., 1998), *B. crocidurae* (Burman et al., 1998) and *B. turicatae* (Cadavid et al., 1997; Pennington et al., 1999a,b) also have *vlp* and *vsp* genes. Serotype 33 of *B. hermsii* has a Vsp-like protein, formerly called 'VmpC' (Barbour et al., 1982) and 'Vmp33' (Carter et al., 1994), that has been renamed 'Vtp' because of its unique signal peptide and expression site (Carter et al., 1994) and because of its association with the tick (Schwan and Hinnebusch, 1998) and laboratory cultivation environments (Carter et al., 1994).

The Vlp family of proteins of *B. hermsii* are divided into four sub-families with less than 60% sequence identity between them: Vlp_α (serotypes 7, 21, 18, 25); Vlp_β (serotypes 9, 10, 12, 14, Vlp_γ (serotypes 5, 19) and Vlp_δ (serotypes 4, 16, 17, 23) (Burman et al., 1990; Restrepo et al., 1992). Different strains of *B. hermsii* from distant geographic origins in western North America had the same

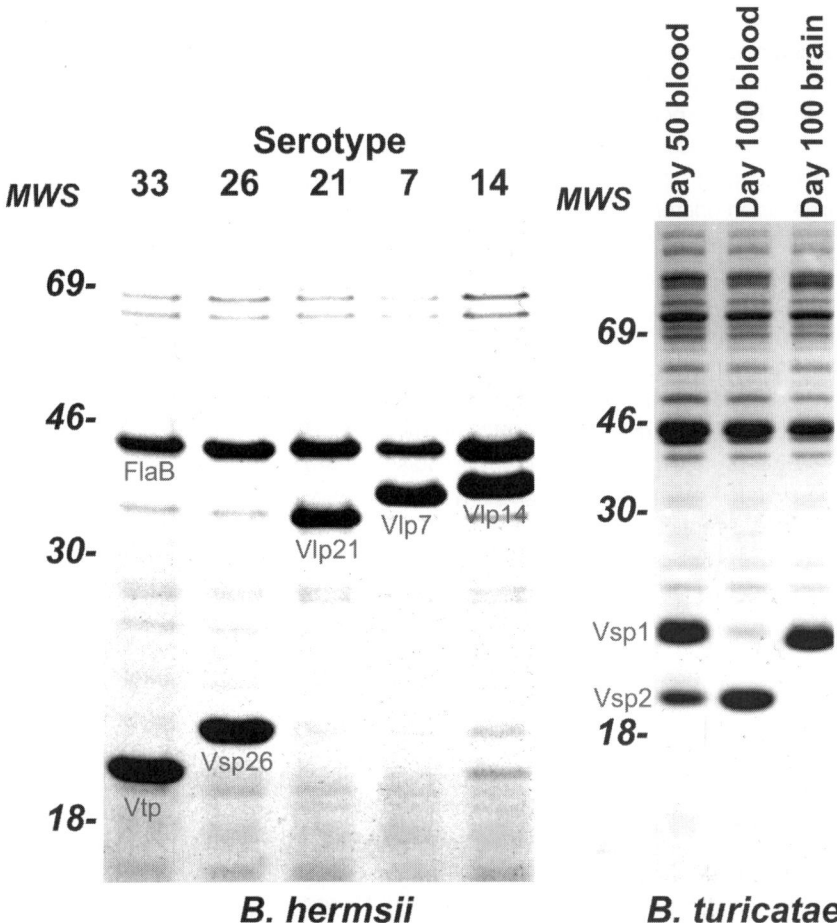

Figure 15.5 Coomasie blue-stained polyacrylamide gels of whole cell lysates of *B. hermsii* or *B. turicatae*. (*Left*) Serotypes 33, 26, 21, 7 and 14 were from strain HS1 of *B. hersmii*; the locations of the Vsp, Vtp and Vlp proteins, as well as the FlaB protein, are indicated (modified from Barbour *et al.*, 1982). (*Right*) Blood (day 50 and 100) and brain (day 100) were collected from an immunodeficient mouse infected with *B. turicatae*; the culture of the samples revealed the changes in the populations at the different sites (modified from Cadavid *et al.*, 1994).

four sub-families (Hinnebusch *et al.*, 1998). The Vsp proteins of *B. turicatae* were originally identified by letters, e.g. 'VspA' (Cadavid *et al.*, 1997; Pennington *et al.*, 1999a), but for compliance with standard genetic nomenclature are now designated with a number to indicate serotype and a subscript to indicate species or strain, e.g. 'Vsp1$_{Bt}$' replaces 'VspA' of *B. turicatae*. Figure 15.5 on the right shows whole cell lysates of *B. turicatae* isolated from the blood or brain of immunodeficient mouse infected with this agent (Cadavid *et al.*, 1994). On day 50 of infection there are two serotypes in the blood, 1 and 2. On day 100 serotype

1 is detectable in a culture of the blood but the predominant type is serotype 2. In contrast, only serotype 1 grew in culture of the brain tissue. The two serotypes are distinguished by the Vsp they produce: Vsp1 and Vsp2.

The LB spirochetes express two lipoproteins, OspA and OspB, in abundance on their surfaces while they are in unfed ticks and in culture but never or infrequently in mammalian hosts. The *ospA* and *ospB* genes exist as a tandem array as an operon in single copy on a linear plasmid in a *B. burgdorferi* genome (Howe *et al.*, 1985; Bergstrom *et al.*, 1989). With *in vitro* selection with antibodies to OspA or OspB proteins, escape variants of *B. burgdorferi* occur at frequencies as high as 10^{-4} (Sadziene *et al.*, 1992). The escape variants have included missense and nonsense mutations in the coding sequence for the epitope for the neutralizing antibody and also chimeric genes that encode proteins that are OspA-like at the N-terminal end and OspB-like at the C-terminal end (Rosa *et al.*, 1992; Sadziene *et al.*, 1992; Fikrig *et al.*, 1993).

OspC proteins, the most polymorphic set of proteins among LB spirochetes, are encoded by genes that, like *ospA*, exist in only one version in each cell and are on a circular plasmid instead of a linear plasmid (Sadziene *et al.*, 1993c). OspC is expressed by the spirochetes in the tick after it begins to feed and when the spirochetes move to the salivary gland and out into the host (Schwan and Piesman, 2000; Gilmore *et al.*, 2001; Ohnishi *et al.*, 2001). During *in vitro* cultivation OspC expression is increased by raising the temperature of incubation or lowering the pH of the medium (Schwan *et al.*, 1995; Carroll *et al.*, 1999; Yang *et al.*, 2000). During the spirochetaemia of early LB infection the spirochetes appear to be expressing OspC (Schwan *et al.*, 1995; Montgomery *et al.*, 1996; de Silva and Fikrig, 1997). Patients with LB and the field mouse, *P. leucopus*, a natural reservoir of *B. burgdorferi*, commonly have antibodies to OspC (Padula *et al.*, 1994; Brunet *et al.*, 1995; Vaz *et al.*, 2001; Bunikis *et al.*, 2003b). Immunity against infection from anti-OspC antibodies is strain-specific (Probert and LeFebvre, 1994; Gilmore *et al.*, 1996; Bockenstedt *et al.*, 1997; Mbow *et al.*, 1999), a finding consistent with the polymorphisms of OspC. In infections of laboratory mice with clonal populations of spirochetes, detectable DNA rearrangements or point mutations of the *ospC* gene were not noted (Persing *et al.*, 1994; Hodzic *et al.*, 2000).

The Vtp protein of *B. hermsii* has a size of about 23 kDa and resembles the homologous OspC protein in its association with transmission from the tick to vertebrate host (Schwan and Piesman, 2002). When a non-infected *Ornithodoros hermsi* tick fed on a mouse infected with *B. hermsii* of serotype 7 or 8, the spirochetes subsequently found in the salivary glands of the tick were a different serotype, 33, and expressed the Vtp protein instead of a Vlp7 or Vsp8 protein (Schwan and Hinnebusch, 1998). When that tick subsequently fed on a second, naive mouse, the spirochetes in the blood of the mouse were of the same serotype, either 7 or 8, that infected the first mouse. Vtp is expressed from a different expression site from the *vsp* and *vlp* genes of *B. hermsii* (Carter *et al.*, 1994; Barbour *et al.*, 2000). The *vtp* gene is in single copy in the *B. hermsii* genome and is located on a 53 kb linear plasmid (Barbour *et al.*, 2000). As described in more detail below, the *vsp* and *vlp* genes are located on other linear plasmids.

While *ospA* and *ospC* are limited to one allele per strain of *B. burgdorferi*, there are several families each with multiple paralogous alleles in the genome, predominantly on the linear and circular plasmids. In some of the paralogous families the alleles are identical or near identical in sequence. Other paralogous genes vary considerably in sequence within a single strain and undergo rearrangements (Sung *et al.*, 1998; Casjens, 1999; Casjens *et al.*, 2000). But for most of these paralogous gene families there is not yet direct experimental evidence of a role in immune evasion or niche selection.

The exception is the *vls* family of sequences of *B. burgdorferi* and other LB agents. The *vlsE* locus encodes a 38 Da lipoprotein on the cell's surface (Zhang *et al.*, 1997; Iyer *et al.*, 2000). The VlsE protein is homologous to Vlp proteins of RF *Borrelia* spp.; it is closest in sequence to members of the δVlp sub-family (Zhang *et al.*, 1997). By a mechanism described below, variants in this protein are generated and then selected for over time from the populations of *B. burgdorferi* in mice but not detectably in populations in cultivation medium or in ticks (Zhang and Norris, 1998a; Indest *et al.*, 2001; Ohnishi *et al.*, 2001). The rate of variation of the *vlsE* locus is difficult to assess, because it has not been detected *in vitro* and *B. burgdorferi* is not readily sampled from the blood. However, *vlsE* variants can be identified as soon as the fourth day of infection, and by a month into the infection many of the spirochetes isolated from mice show substantial changes in the *vlsE* locus (Zhang and Norris, 1998a, Indest *et al.*, 2001, Sung *et al.*, 2001).

Antigen structure

The Vsp and Vtp lipoproteins of RF *Borrelia* species and the OspC proteins of LB *Borrelia* species are homologous. Members of this Vsp–Vtp–OspC protein family are highly alpha-helical and form a compact four-helix bundle (Zückert *et al.*, 2000; Eicken *et al.*, 2001; Kumaran *et al.*, 2001). The processed N-terminus is acylated similarly to other bacterial lipoproteins (Carter *et al.*, 1994) and the lipid moiety anchors them in the outer membrane. The C-terminus is not anchored but is close to the N-terminus near the outer membrane. The loops between alpha-helices are exposed to the environment as ligands or epitopes. Native and recombinant Vsp and OspC proteins exist as dimers in solution and on the surface of the spirochetes (Zückert *et al.*, 2000, 2003). These proteins have conserved signal peptides and N-terminal ends of the processed proteins (Cadavid *et al.*, 1997; Barbour *et al.*, 2000). Although these proteins have numerous lysine residues, the proteins are notably resistant to cleavage by trypsin and other serine proteases (Barbour *et al.*, 1983b; Barbour, 1985; Zückert *et al.*, 2000).

The homologous Vls proteins of LB species and Vlp proteins of RF species are 30–40% identical in sequence across species (Zhang *et al.*, 1997). Although Vsp and Vlp proteins are alternately expressed from the same locus in *B. hermsii* (see below), there is little discernible homology between the two protein families at the amino acid sequence level. However, it is likely that the Vlp–Vls proteins are the result of a duplication of the ancestor of Vsp–Vtp–OspC proteins that occurred before the divergence of LB and RF species. The Vlp proteins are about twice the size of Vsp proteins; both types of proteins have highly alpha-helical secondary structures (Burman *et al.*, 1990; Zückert *et al.*, 2000). While Vsp and OspC proteins are dimers, Vlp proteins are monomers (W. R. Zückert and A. G. Barbour, unpublished findings). Crystallography of a Vls protein revealed a structure similar to Vsp and OspC: the C-terminus was close to the N-terminus and the loops containing the variable regions are localized at the distal surface of the protein (Eicken *et al.*, 2002). Structural predictions indicate that Vls and Vlp have similar overall structures (C. Eicken, unpublished data). The presence of the variable loops over the length of the Vls protein is consistent with the finding that serotype-specific epitopes for Vlp proteins are distributed throughout the protein (Barstad *et al.*, 1985).

Function

Only a few of the numerous lipoproteins of *Borrelia* species have an identified function. In our initial studies of Vsp and Vlp proteins of *B. hermsii* we focused on their characteristics as antigens for this reason: the lack of antigenic cross-reactivity between successively appearing surface proteins was the apparent key to pathogen's success. If the proteins served another purpose, it was perhaps to mask more conserved, critical surface structures, such as a porin (Bunikis and Barbour, 1999). The marked heterogeneity within the families seemed to leave little in common between different Vsps or Vlps at the primary sequence level, especially in the exposed loops between helices.

If there is a function shared between different polymorphic proteins within a family, this may be conferred by the overall conformation of the protein. For instance, we have noted that the Vsp proteins are comparatively resistant to serine proteases (Zückert *et al.*, 2000); although these proteins have high lysine contents, most of the lysines are paired in salt bridges with another amino acid and, thus, are not available for cleavage. Protease resistance of these proteins would affect processing of the proteins for antigen presentation and could direct the immune response to be either T cell independent or predominantly Th1 (Zückert *et al.*, 2000). Another possible shared featured is resistance to serum complement, an important requirement for a blood-borne pathogen. Presence of Osp proteins on the surface increased the minimum concentration of complement for death of *B. burgdorferi* (Sadziene *et al.*, 1995) and Vsp and Vlp protein may have this role for relapsing fever *Borrelia* spp. A general function of these proteins in *B. crocidurae* and other Old World RF species may be adherence to erythrocytes, which is associated with a delayed immune response by the host to the spirochetes (Burman *et al.*, 1998; Shamaei-Tousi *et al.*, 1999).

More recently we have recognized that host-interaction function is associated with the variable regions as well as the conserved overall structure of the protein. Vsp proteins, and perhaps Vlp proteins as well, are determinants of tissue localization or tropisms. Here the polymorphisms serve for niche selection within a given host or between different kinds of vertebrate hosts. This was first observed in *B. turicatae*, in which two serotypes with the same genetic background differed in their neurotropism and virulence (Cadavid *et al.*, 1994). Vsp1 of *B. turicatae*'s serotype 1, which is associated with nervous system invasion, is overall more hydrophobic and less charged than Vsp2, which is associated with retention of high numbers of cells in the blood (Cadavid *et al.*, 1997; Pennington *et al.*, 1999a). On the other hand, Vsp2, but not Vsp1, avidly binds to glycosaminoglycans, such as heparin and dermatan sulphate (Magoun *et al.*, 2000).

There are not isogenic pairs of *B. burgdorferi* expressing different OspC proteins; each strain has only one *ospC* allele. Moreover, allelic replacement of *ospC* to test hypotheses about the role of OspC in infection has not been achieved with infectious isolates. Consequently, little is known of the function of OspC, beyond its known associations with transmission from the tick vector to the vertebrate reservoir (Schwan and Piesman, 2002) and with the early phase of the infection of mammals (Padula *et al.*, 1994; Brunet *et al.*, 1995; Vaz *et al.*, 2001; Bunikis *et al.*, 2003b). There is, however, circumstantial evidence that OspC may confer niche selection similarly to the Vsp proteins of RF *Borrelia* spp. Certain strains of *B. burgdorferi* and other LB species are associated with more invasive disease than other groups of strains (Liveris *et al.*, 1999; Seinost *et al.*, 1999b; Baranton *et al.*, 2001). Strain identity is most easily determined serologically with antibody to OspC (Wilske *et al.*, 1992; Theisen *et al.*, 1993) and OspC proteins are highly polymorphic in sequence (Livey *et al.*, 1995; Theisen *et al.*, 1995; Liveris *et al.*, 1996). Only 10 of 58 known *ospC* groups of *B. burgdorferi*, *B. afzelii* and *B. garinii* strains have been recovered from the blood or cerebrospinal fluid of patients with LB in North America and Europe (Baranton *et al.*, 2001).

EVOLUTION OF VARIABLE ANTIGENS

Borrelia spp. have bacteriophage and numerous plasmids and, accordingly, might be expected to undergo horizontal transfer of genetic material. However, in comparison to many other types of bacteria, LB and RF *Borrelia* spp. are largely clonal organisms. The homoplasy ratio, which provides an estimate of the contribution of recombination to divergence and which runs from 0 (undetectable recombination) to 1.0 (free recombination), is only 0.07 for the *Borrelia fla* genes, in comparison to 0.23 for *E. coli mdh*, 0.59 for *Neisseria recA* and 1.0 for the *fla* gene of the panmictic species *Helicobacter pylori* (reviewed by Dykhuizen and Baranton, 2001). The phylogenetic trees for three different non-contiguous genes, two chromosomal and one plasmid-borne, were essentially identical for several strains of *B. burgdorferi* and related species (Dykhuizen *et al.*, 1993). *B. hermsii* also has a clonal population structure, according to an analysis of the *vsp* and *vlp* gene families and the *fla* and *p66* genes of several geographically dispersed strains (Hinnebusch *et al.*, 1998; Bunikis *et al.*, 2003a).

The few examples of horizontal transfer and inter- and intraspecific recombination in *B. burgdorferi* have been found in plasmid-borne genes. Recombination from horizontal transfer has been most obvious with *ospC*, which is a common target of the immune responses of vertebrate hosts. *ospC* alleles of strains within a species vary by 10–20% and there are about 20 genotypes for each of the three LB species (Jauris-Heipke *et al.*, 1995; Livey *et al.*, 1995; Theisen *et al.*, 1995; Wang *et al.*, 1999; Baranton *et al.*, 2001). Although the rate of horizontal gene transfer between *Borrelia* strains and species is very low, when it did occur with DNA containing *ospC* sequence, the sudden substitution of several amino acids evidently provided an advantage to the clone in the face of selection and the recombinant allele became established in the population (Dykhuizen and Baranton, 2001). Some of the major groups of *ospC* genes seem to have been established in *Borrelia* before further speciation. In a longitudinal study of an area in New York, the *B. burgdorferi* strain frequency distribution changed over time, but it remained homogeneous over space (Qiu *et al.*, 1997). The genetic diversity of *ospC* alleles within *Borrelia* populations is probably a consequence of strong balancing selection on local *Borrelia* clones (Wang *et al.*, 1999; Qiu *et al.*, 2002).

The left hand side of Figure 15.6 shows a distance phylogram of *ospC* alleles of 17 strains of *B. burgdorferi* from a highly endemic area for LB in the northeastern United States. In this rooted tree the outgroup is the *ospC* gene of another North American species, *B. bissettii*. These strains are representative of what is found locally in areas of *B. burgdorferi* transmission. A *P. leucopus* mouse or a human would be at risk of infection with any one of about 10–12 strains in a given area.

On the right of the figure is a comparable tree for a repertoire of 11 *vsp* alleles of each cell of a single species of *B. hermsii*. The outgroup here is the *vsp* gene of *B. turicatae*, another North American species. The divergence between the *vsp* alleles of one strain is almost twice that observed for *ospC* alleles of different strains that are represented. The *vsp* alleles differ by 30–40% in amino acid sequence. While strains of *B. burgdorferi* commonly encounter other strains during mixed infections of ticks and mice, the opportunities for RF *Borrelia* strains to exchange genetic material with another strain are more limited. Unlike the vectors of LB, which attach to a variety of roaming animals, the tick vectors of RF usually remain in one place, infesting the inhabitants of burrows and caves.

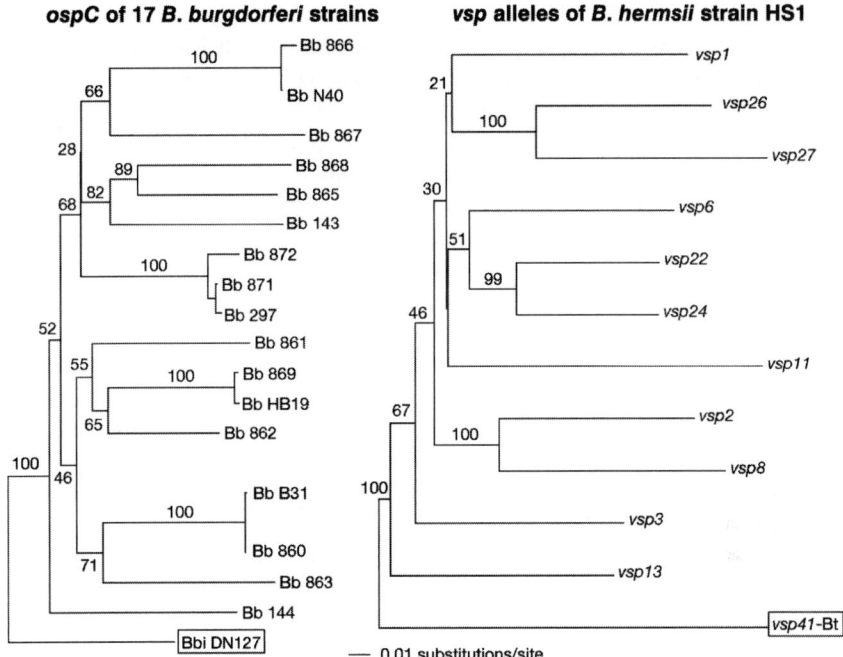

Figure 15.6 Rooted, neighbour-joining phylograms with % bootstrap values (1000 replicates) of nucleotide sequences of 17 *ospC* alleles of different strains of *B. burgdorferi* from the northeastern United States (*left*) and 11 *vsp* alleles of a single strain (HS1) of *B. hermsii* (Rich *et al.*, 2001) (*right*). Sequences Bb 860–872 are Genbank accession numbers AF029860–AF02872; sequences Bb 143 and 144 are accession numbers AF065143 and AF065144 (Wang *et al.*, 1999; Qiu *et al.*, 1997, 2002). The other sequences are from reference strains N40, 297, HB19 and B31. Outgroups are strain DN127 of *B. bissettii* (Bbi DN127) *ospC* for the *B. burgdorferi* and a *B. turicatae vsp* (vsp41-Bt) for the *B. hermsii* tree.

How then could a repertoire as extensive and diverse as the *vsp* family evolve in a clonal microorganism?

Undoubtedly gene duplications occurred, but the diversity of *vsp* and *vlp* alleles of *B. hermsii* is greater than can be explained by accumulation of point mutations. Incremental point mutation would not meet the threshold for antigenic diversity required for escape from the immune response. Mosaicisms of *vsp* alleles indicated the action of intragenic recombination (Restrepo and Barbour, 1994). The evolution of *vsp* alleles within RF *Borrelia* spp. has occurred by punctuated occurrence of allelic differentiation (Rich *et al.*, 2001). For the *vsp* genes, one of the tell-tales of combinatorial recombination is the presence of short sequences that are repeated not only in the same location in different alleles but also in different regions of the genes. For example, a 10-mer found between positions 50 and 60 of four different alleles is also found between positions 200 and 210 of another allele. The original source of another template in RF *Borrelia* spp. may have been the consequence of a horizontal transfer of a second *vsp*-type allele into the cell from another strain or species, but the preponderance of diversity is accountable by intragenic recombination with the clonal population after that event. A present-day, hypothetical model of how this may have happened is the following: a *B. burgdorferi* strain with its single *ospC* allele acquires in a rare event a second *ospC* allele, differing in sequence by 20% from the resident *ospC*, through transduction and then co-integration of the two *ospC*-bearing replicons. Thereafter, several more alleles develop through duplications, mutations and combinatorial recombination between the alleles.

The mechanism of multiple intragenic recombination events within the clonal population appears to have generated the diversity of the *vlp* alleles, which are as divergent within a sub-family, like α, as the *vsp* alleles are (Restrepo *et al.*, 1992). Figure 15.7 shows a distance phylogram of the *vlp* alleles of strain HS1 of *B. hermsii* together with the *vlsE* alleles from several strains of *B. burgdorferi* that have been sequenced to date. For some of the strains, two or more alleles are represented. Overall, the *vlsE* genes of LB *Borrelia* spp. show a lower degree of polymorphism than *vlp* genes. There may be as much diversity of *vlsE* within a strain of *B. burgdorferi*, e.g. strain B31, as there is between strains.

MECHANISMS OF ANTIGENIC VARIATION

There are four general mechanisms for antigenic variation within a clonal population: (1) modification of transcript levels; (2) gene conversion; (3) DNA rearrangement; and (4) multiple point mutations (Deitsch *et al.*, 1997). For the first mechanism two or more loci reversibly alternate in activity and silence; there is no change in the DNA at either locus. With gene conversion the DNA at one locus is changed in sequence, but the pathogen retains a complete repertoire of variable antigen genes. By the mechanism of DNA rearrangement both loci may change in their position in the genome if not in sequence. For example, the first gene in a tandem array of two or more variable antigen genes is deleted moving a previously silent gene up next to a promoter, or there may be a reciprocal recombination between two loci, such as a translocation at the ends of chromosomes. The fourth general mechanism, multiple point mutations, usually occurs in a gene that has already been activated or moved by one of the other three mechanisms. *B. hermsii* and *B. burgdorferi* use all four general mechanisms to change their phenotype of surface protein. The genetic mechanism of antigenic variation of RF *Borrelia* spp. is further detailed elsewhere (Barbour, 2002).

Reciprocal up- and downregulation of transcript levels of the *ospA* and *ospC* in LB borrelias or of *vtp* and *vsp/vlp* in RF borrelias is an example of modification of transcripts by expression-site switching. There are different expression sites with their own promoter for *ospA*, *ospC*, *vtp* and the *vsp/vlp*

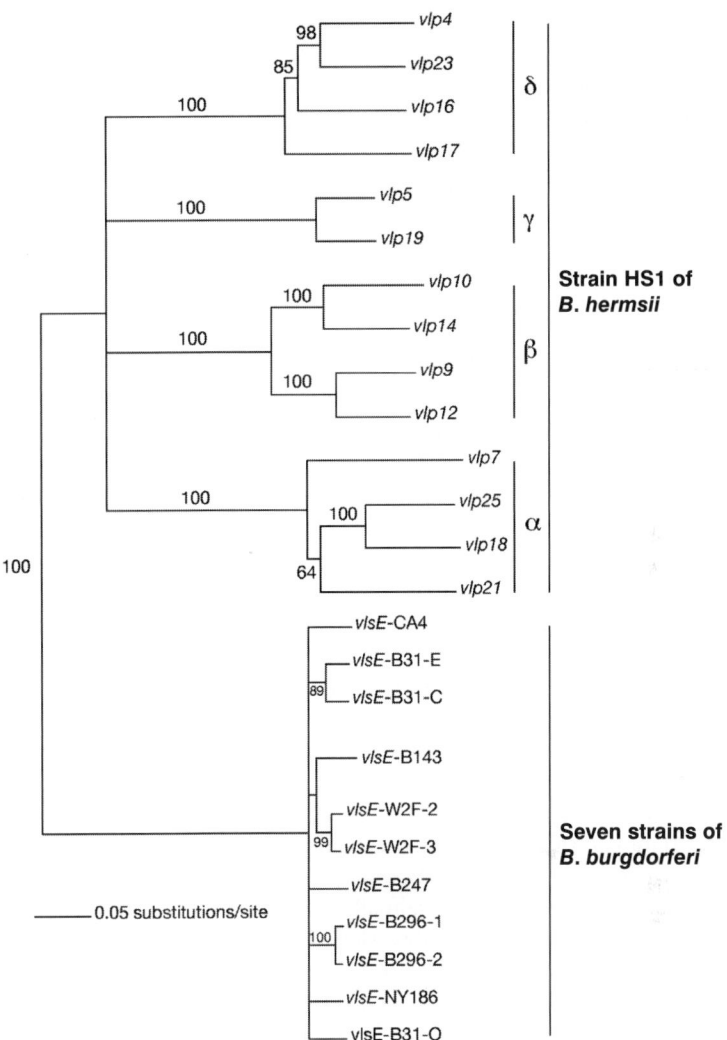

Figure 15.7 Unrooted, neighbour-joining phylogram with % bootstrap values (1000 replicates) of nucleotide sequences of the α, β, γ and δ sub-families of *vlp* alleles of *B. hermsii* (Restrepo *et al.*, 1992) and selected *vlsE* sequences of 11 alleles of 7 strains (CA4, B31, B143, W2F, B247, B296 and NY186) of *B. burgdorferi* in the United States (Zhang *et al.*, 1997; Iyer *et al.*, 2000; Sung *et al.*, 2001).

locus on different plasmids of *B. burgdorferi* or *B. hermsii*. This phenotypic change represents a form of host adaption, but it is not apparently for immune evasion or for niche selection during the course of infection in the same host (Barbour *et al.*, 2000; Schwan and Piesman, 2002) and, consequently, it will not be further considered here.

Relapsing fever

Transposition between a linear plasmid containing a collection of silent *vsp* and *vlp* genes and another linear plasmid with an active *vsp* or *vlp* gene can result in the replacement of one variable antigen gene with another at a site downstream from a promoter (Meier *et al.*, 1985; Plasterk *et al.*, 1985; Burman *et al.*, 1990; Kitten and Barbour, 1990; Barbour *et al.*, 1991; Restrepo *et al.*, 1992). This is a non-reciprocal, unidirectional recombination between two linear plasmids. The recombination is consistent with gene conversion, the second general mechanism of antigenic variation: the unchanged donor sequence is a silent *vsp/vlp* and the target sequence is the *vsp/vlp* gene at the expression site. The boundaries for the recombination are regions of sequence identity between silent and expression sites around and flanking the 5' and 3' ends of the expressed and silent *vsp/vlp* genes. The stimulus for the recombination may be a single-strand or double-strand break at the expression site.

Figure 15.8 shows the arrangement of genes before and after such a switch in *B. hermsii* from serotype 7 to serotype 21. At the top part of the figure are middle (lp32) or right ends (lp28–1 and lp28–2) of three different linear plasmids in a cell. The lp32 and lp28–2 are silent plasmids because they contain the archival, promoter-less copies of several *vsp* and *vlp* genes, arranged in tandem arrays, usually head-to-tail but sometimes head-to-head, e.g. β sub-family gene *vlp41* and the δ sub-family gene *vlp23* on lp32. The *vsp* and *vlp* genes are on either side of a cluster of open reading frames that are homologous to genes of *B. burgdorferi* plasmids and that are probably involved in plasmid replication and partition. The archived *vsp* and *vlp* alleles either are complete open reading frames or lack a few nucleotides at the 5' and/or 3' end.

At the right end of the expression plasmid, lp28–1, is the expression site for the *vsp* and *vlp* genes (Kitten and Barbour, 1990). In the case of serotype 7 there is another variable membrane protein allele, *vsp26*, which is downstream of the expressed allele but is not transcribed (Restrepo *et al.*, 1992, 1994). A duplicate of the *vlp7–vsp26* pair of alleles is also found on lp32 but with different flanking regions. Near to the telomere of the expression plasmid is a repetitive block of about 200 nucleotides, the Downstream Homology Sequence (DHS), that also occurs in at least five locations on the silent plasmids, interspersed among the arrays of *vsp* and *vlp* alleles. In the case of the silent plasmid lp28–2 there are DHS repeats flanking an array of archived *vlp* alleles: *vlp17 vlp25*, *vlp32*, *vlp21* and *vlp34*.

In the switch from serotype 7 to serotype 21 depicted in Figure 15.8, lp28–2 and lp28–1 are shown aligned. The *vlp21* and *vlp34* alleles on lp28–2 convert the two alleles at the expression site. The products of the recombination are an unaltered lp28–2 plasmid and a lp28–1 plasmid that has *vlp21* and *vlp34* at the expression site. Only *vlp21* is transcribed and the cell is phenotypically now serotype 21. There are several copies of the genome, i.e. the chromosome and its accompanying complement of plasmids, in each cell of *Borrelia* spp. (Hinnebusch and Barbour,

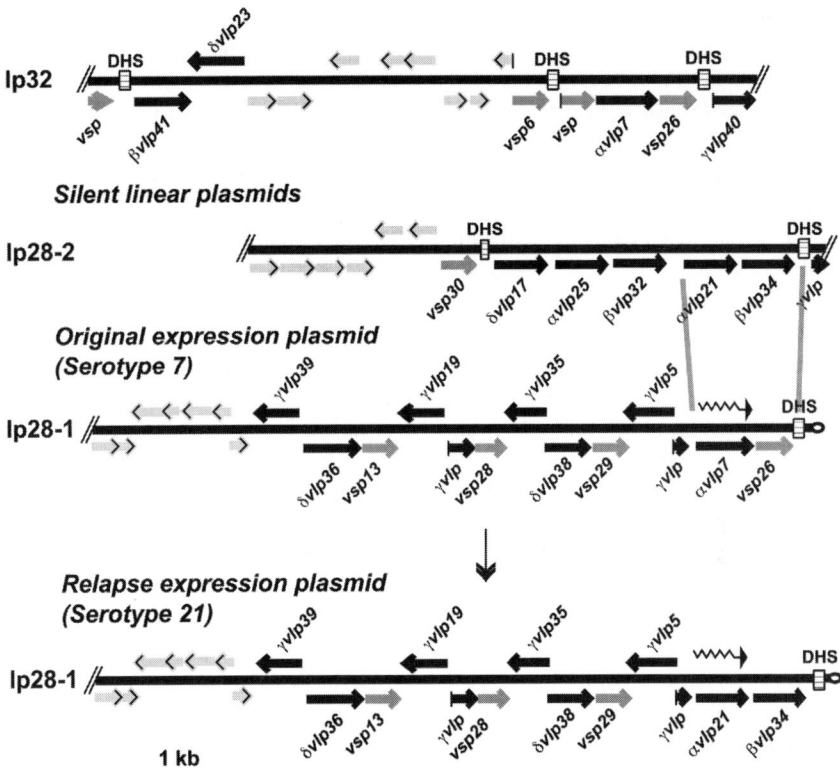

Figure 15.8 Switch in serotype from 7 to 21 of *B. hermsii* (Kitten and Barbour, 1990). Portions of three linear plasmids (lp32, lp28–1 and lp28–2) are represented before the switch. Only lp28–1 is shown after the switch. Transcription and its direction is indicated by a zigzag line and arrow. The locations of the Downstream Homology Sequence (DHS) are shown. The light-grey, thick arrows with black highlights are probable plasmid replication and partition genes.

1992; Kitten and Barbour, 1992). But the recombination probably occurred in only one of the genomes of a cell. For the cell to be 'homozygous' for the new allele at least two rounds of plasmid replication and segregation would have to occur. For the interval, one would predict the occurrence of individual cells expressing two different variable membrane proteins, but such cells have not been detected. It is possible that a newly converted locus on a lp28–1 plasmid is more transcriptionally active than the unconverted expression sites on lp28–1 plasmids in the same cell.

The conversion at the expression site is not always complete (Barbour *et al.*, 2000). We have observed expression by *B. hermsii* of a chimeric *vlp* gene that was half *vlp7* and half *vlp21* in an immunodeficient mouse (Kitten *et al.*, 1993). However, because the epitopes of serotype-specific antibodies to Vlp7 and Vlp21 are distributed throughout the protein (Barstad *et al.*, 1985), replacement of

C-terminal half of a Vlp would retain targets for circulating antibodies in the N-terminal half of the protein. Thus, partial conversions would be selected against in immunocompetent animals. The finding of incomplete conversions demonstrates that the recombinations are not necessarily precise or entirely site-specific.

Figure 15.9 shows a closer view of another switch. In this diagram the serotype and the variable membrane protein that defines the serotype are represented by different patterns. Instead of two alleles at the expression site, only the first of which is transcribed, there is only one allele present, before and after the switch (Restrepo et al., 1992). Downstream of the *vsp* or *vlp* allele is a non-transcribed sequence of different lengths that is specific for the allele. This is represented as similar but not identical patterns between the allele and its non-transcribed 3' flanking region. When the switch is made, both the 3' flanking region and the variable membrane protein allele are converted.

The boundaries for the recombination at the lp28–1 expression site are at its 3' end, the DHS nearest to the telomere and at its 5' end, a ~50 nucleotide region that includes the transcriptional start site (+1) and the start of the gene encoding the conserved signal peptide. Figure 15.10 shows an alignment of sequences with this upstream recombination region. The sequences are lp28–1 with either expressed *vlp7* (Exp 7) or expressed *vsp26* (Exp 26), lp32 with silent *vlp7* (Sil 7) and silent *vsp26* (Sil 26) and lp28–2 with silent *vlp21* (Sil 21). Expressed and silent *vlp7* and *vlp21* are identical from just before the transcription start site to about 30 nucleotides into the open reading frame. The difference between the expression locus and the silent locus in this region is the presence of a σ^{70}-type prokaryotic promoter, represented by the consensus '-10' element of TATAAT in the figure, at the expression site (Meier et al., 1985; Barbour et al., 1991).

In *B. hermsii* what effectively is a double-crossover involving both the upstream and downstream homology sequences is usually observed (Kitten and Barbour, 1992). Less frequent is what appears to be an extensive gene conversion that resembles a single crossover event. Here the entire arm of the silent plasmid after the silent allele translocates to the expression plasmid, yielding an expression plasmid that is longer than previously and in which the expression site is now several kilobases away from the right telomere. In the case of *B. hermsii* it is not known whether the *vlp* or *vsp* alleles at the expression site of these longer plasmids are transcribed or whether the longer expression plasmids are an intermediate in the recombination process. The infrequency of the DHS block among the tandemly arrayed archived alleles may mean that some alleles are activated at the expression only after one or more deletions occurs, as happens with *vsp26*.

In *B. turicatae* an extensive gene conversion that involves 10 or more kilobases downstream of the promoter has been more commonly observed (Pennington et al., 1999b). Most of a 50 kb linear plasmid of *B. turicatae* is converted during a non-reciprocal recombination, the end result of which is change in expression of only one *vsp*. As in *B. hermsii*, the silent *vsp* is activated

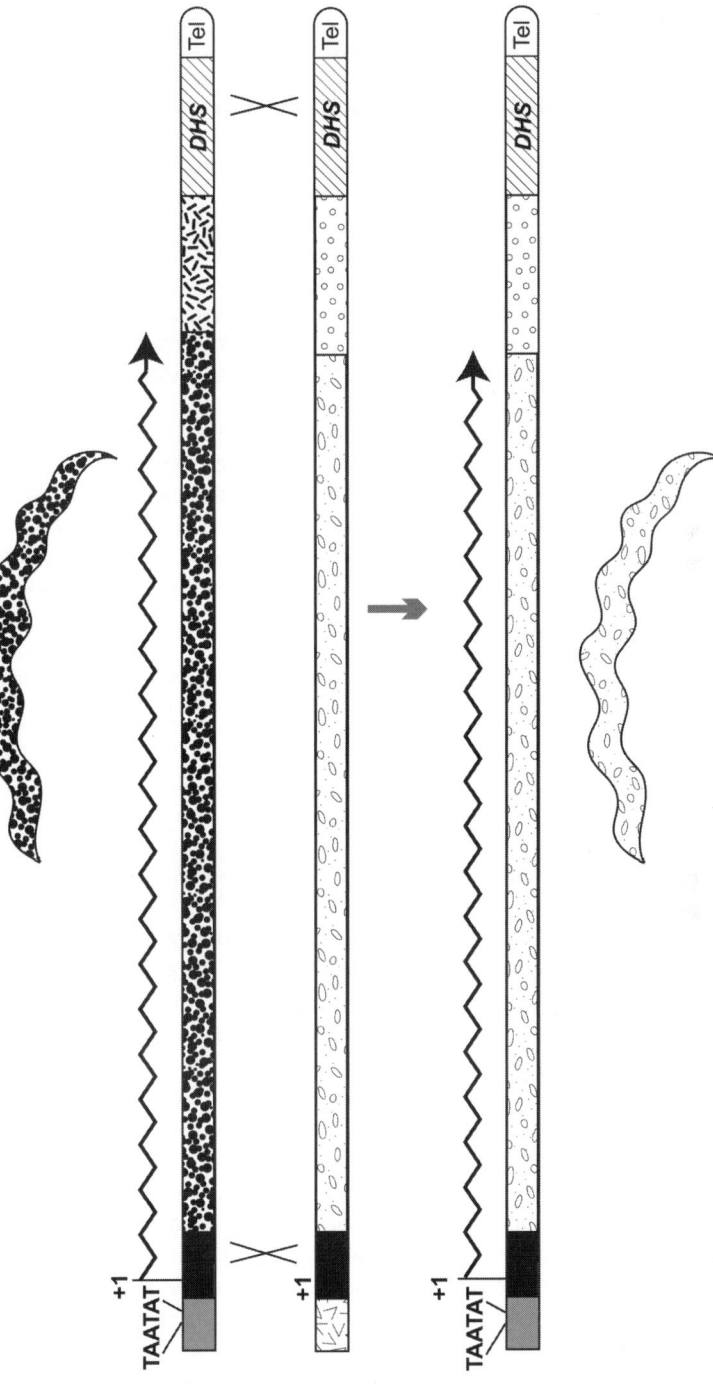

Figure 15.9 Antigenic variation by gene conversion in *B. hermsii*. The zigzag arrow indicates the transcribed variable membrane protein (VMP) gene in the expression plasmid; +1 is the transcriptional start site. TAATA is the '-10' element of the promoter. Between the VMP allele and the DHS and telomere (Tel) of the plasmids are the 3' flanking regions. Surrounding the transcriptional start site is a conserved sequence indicated by the black box. In the figure a recombination between the expression plasmid and a silent plasmid is shown. The boundaries for the recombination are indicated by the crossover sign.

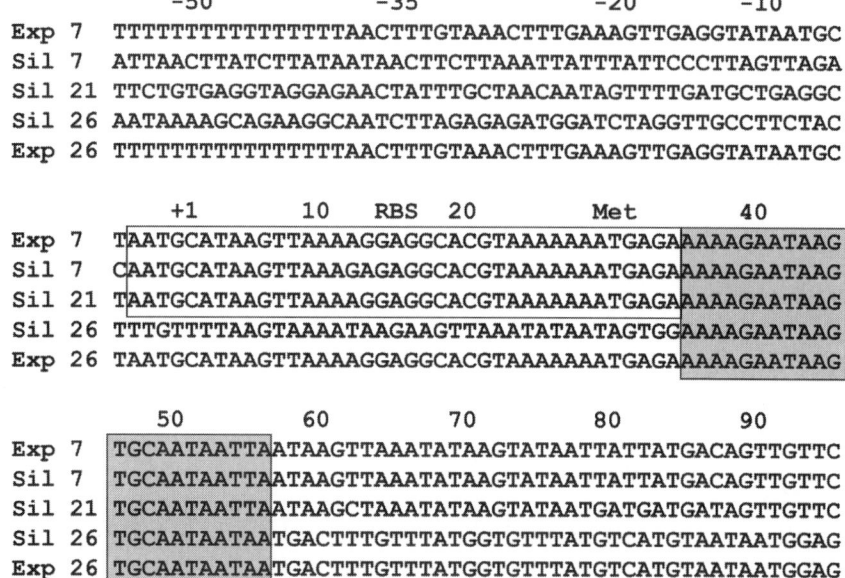

Figure 15.10 Alignment of 5' ends and flanking regions of *vlp7*, *vlp21* and *vsp26* alleles of *B. hermsii* in their expression (Exp) and silent (Sil) loci. The transcriptional start site is +1. The ribosomal binding sequence (RBS) and start codon (Met) are indicated. A transparent box indicates a sequence that is identical between the silent and expression sites for *vlp7* and *vlp21*. The expressed *vsp26* also has the sequence, but the silent *vsp26* does not. All five sequences share a 22 nucleotide sequence in common (light-grey box). The expression site (Exp 7 and Exp 26) is also defined by the run of T's from −56 to −41.

by placing it downstream of a promoter at the expression site, which in *B. turicatae* is not telomeric.

The third mechanism of antigenic variation, DNA rearrangement, occurs in *B. hermsii* when a deletion between direct repeats occurs in the expression plasmid (Restrepo *et al.*, 1994). One such recombination is shown in Figure 15.11. The events start with a serotype 7 cell, whose expression plasmid, lp28–1, has the transcribed *vlp7* and silent *vsp26* at the expression site. The *vsp26* allele at the expression site is silent because there is a transcriptional terminator after *vlp7*; there is no ribosomal binding sequence for *vsp26* and the silent *vsp26* lacks the first two codons for a complete *vsp* (see Figure 15.10). The expressed *vlp7* and silent *vsp26* do share a 22 nucleotide sequence (AAAAGAATAAGTGCAA-TAATTA) right after the first two codons of the open reading frame. As the alleles are tandemly arranged, there is a direct repeat of this sequence. When deletion occurs between these direct repeats the products are a non-replicative circle, which has been experimentally demonstrated (Restrepo *et al.*, 1994), and a new expression plasmid in which the *vsp26* allele has fused to the promoter, ribosomal binding sequence and first two codons of a *vsp* (Figures 15.10, 15.11).

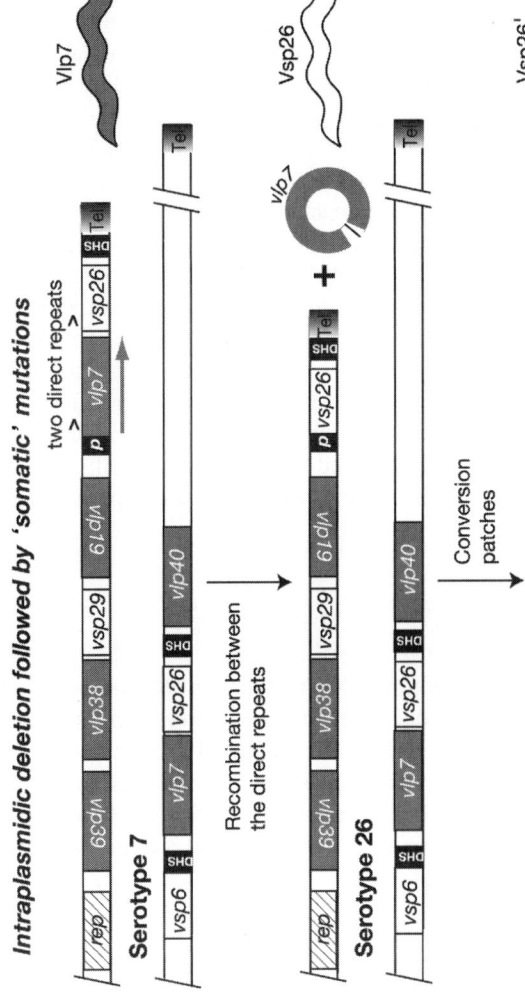

Figure 15.11 Intraplasmidic rearrangement and 'somatic' mutations in *B. hermsii*. In a switch from serotype 7 to 26 a silent *vsp26* moves next to a promoter (*p*) after a deletion of the *vlp7* allele between two direct repeats (>) on the expression plasmid. A product of the recombination is a non-replicative circle with the *vlp7* gene. This is followed by templated conversion patches of the newly expressed *vsp26* allele. There is no change of the silent *vlp7–vsp26* alleles on the other plasmid.

The cell is now serotype 26. Such deletions could also occur at the other silent site for *vsp26* on plasmid lp32 (see Figure 15.8), but there would be no direct selection of this and the cell would have lost the *vlp7* allele from its repertoire.

After this intraplasmidic DNA rearrangement but not after the interplasmidic gene-conversion events shown in Figures 15.8 and 15.9, there is a period of several generations during which frequent point mutations or substitutions in the newly rearranged expression site occur at the 5' end of the gene (Restrepo and Barbour, 1994). This is an example of the fourth mechanism for antigenic variation. The non-synonymous and synonymous base substitutions are usually single nucleotides but dinucleotides are also substituted. We have called these 'somatic' mutations, because they occur in the expressed gene but not in the archived version of the gene. They are lost when the expression site is converted to another allele. The somatic mutations are cumulative; at a rate of 10^{-1} per cell per generation most of the cells in an expanded population may have one or more substitutions in the expressed allele (Restrepo and Barbour, 1994). The templates for what appear to be conversion patches are the *vsp* alleles upstream of the expression locus on lp28-1 (Restrepo and Barbour, 1994). These point mutations in the expressed *vsp* gene have also been noted in *B. turicatae* (Pennington *et al.*, 1999b). These events in a bacterial pathogen are similar to what happens with immunoglobulin genes: first there is a deletion to bring about joining of separate gene segments and then there are somatic mutations at the rearranged locus. The RF agent's defence mirrors the offence the immune system uses against it. But these somatic mutations do not necessarily lead to benefit for the cell; sometimes they result in nonsense mutations, as occurs in *N. gonorrhoeae* pilin genes, or frame shifts (Restrepo and Barbour, 1994).

Lyme borreliosis

Deletions between direct repeats in the *ospA–ospB* operon of *B. burgdorferi* have been documented and these can produce variants that are no longer susceptible to certain anti-OspA or -OspB antibodies (Rosa *et al.*, 1992; Sadziene *et al.*, 1992; Fikrig *et al.*, 1993). However, unlike the situation with *B. hermsii*, whose cells retain the complete archive of the repertoire, the deletion in the *ospA–ospB* operon leads to loss of genetic information, including what appears to be a key protein for tick infection. It could only be reconstituted by a horizontal acquisition from another bacterium. Therefore, this phenomenon is usually a dead-end for further transmission in nature.

A *B. burgdorferi* recombination in which the hereditary material is preserved involves the *vlsE* locus. Figure 15.12 is a schematic representation of the *vlsE* locus of *B. burgdorferi* and the flanking regions. The complete expressed *vlsE* gene of about 1100 nucleotides is located near the end of a 28 kb linear plasmid (Zhang *et al.*, 1997). There are no other complete copies of the locus in the genome. Like the telomeric *vsp/vlp* expression site of *B. hermsii*, the direction of

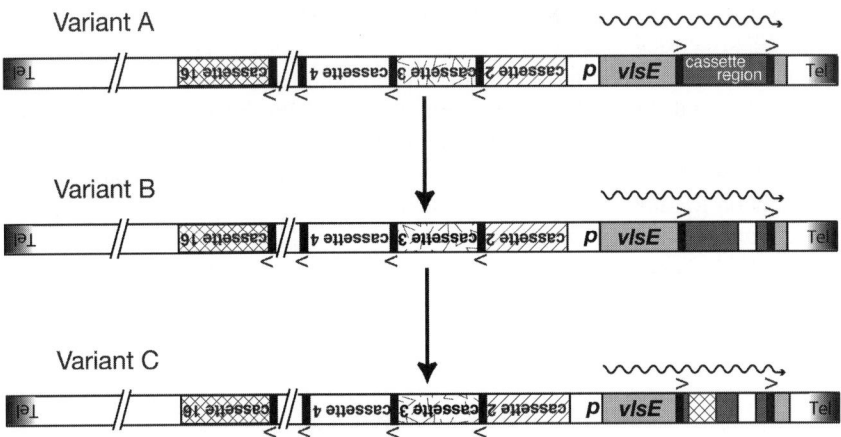

Figure 15.12 Antigenic variation in *B. burgdorferi* through partial gene conversions. Near the telomere (Tel) of a linear plasmid the *vlsE* locus is transcribed (wavy arrow) from a promoter (*p*). Upstream of the promoter on the opposite strand is a tandem array of cassettes, partial *vlsE* sequences. Bordering the cassettes are direct repeats (>) that are also found in the full-length *vlsE* locus. Three variants, A, B and C, are shown.

transcription is toward the plasmid's end. About 200 bp upstream of the *vlsE* promoter and encoded by the complementary strand is a tandem array of 15 or so partial *vls* sequences, called cassettes. These cassettes are about 500 bp in length, unlike the archived *vsp* and *vlp* alleles of *B. hermsii*, which are complete or near-complete copies. The cassettes have an overall sequence identity of ~90%; highly variable regions alternate with conserved regions.

At either end of each cassette, except the last in the series, are 17 bp direct repeats. These repeats are also present at about positions 450 and 1000 in the complete *vlsE* locus. The cassettes correspond to the sequence between the direct repeats in the complete *vlsE* locus. The upstream array of cassettes provides the templates for segmental gene conversions of the cassette in the *vlsE* locus. The previously expressed segment in *vlsE* is apparently degraded but the silent sequences remain unchanged (Zhang and Norris, 1998a). The gene conversions continue as the population expands in the infected mouse; six to eleven recombination events have been documented (Zhang and Norris, 1998a).

The partial gene conversions may be followed by point mutations, the templates for which have not been identified among the upstream cassettes (Indest *et al.*, 2001; Sung *et al.*, 2001). This type of variation is similar to the somatic mutations found in *B. hermsii* after an intraplasmidic deletion (Restrepo and Barbour, 1994). The type of mechanisms that sequentially changes *vlsE* locus most closely resemble the segmental gene conversions of the *pilE* locus of *N. gonorrhoeae* (see Chapter 8). A difference between antigenic variation of VlsE and antigenic variation of pilin in *N. gonorrhoeae* is the more dispersed organization of the silent *pilS* cassettes in the gonococcal chromosome.

ACKNOWLEDGEMENTS

I thank Steve Norris for his thorough and helpful review of the manuscript, Sven Bergström and Steve Rich for their advice, John Swanson for his early encouragement and support and the many students, post-doctoral fellows, research associates and collaborators who have contributed to our laboratory's research on RF and LB since 1981. The laboratory's research on RF and LB has been supported by the National Institutes of Health, National Science Foundation and Arthritis Foundation.

REFERENCES

Alugupalli, K. R., Michelson, A. D., Barnard, M. R., Robbins, D., Coburn, J., Baker, E. K., Ginsberg, M. H., Schwan, T. G. and Leong, J. M. (2001) *Mol. Microbiol.* 39, 330–340.

Baranton, G., Seinost, G., Theodore, G., Postic, D. and Dykhuizen, D. (2001) *Res. Microbiol.* 152, 149–156.

Barbour, A. (1985) In *Pathogenesis of Bacterial Infection* (ed. G. Jackson), Springer-Verlag, Heidelberg, pp. 235–245.

Barbour, A. G. (1987) *Contrib. Microbiol. Immunol.* 8, 125–137.

Barbour, A. G. (1998) *Br. Med. Bull.* 54, 647–658.

Barbour, A. G. (1999a) In *Effects of Microbes on the Immune System* (eds M. W. Cunningham and R. S. Fujinami), Lippincott Williams & Wilkins, Philadelphia, pp. 57–70.

Barbour, A. G. (1999b) In *Tropical Infectious Diseases. Principles, Pathogens, and Practice* (eds R. L. Guerrant, D. H. Walker and P. F. Weller), Churchill Livingstone, Philadelphia, pp. 535–546.

Barbour, A. G. (2001) In *Emerging Infections 5* (eds W. M. Scheld, W. A. Craig and J. Hughes), ASM Press, Washington, DC, pp. 153–174.

Barbour, A. G. (2002) In *Mobile DNA II* (eds N. L. Craig, R. Craigie, M. Gellert and Lambowitz, A.) American Society for Microbiology, Washington, DC, pp. 972–994.

Barbour, A. G. and Bundoc, V. (2001) *Infect. Immun.* 69, 1009–1015.

Barbour, A. G. and Fish, D. (1993) *Science* 260, 1610–1616.

Barbour, A. G. and Garon, C. F. (1987) *Science* 237, 409–411.

Barbour, A. G. and Hayes, S. F. (1986) *Microbiol. Rev.* 50, 381–400.

Barbour, A. G. and Stoenner, H. G. (1985) In *Genome Rearrangement* (eds M. I. Simon and I. Herskowitz), Alan R. Liss, Inc., New York, pp. 123–135.

Barbour, A. G., Barrera, O. and Judd, R. C. (1983b) *J. Exp. Med.* 158, 2127–2140.

Barbour, A. G., Burman, N., Carter, C. J., Kitten, T. and Bergstrom, S. (1991) *Mol. Microbiol.* 5, 489–493.

Barbour, A. G., Carter, C. J., Bundoc, V. and Hinnebusch, J. (1996) *J. Bacteriol.* 178, 6635–6639.

Barbour, A. G., Carter, C. J. and Sohaskey, C. D. (2000) *Infect. Immun.* 68, 7114–7121.

Barbour, A. G., Hayes, S. F., Heiland, R. A., Schrumpf, M. E. and Tessier, S. L. (1986) *Infect. Immun.* 52, 549–554.
Barbour, A. G., Tessier, S. L. and Todd, W. J. (1983a) *Infect. Immun.* 41, 795–804.
Barbour, A. G., Tessier, S. L. and Stoenner, H. G. (1982) *J. Exp. Med.* 156, 1312–1324.
Baril, C., Richaud, C., Baranton, G. and Saint, G. I. (1989) *Res. Microbiol.* 140, 507–516.
Barstad, P. A., Coligan, J. E., Raum, M. G. and Barbour, A. G. (1985) *J. Exp. Med.* 161, 1302–1314.
Barthold, S. W., de Souza, M. S., Janotka, J. L., Smith, A. L. and Persing, D. H. (1993) *Am. J. Pathol.* 143, 959–971.
Berger, B. W. (1989) *Rev. Infect. Dis.* 11, S1475–S1481.
Bergstrom, S., Bundoc, V. G. and Barbour, A. G. (1989) *Mol. Microbiol.* 3, 479–486.
Bockenstedt, L. K., Hodzic, E., Feng, S., Bourrel, K. W., de Silva, A., Montgomery, R. R., Fikrig, E., Radolf, J. D. and Barthold, S. W. (1997) *Infect. Immun.* 65, 4661–4667.
Brandt, M. E., Riley, B. S., Radolf, J. D. and Norgard, M. V. (1990) *Infect. Immun.* 58, 983–991.
Brunet, L. R., Sellitto, C., Spielman, A. and Telford, S. R., 3rd (1995) *Infect. Immun.* 63, 3030–3036.
Bunikis, J. and Barbour, A. G. (1999) *Infect. Immun.* 67, 2874–2883.
Bunikis, J., Rich, S. M. and Barbour, A. G. (2003a) (submitted for publication)
Bunikis, J., Tsao, J., Luke, C. J., Mayeda, K., Mirian, H., Luna, G., Fish, D. and Barbour, A. G. (2003b) (submitted for publication)
Burgdorfer, W. and Gage, K. L. (1987) *Am. J. Trop. Med. Hyg.* 37, 624–628.
Burgdorfer, W. and Schwan, T. G. (1991) *Scand. J. Infect. Dis. Suppl.* 77, 17–22.
Burman, N., Bergstrom, S., Restrepo, B. I. and Barbour, A. G. (1990) *Mol. Microbiol.* 4, 1715–1726.
Burman, N., Shamaei-Tousi, A. and Bergström, S. (1998) *Infect. Immun.* 66, 815–819.
Cadavid, D. and Barbour, A. G. (1998) *Clin. Infect. Dis.* 26, 151–164.
Cadavid, D., Bundoc, V. and Barbour, A. G. (1993) *J. Infect. Dis.* 168, 143–151.
Cadavid, D., Pachner, A. R., Estanislao, L., Patalapati, R. and Barbour, A. G. (2001) *Infect. Immun.* 69, 3389–3397.
Cadavid, D., Pennington, P. M., Kerentseva, T. A., Bergstrom, S. and Barbour, A. G. (1997) *Infect. Immun.* 65, 3352–3360.
Cadavid, D., Thomas, D. D., Crawley, R. and Barbour, A. G. (1994) *J. Exp. Med.* 179, 631–642.
Carroll, J. A., Garon, C. F. and Schwan, T. G. (1999) *Infect. Immun.* 67, 3181–3187.
Carter, C. J., Bergstrom, S., Norris, S. J. and Barbour, A. G. (1994) *Infect. Immun.* 62, 2792–2799.
Casjens, S. (1999) *Curr. Opin. Microbiol.* 2, 529–534.
Casjens, S., Murphy, M., DeLange, M., Sampson, L., van Vugt, R. and Huang, W. M. (1997) *Mol. Microbiol.* 26, 581–596.
Casjens, S., Palmer, N., Van Vugt, R., Mun Huang, W., Stevenson, B., Rosa, P., Lathigra, R., Sutton, G., Peterson, J., Dodson, R. J., Haft, D., Hickey, E., Gwinn, M., White, O. and Fraser, C, M. (2000) *Mol. Microbiol.* 35, 490–516.
Coburn, J., Barthold, S. W. and Leong, J. M. (1994) *Infect. Immun.* 62, 5559–5567.
Coburn, J., Magoun, L., Bodary, S. C. and Leong, J. M. (1998) *Infect. Immun.* 66, 1946–1952.

Coffey, E. M. and Eveland, W. C. (1967a) *J. Infect. Dis.* 117, 23–28.
Coffey, E. M. and Eveland, W. C. (1967b) *J. Infect. Dis.* 117, 29–34.
Comstock, L. E. and Thomas, D. D. (1989) *Infect. Immun.* 57, 1626–1628.
Comstock, L. E. and Thomas, D. D. (1991) *Microbial Pathogen.* 10, 137–148.
Connolly, S. E. and Benach, J. L. (2001) *J. Immunol.* 167, 3029–3032.
Craft, J. E., Fischer, D. K., Shimamoto, G. T. and Steere, A. C. (1986) *J. Clin. Invest.* 78, 934–939.
Cutler, S. J., Fekade, D., Hussein, K., Knox, K. A., Melka, A., Cann, K., Emilianus, A. R., Warrell, D. A. and Wright, D. J. (1994) *Lancet,* 343, 242.
de Silva, A. M. and Fikrig, E. (1997) *J. Clin. Invest.* 99, 377–379.
Deitsch, K. W., Moxon, E. R. and Wellems, T. E. (1997) *Microbiol. Mol. Biol. Rev.* 61, 281–293.
Dunn, J. J., Buchstein, S. R., Butler, L. L., Fisenne, S., Polin, D. S., Lade, B. N. and Luft, B. J. (1994) *J. Bacteriol.* 176, 2706–2717.
Duray, P. H. (1987) *Am. J. Surg. Pathol.* 11, 47–60.
Dykhuizen, D. E. and Baranton, G. (2001) *Trends Microbiol.* 9, 344–350.
Dykhuizen, D. E., Polin, D. S., Dunn, J. J., Wilske, B., Preac-Mursic, V., Dattwyler, R. J. and Luft, B. J. (1993) *Proc. Natl Acad. Sci. USA* 90, 10163–10167.
Eggers, C. H. and Samuels, D. S. (1999) *J. Bacteriol.* 181, 7308–7313.
Eicken, C., Sharma, V., Klabunde, T., Lawrenz, M. B., Hardham, J. M., Norris, S. J. and Sacchettini, J. C. (2002) *J. Biol. Chem.* 277, 21691–21696
Eicken, C., Sharma, V., Klabunde, T., Owens, R. T., Pikas, D. S., Hook, M. and Sacchettini, J. C. (2001) *J. Biol. Chem.* 276, 10010–10015.
Escudero, R., Halluska, M. L., Backenson, P. B., Coleman, J. L. and Benach, J. L. (1997) *Infect. Immun.* 65, 1908–1915.
Felsenfeld, O. (1971) *Borrelia. Strains, Vectors, Human and Animal Borreliosis.* Warren H. Greene, Inc., St Louis, MO.
Ferdows, M. S. and Barbour, A. G. (1989) *Proc. Natl Acad. Sci. USA,* 86, 5969–5973.
Ferdows, M. S., Serwer, P., Griess, G. A., Norris, S. J. and Barbour, A. G. (1996) *J. Bacteriol.* 178, 793–800.
Fikrig, E., Tao, H., Kantor, F. S., Barthold, S. W. and Flavell, R. A. (1993) *Proc. Natl Acad. Sci. USA* 90, 4092–4097.
Fraser, C. M., Casjens, S., Huang, W. M., Sutton, G. G., Clayton, R., Lathigra, R., White, O., Ketchum, K. A., Dodson, R., Hickey, E. K., Gwinn, M., Dougherty, B., Tomb, J. F., Fleischmann, R. D., Richardson, D., Peterson, J., Kerlavage, A. R., Quackenbush, J., Salzberg, S., Hanson, M., van Vugt, R., Palmer, N., Adams, M. D., Gocayne, J., Venter, J. C. et al. (1997) *Nature* 390, 580–586.
Fukunaga, M., Okada, K., Nakao, M., Konishi, T. and Sato, Y. (1996) *Int. J. Syst. Bacteriol.* 46, 898–905.
Garcia-Monco, J. C., Miller, N. S., Backenson, P. B., Anda, P. and Benach, J. L. (1997) *J. Infect. Dis.* 175, 1243–1245.
Gilmore, R. D., Jr, Kappel, K. J., Dolan, M. C., Burkot, T. R. and Johnson, B. J. (1996) *Infect. Immun.* 64, 2234–2239.
Gilmore, R. D., Jr, Mbow, M. L. and Stevenson, B. (2001) *Microbes Infect.* 3, 799–808.
Guo, B. P., Norris, S. J., Rosenberg, L. C. and Hook, M. (1995) *Infect. Immun.* 63, 3467–3472.

Guttman, D. S., Wang, P. W., Wang, I. N., Bosler, E. M., Luft, B. J. and Dykhuizen, D. E. (1996) *J. Clin. Microbiol.* 34, 652–656.
Hinnebusch, J. and Barbour, A. G. (1991) *J. Bacteriol.* 173, 7233–7239.
Hinnebusch, J. and Barbour, A. G. (1992) *J. Bacteriol.* 174, 5251–5257.
Hinnebusch, B. J., Barbour, A. G., Restrepo, B. I. and Schwan, T. G. (1998) *Infect. Immun.* 66, 432–440.
Hinnebusch, J., Bergstrom, S. and Barbour, A. G. (1990) *Mol. Microbiol.* 4, 811–820.
Hodzic, E., Feng, S. and Barthold, S. W. (2000) *J. Infect. Dis.* 181, 750–753.
Hofmeister, E. K., Ellis, B. A., Glass, G. E. and Childs, J. E. (1999) *Am. J. Trop. Med. Hyg.* 60, 598–609.
Holt, S. C. (1978) *Microbiol. Rev.* 38, 114–160.
Howe, T. R., Mayer, L. W. and Barbour, A. G. (1985) *Science* 227, 645–646.
Hughes, C. A. and Johnson, R. C. (1990) *J. Bacteriol.* 172, 6602–6604.
Indest, K. J., Howell, J. K., Jacobs, M. B., Scholl-Meeker, D., Norris, S. J. and Philipp, M. T. (2001) *Infect. Immun.* 69, 7083–7090.
Iyer, R., Hardham, J. M., Wormser, G. P., Schwartz, I. and Norris, S. J. (2000) *Infect. Immun.* 68, 1714–1718.
Jauris-Heipke, S., Liegl, G., Preac-Mursic, V., Rossler, D., Schwab, E., Soutschek, E., Will, G. and Wilske, B. (1995) *J. Clin. Microbiol.* 33, 1860–1866.
Kazragis, R. J., Dever, L. L., Jorgensen, J. H. and Barbour, A. G. (1996) *Antimicrob. Agents Chemother.* 40, 2632–2636.
Kitten, T. and Barbour, A. G. (1990) *Proc. Natl Acad. Sci. USA* 87, 6077–6081.
Kitten, T. and Barbour, A. G. (1992) *Genetics* 132, 311–324.
Kitten, T., Barrera, A. V. and Barbour, A. G. (1993) *J. Bacteriol.* 175, 2516–2522.
Kumaran, D., Eswaramoorthy, S., Luft, B. J., Koide, S., Dunn, J. J., Lawson, C. L. and Swaminathan, S. (2001) *EMBO J.* 20, 971–978.
Leong, J. M., Robbins, D., Rosenfeld, L., Lahiri, B. and Parveen, N. (1998) *Infect. Immun.* 66, 6045–6048.
Liveris, D., Varde, S., Iyer, R., Koenig, S., Bittker, S., Cooper, D., McKenna, D., Nowakowski, J., Nadelman, R. B., Wormser, G. P. and Schwartz, I. (1999) *J. Clin. Microbiol.* 37, 565–569.
Liveris, D., Wormser, G. P., Nowakowski, J., Nadelman, R., Bittker, S., Cooper, D., Varde, S., Moy, F. H., Forseter, G., Pavia, C. S. and Schwartz, I. (1996) *J. Clin. Microbiol.* 34, 1306–1309.
Livey, I., Gibbs, C. P., Schuster, R. and Dorner, F. (1995) *Mol. Microbiol.* 18, 257–269.
Magoun, L., Zuckert, W. R., Robbins, D., Parveen, N., Alugupalli, K. R., Schwan, T. G., Barbour, A. G. and Leong, J. M. (2000) *Mol. Microbiol.* 36, 886–897.
Mbow, M. L., Gilmore, R. D., Jr and Titus, R. G. (1999) *Infect. Immun.* 67, 5470–5472.
McKisic, M. D. and Barthold, S. W. (2000) *Infect. Immun.* 68, 5190–5197.
Meier, J. T., Simon, M. I. and Barbour, A. G. (1985) *Cell* 41, 403–409.
Meleney, H. E. (1928) *J. Exp. Med.* 48, 65–82.
Montgomery, R. R., Malawista, S. E., Feen, K. J. and Bockenstedt, L. K. (1996) *J. Exp. Med.* 183, 261–269.
Newman, K., Jr and Johnson, R. C. (1984) *Infect. Immun.* 45, 572–576.
Ohnishi, J., Piesman, J. and de Silva, A. M. (2001) *Proc. Natl Acad. Sci. USA* 98, 670–675.

Pachner, A. R., Delaney, E. and O'Neill, T. (1995) *Ann. Neurol.* 38, 667–669.
Padula, S. J., Dias, F., Sampieri, A., Craven, R. B. and Ryan, R. W. (1994) *J. Clin. Microbiol.* 32, 1733–1738.
Paster, B., Dewhirst, F., Weisburg, W., Tordoff, L., Fraser, G., Hespell, R., Stanton, T., Zablen, L., Mandelco, L. and Woese, C. (1991) *J. Bacteriol.* 173, 6101–6109.
Pennington, P. M., Allred, C. D., West, C. S., Alvarez, R. and Barbour, A. G. (1997) *Infect. Immun.* 65, 285–292.
Pennington, P. M., Cadavid, D. and Barbour, A. G. (1999a) *Infect. Immun.* 67, 4637–4645.
Pennington, P. M., Cadavid, D., Bunikis, J., Norris, S. J. and Barbour, A. G. (1999b) *Mol. Microbiol.* 34, 1120–1132.
Persing, D. H., Mathiesen, D., Podzorski, D. and Barthold, S. W. (1994) *Infect. Immun.* 62, 3521–3527.
Picardeau, M., Lobry, J. R. and Hinnebusch, B. J. (1999) *Mol. Microbiol.* 32, 437–445.
Picardeau, M., Lobry, J. R. and Hinnebusch, B. J. (2000) *Genome Res.* 10, 1594–1604.
Piesman, J. (1993) *J. Infect. Dis.* 167, 1082–1085.
Plasterk, R. H., Simon, M. I. and Barbour, A. G. (1985) *Nature* 318, 257–263.
Pride, M. W., Brown, E. L., Stephens, L. C., Killion, J. J., Norris, S. J. and Kripke, M. L. (1998) *J. Leukoc. Biol.* 63, 542–549.
Probert, W. S. and LeFebvre, R. B. (1994) *Infect. Immun.* 62, 1920–1926.
Qiu, W. G., Bosler, E. M., Campbell, J. R., Ugine, G. D., Wang, I. N., Luft, B. J. and Dykhuizen, D. E. (1997) *Hereditas* 127, 203–216.
Qiu, W. G., Dykhuizen, D. E., Acosta, M. S. and Luft, B. J. (2002) *Genetics* 160, 833–849.
Radolf, J. D., Bourell, K. W., Akins, D. R., Brusca, J. S. and Norgard, M. V. (1994) *J. Bacteriol.* 176, 21–31.
Rao, T. D. and Frey, A. B. (1995) *Cell Immunol.* 162, 225–234.
Ras, N. M., Lascola, B., Postic, D., Cutler, S. J., Rodhain, F., Baranton, G. and Raoult, D. (1996) *Int. J. Syst. Bacteriol.* 46, 859–865.
Restrepo, B. I. and Barbour, A. G. (1994) *Cell* 78, 867–876.
Restrepo, B. I., Carter, C. J. and Barbour, A. G. (1994) *Mol. Microbiol.* 13, 287–299.
Restrepo, B. I., Kitten, T., Carter, C. J., Infante, D. and Barbour, A. G. (1992) *Mol. Microbiol.* 6, 3299–3311.
Rich, S. M., Sawyer, S. A. and Barbour, A. G. (2001) *Proc. Natl Acad. Sci. USA* 98, 15038–15043.
Rosa, P. A., Schwan, T. and Hogan, D. (1992) *Mol. Microbiol.* 6, 3031–3040.
Russell, H. (1936) *Trans. R. Soc. Trop. Med. Hyg.* 30, 179–190.
Sadziene, A., Barbour, A. G., Rosa, P. A. and Thomas, D. D. (1993a) *Infect. Immun.* 61, 3590–3596.
Sadziene, A., Jonsson, M., Bergstrom, S., Bright, R. K., Kennedy, R. C. and Barbour, A. G. (1994) *Infect. Immun.* 62, 2037–2045.
Sadziene, A., Rosa, P. A., Thompson, P. A., Hogan, D. M. and Barbour, A. G. (1992) *J. Exp. Med.* 176, 799–809.
Sadziene, A., Thomas, D. D. and Barbour, A. G. (1995) *Infect. Immun.* 63, 1573–1580.
Sadziene, A., Thomas, D. D., Bundoc, V. G., Holt, S. C. and Barbour, A. G. (1991) *J. Clin. Invest.* 88, 82–92.
Sadziene, A., Thompson, P. A. and Barbour, A. G. (1993b) *J. Infect. Dis.* 167, 165–172.

Sadziene, A., Wilske, B., Ferdows, M. S. and Barbour, A. G. (1993c) *Infect. Immun.* 61, 2192–2195.
Schuhardt, V. T. and Wilkerson, M. (1951) *J. Bacteriol.* 62, 215–219.
Schwan, T. G. and Hinnebusch, B. J. (1998) *Science* 280, 1938–1940.
Schwan, T. G. and Piesman, J. (2000) *J. Clin. Microbiol.* 38, 382–388.
Schwan, T. G. and Piesman, J. (2002) *Emerg. Infect. Dis.* 8, 115–121.
Schwan, T. G., Piesman, J., Golde, W. T., Dolan, M. C. and Rosa, P. A. (1995) *Proc. Natl Acad. Sci. USA* 92, 2909–2913.
Seiler, K. P. and Weis, J. J. (1996) *Curr. Opin. Immunol.* 8, 503–509.
Seinost, G., Dykhuizen, D. E., Dattwyler, R. J., Golde, W. T., Dunn, J. J., Wang, I. N., Wormser, G. P., Schriefer, M. E. and Luft, B. J. (1999b) *Infect. Immun.* 67, 3518–3524.
Seinost, G., Golde, W. T., Berger, B. W., Dunn, J. J., Qiu, D., Dunkin, D. S., Dykhuizen, D. E., Luft, B. J. and Dattwyler, R. J. (1999a) *Arch. Dermatol.* 135, 1329–1333.
Shamaei-Tousi, A., Burns, M. J., Benach, J. L., Furie, M. B., Gergel, E. I. and Bergstrom, S. (2000) *Cell Microbiol.* 2, 591–599.
Shamaei-Tousi, A., Collin, O., Bergh, A. and Bergstrom, S. (2001) *J. Exp. Med.* 193, 995–1004.
Shamaei-Tousi, A., Martin, P., Bergh, A., Burman, N., Brännström, T. and Bergström, S. (1999) *J. Infect. Dis.* 180, 1929–1938.
Shih, C. M., Pollack, R. J., Telford, S. R. d. and Spielman, A. (1992) *J. Infect. Dis.* 166, 827–831.
Sigal, L. H. (1997) *Annu. Rev. Immunol.* 15, 63–92.
Southern, P. and Sanford, J. (1969) *Medicine* 48, 129–149.
Steere, A. C. (1997) *Med. Clin. North Am.* 81, 179–194.
Steere, A. C. (2001) *N. Engl. J. Med.* 345, 115–125.
Stoenner, H. G., Dodd, T. and Larsen, C. (1982) *J. Exp. Med.* 156, 1297–1311.
Sung, S. Y., Lavoie, C. P., Carlyon, J. A. and Marconi, R. T. (1998) *Infect. Immun.* 66, 4656–4668.
Sung, S. Y., McDowell, J. V. and Marconi, R. T. (2001) *J. Bacteriol.* 183, 5855–5861.
Szczepanski, A., Furie, M. B., Benach, J. L., Lane, B. P. and Fleit, H. B. (1990) *J. Clin. Invest.* 85, 1637–1647.
Takayama, K., Rothenberg, R. J. and Barbour, A. G. (1987) *Infect. Immun.* 55, 2311–2333.
Theisen, M., Borre, M., Mathiesen, M. J., Mikkelsen, B., Lebech, A. M. and Hansen, K. (1995) *J. Bacteriol.* 177, 3036–3044.
Theisen, M., Frederiksen, B., Lebech, A. M., Vuust, J. and Hansen, K. (1993) *J. Clin. Microbiol.* 31, 2570–2576.
Ulbrandt, N. D., Cassatt, D. R., Patel, N. K., Roberts, W. C., Bachy, C. M., Fazenbaker, C. A. and Hanson, M. S. (2001) *Infect. Immun.* 69, 4799–4807.
Vaz, A., Glickstein, L., Field, J. A., McHugh, G., Sikand, V. K., Damle, N. and Steere, A. C. (2001) *Infect. Immun.* 69, 7437–7444.
Vidal, V., Scragg, I. G., Cutler, S. J., Rockett, K. A., Fekade, D., Warrell, D. A., Wright, D. J. and Kwiatkowski, D. (1998) *Nat. Med.* 4, 1416–1420.
Walker, E. M., Borenstein, L. A., Blanco, D. R., Miller, J. N. and Lovett, M. A. (1991) *J. Bacteriol.* 173, 5585–5588.
Wang, I. N., Dykhuizen, D. E., Qiu, W., Dunn, J. J., Bosler, E. M. and Luft, B. J. (1999) *Genetics* 151, 15–30.

Wilske, B., Barbour, A. G., Bergstrom, S., Burman, N., Restrepo, B. I., Rosa, P. A., Schwan, T., Soutschek, E. and Wallich, R. (1992) *Res. Microbiol.* 143, 583–596.

Yang, X., Goldberg, M. S., Popova, T. G., Schoeler, G. B., Wikel, S. K., Hagman, K. E. and Norgard, M. V. (2000) *Mol. Microbiol.* 37, 1470–1479.

Zhang, J. R. and Norris, S. J. (1998a) *Infect. Immun.* 66, 3689–3697.

Zhang, J. R. and Norris, S. J. (1998b) *Infect. Immun.* 66, 3698–3704.

Zhang, J. R., Hardham, J. M., Barbour, A. G. and Norris, S. J. (1997) *Cell* 89, 275–285.

Zuckert, W. R. and Meyer, J. (1996) *J. Bacteriol.* 178, 2287–2298.

Zückert, W., Kerentseva, T. A., Lawson, C. L. and Barbour, A. G. (2000) *J. Biol. Chem.* 276, 457–463.

Zückert, W. R., Stewart, P. E., Rosa, P. A. and Barbour, A. G. (2003) (submitted for publication).

16

SURFACE ANTIGENIC VARIATION IN *GIARDIA LAMBLIA*

THEODORE E. NASH

INTRODUCTION

Giardia lamblia is a binucleated, intestinal-dwelling flagellate that is a cause of diarrhoea and intestinal upset worldwide. It is also one of the earliest-diverging eukaryotes (Sogin *et al.*, 1989) and has a simple faecal–oral life cycle. Infection occurs after ingestion of a cyst, which excysts and multiplies in the small intestine as motile trophozoites. It is this stage that is responsible for symptoms and disease. As the trophozoites move down the intestines they differentiate into infectious cysts, which are then excreted in the faeces. Because cyst excretion may approach 10^7 per gram and 10–100 cysts can infect 100% of individuals (Rendtorff, 1954; Rendtorff and Holt, 1954), ingestion of only a very small amount of faecal contamination from food, water or hands is likely to result in infection. Understandably, *Giardia* infections are among the most common parasitic infections in developed countries, and not surprisingly they are almost universal by the age of 2 in regions where faecal contamination is high (Mata, 1978).

Unlike other organisms where surface antigenic variation was suggested by the biology of the organism in the host, antigenic variation was unsuspected in *G. lamblia* and was discovered as an *in vitro* phenomenon (Adam *et al.*, 1988; Nash *et al.*, 1988). Although there are many unanswered questions concerning antigenic variation in *Giardia*, perhaps the most vexing are:

1 What is its biological role in infection?
2 How do the unique structural features of the variant-specific surface proteins (VSPs), the family of surface-varying proteins in *Giardia*, contribute to its biology?
3 What molecular mechanisms are involved in controlling VSP expression?

CHARACTERISTICS OF ANTIGENIC VARIATION IN VITRO

Antigenic variation occurs spontaneously *in vitro* (Nash *et al.*, 1988) (Table 16.1). It is a general property of all *G. lamblia* isolates (Nash, 1992; Bruderer *et al.*, 1993) and probably occurs in other *Giardia* species as well because VSP gene fragments have been detected in *Giardia muris* (T. E. Nash and M. F. Heyworth, unpublished results). Antigenic variation *in vitro* was initially suggested by varying surface-labelling patterns of cultures over time but was definitively demonstrated with the development of VSP-specific cytotoxic monoclonal antibodies (Mabs) (Nash *et al.*, 1988). Analyses of clones exposed to specific cytotoxic Mabs that recognize the predominantly expressed VSP resulted in elimination of all but a few organisms that repopulated the culture with trophozoites expressing other VSPs. The process was repeated in one of the subsequent clones, which again resulted in killing of most of the trophozoites and repopulation with trophozoites expressing VSPs unrecognized by that particular Mab. Except at the time of switching, only one VSP is present on the surface. During switching, one of the two residing surface VSPs is lost after 12–36 hours and if linearity of the process is assumed then the half-life is 17.3 hours (Nash *et al.*, 2001). The dynamics of VSP switching are superficially and perhaps mechanistically similar to other parasites undergoing antigenic variation. The propensity to switch VSPs in culture is strain- and VSP-dependent and ranges from about 6.5 generations to 13 generations until VSP switching occurs (Nash *et al.*, 1990a). Therefore, most cultures consist of a mixture of expressed VSPs.

Encystation–excystation-induced switching has been documented in one instance but does not seem to occur in all *Giardia*. Following encystation and subsequent excystation *in vitro* the expressed VSP of the Group 1 (Assemblage A) WB isolate changes but the expressed VSP of Group 3 (Assemblage B) GS isolate remains unchanged (Svard *et al.*, 1998).

Table 16.1 Characteristics of antigenic variation in *G. lamblia*

Occurs in all *G. lamblia* isolates

The varying-specific surface proteins(VSPs) are a family of related proteins; VSP repertoires differ among *Giardia* groups

Switching occurs spontaneously in culture but in some isolates can also occur during encystation–excystation

Rate of VSP change is isolate- and VSP-dependent and ranges from one switch every 6.5 to 13 generations

Occurs *in vivo* in humans, mice and gerbils

Monoclonal antibodies to VSPs are cytotoxic or inhibit growth

Switching during encystation–excystation occurs in some *Giardia*

Table 16.2 Characteristics of variant-specific surface proteins of *G. lamblia*

VSPs vary in size from about 20 kDa to 200 kDa
Cysteine-rich with numerous CXXC motifs
Conserved hydrophobic tail that probably anchors the VSP to the plasma membrane
Hydrophilic terminal 5 amino acids are CRGKA
One or more GGCY motifs of unknown function
Only surface-residing Zn finger motif known
O-linked glucose and N-acetylglucosamine
Palmitate residue at carboxyl terminus
Amino terminal most peripheral at parasite–host interface

CHARACTERISTICS OF VARIANT-SPECIFIC SURFACE PROTEINS

VSPs consist of a family of related cysteine-rich proteins that vary in molecular weight from about 20 kDa (Ey and Darby, 1998) to over 200 kDa (Nash, 1992) (Table 16.2). VSPs cover the entire surface of the organism, including the flagella. Using indirect fluorescent microscopy (IFA) and VSP-specific Mabs, a bright outline of the organism is appreciated (Figure 16.1) (Nash and Aggarwal, 1986). By immunoelectronmicroscopy VSPs are present in an amorphous layer at the surface of the trophozoite (Figure 16. 2) (Pimenta *et al.*, 1991). The cysteine content is usually about 11–12% (Nash, 1992). The carboxyl terminus is

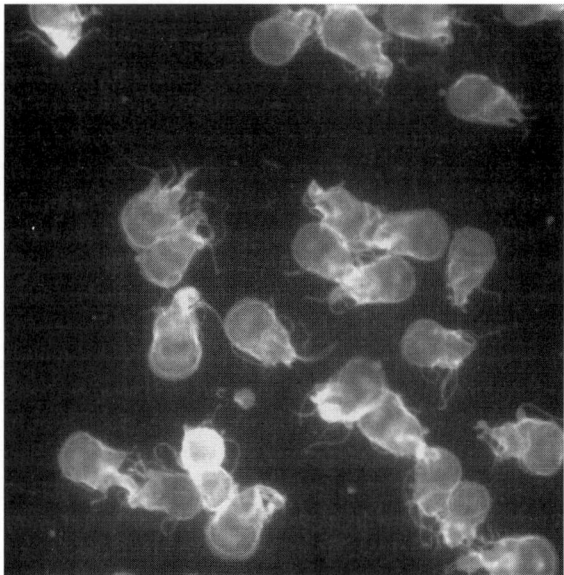

Figure 16.1 Indirect immunofluorescence of native VSPH7 using Mab G10/4. All the visualized trophozoites represent fluorescing parasites. VSP localizes to the entire surface of the trophozoites including the flagella.

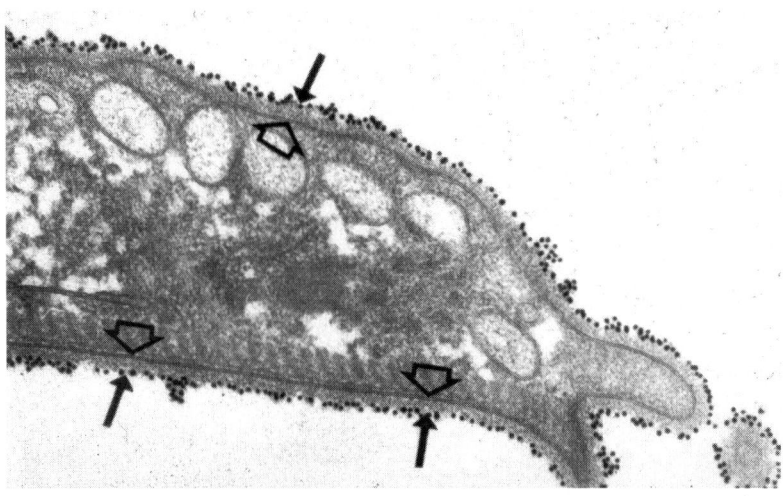

Figure 16.2 Immunoelectromicrograph Mab 6E7 reacting with the surface of viable parasites expressing VSPA6 and then fixed (adapted from Pimenta, de Silva and Nash (Pimenta *et al.*, 1991)). There is intense labelling of the entire surface of the parasite, including the flagella. The large arrows point to plasma membrane and the small arrows to the gold particles that correspond to VSP reactivity.

well conserved and consists of a hydrophobic, presumably intramembranous, domain that anchors the VSP to the parasite surface and a perfectly conserved terminal of 5 amino acids, CRGKA, that are hydrophilic and probably located in the cytoplasm (Mowatt *et al.*, 1991). All VSPs have a GGCY-containing motif in the carboxyl terminal end, which have been shown to be immunodominant in a *Giardia lamblia* mouse model (Bienz *et al.*, 2001a), and some have an additional GGCY-containing motif in the amino terminal region as well (Nash *et al.*, 1995). Most cysteines in VSPs are present in numerous CXXC motifs (Adam *et al.*, 1988), occurring 15–30 times or more depending primarily on the size of the VSP (Gillin *et al.*, 1990; Nash, 1992). The presence of numerous CXXC motifs is not unique to VSPs. They are found uncommonly in other proteins, including the surface proteins of ciliates that also undergo antigenic variation or have surface antigen variability, including *Tetrahymena* species (Bannon *et al.*, 1986) and *Ichthophthirius multifliis* (Clark *et al.*, 1992), and the surface protein of *Lembadion bullinum* (Peters-Regehr *et al.*, 1997) as well as in a class of serine proteases (Nakagawa *et al.*, 1993), furin-like proteases (AJ{84} *et al.*, 1992). Interestingly, a 150 kDa intermediate sub-unit of the Gal/GalNAc lectin of *Entamoeba histolyica* was recently characterized and found to have multiple CXXC motifs (Cheng *et al.*, 2001). Inexplicably, all VSPs contain a combination of the RING and LIM Zn finger motifs, which is the only surface-residing Zn finger motif thus far described in any organism (Nash, 1992; Nash and Mowatt, 1993; Zhang *et al.*, 1993). The core Zn finger motif, CXXCHXXCXXC, contains a critical histidine residue, which is replaced with an aspartic acid residue in some VSPs. The actual presence of Zn or other metals in the VSP Zn finger is disputed. While one laboratory documented the presence of Zn in one particular purified VSP7 (Nash and Mowatt, 1993), a second laboratory did not detect Zn in another VSP (Papanastasiou et al., 1997a). Both laboratories could not incorporate radioactive Zn into the VSP *in vitro* under conditions that limited growth (Papanastasiou et al., 1997a) or where the radioactive Zn was probably of low specific activity (T. E. Nash, unreported results). Curiously, in-depth studies of Zn binding of VSP fragments showed that Zn binding was not limited to the Zn finger motif (Nash and Mowatt, 1993). Presumably, $CXXCX_nCXXC$ motifs, where $n = 13$, 19, or 20, are functioning as metal-binding regions. The functions of Zn finger motifs in proteins of higher eukaryotes are varied but commonly contribute to protein-to-protein interactions. It is tempting to speculate that this motif contributes to increased surface-coat stability via VSP-to-VSP interactions resulting in enhanced protease resistance. Certain Zn fingers function as transcription factors, but there is at present no evidence to support this function of VSPs in *Giardia*. In a compelling series of experiments, one laboratory has demonstrated the fatty acid palmitate at or near the conserved carboxyl terminus and the presence of O-linked sugars consisting of N-acetylgluscosamine and glucose (Papanastasiou et al., 1997b; Hiltpold et al., 2000). A recent undocumented comment in a publication from the same laboratory suggested that the carbohydrates may not be covalently bound in VSPH7 (Marti et al., 2002). The presence of the palmitate explains the high hydrophobicity of VSPs, which has made purification challenging. In addition to the common motifs in the coding region, an extended polyadenylation signal is also relatively well conserved in VSPs (Svard et al., 1998).

A number of closely related VSP families have been described, each consisting of a varying number of identical or almost identical repeating peptide units that are found in the amino terminus (Adam et al., 1988; Mowatt et al., 1994; Chen et al., 1995, 1996; Yang et al., 1994). In one example studied most intensively there are 21 repeating units of 195 bp or about 4000 bp in all (Yang et al., 1994), encompassing over two-thirds of the size of the VSP. The repeating peptide is immunodominant and antibodies or Mabs to this peptide were immobilizing and cytotoxic (Nash and Aggarwal, 1986; Mowatt et al., 1994). Surface-localized repeating units may impart a unique biological function to specific VSPs and some have suggested particular functions based on the presence of low-level homologies (Chen et al., 1995, 1996), but since no biological functions have been directly demonstrated this idea is at best a suggestion.

VSPs are secreted from the surface of the parasite and are highly abundant in the culture supernatant and presumably the intestinal lumen (Nash et al., 1983; Papanastasiou et al., 1996). While the membrane-associated VSP is hydrophobic, the secreted form is hydrophilic and lacks both the hydrophobic tail and the palmitic acid residue (Papanastasiou et al., 1996). A presumed proteolytic cleavage is proposed by an as yet unidentified protease as the likely mechanism of release (Papanastasiou et al., 1996).

The three-dimensional structure of VSPs is not known and is particularly difficult to predict. All the cysteines are either bound to each other as disulphide bonds or otherwise non-reactive (Papanastasiou *et al.*, 1997a; T. E. Nash, unreported results). Because of the large number of cysteines, it is not possible to predict how cysteines are bound to each other and therefore predict structure. Inspection of the CXXC spacing (Nash and Mowatt, 1993) and location of the GGCY and Zn finger motifs of nine published VSPs and one unpublished VSP (A5 from the GS Group 3 isolate) is shown in Figure 16.3. The location of the carboxyl GGCY motif is well conserved in all the VSPs and demarcates a region relatively devoid of CXXC motifs. The amino terminal GGCY motifs are

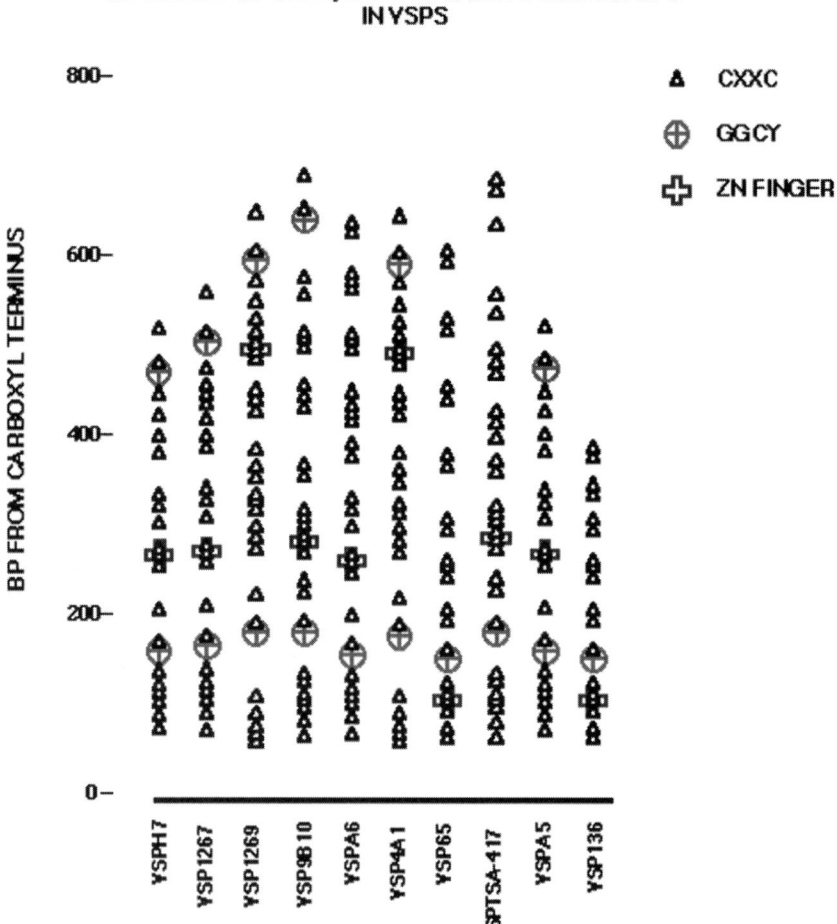

Figure 16.3 Localization of CXXC, GGCY and Zn finger motifs in various VSPs.

variably present. All the VSPs have a Zn finger motif but its location is surprisingly variable. While most are located between 250 and 300 amino acids from the carboxyl terminus and proximal to the carboxyl GGCY motif, others are found in the amino terminus or unusually between the carboxyl GGCY motif and the end of the protein. Inspection of the location of these motifs reveals obvious families of VSP proteins. For instance, VSP1269 (Mowatt et al., 1991) and VSP4A1 (Papanastasiou et al., 1997a), or VSPA5 (L. Kalakova and T. E. Nash, unreported results), VSPH7 (Nash and Mowatt, 1992a) and VSP 1267 (Mowatt et al., 1991). A number of studies have shown some VSPs are more closely related compared to others (Yang and Adam, 1995a,b; Nash et al., 1995; Ey and Mayrhofer, 1993; Ey et al., 1993a,b, 1998, 1999; Ey and Darby, 1998).

The number of *vsp* genes in the *Giardia* genome has been estimated at between 133 and 151 in the Group 3 isolate GS (Nash et al., 1990a). However, the calculation of the number of unique *vsp* genes depends on several assumptions. Surprisingly, another estimate that employed analysis of partial sequencing of *Giardia* cosmid genomic libraries found 2.4% of the *Giardia* genome or 144 *vsps* in one genome (Smith et al., 1998). These two estimates are in surprisingly close agreement even though the methods used were different. The genomic organization of VSP1267 consists of two identical *vsp* genes in a tail-to-tail arrangement, suggesting that more than one copy of a particular *vsp* gene may be at a single locus (Mowatt et al., 1991). The genomic organization of the few other *vsp* genes studied did not show a similar organization, so it is unknown how often two copies of the same *vsp* are found close to one another. However, if common, it would obviously alter the estimate of the number of different VSPs in a genome. A clearer understanding of the organization of the *vsp* genes should be obtained from the *Giardia* sequencing project.

The amino terminus is probably the region where the VSP makes direct contact with the environment, which is implied by the progressive decrease in conservation of CXXC motifs from the carboxyl terminus to the amino terminus (Nash and Mowatt, 1993). This suggests that the most variable region, and hence the region that would most profitably expose the host to biologic variability at the interface between the VSP and host, lies in the amino terminus. This is also one of the most antigenic regions of the molecule. Three Mabs that react with the surface of viable trophozoites locate to the amino terminus, supporting this view (Mowatt et al., 1994; Stager et al., 1997; Nash et al., 2001). In addition, influenza haemagglutinin epitopes inserted in the amino terminus portion of one VSP are detected on the surface of viable transfected parasites, confirming that this region is at the surface interface (L. Kulakova and T. E. Nash, unreported results).

The presence or absence of a certain VSP in any trophozoite is dependent upon its genetic groupings (Nash and Mowatt, 1992b). In other words, different genetic groups of *Giardia* seem to have their own VSP repertoires, which are more closely related to themselves than to other genetic groupings. The first meaningful analysis by Nash et al. using RFLPs placed *G. lamblia* isolates into three groups (Nash et al., 1985). Group 1 was similar but measurably different from Group 2, but both differed rather dramatically from Group 3. This grouping was validated and expanded using a panel of molecular and phenotypic characteristics, including the presence of vsp genes and the ability to express them (Nash and Mowatt, 1992b). Other studies have validated this grouping (Weiss *et al.*, 1992; Baruch *et al.*, 1996). Subsequently other more quantitative analyses confirmed the above groupings but combined the Group 1 and Group 2 isolates into Assemblage A and Group 3 into Assemblage B (Adam, 2001). Additional Assemblages, mostly from animals, have more recently been suggested (Adam, 2001). The limitation of VSPs to specific genetic groups is readily shown by reactivity using VSP-specific Mabs. Analysis of the reactivity of three Mabs specific to Group 1 *Giardia* isolates (Mabs 6E7, 5C1 and 3F6) and six Mabs specific to Group 3 showed no cross-reactivity between Groups 1 and 3 (S. M. Singer and T. E. Nash, unreported results) (Nash *et al.*, 1990b; Nash and Mowatt, 1992b). Similarly, *vsp* genes were also limited to their own genetic group (Nash and Mowatt, 1992b). Other studies showed more subtle differences between VSPs in Groups 1 and 2 (Ey *et al.*, 1998). A recent analysis of 30 Group 3 or Assemblage B partial *vsp* sequences found that all but four were most closely related to Group 3 or Assemblage B, confirming segregation of *vsp* genes to their own Group (Bienz *et al.*, 2001b).

BIOLOGY OF VARIANT-SPECIFIC PROTEINS

VSPs appear to be important to the organism even without knowing their biological role(s). As noted above, a considerable amount of its genome, estimated at 9%, encodes for VSPs and a considerable effort is devoted to producing them. Radioactive incorporation of cysteine into trophozoites leads mostly to incorporation into VSPs (Aggarwal *et al.*, 1989; Bruderer *et al.*, 1993) and the entire trophozoite surface including the flagella are coated by VSP (Nash and Aggarwal, 1986; Nash *et al.*, 1990b). Immunoelectronmicroscopy revealed remarkably heavy localization of VSP to the surface (Gillin *et al.*, 1990; Pimenta *et al.*, 1991), and one unsubstantiated analysis estimated that there are VSP 5 $\times 10^6$ copies/trophozoite (Papanastasiou *et al.*, 1997a).

The transport and processing of VSPs appear similar to that of higher eukaryotes. Brefeldin A inhibits VSP secretion, suggesting similar mechanisms of vesicular transport compared to higher eukaryotes (Lujan *et al.*, 1995). Additionally, VSPH7 is secreted and placed on the surface of transfected COS cells (T. E. Nash, unpublished results). Therefore, the VSP's signal peptide is able to direct it to the secretory pathway in mammalian cells, indicating that the signal peptide operates similarly in *Giardia* and higher eukaryotes. Although trophozoites do not have an easily visualized Golgi (Lujan *et al.*, 1995; Lanfredi-Rangel *et al.*, 1999) and no VSP has been localized to it, VSPH7 clearly is processed through the Golgi in COS cells and therefore the same is likely to occur in *Giardia* (T. E. Nash, unpublished results). VSPs are released from the surface as a somewhat shorter, less hydrophobic species, indicating a protease cleavage that releases the VSP from the surface, suggesting it is separated from its hydrophobic tail (Papanastasiou et al., 1996). Interestingly, VSPH7 could not be surface-radiolabelled when present on the surface of COS cells nor is it released from the surface in any appreciable amount (T. E. Nash, unpublished results). In contrast, surface VSPH7 is readily surface-radiolabelled and released from the surface of Giardia (Nash et al., 1983). Not unexpectedly, this indicates that there are specific processing events relating to VSP surface placement and secretion that are unique to Giardia and that cannot be carried out in other cells. VSPs have also been localized to the peripheral vacuoles during encystation, suggesting that at certain times VSPs may be internalized (Mccaffery et al., 1994; Svard et al., 1998).

ROLE OF IMMUNE AND BIOLOGICAL SELECTION

Because there is spontaneous change in VSP expression *in vitro*, biological host factors *probably* determine which VSPs persist *in vivo*. There is evidence that selective pressures by both immunological and non-immunological mechanisms determine which VSPs are preferred or not preferred in a particular circumstance.

The time-honoured rationale of the purpose of antigenic variation is immune escape from the host. The expected pattern of infection is one of waves of parasites expressing one dominant surface variant followed by immune elimination and a decrease in the numbers of parasites followed by a second wave of parasites expressing another variant antigen and so on. In humans experimentally infected with the Group 3 isolate GS expressing VSPH7, the same VSP is expressed until antibodies to VSPH7 appear at about 10–14 days, when organisms expressing VSPH7 are gradually replaced by trophozoites expressing other VSPs by day 21 (Nash et al., 1990c). Thereafter, the pattern of VSP expression remains uncharacterized in humans. In adult (Byrd et al., 1994; Singer et al., 2001) and neonatal mice (Gottstein et al., 1990) infected with the same clone, expression of VSPH7 continues for 10–14 days, at which time humoral responses are first detected and the number of trophozoites decreases so that after day 21 barely perceptible numbers persist for months, if not for the life of the animal (Muller and Stager, 1999). In one experiment using neonatal mice whose mothers were previously infected or not with VSPH7 while gravid, there was a suggestion of sequential waves of intestinal trophozoites (Stager et al., 1998). In the control mice from uninfected mothers, there was a decrease in the number of intestinal parasites only on day 8, while in the maternally exposed neonatal mice there were decreases in the number of intestinal parasites on days 8, 10 and 14. The decrease noted in the control mice in day 8 occurred at the same time that antibodies to VSPH7 were first detected and is consistent with a transient effect of antibodies. In the maternally exposed mice the decreases in parasite burdens were transient and occurred within a very short time frame, suggesting that the changes were perhaps not due to humoral or immunological mechanisms. Nevertheless, the pattern of infection in these models clearly differs from the idealized sequential waves of parasites each expressing predominately one VSP. The overall pattern is characterized by a relatively short time of increased parasite intestinal burden followed by a profound suppression. Analysis of VSPs during this suppressive period appears to show a diversification of VSPs perhaps followed by a limited selection of expressed VSPs. In the neonatal mouse model analysis of the VSPs expressed over time in separate mice showed different sets of VSPs expressed at 14, 21 and 42 days after inoculation (Bienz et al., 2001b); but at 21 and 42 days there was over representation of one VSP. Because of the techniques employed, these analyses were not quantitative, so definitive conclusions are not possible.

Immune responses to VSPs are important in determining what VSPs are expressed. There are consistent data indicating that specific anti-VSP humoral responses are important in the immune elimination of *Giardia* expressing those VSPs. On the other hand, it is unclear if humoral responses are primarily responsible for control of the infection and the profound decrease in intestinal-dwelling trophozoites detected in the intestine over time. Switching *in vivo* occurs at the time antibodies are first detected (Gottstein et al., 1990; Nash et al., 1990c; Stager et al., 1997). Humoral responses in experimental mice (Gottstein et al., 1990; Stager and Muller, 1997; Stager et al., 1997, 1998), human (Nash et al., 1990c) and gerbil infections (Aggarwal and Nash, 1987) are directed to the VSPs and result in growth inhibition and/or death of the trophozoite (Nash and Aggarwal, 1986; Hemphill et al., 1996). In the adult and neonatal mouse models of *G. lamblia* infection, B cell-deficient (Singer and Nash, 2000; Stager and Muller, 1997) and SCID mice (Gottstein and Nash, 1991; Singer and Nash, 2000; Singer et al., 2001) do not undergo antigenic variation although they may undergo non-antibody-mediated changes. In contrast, T cell-deficient animals have a markedly prolonged and enhanced intensity of infection but still undergo antigenic variation (Gottstein and Nash, 1991; Singer and Nash, 2000). These animals are able to produce humoral antibodies and immune selection and yet are unable to control infections.

There are conflicting results concerning the role of humoral responses in controlling *Giardia* infections mice. Using a limited number of *G. lamblia*-infected, B cell-deficient mice, Stager et al. found low numbers of intestinal trophozoites in B cell-deficient mice and even lower numbers in immunocompetent mice, hinting at B cell-mediated suppressive effects (Stager and Muller, 1997). No statistical evaluation was possible in these experiments. More recently, Langford et al. (2002), using *G. muris* infections in B cell- and IgA-deficient mice, showed that humoral responses were paramount in the control *G. muris* infections. They also found that the control *G. lamblia* infections in mice were also IgA-dependent. In contrast, studies comparing the roles of T cell-, B cell- and IgA-deficient mice in *G. lamblia* infections at the same time found that acute infections were primarily controlled by T cell-mediated mechanisms (Singer and

Nash, 2000). Although there are a number of differences in the methods that may explain some of these disparate results, nevertheless, at the present time it is difficult to reconcile these contrasting results.

A detailed study of the humoral responses to VSPH7 in neonatal mice showed that cytotoxic activity was directed to the first 119 amino acids similar to the cytotoxic reactivity of Mab G10/4 (Muller *et al.*, 1996; Stager *et al.*, 1997). Antibodies appearing during the chronic phase were transiently inhibitory and mostly reactive to the carboxyl terminal GGCY motif (Bienz *et al.*, 2001a) and not to the conserved hydrophobic tail (Muller *et al.*, 1996). No antibodies were formed to the latter. T cell proliferative responses to *Giardia* homogenates have been detected in mice (Gottstein *et al.*, 1990) and humans (Gottstein *et al.*, 1991), but specific responses to VSPs were not detected.

In addition to the adaptive immune responses, biological selection apart from immunologically based mechanisms is also important in antigenic variation. Although VSPs have a number of characteristics and features in common, they are distinct enough to possess unique biochemical–physical characteristics. They differ in size, presence and number of repeating units, number or make up or location of common motifs, and composition of the intervening sequence. As noted above, the locations of the CXXC and Zn finger motifs are not the same in all VSPs, adding to the variability of these proteins. The first indication that VSPs differed biochemically came from studies of the *in vitro* effects of proteases on *Giardia* expressing different VSPs. When different VSP-expressing *Giardia* clones were cultured in high concentrations of trypsin or chymotrypsin, some clones were relatively unaffected while in others almost all the organisms were killed. However, a few trophozoites expressing other VSPs resistant to these proteases survived and continued to grow, repopulating the culture (Nash *et al.*, 1991). Since trophozoites survive and multiply in the small intestines, where proteases are present in high concentrations, resistance to intestinal proteases is necessary for survival. Presumably the large number of disulphide bonds contributes to protease resistance, as reduction of VSPs leads to sensitivity to digestion at the same enzyme concentration (Papanastasiou *et al.*, 1997a). Why VSPs that are sensitive to intestinal proteases exist is not obvious, but VSPs may have unique and/or different properties under more physiological conditions. It is likely that there are a number of as yet unknown biological features of VSPs that distinguish them. Nevertheless, these *in vitro* experiments amply demonstrate that VSPs may vary in their biochemical–physical properties and therefore can potentially be positively or negatively selected for *in vivo* on the basis of these differences.

Biological selection occurs *in vivo* in experimentally infected mice and there is evidence that it also occurs in humans and gerbils as well. Adult immunocompetent and immunoincompetent SCID mice lacking B and T cell-mediated immunity were infected with clones of the GS isolate expressing VSPs that reacted with specific Mabs (Singer *et al.*, 2001). As mentioned above, adult immunocompetent mice eliminated the predominant VSP in the inocula. In contrast, infections in SCID mice revealed that organisms expressing certain VSPs were maintained and even enhanced, such as those expressing VSPH7, while others were eliminated and replaced by organisms expressing other VSPs. These experiments were repeated in gerbils made immunodeficient by irradiation. Although certain VSP-expressing clones were favoured or eliminated, similar to infections in SCID mice, there were differences. Some VSP-expressing clones were eliminated in mice but survived in gerbils while others were favoured in both or eliminated in gerbils and favoured in mice. *In vitro* cultures of these clones carried out as controls showed gradual loss of the initial VSP in each, which is the expected replacement by organisms expressing many different VSPs over the same time frame. These results clearly demonstrate that biological selection of VSPs occurs in the absence of an adaptive immune response. Differences in host VSP preferences suggest that one of the benefits of this is to broaden the host range of *G. lamblia* in nature.

Biological selection was also suggested in human experimental infections (Nash *et al.*, 1987, 1990c). Humans were inoculated with a heterologous GS culture consisting of trophozoites expressing many different VSPs, including VSPH7 (from 0.1% to 1.2%) and another VSP recognized by Mab 3F6 (from 1.0% to 2.6%). Studies of the surface expression of organisms obtained between 7 and 14 days post-inoculation showed a change in the expression pattern. Despite being only one of many VSPs present in the inoculum, VSPH7 became the predominant VSP expressed in five of six volunteers, while the numbers of organisms expressing the VSP recognized by Mab 3F6 decreased to 0% in five of six. Although the exact date the organisms were obtained was not recorded, all were obtained on or before day 14, when humoral responses were first detected. Subsequently, as noted above, experimental infections of another group of volunteers with VSPH7-expressing organisms showed that it remained the predominant expressed VSP, and therefore preferred, until day 17, when the numbers of VSPH7-expressing organisms gradually decreased until day 21, when they were completely replaced by other VSP-expression organisms. These results demonstrate biological selection because VSPH7-expressing organisms were preferred and were only eliminated with the initiation of an immune response directed against it. A clear demonstration of VSP-expression changes before the presence of an immune response was seen in experimental infection of gerbils with clones of the Group 1 isolate WB. Analysis of surface VSP expression over the first week demonstrated loss of the predominant VSP that was present in the inoculum (Aggarwal and Nash, 1988).

The host immune responses are of major importance. However, infections in mice, humans and gerbils clearly demonstrate that VSP expression *in vivo* is not random and in the absence of an immune response some VSPs are preferred while others are not. Consequently, for a particular VSP to be highly expressed it first must be able to survive and grow in the host and, second, not be recognized by the host. One would predict that the numbers of VSPs that can survive in a particular host or environment are more limited than the entire repertoire of VSPs. This is likely to further decrease the number of VSPs that can exist in one particular host. Exactly how the parasite is able to maintain a chronic infection is still unclear. The intensity of infection may be so low that an immune response is not initiated. Conversely, immune escape may be predominant during this phase of the infection. Alternatively, the mechanisms may be more complex and include specific immune suppression for instance.

MOLECULAR MECHANISMS

The molecular mechanisms involved in antigenic variation are not known but do not appear to be very similar to those employed by more well-studied parasites that undergo antigenic variation, such as trypanosomes. Steady-state transcripts from clones expressing predominantly one VSP reveals one major transcript. Immediate progeny of a clone not expressing the first VSP but expressing a second VSP shows loss of the first transcript and gain of the second transcript (Adam *et al.*, 1988; Mowatt *et al.*, 1991; Yang *et al.*, 1994). There is no indication that post-transcriptional processing occurs, although this has not been vigorously studied. To date there are no studies that suggest the production of non-productive transcripts where only one of many transcripts is permitted to produce a single surface protein (L. Kulakova and T. E. Nash, unreported results). Analyses of *vsp* genes from expressing and non-expressing clones has not shown rearrangements of these genes as would be expected in duplicative transposition events, so movement into telomere-associated expression sites as occurs in trypanosomes appears unlikely. Furthermore, *vsp* genes, except in two instances reported from one laboratory (Upcroft *et al.*, 1997), are not associated with telomeres (Yang and Adam, 1995a; Adam, 2001). Studies of closely related *vsp* genes indicated that they occupy the same position in homologous chromosomes, suggesting that they are analogous to alleles (Yang and Adam, 1994). These closely related *vsp*s as well as other related *vsp* genes have upstream and downstream sequences that are practically identical among them (Mowatt *et al.*, 1991; Yang and Adam, 1994, 1995a; Nash *et al.*, 1995). Analysis from alleles from three genes suggests that only one of the homologous genes is transcribed and expressed (Yang and Adam, 1994; Nash *et al.*, 1995; Adam, 2001). If upstream and/or downstream sequence controls antigenic variation then there must be a mechanism to differentiate these genes and allow expression of only one. For these reasons epigenetic mechanisms may play an essential role in antigenic variation.

Chromosomal organization and location of *vsp* genes has been studied to see if the organization sheds some light on the mechanisms involved in antigenic variation. *Vsp* genes apparently are present on most chromosomes and appear dispersed throughout the genome (Adam, 2001). The first gene for which the genomic organization was studied, *vsp1267*, consisted of two identical tail-to-tail copies with identical 5' and 3' UTRs up to several hundred base pairs (Mowatt *et al.*, 1991). Other *vsp* genes have not shown this type of organization.

Two *vsp* genes have been extensively studied (Adam *et al.*, 1988). The *vspA6* family consists of four closely related but different genes, three of which clearly localize to the identical region of chromosome 4, indicating that they are alleles. The signature changes of *vspA6* alleles are in the number of repeating units and in a small number of nucleotide changes (Yang and Adam, 1994; Yang *et al.*, 1994). In addition, three of these alleles were identical to the 5' UTR of the expressed gene up to −372 bp (Adam *et al.*, 1988). The *vspC5* is another *vsp* gene with closely related alleles. (Yang and Adam, 1995a,b). Extended maps around some of these *vsp* genes fail to include the telomeres, and the *vspA6* gene was not found in the 150 kb and 240 kb telomere fragments of chromosome 4 (Yang and Adam, 1995a,b; Adam, 2001). Another extended but shorter restriction map of the *vspH7* gene did not include the end of the chromosome (Nash *et al.*, 1995). In contrast, two *vsp* genes were proximal to a protein kinase gene and a gene consisting of ankryin repeats which abutted on the large sub-unit of rDNA near the telomere (Upcroft *et al.*, 1997). Therefore, some *vsp* genes are associated with the telomere while others, including *vsps* that were being expressed, were not. Although telomere association is of potential importance, *vsps* do not seem to be located near the telomere to be expressed.

Although there is a general plasticity of the genome, particularly around the rDNA at the ends of some of the chromosomes (Le Blancq and Adam, 1998), analysis of the same *vsp* genes studied in clones expressing or not expressing the gene have only rarely demonstrated any changes in the sequence or location of the gene that correlated with change of VSP expression (Adam *et al.*, 1988, 1992). In one case the loss of a *vsp* gene was found in a non-expressing clone (Adam *et al.*, 1992). Presumably this represented loss of the expressed allele.

Control of VSP transcription differs compared to transcription of constitutively expressed proteins. Expression of constitutive and regulated proteins in *Giardia* is controlled by a relatively small promoter situated just upstream of the translation start site (Yee and Nash, 1995; Sun *et al.*, 1998; Knodler *et al.*, 1999; Sun and Tai, 1999, 2000; Yee *et al.*, 2000; Elmendorf *et al.*, 2001; Hayman and Nash, 2002). Commonly, all the controlling elements reside within 60 bp, and mutation of certain regions, as expected, severely alters transcription. Many of these upstream regions consists of stretches of A's and T's where transcription is initiated. Upstream regions of VSPs differ from those of constitutive proteins.

Using the newly developed transcription system, we have been able to study the 5' UTR of VSPH7. When the alpha tubulin gene 5' and 3' UTRs is used to drive VSP transcription, VSPH7 is expressed and appropriately placed on the surface of the heterologous Group 1 isolate WB (Elmendorf *et al.*, 2001). This organism does not have the *vspH7* gene and therefore any VSPH7 on its surface is evidence of expression and proper transport. Use of the 5' UTR up to 400 bp upstream and its own 3' UTR resulted in the intermittent expression of VSPH7 similar to antigenic variation (T. E. Nash, S. M. Singer and L. Kulakova, unpublished results). This occurred similarly in constructs that were either integrated or maintained as plasmids. However, transcription initiation began spuriously at varying sites upstream of the usual start site, which was only rarely employed in these constructs (L. Kulakova and T. E. Nash, unreported results) Presumably most of these were sterile transcripts. In GS clones not expressing VSPH7 (e.g. the native state) no transcription of vspH7 was detected, indicating that spurious transcription does not occur normally in VSPs that are not expressed (L. Kalakova. and T. E. Nash, unreported results). Additionally, replacement of the immediate 5' 73 bp that allowed intermittent VSPH7 expression with an unrelated sequence did not result in loss of the ability to allow intermittent expression. These results indicate that the upstream 73 bp is not essential, not in its proper genomic context, or cannot be appropriately altered to promote antigenic variation (L. Kalakova and T. E. Nash, unreported results). Another possibility is that the heterologous isolate, WB, is unable to support antigenic variation appropriately as would occur in the homologous isolate.

CONCLUSION

Surface antigenic variation in *G. lamblia* is unique. However, the major questions about the phenomenon remain unanswered. What is the biologic role of antigenic variation? How does the unique structure of VSPs help the parasite and does it contribute to the pathophysiology of disease? What are the molecular mechanisms involved in antigenic variation? The ability to transfect *Giardia*, use of appropriate animal models, detailed studies of human infection and knowledge of VSP genomic organization will probably lead to a better understanding of this process.

REFERENCES

Adam, R. D. (2001) *Clin. Microbiol. Rev.* 14, 447.
Adam, R. D., Aggarwal, A., Lal, A. A., Cruz, d. l. V. F., McCutchan, T. and Nash, T. E. (1988) *J. Exp. Med.* 167, 109–118.
Adam, R. D., Yang, Y. M. and Nash, T. E. (1992) *Mol. Cell Biol.* 12, 1194–1201.
Aggarwal, A. and Nash, T. E. (1987) *Am. J. Trop. Med. Hyg.* 36, 325–332.
Aggarwal, A. and Nash, T. E. (1988) *Infect. Immun.* 56, 1420–1423.
Aggarwal, A., Merritt, J. J. and Nash, T. E. (1989) *Mol. Biochem. Parasitol.* 32, 39–47.
Bannon, G. A., Perkins, D. R. and Allen, N. A. (1986) *Mol. Cell Biol.* 6, 3240–3245.
Baruch, A. C., Isaacrenton, J. and Adam, R. D. (1996) *J. Infect. Dis.* 174, 233–236.
Bienz, M., Siles-Lucas, M., Wittwer, P. and Muller, N. (2001b) *Infect. Immun.* 69, 5278–5285.
Bienz, M., Wittwer, P., Zimmermann, V. and Muller, N. (2001a) *Int. J. Parasitol.* 31, 827–832.

Bruderer, T., Papanastasiou, P., Castro, R. and Kohler, P. (1993) *Infect. Immun.* 61, 2937–2944.
Byrd, L. G., Conrad, J. T. and Nash, T. E. (1994) *Infect. Immun.* 62, 3583–3585.
Chen, N., Upcroft, J. A. and Upcroft, P. (1995) *Parasitology* 111, 423–431.
Chen, N., Upcroft, J. A. and Upcroft, P. (1996) *Gene* 169, 33–38.
Cheng, X. J., Hughes, M. A., Huston, C. D., Loftus, B., Gilchrist, C. A., Lockhart, L. A., Ghosh, S., Miller-Sims, V., Mann, B. J., Petri, W. A. and Tachibana, H. (2001) *Infect. Immun.* 69, 5892–5898.
Clark, T. G., McGraw, R. A. and Dickerson, H. W. (1992) *Proc. Natl Acad. Sci. USA* 89, 6363–6367.
Davis-Hayman, S. R. and Nash, T. E. (2002) *Mol, Biochem. Parasitol.* 122, 1–7.
Elmendorf, H. G., Singer, S. M., Pierce, J., Cowan, J. and Nash, T. E. (2001) *Mol. Biochem. Parasitol.* 113, 157–169.
Ey, P. L. and Darby, J. M. (1998) *Exp. Parasitol.* 90, 250–261.
Ey, P. L. and Mayrhofer, G. (1993) *Gene* 129, 257–262.
Ey, P. L., Darby, J. M., Andrews, R. H. and Mayrhofer, G. (1993a) *Int. J. Parasitol.* 23, 591–600.
Ey, P. L., Darby, J. M. and Mayrhofer, G. (1998) *Parasitology,* 117, 445–455.
Ey, P. L., Darby, J. M. and Mayrhofer, G. (1999) *Mol. Biochem. Parasitol.* 99, 55–68.
Ey, P. L., Khanna, K. K., Manning, P. A. and Mayrhofer, G. (1993b) *Mol. Biochem. Parasitol. 58,* 247–257.
Gillin, F. D., Hagblom, P., Harwood, J., Aley, S. B., Reiner, D. S., McCaffery, M., So, M. and Guiney, D. G. (1990) *Proc. Natl Acad. Sci. USA* 87, 4463–4467.
Gottstein, B. and Nash, T. E. (1991) *Parasite Immunol.* 13, 649–659.
Gottstein, B., Harriman, G. R., Conrad, J. T. and Nash, T. E. (1990) *Parasite Immunol.* 12, 659–673.
Gottstein, B., Stocks, N. I., Shearer, G. M. and Nash, T. E. (1991) *Infection* 19, 421–426.
Hemphill, A., Stager, S., Gottstein, B. and Muller, N. (1996) *Parasitol. Res.* 82, 206–210.
Hiltpold, A., Frey, M., Hulsmeier, A. and Kohler, P. (2000) *Mol. Biochem. Parasitol.* 109, 61–65.
Knodler, L. A., Svard, S. G., Silberman, J. D., Davids, B. J. and Gillin, F. D. (1999) *Mol. Microbiol.* 34, 327–340.
Lanfredi-Rangel, A., Kattenbach, W. M., Diniz, J. A., Jr and de Souza, W. (1999) *FEMS Microbiol. Lett.* 181, 245–251.
Langford, T. D., Housley, M. P., Boes, M., Chen, J., Kagnoff, M. F., Gillin, F. D. and Eckmann, L. (2002) *Infect. Immun.* 70, 11–18.
Le Blancq, S. M. and Adam, R. D. (1998) *Mol. Biochem. Parasitol.* 97, 199–208.
Lujan, H. D., Marotta, A., Mowatt, M. R., Sciaky, N., Lippincott, S. J. and Nash, T. E. (1995) *J. Biol. Chem.* 270, 4612–4618.
Marti, M., Li, Y. J., Kohler, P. and Hehl, A. B. (2002) *Infect. Immun.* 70, 1014–1016.
Mata, L. J. (1978) *The Children of Santa Maria Cauque: a Prospective Field Study of Health and Growth.* MIT Press, Cambridge, MA.
Mccaffery, J. M., Faubert, G. M. and Gillin, F. D. (1994) *Exp. Parasitol.* 79, 236–249.
Mowatt, M. R., Aggarwal, A. and Nash, T. E. (1991) *Mol. Biochem. Parasitol.* 49, 215–227.
Mowatt, M. R., Nguyen, B. Y., Conrad, J. T., Adam, R. D. and Nash, T. E. (1994) *Infect. Immun.* 62, 1213–1218.

Muller, N. and Stager, S. (1999) *Int. J. Parasitol.* 29, 1917–1923.
Muller, N., Stager, S. and Gottstein, B. (1996) *Infect. Immun.* 64, 1385–1390.
Nakagawa, T., Murakami, K. and Nakayama, K. (1993) *FEBS Lett.* 327, 165–171.
Nash, T. (1992) *Parasitol. Today* 8, 229–234.
Nash, T. E. and Aggarwal, A. (1986) *J. Immunol.* 136, 2628–2632.
Nash, T. E. and Mowatt, M. R. (1992a) *Mol. Biochem. Parasitol.* 51, 219–227.
Nash, T. E. and Mowatt, M. R. (1992b) *Exp. Parasitol.* 75, 369–378.
Nash, T. E. and Mowatt, M. R. (1993) *Proc. Natl Acad. Sci. USA* 90, 5489–5493.
Nash, T. E., Aggarwal, A., Adam, R. D., Conrad, J. T. and Merritt, J. W., Jr (1988) *J. Immunol.* 141, 636–641.
Nash, T. E., Banks, S. M., Alling, D. W., Merritt, J. J. and Conrad, J. T. (1990a) *Exp. Parasitol.* 71, 415–421.
Nash, T. E., Conrad, J. T. and Merritt, J. W., Jr (1990b) *Mol. Biochem. Parasitol.* 42, 125–132.
Nash, T. E., Conrad, J. T. and Mowatt, M. R. (1995) *J. Eukaryot. Microbiol.* 42, 604–609.
Nash, T. E., Gillin, F. D. and Smith, P. D. (1983) *J. Immunol.* 131, 2004–2010.
Nash, T. E., Herrington, D. A., Levine, M. M., Conrad, J. T. and Merritt, J. J. (1990c) *J. Immunol.* 144, 4362–4369.
Nash, T. E., Herrington, D. A., Losonsky, G. A. and Levine, M. M. (1987) *J. Infect. Dis.* 156, 974–984.
Nash, T. E., Lujan, H. T., Mowatt, M. R. and Conrad, J. T. (2001) *Infect. Immun.* 69, 1922–1923.
Nash, T. E., McCutchan, T., Keister, D., Dame, J. B., Conrad, J. D. and Gillin, F. D. (1985) *J. Infect. Dis.* 152, 64–73.
Nash, T. E., Merritt, J. J. and Conrad, J. T. (1991) *Infect. Immun.* 59, 1334–1340.
Papanastasiou, P., Bruderer, T., Li, Y., Bommeli, C. and Kohler, P. (1997a) *Mol. Biochem. Parasitol.* 86, 13–27.
Papanastasiou, P., Hiltpold, A., Bommeli, C. and Kohler, P. (1996) *Biochemistry* 35, 10143–10148.
Papanastasiou, P., McConville, M. J., Ralton, J. and Kohler, P. (1997b) *Biochem. J.* 49–56.
Peters-Regehr, T., Kusch, J. and Heckmann, K. (1997) *Eur. J. Parasitol.* 33, 389–395.
Pimenta, P. F., da Silva, P. P. and Nash, T. (1991) *Infect. Immun.* 59, 3989–3996.
Rendtorff, R. C. (1954) *Am. J. Hyg.* 59, 209–220.
Rendtorff, R. C. and Holt, J. C. (1954) *Am. J. Hyg.* 60, 327–338.
Roebrock, A. J., Creemers, J. W., Pauli, I. G., Kurzik-Dumke, U., Rentrop, M., Gataff, E. A., Leunissen, J. A. and Van de Ven, W. J. (1992) *J. Biol. Chem.* 267, 17208–17215.
Singer, S. M. and Nash, T. E. (2000) *Infect. Immun.* 68, 170–175.
Singer, S. M., Elmendorf, H. G., Conrad, J. T. and Nash, T. E. (2001) *J. Infect. Dis.* 183, 119–124.
Smith, M. W., Aley, S. B., Sogin, M., Gillin, F. D. and Evans, G. A. (1998) *Mol. Biochem. Parasitol.* 95, 267–280.
Sogin, M. L., Gunderson, J. H., Elwood, H. J., Alonso, R. A. and Peattie, D. A. (1989) *Science* 243, 75–77.
Stager, S. and Muller, N. (1997) *Infect. Immun.* 65, 3944–3946.
Stager, S., Felleisen, R., Gottstein, B. and Muller, N. (1997) *Mol. Biochem. Parasitol.* 85, 113–124.

Stager, S., Gottstein, B., Sager, H., Jungi, T. W. and Muller, N. (1998) *Infect. Immun.* 66, 1287–1292.

Sun, C. H. and Tai, J. H. (1999) *J. Biol. Chem.* 274, 19699–19706.

Sun, C. H. and Tai, J. H. (2000) *Mol. Biochem. Parasitol.* 105, 51–60.

Sun, C. H., Chou, C. F. and Tai, J. H. (1998) *Mol. Biochem. Parasitol.* 92, 123–132.

Svard, S. G., Meng, T. C., Hetsko, M. L., McCaffery, J. M. and Gillin, F. D. (1998) *Mol. Microbiol.* 30, 979–989.

Upcroft, P., Chen, N. H. and Upcroft, J. A. (1997) *Genome Res.* 7, 37–46.

Weiss, J. B., van Kewen, H. and Nash, T. E. (1992) *Mol. Biochem. Parasitol.* 54, 73–86.

Yang, Y. M. and Adam, R. D. (1994) *Nucleic Acids Res.* 22, 2102–2108.

Yang, Y. and Adam, R. D. (1995a) *Mol. Biochem. Parasitol.* 75, 69–74.

Yang, Y. M. and Adam, R. D. (1995b) *J. Eukaryot. Microbiol.* 42, 439–444.

Yang, Y. M., Ortega, Y., Sterling, C. and Adam, R. D. (1994) *Mol. Biochem. Parasitol.* 68, 267–276.

Yee, J. and Nash, T. E. (1995) *Proc. Natl Acad. Sci. USA* 92, 5615–5619.

Yee, J., Mowatt, M. R., Dennis, P. P. and Nash, T. E. (2000) *J. Biol. Chem.* 275, 11432–11439.

Zhang, Y. Y., Aley, S. B., Stanley, S. L. and Gillin, F. D. (1993) *Infect. Immun.* 61, 520–524.

17

FREE-LIVING AND PARASITIC CILIATES

THEODORE G. CLARK AND JAMES D. FORNEY

INTRODUCTION

Nearly one hundred years ago, Rössle (1905) observed that *Paramecium* exposed to the appropriate dilution of homologous antiserum are immobilized and killed. Jollos (1921) later demonstrated that resistant cell lines would arise from the initial culture, thus establishing the principle of antigenic variation in *Paramecium*. The following decades of research have sought to define the antigens involved in this phenomenon and the regulatory mechanisms responsible for switching. Despite progress in these areas (reviewed here), the biological significance of antigenic variation in free-living ciliates remains mysterious. Unlike parasitic protozoa, *Paramecium* has no need to avoid an immune response – it has no host. Nevertheless, its ability to switch surface coats is not unique, since *Tetrahymena*, a related free-living ciliate, exhibits antigenic variation as well (Smith *et al.*, 1992). Speculation regarding the function of variable surface antigens in the ciliates and the need for antigen switching has been presented, yet few studies that directly address these issues have been performed (for an exception see Harumoto and Miyake, 1993). The mystery is unlikely to be resolved solely with molecular investigations and will require a better understanding of the natural history and ecology of the ciliates as a whole. In the absence of a clear explanation the authors have sought common themes within the ciliate phylum. Here we review our present knowledge of antigenic variation in *Paramecium*, and contrast this with the current understanding of the i-antigens of *Ichthyophirius*, a parasitic ciliate that represents a bridge between the free-living species and more distantly related parasitic protozoa. We highlight the similarities and differences between the i-antigens of these organisms, and offer speculation regarding the role of major surface antigens in ciliates.

PARAMECIUM VARIABLE SURFACE ANTIGENS

Paramecium variable surface antigens were originally identified using antisera that immobilized and killed cells expressing the corresponding antigen, therefore the proteins are commonly referred to as immobilization antigens (i-antigens). In this review the term 'i-antigen' will be used interchangeably with 'variable surface antigen' or 'variable surface protein'. A pure culture expressing one i-antigen defines a serotype and each serotype (along with its corresponding antigen) is given a letter name (A, B, C etc.). The full repertoire of variable antigens in *Paramecium* is not known. The most extensively studied cell line, *Paramecium tetraurelia*, stock 51, can express at least 11 different antigens on the cell surface (Forney *et al.*, 1983). However, the expression of some i-antigens is unstable, making it difficult to maintain a pure culture. This leaves open the possibility that rare antigens remain undetected. Indeed, even the limited genome sequence obtained from randomly selected plasmid clones of *P. tetraurelia* (less than 2% of the genome) has revealed several i-antigen genes that do not correspond to already known antigen sequences (Dessen *et al.*, 2001). This suggests that a large number of i-antigen genes or gene fragments may be present in the genome.

The most remarkable feature of *Paramecium* i-antigens is their mutually exclusive expression; only one antigen type is present on the cell surface at any given time. Expression is generally stable and most serotypes can be maintained in pure cultures for an extended period. Unlike the stochastic switching of some protozoan parasites, switching in *Paramecium* can be directed. For example, high temperature (34 °C) and abundant food will induce essentially all stock 51 cells to express the A51 antigen within a few cell divisions. Alternatively, low temperature or special culture conditions can induce other serotypes. Despite these observations mutual exclusion cannot be explained merely by a series of alternate environmental conditions for each antigen type. In stock 51 the A, B, C and D serotypes all show stable expression at 27 °C under standard culture conditions. This stable yet exclusive expression of *Paramecium* variable surface antigens has drawn the interest of investigators for several decades. This review will provide some background on early studies, and then focus on investigations at the molecular and genetic levels conducted over the past 10 years. For a thorough description of prior research the reader may consult other reviews, including Caron and Meyer (1989), Preer (1968) or Preer (1986).

Genetics and molecular biology of *Paramecium*

There are several unusual features of *Paramecium* genetics and molecular biology that require an introduction before launching into a description of surface antigen expression. The following is a brief account of the salient features relevant to this review.

Each cell contains two identical diploid micronuclei and a single macronucleus with about 1000 copies of each gene (McTavish and Sommerville, 1980). The micronuclei are transcriptionally silent and therefore gene expression is generally a function of the transcribed macronucleus. The genome size is roughly 9×10^7 bp (Soldo and Godoy, 1972) and the haploid micronuclear chromosome number is estimated at between 41 and 45 in the commonly studied stock 51 (Dippell, 1954). Vegetative reproduction occurs by binary fission in which the micronuclei undergo mitosis and macronuclei divide by an amitotic mechanism. Sexual reproduction, either conjugation or a self-fertilization process called autogamy, results in the formation of a new macronuclear genome from the micronuclear DNA. Since the genetic consequence of autogamy is a completely homozygous genome and autogamy occurs after a defined number of fissions since the previous sexual event, cell lines in the *Paramecium aurelia* complex are normally homozygous. In fact it is difficult to maintain heterozygous cell lines for long periods of time.

The macronucleus contains hundreds of different linear DNA molecules that result from fragmentation of the micronuclear chromosomes. Based on the average size of these linear fragments (300 kb) each micronuclear chromosome must be broken into about seven pieces (Preer and Preer, 1979; Phan *et al.*, 1989). The ends of *Paramecium* macronuclear chromosomes show considerable heterogeneity. For example, the A51 gene can be located 8, 13 or 26 kb from the end of a macronuclear chromosome (Forney and Blackburn, 1988). Similarly the G gene of *Paramecium primaurelia*, stock 156 (G156) is located about 5 kb from a chromosome end, but minor telomere locations have been identified 2.8 kb closer to the gene (Amar, 1994). These telomere variations occur as a result of differential DNA processing during formation of the macronuclear genome from the micronuclear DNA. Antigen switching does not require the formation of a new macronucleus therefore these chromosomal variations are not the primary mode of regulation. Nevertheless, there is evidence that variations in the macronuclear genome can alter the stability of i-antigen expression (see Regulation of i-Antigen Expression section).

Perhaps the most unusual feature of ciliates is their use of a non-standard genetic code. Although different variations are known within the phylum, *Paramecium*, *Tetrahymena* and *Ichthyopthirius* all appear to use two of the three universal stop codons (UAA and UAG) as glutamine codons (Caron and Meyer, 1985; Horowitz and Gorovsky, 1985; Preer *et al.*, 1985; Clark *et al.*, 1992). Although this does not present a problem for *in vivo* investigations, it does limit the use of heterologous systems such as yeast two-hybrid, protein production in *Escherichia coli* and expression-based cloning.

Despite these limitations, *Paramecium* offers some unique advantages for DNA-mediated transformation. Foreign DNA can be injected directly into the macronucleus, where it is maintained as extrachromosomal linear molecules after telomeres are added *in vivo* (Godiska *et al.*, 1987). No specialized vectors are required and copy number is maintained for indefinite periods at roughly the level that was injected.

Antigen structure and variation

In *Paramecium* the i-antigens are abundant, high molecular weight polypeptides ranging from 240 to 310 kDa (Forney *et al.*, 1983). They are located on the outer membrane of the cilia and the cell body and account for roughly 3.5% of the total cellular protein (Mott, 1965; Macindoe and Reisner, 1967). Each protein is encoded by a separate unlinked genetic locus. Our current understanding of variable surface antigen variation and structure is based on a combination of older biochemical studies along with the deduced amino acid sequence from several genes. Complete coding sequences are available for the A51, B51, C51 and D51 genes of *P. tetraurelia* stock 51, and three genes from *P. primaurelia*: G156, G168 and D156 (Prat *et al.*, 1986; Prat 1990; Nielsen *et al.*, 1991; Scott *et al.*, 1993; Bourgain-Guglielmetti and Caron 1996; Breuer *et al.*, 1996). Prat *et al.* (1986) first determined the G156 sequence, and their primary observations regarding the structure of i-antigens have remained fundamentally correct for the sequences that have followed.

The predominant features of the *Paramecium* i-antigens are the regular spacing of cysteine residues and the strong degree of internal homology. The central portion of the G156 sequence contains four almost identical repeats of 74 amino acids each contain eight cysteine residues. Using cysteine residues as the basis for alignment it is possible to display almost the entire sequence as a series of 37 periods containing eight cysteine residues each (along with four half-periods). The A51 gene product also contains internal repeats that can be displayed with a similar cysteine periodicity (Figure 17.1). Thus far, all i-antigen sequences display cysteine periodicity, though some (such as C51, D51 and D156) do not contain tandem repeats. There are also variations in the number of cysteines per period. For example, half-periods (four cysteines) occur in all the antigens sequenced to date. In addition, the D gene product from stock 156 has one cysteine period that contains only three cysteines (Bourgain-Guglielmetti and Caron, 1996). It is worth noting that the most common and stable antigens (A51 and G156) have the most prominent internal repeats. Thus, one feature of the less stable antigenic types may be fewer internal repeats with less ordered cysteine arrangements (although additional sequencing will be required to bear this out).

Comparisons between the G156, G168, A51 and C51 sequences show that the amino-terminal and carboxy-terminal third of the polypeptides are more conserved than the central third of the molecule (Nielsen *et al.*, 1991). The current hypothesis is that the central variable portion of the molecule is exposed to the aqueous medium and the N- and C-terminal regions are tightly packed and inaccessible on the cell surface. This is consistent with the observation that antiserum specific for *in vivo* immobilization cross-reacts with other i-antigens on western blots (for an example see Deregnaucourt 1996). Presumably the polyclonal antiserum reacts only with the exposed variable central region on the intact cell, but after extraction the exposed amino- and carboxy-terminal regions are detected by the antiserum.

	MNQKFPIILSLMLALAASQTYSLTSCTCAQLLSEGDCTKNASLGCSWDSTKKACAVSTTPVTPVMTYAAYCDTFAETDCPKAKPCTDCGSYAACAWVDSKCTYFTG				
1	CTAFAKTTDSD	CQAISNR	CIITDGTH	CVEVDA	CNTYKKQLP	CVKNTAGSL	CVDANT
2	CDKLPVNLATDSD	CRALIST	CTTKTGGG	CVDSGNN	CSDQTLEIQ	CVWNKLKTTA	CKDRI
3	CDNAPTSLTTDDA	CKLFRVDGS	CTTKANGG	CVTRTT	CSAATIQAS	CVKNSSGGD	CVDKN
4	CANAPVTMTNSA	CAGFVTG	CIITKSGGG	CVANGA	CSVANVQAA	CVKNSSNQD	CKEKT
5	CANAPTTNNTHDL	CTSYLST	CIVKTGGG				CQNRT
6	CANAPVTLLTNDA	CEAYLTGNNN	CIITKSGGG	CVTNTT	CAAITLEAA	CVFNSSGQT	CKDKT
7	CLNAPSTNTTHDL	CQAFLNT	CTVNSTSAG	CVEKT	CENSLVLAI	CDKDTSNRA	CIWKGK
8	CVLASSATTSHAD	CQTYSSG	CTLSNTGSG	CVTLPLK	CEAITIEAA	CQMKSNGQP	CGWTGTQ
9	CSTTASKTFSTTTQ	CQTHLSS	CVANNPATVNGSVTIQG	CQDLPTT	CAGRKSSEN	CEITRSGFPT	CLWVSSSST
10	CTTASTVGTTGALSTGQFSFAN	CQNYISI	CTATNAQGVCTATSYPCISNNAGDG	CIAKPTS	CSSLIQSN	CQAGSKSTGD	CVWNGAA
11	CANIALTTHNS	CNTTLNT	CTVNNGATA	CQPLATA	CTSYTTSEN	CKLTSANKK	CVWTGLA
12	CADAPDDTTSDTDGE	CFNYQTPSET	CTVVKVGALG	CVPRSAN	CDYMTQAQ	CHKTLTNLTANDD	CKWIVDK
13	CTSFKGTQTM	CQSYRVG	CINTNSATSSTA				CYATSFASGA
14	CTLKTGTGLAFSD	CQAVDTT	CSVNSTGTG	CIAIQSA	CTGYGQTAAN	CFRSTAGL	CTLD
15	CVLVTGKTGLDHAK	CQAYHTS	CTSLNDGTG	CQEFRAT	CPAFTGDAHAK	CTASQQGK	CQAVTQASE
16	CAAISGIGLTDAK	CAGYNAD	CTVNAAGTA	CQEQKAT	CAVTLTQDS	CTTSKDTATADK	CVRFST
17	CAFVTGSGLDDTQ	CATYNAG	CVANATGTA	CQEKKAA	CTDYTTSTA	CATSTAAK	CAVTTVATQ
18	CLKVTGTGLDDAK	CIAYNAG	CVANSTGTA	CQEKKAA	CTDYTTSTA	CGTSTAAK	CIQISTVGTD
19	CLKVTGTGLDDTK	CIAYNAG	CVANSTGTA	CQEKKAA	CTDYADSTA	CGTSTAAK	CIQISTVGTD
20	CLKVTGTGLDDAK	CIAYNAG	CVANSTGTA	CQEKKAA	CTDYTTSTA	CGTSTAAK	CIQISTVATD
21	CQLVLGSGLNDSK	CSAYNAA	CTSLVDGTA	CQEKKAN	CKDYTTQNK	CISTSSVT	CIQISTVATD
22	CSSITGTSLDHTK	CQAYNTK	CTSITDGTT	CQDPKTT	CEQYPGTAAG	CTKTATSK	CYSITAVS
23	CAKITGSNLTYDT	CQSYTG	CSVNRAKTA	CVQQAL	CSGYSSAMTS	CYKSAAGL	CITISNVATD
24	CDAVFLGAGNYNDAN	CSSFKTG	CTVNGSA				CVAASTAST
25	CANATGITFNHAN	CNSWLNT	CTVNSGSS	CQAMASK	CADQSTA	CQYSVEGE	CTART
26	CDTADADASPDSDSE	CGTYQGS	CTVARLGA	CQARTA	CASYKSLLQ	CKFNTSGGK	CVRKT
27	CGNIEATTTYDSHSE	CVAVDSTLL	CTVRATNGAAVPG	CMARGS	CSSYTIEDQ	CRTNSNGGE	CVDLN
28	CSAFTSTPATHND	CFAYYTSAITVK	CTVAATPSNSGGAATLGG	CQQTAA	CSSYIDKEQ	CQINANGDP	CQDKS
29	CATAPATADYDDTK	CRAYFSNK	CTVSETGQG	CVDIPAT	CETMTQKQ	CNLNKNGDP	CADKS
30	CDNAPDSTATADD	CNTYLSG	CTLDSVK				CITKS
31	CEDFAFATDAL	CKQAIST	CTTNGTN	CVTRGT	CFQALSQSG	CVTSSTGQQ	CKTKV
32	CSTAPLTLLTSEAA	CASYFTN	CTTKNGGG	CVTKSS	CSAVTIVA	CTTALNGTV	CEWIPAVLNAQNVTSPAY
33	CQDFSGTTHAA	CQTQRAG	CTAGANGK	CARVQN	CEQTTVRAA	CIEGSNGP	CTIKT
34	CKSLQMNNDAS	CKMISNK	CTTNGSN	CVGITL	CSETNTDGQ	CVTGYDGA	CRDKD
35	CADAFYTTHSD	CQIASNK	CTTNGTTG	CIALGA	CSSYTSQAG	CYFNDKGAVLQSGAIVSTGL CTWDTTASS	CAWDSAQNK
36	CADLTGTTHAT	CSSQLST	CTSDGTT	CLLKGA	CSSYSTTA	CTTAVGSDGI	CFRYTS
37	CADIQNGTSTSV	CVVALST	CVSNGTA	CIPKAN	CSTYTNKIA	CNSGGLDGI	CKPYTS
	CTTAANDGIACQAARDRCSWTPSGSGATAVASKCATHTCATNQATNGACTRFLMWDKKTQQVCTLVSGTCTATDPSTLSSNDCFLVSGYTYTWNASTSKCGVCTAVVVQPNTSDNNTNNTTTDSGYILGMSIVLGYLMF					CVFTQSTATGAVAGTGT	CLMIDKYPNNDGTKGA
							CIQSVPALNSSDPKV
							CWELASATNNNTAK
							CALMTS
							CRDQT
							CRLLT

Figure 17.1 Repeated cysteine motif displayed for the A51 i-antigen from *P. tetraurelia*. The entire amino acid sequence of A51 can be read left to right starting with the N-terminus in the upper left corner and ending with the C-terminus in the lower right corner. Each numbered cysteine period contains eight cysteine residues with the exception of four half-periods and one 10-cysteine period (period 10). The underlined region in period 35 is essential for antigen expression as shown by Thai and Forney (2000) (Adapted with permission from Nielsen, You and Forney (1991) *J. Mol. Biol.* 222, 835–841.)

Cysteine repeats are a common feature among protozoan surface proteins yet few studies have been performed to analyse the importance of the cysteine arrangement. Thai and Forney (2000) analysed a series of coding region deletions to investigate the flexibility of altering the cysteine pattern. They found that entire periods or portions of periods could be deleted from a region in the N-terminal third of the molecule. Stable expression of the A51 surface protein was used as the assay after transformation of the altered gene into an A- mutant cell line. Surprisingly, a similar set of deletions in the C-terminal region prevented expression of the protein on the cell surface. The 20 amino acid region required for expression had an unusually long cysteine segment when compared with other segments in row 6 of the period arrangement. A stable mRNA was produced from the construct so it is presumed that the defect acts at the protein level. Although the function of this region is not understood, it illustrates that not all parts of the sequence are equivalent despite its repetitive nature. The systematic deletion of areas within the coding region may identify other critical portions of the protein.

There is little information regarding the three-dimensional structure of i-antigens. Computer secondary structure predictions of a low alpha helix content and high beta sheet (Prat *et al.*, 1986) are consistent with 7% alpha helix content measured by optical rotary dispersion (Reisner *et al.*, 1969). Electron microscopy along with physical studies concluded that the i-antigens are ellipsoid spheres roughly 20 by 1.5 nm (Reisner *et al.*, 1969). Early studies showed that few of the sulphydryl groups react with iodoacetate, suggesting that the abundant cysteine residues are involved in disulphide bonds (Jones, 1965).

The i-antigens are attached to the membrane via a glycosylphosphatidylinositol (GPI) lipid anchor (Capdeville *et al.*, 1987). This modification generally requires the cleavage of a hydrophobic N-terminus after entering the ER as well as cleavage of a hydrophobic C-terminus followed by attachment of the glycan in an amide linkage to the new C-terminus. Recent studies have shown that the lipid component for the *Paramecium* variable surface antigens is a ceramide, and the amide-linked fatty acid (linked to the ceramide) varies when paramecia are grown at 23 °C verses 32 °C (Benwakrim *et al.*, 1998). The authors suggest the possibility that variation in fatty acid composition of the GPI anchor could influence surface antigen stability. Interestingly, despite an extensive knowledge of the GPI anchor neither the mature N-terminus nor C-terminus of any of the *Paramecium* i-antigens is known. Newer technical approaches using protein mass spectrometry may provide a relatively straightforward method for obtaining this information.

An association of carbohydrate with the i-antigens was originally reported by Preer (1959b) and later confirmed by Merkel *et al.* (1981). Both lentil lectin and conconavalin A, which recognize mannose and glucose residues, bind specifically to the proteins. Estimates of the molecular mass of the i-antigens combined with the known amino acid sequence suggest that the carbohydrate content is not more than 3% of the molecular mass, and it remains unclear whether the sugar modifications exist apart from the glycolipid anchor.

Regulation of i-antigen expression

Since the discovery of variable surface antigens, the primary effort has been to understand the regulation of mutually exclusive expression. Although our understanding remains incomplete, substantial progress has been made using a combination of molecular biology and genetics. We will review some of the major observations and experimental approaches over the past several years and then provide a general model that can account for the major experimental results.

Stable genes without introns

The cloning of the first *Paramecium* i-antigen genes followed shortly after the observation that DNA rearrangements were associated with trypanosome surface antigen switching (reviewed in Pays *et al.*, 1994). As a result there were several attempts to identify DNA rearrangements associated with *Paramecium* antigen switching. These experiments relied on the analysis of genomic DNA via Southern hybridization at a fine scale using frequent cutting restriction enzymes and on a larger scale using pulse field gels (Forney *et al.*, 1983; Meyer *et al.*, 1985). No evidence for DNA rearrangements was found and in retrospect perhaps this was a naive idea considering the presence of 1000 copies of each gene in the macronucleus.

Two other observations regarding the genomic structure of surface antigen genes may be relevant to their regulation. The first is that some of the surface antigen genes are located near macronuclear telomeres. These include the *P. tetraurelia* A51 and B51 genes as well as the G156 and G168 alleles of *P. primaurelia* (Meyer *et al.*, 1985; Forney and Blackburn, 1988; Scott *et al.*, 1993). The significance of this observation is not entirely clear since other surface antigen genes apparently have more proximal locations. Nevertheless, an analysis of allelic exclusion between G156 and G168 provides a correlation between the location of the telomere downstream of the G156 gene and the expression of the G168 allele in 156/168 heterozygous cells (Keller *et al.*, 1992). The observation is not directly relevant to mutual exclusion between different genes since telomere location is determined during formation of the macronucleus (sexual reproduction) and switching occurs during vegetative growth. It does show that genomic location can influence antigen expression and it may be relevant to understanding transcription of i-antigen genes.

A second potentially important observation is the absence of introns in surface antigen genes, despite the common occurrence of introns in the *Paramecium* genome. Of seven genes completely sequenced none contains an intron. This might be dismissed were it not for the large size of the coding regions (7–8 kb). The significance may be related to the evolutionary history of the genes and the observation that they most likely arose from tandem duplications of a cysteine-rich sequence. Alternatively, the lack of introns might reflect the presence of regulatory regions within the coding sequence. The introns would disrupt those elements and alter gene regulation.

Isolation and analysis of mutants

One advantage *Paramecium* can provide over many of the other protozoa that exhibit antigen variation is the ability to isolate and analyse mutants. In the mid-1950s Reisner isolated mutants unable to express the H antigen (Reisner, 1955). After the A51 gene was cloned, Epstein and Forney (1984) used X-rays to induce mutations and then selected for cell lines that failed to express A (A- lines). In the absence of A, cells tend to express B, D, but other types are observed under the appropriate conditions. Three of the initial A- mutants (d12, d16 and d48) were large deletions in or near the A51 structural gene. The d48 cell line has been studied extensively because of its unusual inheritance resulting from defects in DNA processing during formation of the macronuclear genome (reviewed in Forney *et al.*, 1996). The d12 cell line has a deletion of the A51 gene in both the macronucleus and the micronucleus and has been used extensively for DNA transformation experiments. Two cell lines (d8 and d29) contained mutations that were unlinked to the A51 gene. Interestingly, at high temperature these cell lines can express the A51 antigen, yet they quickly revert to C51 and E51 antigens at 27 °C, which is unusual for wild-type cells. Thus, rather than being defective for A51 expression these mutants appear to be more stable for C and E expression. Additional mutagenesis in John Preer's laboratory resulted in deletions of the 51B and 51D genes (Preer *et al.*, 1987).

Thus far, the analysis of mutants has shown that:

1 there is no obvious deleterious phenotype of deletions of individual surface antigen genes;
2 the deletion of the A51 or B51 surface antigen genes does not significantly alter the expression pattern of other antigen genes;
3 loci unlinked to the structural gene influence surface antigen expression.

Although it remains possible that the unlinked regulatory loci are other surface antigens genes, this seems unlikely based on the phenotype of the A51 and B51 gene deletions. The analysis of unlinked genes has tremendous potential to increase our understanding of surface antigen regulation. The recent development of methods for cloning genes by complementation in *Paramecium* provides new incentives to pursue this approach in the future (Haynes *et al.*, 1998).

Transcriptional control of antigen expression

Although evidence for multiple levels of i-antigen regulation has been presented (Gilley *et al.*, 1990), it is apparent that under most conditions of stable expression the primary control is at the level of transcription. This was initially suggested by Preer *et al.* (1981), who showed that cultures expressing a single antigen type contained the mRNA corresponding to that antigen and no others. The Preer laboratory later extended the observation to all 11 serotypes in *P. tetraurelia* stock 51 (Forney *et al.*, 1983). The first direct demonstration of transcriptional control was provided by Gilley *et al.* (1990) using run-on transcription assays. An examination of the A51, C51 and H51 genes showed that A51 and H51 were controlled at the level of transcription. Yet the C51 gene was transcribed along with the H51 gene in an H-expressing culture (Gilley et al., 1990). Leeck and Forney (1994) later confirmed the transcriptional regulation of A51 and showed this was true for the B51 gene also. Two studies have examined the sequence requirements for A51 expression (Martin et al., 1994; Leeck and Forney, 1994). The results from both are consistent with the conclusion that 260–270 bp of sequence upstream of the A51 gene (relative to the translation start) are sufficient for normal antigen expression. The surface antigen transcripts have very short 5' non-translated regions of 8–12 nucleotides. Comparisons of gene sequences have identified some candidate regulatory elements, including a long poly (dA-dT) stretch and a short 12 bp sequence that overlaps the transcription start site (Preer et al., 1987), but there has been no detailed experimental analysis of the upstream region. More attention has been focused on the problem of mutual exclusion, and although the upstream region is required for transcription, in the one case examined it does not regulate mutual exclusion (see next section).

Role of the antigen in regulation of mutual exclusion
Over the years a number of models for *Paramecium* i-antigen expression have postulated that the surface antigen itself is involved in the regulation of mutual exclusion (Kimball, 1947; Finger, 1967; Capdeville, 1979). The idea is attractive because it can account for the relatively stable expression of a single antigen type in the face of many alternatives. Although a specific regulatory role for the antigen has not been established there is increasing evidence for some involvement of the antigen. Previous observations along with more recent molecular studies provide support for this view.

An early genetic analysis showed that crosses between cell lines expressing different surface antigens appear to follow cytoplasmic inheritance. For example, when A51 and B51 cells are mated the exconjugant from the A51 parent continues to express A and the exconjugant from the B51 parent continues to express B. This occurs despite the fact that both F1 cell lines are genetically identical. If cell separation is delayed so that cytoplasm is exchanged between mating cells, the resulting exconjugants tend to switch antigen type to that of their partner (Sonneborn, 1948). Historically, this experiment has been interpreted as evidence for a cytoplasmic determinant, but delayed cell separation may also result in membrane exchange. The possibility of a membrane signal involved in regulating antigen expression is relevant to recent molecular observations discussed below.

Capdeville (1979) made a series of observations regarding the expression of the 156G and 168G alleles in *P. primaurelia*. These two alleles exhibit intergenic exclusion as well as interallelic exclusion (156G is dominant over 168G). Comparison of expression patterns of homozygous, heterozygous and G156 isogenic strains (G156 allele in a 168 background) under different temperature and environmental conditions led to the conclusion that the antigen expression pattern follows the gene encoding the antigen rather than unlinked regulatory genes. Capdeville proposed that the antigen expressed on the cell surface has a self-regulatory role acting as a 'primer' for its continued synthesis. She also recognized that the previous interpretation of 'cytoplasmic' inheritance of antigen type could also be a result of exchanging small portions of membrane that prime the synthesis of an alternative antigen (Capdeville, 1979).

Although older genetic studies support the hypothesis of self-regulation by the antigen, they fall short of direct evidence. More recent investigations have tried to establish which regions of the surface antigen gene (or protein) are involved in the regulation of mutual exclusion. These experiments are based on the use of gene chimeras: combining portions of two surface antigen genes that normally show mutual exclusion and observing whether they show co-expression or exclusion when transformed with the wild-type genes. Thus far, all experiments have involved the cloned A51 and B51 genes introduced into a cell line that has macronuclear deletions of the endogenous A51 and B51 genes. The first study by Leeck and Forney (1994) showed that although the 5' upstream sequence is required for transcription, it is not sufficient to control mutually exclusive expression. Substituting the B51 5' upstream region on the A51 gene did not alter the normal A51 expression pattern; the chimera was expressed when injected alone and was not expressed when co-injected with the B51 gene (Figure 17.2). Further experiments (Leeck and Forney, 1996) showed that a region downstream of the start codon in A51 regulates mutually exclusive transcription. Substituting nucleotides −1649 to +885 of the B51 gene onto the A51 gene allowed co-expression of the gene chimera with B51 but excluded expression of the A51 gene (as measured by transcription). A smaller substitution from −26 to +885 had the same effect on antigen expression and transcription (Figure 17.2). As a control, roughly 600 nucleotides from the B51 gene (+2375 to +2971) were substituted for the corresponding region of A51. The results showed that the gene chimera was transcribed along with the wild-type A51 gene. Other experiments showed that the 3' coding region (1.2 kb) of 51B does not regulate mutual exclusion (Thai and Forney, 1999). Although every region of the A51 and B51 genes has not been tested, it seems clear that the 5' coding region has a special role in the regulation of mutual exclusion. Whether the same conclusions will be true for other antigen genes remains to be determined.

Thai and Forney (1999) recognized that the ability to co-express two antigen genes allows a direct test of whether the expression of the i-antigen on the cell surface influences transcription of the corresponding gene. Since the pSB/A construct can be co-expressed with the wild-type B51 gene (pSB) a frame-shift mutation was introduced into the coding region of pSB/A, creating a construct called pSB/A(fs) that is incapable of producing mature protein. Co-transformation of pSB/A(fs) and pSB was compared with co-transformation of pSB/A and pSB (Figure 17.2). Analysis of run-on transcription from transformants showed a five- to sixfold decrease in transcription from the gene with a frame-shift mutation [pSB/A(fs)] compared to the wild-type B gene (pSB). Although low

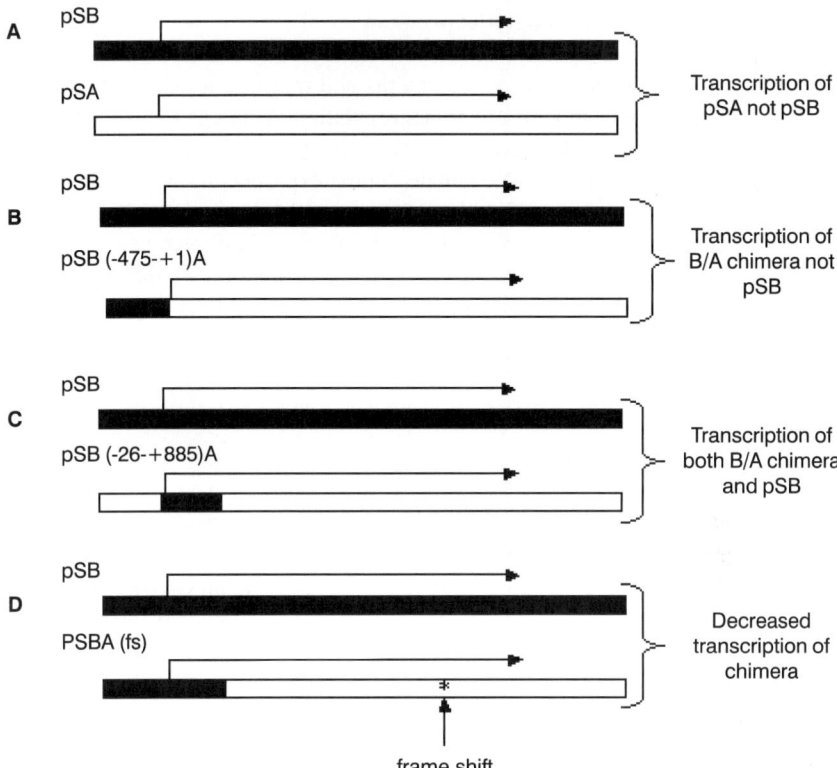

Figure 17.2 Summary of results from *Paramecium* i-antigen gene chimeras. (A) Co-transformation of wild-type A51 and B51 genes generally results in transcription of pSA. (B) Substituting the 5' upstream region of B51 on the A51 gene does not alter the expression pattern. The chimeric gene is transcribed and A51 expressed on the cell surface. (C) Substitution of the coding region near the translation start allows co-expression of the chimera and B51. (D) Introducing a frame-shift mutation into the chimeric gene reduces transcription when co-transformed with pSB (wild-type B51 gene) as compared with the same chimera with no frame-shift mutation.

steady-state RNA levels might be the result of nonsense-mediated decay it is difficult to argue that transcription could be similarly affected. The authors conclude that some portion of the i-antigen transcriptional pathway is self-regulatory (Thai and Forney, 1999).

The presence or absence of the protein encoded by pSB/A must be signalled from the outer membrane to the transcriptional machinery in the nucleus. Direct interaction between the GPI-anchored protein and a repressor/activator in the nucleus seems unlikely, but evidence for signal transduction pathways involving GPI-anchored proteins has been established in the immune system (Brown, 1993; Lund-Johansen, 1993). The possibility of GPI protein signalling is particularly interesting in regard to antibody-induced exit of *Ichthyophthirius* as described in the second section of this chapter.

Model for regulation of antigen expression

The recent data obtained from gene chimeras and transcription analysis provide the basis for speculative models of antigen regulation. Although the lack of *trans*-acting factors prevents a detailed hypothesis, this model may be useful to place the current data in context and evaluate future experimental approaches. The current model is built around the idea of a silenced (transcriptionally inactive) chromatin configuration that prevents expression of i-antigen genes (Figure 17.3). Activation of one gene allows the stable expression of that gene until conditions (temperature, stress) alter the balance between repressors and activators and cause switching.

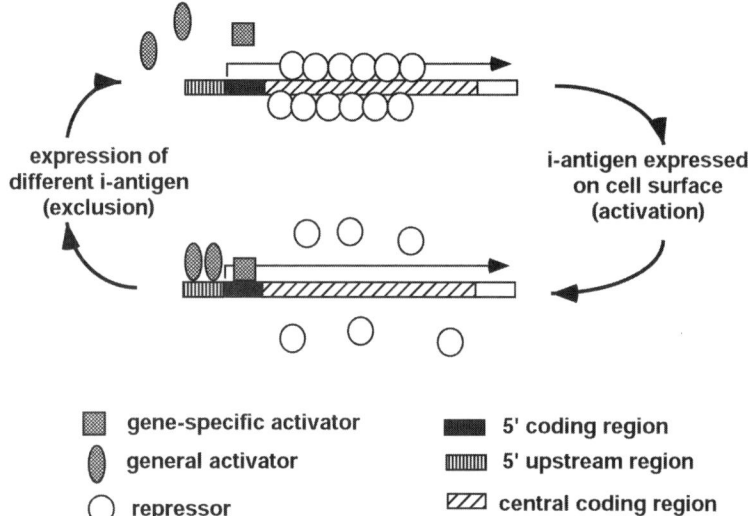

Figure 17.3 Model for regulation of *Paramecium* i-antigen expression. Repressors bind to the coding region of the i-antigen gene and maintain a silent chromatin configuration. Altered conditions allow competition for limiting amounts of transcription activators. The activated gene generally continues stable expression due to its open (transcribed) configuration. See text for details. (Adapted with permission from Thai and Forney (1999) *Gene Expresssion* 8, 263–272.)

The model has the following features:

1 Surface antigen gene transcription is the result of competition between transcription activators and repressors. The genes are normally in the repressed (inactive) state. Although there is no direct experimental data for this assumption, conceptually we favour a ground state of repression, which is overcome by a limited set of activators.

2 Upstream regions are transcriptional promoters, yet they bind general activators not gene-specific activators. Gene chimeras that contain the upstream region of B51 attached to the coding region of A51 support this statement since they do not alter the pattern of mutual exclusion (Leeck and Forney, 1994).
3 The 5' coding region is part of the promoter and binds gene-specific activators. This conclusion is based on the dramatic effect of the 5' coding region on mutually exclusive transcription (Leeck and Forney, 1996).

The central region of the gene binds repressor molecules. No effect on transcription is seen when the 3' regions of A51 and B51 are exchanged. In addition, transcription of the pSB/A frame-shift construct is decreased sixfold relative to wild-type B51 even though it contains the B51 gene upstream sequence and 5' coding region (Thai and Forney, 1999). From this we indirectly infer that the central region binds repressors that bind tightly when the B/A protein chimera is not detected on the cell surface. The repressor-binding hypothesis could also explain the lack of introns in i-antigen genes since they would disrupt the normal regulatory pathway.

The activator–repressor competition resulting in a repressed and silent chromatin conformation versus an open and transcriptionally active form can explain the stable but exclusive nature of *Paramecium* antigen expression. Some aspects of this model incorporate known features of chromatin silencing, particularly yeast studies (Lustig, 1998). The identification of *trans*-acting molecules will be critical to the elucidation of the molecular regulation of this remarkable system.

I-ANTIGENS IN THE CONTEXT OF INFECTION AND IMMUNITY

Antigenic variation in the free-living ciliates is interesting from both a functional and an evolutionary standpoint. The ability to vary surface antigens is generally associated with evasion strategies used by pathogens to escape the adaptive immune response. Nevertheless, as indicated above, *Paramecium* and *Tetrahymena* are primarily free-living organisms that fail to 'see' antibodies in their natural environments. Given the energy costs of replacing their surface coats (and in the absence of selective pressure by the adaptive immune system), why evolve and maintain mechanisms for doing so? Of course not all ciliates are free-living and a number of parasitic forms are known to exist. These include *Ichthyophthirius multifiliis*, a commercially important pathogen of freshwater fish. In the case of *Ichthyophthirius*, i-antigens are major targets of the host immune system, and at the same time, play a critical role in evasion of the humoral response. Recent studies would nevertheless suggest that their evasion strategy involves an alteration in behaviour rather than antigenic switching *per se* (Cross and

Matthews, 1992; Clark *et al.*, 1996; Lin *et al.*, 1996; Clark and Dickerson, 1997; Wahli and Matthews, 1999).

The i-antigens as targets of humoral immunity

As the causative agent of 'white-spot' (Figure 17.4), *Ichthyophthirius*, occurs worldwide and is among the most significant pathogens of farm-raised fish (Matthews, 1994; Dickerson and Dawe, 1995). It has an extremely broad host range and infects virtually all freshwater species. Although taxonomically related to both *Paramecium* and *Tetrahymena* (order *Hymenostomatida*), *Ichthyophthirius* is a strict obligate parasite and has only limited ability to survive outside the host. Parasites numbers increase by two to three orders of magnitude with each round of the life cycle (Figure 17.5), and in closed aquatic systems quickly overwhelm fish. Nevertheless, animals that survive

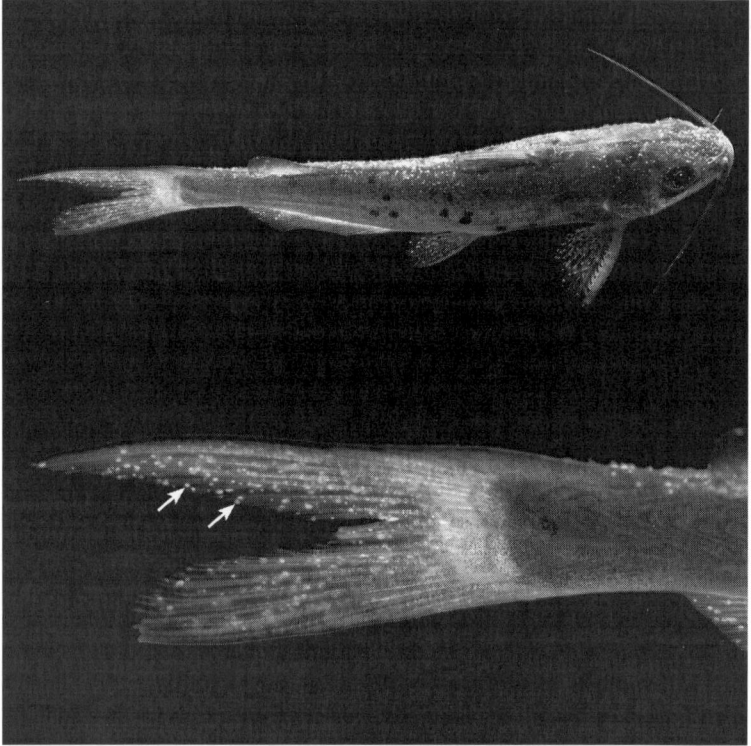

Figure 17.4 Juvenile channel catfish infected with *I. multifiliis*. Refractile white spots visible on the fins and the upper body surface of the fish are individual parasites in the trophont stage. A higher magnification image of the tail fin is shown at the bottom of the panel. Arrows point to single trophonts. The fish shown here is approximately 10 cm in length.

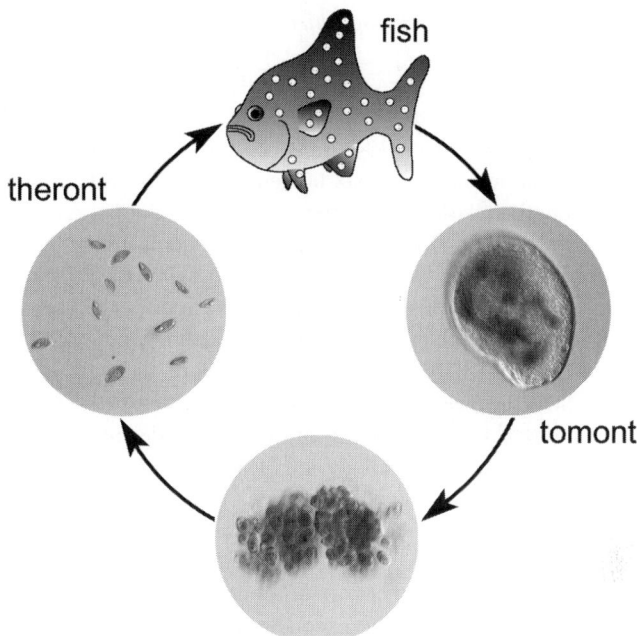

Figure 17.5 Parasite life cycle. Free-swimming theronts invade the skin and gill epithelia of freshwater fish. Soon after invasion, theronts transform into trophonts, which feed on host tissue. Over the course of a week (20 °C), trophonts grow to several hundred microns in diameter and then exit the fish (as tomonts) to complete the life cycle. Tomonts swim briefly, then attach to an inert support and divide within a gelatinous cyst to form 100–1000 infectious theronts. Theronts have a limited lifespan, and lose viability within about 48 hours in the absence of a host.

epizootics (as well as fish exposed to controlled infections) can develop a solid acquired immunity against subsequent challenge. Fish produce antibodies in response to infection that are directed primarily against a class of abundant GPI-anchored surface membrane proteins that are structurally analogous to the i-antigens of the free-living ciliates (Clark *et al.*, 1995, 2001). Sera from immune fish rapidly immobilize *Ichthyophthirius* in a manner almost identical to that seen following treatment of *Paramecium* and *Tetrahymena* with homologous rabbit antisera (Clark *et al.*, 1987; Cross, 1993). Indeed, high-titre sera that strongly immobilize *Ichthyophthirius* can be generated in both rabbits and mice following injection with whole parasites, isolated cilia, or the purified proteins themselves (Dickerson *et al.*, 1989; Lin and Dickerson, 1992). While it is clear that fish produce antibodies against i-antigens in response to infection, a role for these proteins in protective immunity is based on two further observations. First, naive fish are completely resistant to parasite infection following passive antibody transfer (Clark *et al.*, 1996; Lin *et al.*, 1996), and second, the purified i-antigens can act as effective sub-unit vaccines (Wang *et al.*, 2002).

Structural features of the *Ichthyophthirius* i-antigens

Based on sequences obtained from three recently cloned genes, the i-antigens of *Ichthyophthirius* share the same basic structure, which consists of hydrophobic N- and C-termini that direct ER translocation and GPI-anchor addition, respectively, and a series of five to six tandem repeats of ~80 amino acids each spanning their length (Clark *et al.*, 1999; Lin *et al.*, 2002). As is true of the i-antigens of the free-living ciliates, the repeats are characterized by the presence of periodic cysteine residues, which fall into register when the repeats are aligned (Figure 17.6). The parasite antigens contain six cysteines per repeat, which occur as -C-$X_{2,3}$-C- motifs within the larger-order framework -C-X_{19-21}-C-X_2-C-X_{16-47}-C-X_2-C-X_{20-22}-C-X_3. While the spacing of the cysteine residues is consistent with the i-antigens being metal-binding proteins (Clark *et al.*, 1999), the epitopes recognized by immobilizing antibodies are destroyed by sulphhydryl reducing agents (X. Wang, T. G. Clark and H. W. Dickerson, unpublished), and based on studies of the *Paramecium* i-antigens (Reisner *et al.*, 1969), there is good reason to believe that some or all of the cysteine residues within the ciliate i-antigens are disulphide-linked. In addition to the -C-$X_{2,3}$-C- motifs, a number of other structural elements appear to be conserved among the *I. multifiliis* antigens, most notably, CP-X-G(T/A) sequences at the start of each repeat, and a KKLTSGA domain just upstream of the C-termini in each protein (Lin *et al.*, 2002). The latter element may be a recognition sequence for the transamidase that adds the GPI-anchor to the C-terminal residue of the mature protein (Udenfriend and Kodukula, 1995; Clark *et al.*, 2001).

Homology searches of existing protein databases place the i-antigens of *Ichthyophthirius* closest to the SerL paralogues of *Tetrahymena thermophila*. The SerL proteins are a group of three to five polypeptides that are coordinately expressed below 20 °C, and like the parasite i-antigens, contain variable numbers of tandem repeats with six cysteines per repeat (Doerder and Gerber, 2000). The interval between nearest-neighbour cysteines is nevertheless different in the *Tetrahymena* proteins, with the SerL paralogues having C-$X_{1,2}$-C, rather than -C-$X_{2,3}$-C- spacing. Along with SerL, additional homologies have been reported between the parasite proteins and i-antigens specified by the *SerH* alleles of *T. thermophila* (*SerH* is expressed in a mutually exclusive fashion in cells cultured between 20 and 37 °C) (Smith *et al.*, 1992). Although BLAST searches reveal further weak homologies with the i-antigens of *Paramecium*, the *Tetrahymena* and *Ichthyophthirius* proteins are far more similar with respect to both primary sequence and overall size (40–60 kDa as opposed to 250–300 kDa in *Paramecium*). This is entirely consistent with the closer taxonomic relationship between these organisms (Wright and Lynn, 1995).

Quite interestingly, similarities between the i-antigens of *Ichthyophthirius* and the variant-specific surface proteins (VSPs) of *Giardia lamblia* have also been noted (Clark *et al.*, 1999). As in the case of the i-antigens, the VSPs contain tandemly repetitive amino acid sequence domains with distinct cysteine

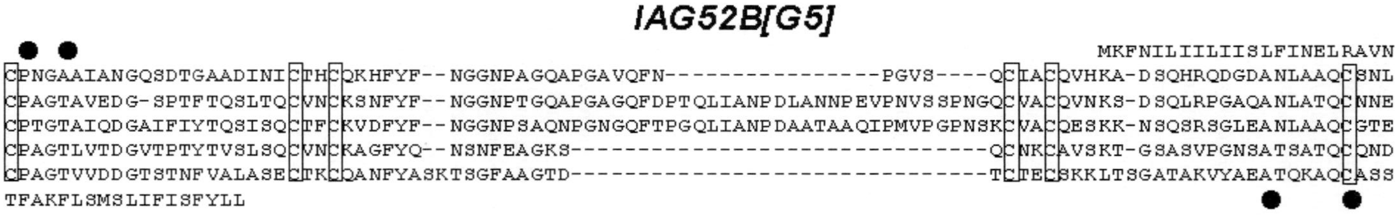

IAG52B[G5]

```
                                                          MKFNILIIIISLFINELRAVN
CPNGAATANGQSDTGAADINI CTHC QKHFYF--NGGNPAGQAPGAVQFN-------------PGVS----QC IAC QVHKA-DSQHRQDGDANLAAQ CSNL
CPAGTAVEDG-SPTFTQSLTQ CVNC KSNFYF--NGGNPTGQAPGAGQFDPTQLIANPDLANNPEVPNVSSPNGQ CVAC QVNKS-DSQLRPGAQANLATQ CNNE
CPTGTAIQDGAIFIYTQSISQ CTFC KVDFYF--NGGNPSAQNPGNGQFTPGQLIANPDAATAAQIPMVPGPNSKC VAC QESKK-NSQSRSGLEANLAAQ CGTE
CPAGTLVTDGVTPTYTVSLSQ CVNC KAGFYQ--NSNFEAGKS-------------------QC NKC AVSKT-GSASVPGNSATSATQ CQND
CPAGTVVDDGTSTNFVALASE CTKC QANFYASKTSGFAAGTD-----------------------TC TEC SKKLTSGATAKVYAEATQKAQ CASS
TFAKFLSMSLIFISFYLL
```

Figure 17.6 Arrangement of tandem repeats within the i-antigens of *Ichthyophthirius*. The predicted sequence of i-antigen gene *IAG52B[G5]* is shown. The organization of cysteine residues and tandem repeats is typical of the i-antigens of *Ichthyophthirius*. The repeats span the length of the protein and are flanked by hydrophobic signal peptides. Six invariant cysteines (with the closest spacing -C-$X_{2,3}$-C-) are present within each repeat (boxed). Dark circles indicate additional amino acid residues that are conserved within every repeat of the *I. multifiliis* antigens characterized to date. The regions of greatest variability among these proteins lie between adjacent -C-X_2-C- motifs in the central (2–3) repeats.

periodicity. Sequence alignments between the 48 kDa i-antigen of parasite isolate G1 and the *Giardia vspA6-S1* gene product have shown that 29 of a possible 30 cysteines in the *I. multifiliis* antigen overlap with the same amino acid in the *Giardia* protein (Clark *et al.*, 1999). This is due primarily to the existence of tandem repeats, along with comparable $-C-X_{2,3}-C-$ motifs within the predicted sequences. While *Giardia* and *Ichthyophthirius* are highly diverged from an evolutionary standpoint, such structural similarities have suggested that these proteins may be functionally related (Clark *et al.*, 1999). In this regard, the $-C-X_{2,3}-C-$ motifs within the *Giardia* proteins were recognized as having the potential for metal binding, and it has been shown that the VSPs are capable of binding zinc *in vitro* (Nash and Mowatt, 1993; Zhang *et al.*, 1993). Clearly, the i-antigens have similar potential, although whether any of these proteins bind metal ions *in vivo* has yet to be established.

I-antigen variation in natural parasite populations

Parasite strains collected from the wild and maintained as clonal isolates coordinately express between 1 and 3 i-antigen polypeptides with apparent M_rs of 40–60 kDa on one-dimensional SDS-PAGE (Clark *et al.*, 1995). These isolates can be distinguished from one another based on immobilization with specific antisera, and often fall into one of a number of defined serotypes. At least five such serotypes are presently known to exist (designated A–E) (Clark *et al.*, 1995; Dickerson and Clark, 1998). The most common of these, namely serotype D, has been identified in isolates ranging from China to the United States (Dickerson and Clark, 1998). Cells that express the D serotype show a single i-antigen band of ~52/55 kDa following 1-D SDS-PAGE and western blotting. Recently, the 52/55 kDa band has been shown to resolve into four or more isoelectric variants on two-dimensional gels, and genes encoding two distinct i-antigens have now been isolated from the G5 parasite strain (a representative of serotype D) (Lin *et al.*, 2002). The two genes (designated *IAG52A[G5]* and *IAG52B[G5s]*) are both transcribed (although at widely different levels), and their products share roughly 40% identity in terms of primary sequence (Lin *et al.*, 2002). Aside from D, the most well-characterized parasite i-antigen serotype is A. Natural isolates obtained from Georgia (G1) and New York (NY1) are members of this group. These isolates express either two (G1) or three (NY1) i-antigen polypeptides based on 1-D SDS-PAGE and western blotting, with the G1 proteins migrating at ~48 and 60 kDa (Clark *et al.*, 1995; Wang *et al.*, 2002). The gene for the 48 kDa antigen (designated *IAG48[G1]*) has been sequenced, and predicts a protein that shares 44% and 57% homology with the *IAG52A[G5]* and *IAG52B[G5]* gene products, respectively. When the predicted sequences of all three proteins are compared, they differ most in the region between adjacent $-C-X_2-C-$ motifs (particularly in the central repeats), and are most conserved at their carboxy-termini. As suggested for the *Paramecium* antigens (Prat, 1990), the central repeats may harbour epitopes recognized by immobilizing antibodies.

Exact homologues of the *IAG52A[G5]* and *IAG52B[G5]* genes have been isolated from parasite strain G3, an independent isolate belonging to serotype D (T. G. Clark, unpublished). Similarly, a gene with the identical coding sequence of *IAG48[G1]* has been obtained from NY1, an independent isolate belonging to serotype A (T. G. Clark, unpublished). In one study, attempts to amplify the gene for the 48 kDa antigen using gene-specific primers and genomic DNA from heterologous serotypes proved unsuccessful (Dickerson et al., 1993). Furthermore, *IAG48[G1]* failed to hybridize with genomic DNA of heterologous i-antigen serotypes in southern blots probed under conditions of high stringency (Dickerson et al., 1993). When probed under conditions of low stringency, the same gene recognized only a limited number of bands in southern blots of genomic DNA from homologous or heterologous strains, suggesting that only a small number of i-antigen genes exists in any given isolate. One would have to assume that this limited repertoire arose from gene duplication and genetic drift involving point mutations and/or intra- or intergenic recombination. With regard to the latter, alignments of the *IAG52A[G5]* and *IAG48[G1]* gene products split the first repeat of the 48 kDa protein in two, and place the first half with the first repeat, and the second half with the second repeat of the 52 kDa antigen (Lin et al., 2002).

Regardless of the origin of these genes, the presence of only a limited set of i-antigen coding sequences in *Ichthyophthirius* would suggest that the differences in i-antigen expression among serotypes results from allelic variation rather than antigenic shift. Of course, even a small number of genes could give rise to different serotypes if they are differentially expressed within a given isolate. While this is clearly true in the free-living ciliates, i-antigen serotypes in *Ichthyophthirius* tend to be highly stable when parasites are cultured for prolonged periods on fish. In some instances, clonal isolates have been shown to maintain a given serotype for >100 generations (T. G. Clark and H. W. Dickerson, unpublished). This in itself would suggest that variation in serotype is not a frequent occurrence in *I. multifiliis*. It should also be noted that, in contrast with the free-living ciliates (in particular, *Tetrahymena*), obvious changes in the steady-state levels of i-antigen transcripts have not been detected when infective theronts are maintained at different temperatures (Clark et al., 1992). Still, parasites have a relatively short lifespan outside the host and it is difficult, from a technical standpoint, to look for antigenic variation during the theront stage. Furthermore, spontaneous alterations in serotype have occasionally been seen in parasites cultured on fish (Dickerson et al., 1993). Unfortunately, in such cases, it has not been possible to rule out contamination by heterologous strains. Finally, depending on the challenge dose, immunity against *Ichthyophthirius* is not sterile, and it would be attractive to think that parasites that can establish on fish that are actively immune have switched expression of their surface antigens. This has yet to be formally demonstrated, however, and it should be mentioned that active immunity in response to infection appears to extend across serotypes (Leff et al., 1994).

While the capacity for antigenic variation in *Ichthyophthirius* is still unclear, the expression of i-antigen genes is highly regulated from a developmental standpoint. As parasites transit from the host-associated trophont to the infective theront stage, steady-state levels of i-antigen transcripts increase dramatically, reaching levels as high 6% of polyA$^+$ RNA (Clark et al., 1992). Moreover, in cases where multiple antigens are expressed in a given serotype, the genes themselves appear to be independently regulated, with levels of expression varying by nearly 100-fold (Lin et al., 2002). The consequences of this, relative to alterations in serotype, are not known.

I-antigens in immune evasion

In the early 1970s, Hines and Spira demonstrated that sera from immune fish will immobilize *I. multifiliis* theronts in culture (Hines and Spira, 1974). These initial observations suggested a role for i-antigens in protective immunity, and a series of passive immunization studies was undertaken to test this idea. In these studies, murine monoclonal antibodies that immobilize *Ichthyophthirius in vitro* were found to confer virtually complete protection against an otherwise lethal parasite challenge when injected intraperitoneally into naive channel catfish. As was the case for immobilization, the protection afforded by these antibodies was serotype-specific (Lin et al., 1996). While these studies provided clear evidence of a role for i-antigens in protective immunity, experiments carried out by Cross and Matthews with actively immune fish generated conflicting results. When carp that had been immunized by previous exposure to *Ichthyophthirius* were re-exposed to the infectious theronts, parasites readily penetrated the skin (Cross and Matthews, 1992). Nevertheless, within 2 hours of exposure, ~ 80% of those initially present had disappeared with little or no trace. The conclusion was that parasites were being forced to exit fish prematurely in response to host factors (presumably antibodies) within the skin (Cross and Matthews, 1992). Although a remarkable finding, the ability of parasites to invade and then exit the skin was clearly inconsistent with the presence of immobilizing antibodies within tissues, and raised obvious questions regarding the role of i-antigens in forced exit and immunity overall. To reconcile these findings, a further series of passive immunization trials was conducted, this time using fish that were already infected with *Ichthyophthirius* (Lin et al., 1996). In this case, passive transfer of immobilizing antibodies forced rapid, premature exit of parasites from the host exactly as described by Cross and Matthews (1992) for actively immune fish. Trophonts that exited fish in response to murine antibodies were viable and could divide to form new theronts (Lin et al., 1996). Indeed, subsequent studies with actively immune fish showed that parasites that escape the host can directly reinfect naive animals (Wahli and Matthews, 1999). Thus, while premature exit represents an entirely novel mechanism of humoral immunity, it can act as an evasion strategy for the parasite as well.

I-antigens and transmembrane signalling

While the mechanisms responsible for premature exit are unknown, escape from the host involves a physiological response by the parasite, rather than a blocking of receptor-ligand interactions within the skin. This is most clearly demonstrated by the fact that monovalent Fab subfragments of immobilizing mouse monoclonal antibodies fail to elicit premature exit following passive transfer, despite the fact that they can bind to the parasite surface *in vivo*. By contrast, bivalent Fab$_2$ has the same effect as intact antibody, and Fab subfragments can be reactivated by subsequent administration of bivalent goat anti-mouse IgG to passively immunized fish (Lin *et al.*, 1996). Such findings clearly indicate that forced exit requires antigen cross-linking at the parasite surface rather than antibody binding *per se*, and suggests that lateral movements within the membrane trigger the response (Clark and Dickerson, 1997; Lin *et al.*, 1996). Since clustering of membrane receptors is a hallmark of signal transduction phenomena, such results are consistent with the idea that ciliate i-antigens can initiate transmembrane-signalling events *in vivo*.

That the i-antigens can, in fact, transduce signals is supported by two additional observations. First, antibody concentrations below threshold for immobilization induce a classical 'avoidance' response in *Paramecium* (Preer, 1959a). This type of behaviour is a manifestation of backward swimming caused by alterations in voltage-dependent calcium flux across the membrane, leading to a reversal in ciliary beat (Machemer and Teunis, 1996; Plattner and Klauke, 2001). Second, i-antigen antibodies trigger rapid discharge of cortical secretory organelles in both *Ichthyophthirius* and the free-living ciliates. As in all eukaryotic cells, regulated secretion in the ciliates involves fusion of dense-core vesicles (mucocysts in *Tetrahymena* and *Ichthyophthirius*, and trichocysts in *Paramecium*) with the plasma membrane in response to elevated intracellular [Ca^{++}] (Turkewitz *et al.*, 1999; Plattner and Klauke, 2001). Thus, cross-linking of i-antigens at the cell surface must initiate a signalling cascade that triggers calcium mobilization, and in the case of *I. multifiliis*, the downstream events leading directly to premature exit.

As GPI-anchored proteins, the i-antigens are restricted to the outer leaflet of the plasma membrane. How such proteins transduce signals to the interior of the cell is an interesting question in itself. Indeed, while GPI-anchored proteins are known to transduce signals in metazoan cells, they are generally thought of as having exclusively extracellular functions in protozoa (Ferguson, 1994, 1999). In keeping with this view, the shape and presumed packing density of the *Paramecium* i-antigens has suggested that their primary function is to shield the plasma membrane (and associated transmembrane receptors) from potentially damaging effects of the environment (Reisner *et al.*, 1969). Similar ideas have been advanced for the variant surface glycoproteins (VSGs) of African trypanosomes, and for surface-coat proteins of other parasitic protozoa (Ferguson, 1994, 1999). The fact that i-antigens can initiate signalling events in both *Ichthyophthirius* and the free-living ciliates clearly challenges this view. Based on studies of metazoan cells, GPI-linked proteins associate with detergent-resistant microdomains within the plasma membrane that are enriched in sphingolipids and cholesterol (so-called lipid rafts) (Brown and London, 1998; Friedrichson and Kurzchalia, 1998). Current models would suggest that protein clustering in response to antibody-mediated cross-linking leads to a coalescence of such domains, and an association of key signalling molecules, in particular, acylated src-family tyrosine kinases and their respective target receptors, to form activated complexes (Brown and London, 1998; Friedrichson and Kurzchalia, 1998). Although biochemical evidence would suggest that lipid rafts, in fact, exist in ciliates (Zhang and Thompson, 1997), true tyrosine kinases are thought to have evolved at the protozoan:metazoan boundary (Darnell, 1997) and it is not at all clear how such signalling would occur in ciliated protozoa. Recent work with African trypanosomes might nevertheless speak to this question (O'Beirne *et al.*, 1998). In the presence of anti-VSG antibodies, bloodstream forms of *Trypanosoma brucei* undergo transient aggregation for periods lasting up to ~30 minutes, at which point they disaggregate. This response is remarkably similar to that reported for *Tetrahymena* in the presence of i-antigen-specific antibodies (that is, transient aggregation and a resumption of normal swimming within 30 minutes to several hours) (Margolin *et al.*, 1959). This does not appear to result from antigenic switching, and while it remains to be determined whether the mechanisms responsible for aggregation/disaggregation in these instances are related, drug studies provide indirect evidence that cyclic nucleotide metabolism and protein kinase A activities are involved in the case of *T. brucei* (O'Beirne *et al.*, 1998).

I-antigen clustering and parasite behaviour

The potential involvement of i-antigens in transmembrane signalling raises interesting questions regarding the mechanisms responsible for both premature exit (in the case of *Ichthyophthirius*) and immobilization of cells in culture. As with premature exit, immobilization is strictly dependent on i-antigen cross-linking, and does not occur in the presence of monovalent Fab (Barnett and Steers, 1984; Clark *et al.*, 1996). The requirement for multivalent antibodies in loss of motility has suggested that immobilization results from a physical restraint imposed by the cross-linking of cilia, or the tethering of individual cells. Immobilization is generally accompanied by cell-to-cell aggregation in all ciliates studied thus far. Furthermore, ciliary membranes appear to fuse laterally following antibody binding in the case of *Paramecium* (Barnett and Steers, 1984). The ability of i-antigens to transduce signals across the membrane would nevertheless suggest an alternative hypothesis, namely, that loss of motility results from aberrant signalling across the membrane. Inappropriate signalling could lead to (1) interference with coordinated ciliary beat, or (2) mucus secretion, which, in the case of *Tetrahymena* and *Ichthyophthirius*, could simply entrap cells (Turkewitz *et al.*, 1999). In such cases, cell-to-cell aggregation would be a secondary consequence of a loss in motility, with immobilized cells being held together by antibody, or mucus, to form a lattice.

The mechanisms responsible for premature exit are no less clear. Three hypotheses have been advanced to explain this behaviour (Clark and Dickerson, 1997). As indicated above, *Paramecium* undergoes backward swimming in response to antibody concentrations that are subthreshold for immobilization. It is possible that *Ichthyophthirius* responds in a similar manner *in vivo*, moving out of the skin in order to avoid antibody. Alternatively, *I. multifiliis* normally exits the skin to complete the life cycle. I-antigen cross-linking could short-circuit the normal developmental pathway that leads to this event, and result in premature exit from the host. Finally, uncontrolled release of tissue proteases by the parasite could dissolve tissue locally and release trophonts directly into the surrounding water column. This last idea is consistent with the fact that (1) parasites undergo regulated secretion in response to i-antigen cross-linking; (2) large amounts of melanin are released from the skin following passive antibody transfer (T. G. Clark, unpublished); and (3) parasites that exit the skin in response to passive immunization are invariably associated with clumps of epidermal tissue (Clark and Dickerson, 1997).

Antigenic variation and i-antigen function

Regardless of the precise mechanisms responsible for these phenomena, a role for i-antigens in transmembrane signalling has a number of important implications with regard to antigenic variation, and the function of the i-antigens overall. First and foremost, clustering of antigens at the cell surface could provide the stimulus for antigenic variation itself. Indeed, there is some evidence for this in both *Paramecium* and *Tetrahymena*, where antibodies specific for particular i-antigens can trigger antigenic shift (Margolin *et al.*, 1959; Juergensmeyer, 1969; Preer, 1986). Nevertheless, in the case of the free-living ciliates, antibodies are non-physiological ligands, and if antigen clustering at the cell surface is the trigger for antigenic switching, one can only guess as to the natural ligands that initiate this process in a pond-water setting. Of course, an answer to this question could be highly informative with regard to i-antigen function, and the reason(s) for antigenic switching in free-living species. In the case of *Ichthyophthirius*, the situation is somewhat different. While it is not entirely clear whether antigenic variation (*sensu stricto*) occurs in this system, the i-antigens are immunodominant in fish, and their interaction with host antibodies has a profound effect on parasite behaviour, leading to forced exit. In this case, antibodies are highly relevant in physiological terms, and one can plausibly argue that the i-antigens play a role in defence. Interestingly, while antibodies have little relevance for free-living species, interactions between *Paramecium* spp. and predatory ciliates such as *Didinium nasutum* have led to similar conclusions as well.

Paramecium and *Tetrahymena* are routinely eaten by predatory ciliates encountered in their natural environments (Figure 17.7). As prey, however, they are not defenceless. A variety of studies have suggested that extrusomes (trichocysts, pigment granules, etc.) are among the primary weapons used by

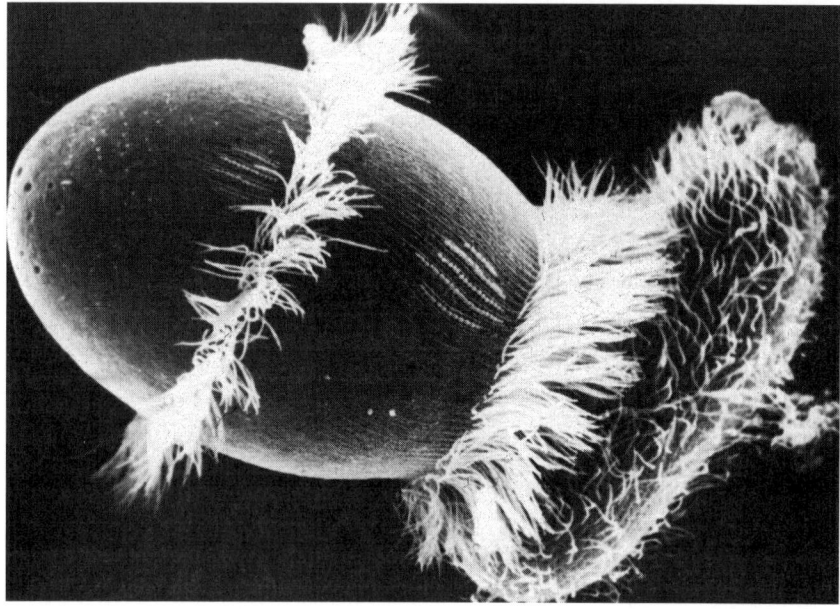

Figure 17.7 Prey–predator interactions among ciliates. While free-living species such as *Paramecium* and *Tetrahymena* need not contend with the adaptive immune response, they are eaten by predatory ciliates in their natural environments. The panel shows a scanning electron micrograph of *Paramecium* being eaten by *Didinium* (Courtesy of Dr Steve L'Hernault, Emory University.)

these organisms to escape engulfment (Harumoto and Miyake, 1991; Harumoto, 1994). Moreover, experiments with *Paramecium* and *Didinium* have suggested that either masking or removing i-antigens from the prey interferes with recognition by the predator during the course of an attack (Harumoto and Miyake, 1993; Harumoto, 1994). Based on these studies, Harumoto and Miyake (1993) have argued that surface antigens not only participate in prey recognition, but that antigenic variation is used as a means of escape. Indeed, in the case of *P. tetraurelia*, cells that express different surface antigens appear to be eaten at different rates depending on the antigens they express (Harumoto and Miyake,

1993; Harumoto, 1994). Assuming that predator populations vary at different times of the year, this could account for seasonal variations in the prevalence of i-antigen serotypes expressed by some ciliates in the wild (Saad and Doerder, 1995; Doerder *et al.*, 1996). While antigenic switching could play a role in this regard, a second hypothesis might argue that direct interactions between the i-antigens and ligands associated with predatory species results in discharge of cortical secretory organelles from the prey, and provides yet another means of escape. Although clearly speculative, the parallel between prey–predator and host–pathogen interactions provides an alternative framework in which to consider these proteins. If the i-antigens indeed play a role in defence, one might argue that this function arose initially in the free-living ciliates. With the evolution of pathogenic species (e.g. *Ichthyophthirius*), the same proteins might then have been adapted for use in evasion of the humoral immune response. Whether similar forces were at play in the evolution of surface antigens (and antigenic variation) in more distant phyla remains an open question.

REFERENCES

Amar, L. (1994) *J. Mol. Biol.* 236, 421–426.
Barnett, A. and Steers, E., Jr (1984) *J. Cell Sci.* 65, 153–162.
Benwakrim, A., Tremoliere, A., Labarre, J. and Capdeville, Y. (1998) *Protist.* 149, 39–50.
Bourgain-Guglielmetti, F. and Caron, F. (1996) *J. Eukaryot. Microbiol.* 43, 303–313.
Breuer, M., Schulte, G., Schwegmann, K. J. and Schmidt, H. J. (1996) *J. Eukaryot. Microbiol.* 43, 314–322.
Brown, D. (1993) *Curr. Opin. Immunol.* 5, 349–354.
Brown, D. A. and London. E. (1998) *Annu. Rev. Cell Dev. Biol.* 14, 111–136.
Capdeville, Y. (1979) *J. Cell. Physiol.* 99, 383–394.
Capdeville, Y., Cardoso de Almeida, M. L. and Deregnaucourt, C. (1987) *Biochem. Biophys. Res. Commun.* 147, 1219–1225.
Caron, F. and Meyer, E. (1985) *Nature* 314, 185–188.
Caron, F. and Meyer, E. (1989) *Ann. Rev. Microbiol.* 43, 23–42.
Clark, T. G. and Dickerson, H. W. (1997) *Parasitol. Today* 13, 477–480.
Clark, T. G., Dickerson, H. W., Gratzek, J. B. and Findly, R. C. (1987) *J. Fish Biol.* 31 (Supplement A), 203–208.
Clark, T. G., Gao, Y., Gaertig, J. and Cheng, G. (2001) *J. Eukaryot. Microbiol.* 48, 332–337.
Clark, T. G., Lin, T. L. and Dickerson, H. W. (1995) *Annu. Rev. Fish Dis.* 5, 113–131.
Clark, T. G., Lin, T. L. and Dickerson, H. W. (1996) *Proc. Natl Acad. Sci. USA* 93, 6825–6829.
Clark, T. G., Lin, T. L., Jackwood, D. A., Sherrill, J., Lin, Y. and Dickerson, H. W. (1999) *Gene* 229, 91–100.
Clark, T. G., McGraw, R. A. and Dickerson, H. W. (1992) *Proc. Natl Acad. Sci. USA* 89, 6363–6367.
Cross, M. L. (1993) *Dis. Aqua. Org.* 17, 159–164.

Cross, M. L. and Matthews, R. A. (1992) *J. Fish Dis.* 15, 497–505.
Darnell, J. E. (1997) *Proc. Natl Acad. Sci. USA* 94, 11767–11769.
Deregnaucourt, C. (1996) *Eur. J. Protistol.* 32, 380–388.
Dessen, P., Zagulski, M., Gromadaka, R. *et al.* (2001) *Trends Genet.* 17, 306–308.
Dickerson, H. W. and Clark, T. G. (1998) *Immunol. Rev.* 166, 377–384.
Dickerson, H. W. and Dawe, D. L. (1995) In *Fish Diseases and Disorders* (ed. P. T. K. Woo), CAB International, Wallingford, pp. 181–227.
Dickerson, H. W., Clark, T. G. and Findly, R. C. (1989) *J. Protozool.* 36, 159–164.
Dickerson, H. W., Clark, T. G. and Leff A. A., (1993) *J. Eukaryot. Microbiol.* 40, 816–820.
Dippell, R. V. (1954) *Caryologia* 6, 1109–1111.
Doerder, F. P. and Gerber, C. A. (2000) *Biochem. Biophys. Res. Commun.* 278, 621–626.
Doerder, F. P., Arslanyolu, M., Saad, Y., Kaczmarek, M., Mendoza, M. and Mita, B. (1996) *J. Eukaryot. Microbiol.* 43, 95–100.
Epstein, L. M. and Forney, J. D. (1984) *Mol. Cell. Biol.* 4, 1583–1590.
Ferguson, M. A. J. (1994) *Parasitol. Today* 10, 48–52.
Ferguson, M. A. J. (1999) *J. Cell Sci.* 112, 2799–2809.
Finger, I. (1967) In *The Control of Nuclear Activity* (ed. L. Goldstein), Prentice Hall, Englewood Cliffs, NJ, pp. 377–411.
Forney, J. D. and Blackburn, E. H. (1988) *Mol. Cell. Biol.* 8, 251–258.
Forney, J. D., Epstein, L. M., Preer, L. B., Rudman, B. M., Widmayer, D. J., Klein, W. H. and Preer, J. R., Jr (1983) *Mol. Cell. Biol.* 3, 466–474.
Forney, J. D., Yantiri, F. and Mikami, K. (1996) *J. Eukaryot. Microbiol.* 43, 462–467.
Friedrichson, T. and Kurzchalia, T. V. (1998) *Nature* 394, 802–805.
Gilley, D., Rudman, B. M., Preer, J. R., Jr and Polisky, B. (1990) *Mol. Cell. Biol.* 10, 1538–1544.
Godiska, R., Aufderheide, K. J., Gilley, D., Hendrie, P., Fitzwater, T., Preer, L. B., Polisky, B. and Preer, J. R., Jr (1987) *Proc. Natl Acad. Sci. USA* 84, 7590–7594.
Harumoto, T. (1994) In *Progress in Protozoology. Proceedings of the IX International Congress of Protozoology* (eds K. Hausman and N. Hülsmann), Gustav Fischer Verlag, Stuttgart, pp. 55–56.
Harumoto, T. and Miyake, A. (1991) *J. Exp. Zool.* 260, 84–92.
Harumoto, T. and Miyake, A. (1993) *J. Eukaryot. Microbiol.* 40, 27A.
Haynes, J. W., Vaillant, B., Preston, R. R., Yoshiro, S. and Kung C. (1998) *Genetics* 149, 947–957.
Hines, R. S. and Spira, D. T. (1974) *J. Fish Biol.* 6, 373–378.
Horowitz, S. and Gorovsky, M. A. (1985) *Proc. Natl Acad. Sci. USA* 82, 2452–2455.
Jollos, V. (1921) *Arch. F. Protistenk.* 43, 1–222.
Jones, I. G. (1965) *Biochem. J.* 96, 17–23.
Juergensmeyer, E. B. (1969) *J. Protozool.* 16, 344–352.
Keller, A. M., Le Mouel, A., Caron, F., Katinka, M. and Meyer, E. (1992) *Dev. Genet.* 13, 306–317.
Kimball, R. F. (1947) *Genetics* 32, 486–499.
Leeck, C. L. and Forney, J. D. (1994) *J. Biol. Chem.* 269, 31283–31288.
Leeck, C. L. and Forney, J. D. (1996) *Proc. Natl Acad. Sci. USA* 93, 2838–2843.
Leff, A. A., Yoshinaga, T. and Dickerson, H. W. (1994) *J. Fish Dis.* 17, 429–432.
Lin, T. L. and Dickerson, H. W. (1992) *J. Protozool.* 39, 457–463.

Lin, T. L., Clark, T. G. and Dickerson, H. W. (1996) *Infect. Immun.* 64, 4085–4090.
Lin, Y., Lin, T. L., Wang, C. C., Wang, X., Klobfleisch, R., Stieger, K. and Clark, T. G. (2002) *Mol. Biochem. Parasitol.* 120, 93–106.
Lund-Johansen, F., Olweus, J., Symington, F. W., Arli, A., Thompson, J. S., Vilella, R., Skubitz, K. and Horejsi, V. (1993) *Eur. J. Immunol.* 23, 2782–2791.
Lustig, A. (1998) *Curr. Opin. Gen. Devel.* 8, 233–239.
Machemer, J. and Teunis, P. F. M. (1996) In *Ciliates. Cells and Organisms* (eds K. Hausman and P. C. Bradbury), Gustav Fischer Verlag, Stuttgart, pp. 379–402.
Macindoe, H. and Reisner, A. H. (1967) *Aust. J. Biol. Sci.* 20, 141–152.
Margolin, P., Loefer, J. B. and Owen, R. D. (1959) *J. Protozool.* 6, 207–215.
Martin, L. D., Pollack, S., Preer, J. R., Jr and Polisky, B. (1994) *Dev. Genetics* 15, 443–451.
Matthews, R. A. (1994) In *Parasitic Diseases of Fish* (eds A. W. Pike and J. W. Lewis), Samara, Dyfed, pp. 17–42.
McTavish, C. and Sommerville, J. (1980) *Chromosoma* 78, 147–164.
Merkel, S. J., Kaneshiro, E. S. and Gruenstein, E. I. (1981) *J. Cell Biol.* 89, 206–215.
Meyer, E., Caron, F. and Baroin, A. (1985) *Mol. Cell. Biol.* 5, 2414–2422.
Mott, M. R. (1965) *J. Gen. Microbiol.* 41, 251–261.
Nash, T. E. and Mowatt, M. R. (1993) *Proc. Natl Acad. Sci. USA* 90, 5489–5493.
Nielsen, E., You, Y. and Forney, J. (1991) *J. Mol. Biol.* 222, 835–841.
O'Beirne, C., Lowry, C. M. and Voorheis, H. P. (1998) *Mol. Biochem. Parasitol.* 91, 165–193.
Pays, E., Vanhamme, L. and Berberof, M. (1994) *Annu. Rev. Microbiol.* 48, 25–52.
Phan, H. L., Forney, J. D. and Blackburn, E. H. (1989) *J. Protozool.* 36, 402–408.
Plattner, H. and Klauke, N. (2001) *Int. Rev. Cytol.* 201, 115–208.
Prat, A. (1990) *J. Mol. Biol.* 211, 521–535.
Prat, A., Katinka, M., Caron, F. and Meyer, E. (1986) *J. Mol. Biol.* 189, 47–60.
Preer, J. R., Jr (1959a) *J. Immunol.* 83, 276–283.
Preer, J. R., Jr (1959b) *J. Immunol.* 83, 385–391.
Preer, J. R., Jr (1968) In *Research in Protozoology* (ed. T. T. Chen), Pergamon Press, New York, pp. 130–278.
Preer, J. R., Jr (1986) In *The Molecular Biology of Ciliated Protozoa* (ed. J. G. Gall), Academic Press, New York, pp. 301–339.
Preer, J. R., Jr and Preer, L. B. (1979) *J. Protozool.* 26, 14–18.
Preer, J. R., Jr, Preer, L. B. and Rudman, B. M (1981) *Proc. Natl Acad. Sci. USA* 78, 6776–6778.
Preer, J. R., Jr, Preer, L. B., Rudman, B. M. and Barnett, A. J. (1985) *Nature* 314, 188–190.
Preer, J. R., Jr, Preer, L. B., Rudman, B. and Barnett, A. (1987) *J. Protozool.* 34, 418–423.
Reisner, A. (1955) *Genetics* 40, 591–592.
Reisner, A. H., Rowe, J. and Sleigh, R. (1969) *Biochemistry* 8, 4637–4644.
Rossel, R. (1905) *Arch. Hyg. Berl.* 54, 1–31.
Saad, Y. and Doerder, F. P. (1995) *Eur. J. Protistol.* 31, 45–53.
Smith, D. L., Berkowitz, M. S., Potoczak, D., Krause, M., Raab, C., Quinn, F. and Doerder, F. P. (1992) *J. Protozool.* 39, 420–428.
Scott, J., Leeck, C. L. and Forney, J. D. (1993) *Genetics* 133, 189–198.
Soldo, A. T. and Godoy, G. A. (1972) *J. Protozool.* 19, 673–678.

Sonneborn, T. M. (1948) *Proc. Natl Acad. Sci. USA* 34, 413–418.
Thai, K. Y. and Forney, J. D. (1999) *Gene Expresssion* 8, 263–272.
Thai, K. Y. and Forney, J. D. (2000) *J. Eukaryot. Microbiol.* 47, 242–248.
Turkewitz, A. P., Chilcoat, N. D., Haddad, A. and Verbsky, J. W. (1999) *Meth. Cell Biol.* 62, 347–362.
Udenfriend, S. and Kodukula, K. (1995) *Meth. Enzymol.* 250, 571–582.
Wahli, T. and Matthews, R. A. (1999) *Dis. Aquat. Org.* 36, 201–207.
Wang, X., Clark, T. G., Noe, J. and Dickerson, H. W. (2002) *Fish Shellfish Immunol.* 13, 337–350.
Wright, A. D. and Lynn, D. H. (1995) *Mol. Bio. Evol.* 12, 285–290.
Zhang, X. and Thompson, G. A., Jr (1997) *Biochem. J.* 323, 197–206.
Zhang, Y-Y., Aley, S. B., Stanley, S. L., Jr and Gillin, F. D. (1993) *Infect. Immun.* 61, 520–524.

18

THE IMPACT OF ANTIGENIC VARIATION ON PATHOGEN POPULATION STRUCTURE, FITNESS AND DYNAMICS

NEIL M. FERGUSON AND ALISON P. GALVANI

Given the ubiquity of antigenic variability in many parasite species, it is perhaps surprising that theoretical analysis of the impact of variation on parasite population biology has been limited until recent years. In part this reflects the Western public health priority in the post-war years on controlling the classic childhood diseases through mass vaccination, and consequent focus of much theoretical research on those diseases as archetypes of all infectious agents. The breakthroughs in our understanding of the molecular basis of vertebrate immunity and disease pathogenesis in recent decades have been key to challenging this view – and have revealed the antigenic stability of diseases such as measles, mumps and rubella to be very much the exception rather than the rule. Indeed, the volume of sequence, immunological and epidemiological data now being generated is outpacing our ability to develop the new theoretical tools necessary for a rigorous analysis of its implications for our understanding of disease population biology. Mathematical studies of antigenic variation are thus still largely focused on using simple models to explore the impact of diversity on pathogen fitness, transmission and evolution – with the correspondence between models and data typically remaining qualitative or conceptual rather than statistically rigorous.

This chapter therefore provides an introduction into the conceptual insights gained from this recent work, whilst highlighting the many research challenges remaining, particularly in relation to improving model realism. Additionally, some new results examining the relationship between rate of variation and pathogen fitness are presented. We begin by reviewing the mechanisms by which antigenic variation can influence pathogen fitness, placing the discussion within a framework of simple models of disease replication and spread. In this context we consider some of the ways antigenic variation can influence disease pathogenesis, though we do not attempt to be exhaustive in our coverage of this complex subject (for an excellent overview of this area, and many other aspects of antigenic variation, see Frank, 2002). We then consider in more detail how antigenic variation shapes pathogen population structure at the level of the host population, focusing on the role of intra-strain competition mediated by host cross-immunity. Next, we review the implications of antigenic variation for infectious disease control strategies revealed by recent theoretical studies in this area, before concluding with a discussion on avenues for future research. Whilst many of the concepts discussed are illustrated with results from simple theoretical models, for accessibility we have tried to keep the main text as non-mathematical as possible. Technical descriptions of models used are provided in an Appendix for the interested reader.

INFLUENCE OF ANTIGENIC VARIATION ON PATHOGEN FITNESS

The central advantage of antigenic variation to the replicative success of a parasite is in escaping antigen-specific host immune responses (Nowak and May, 1991; Phillips *et al.*, 1991; Koenig *et al.*, 1995). The simplicity of this statement belies its underlying complexity, however. To understand its implications for pathogen population dynamics requires precise consideration of two key factors:

1 *The appropriate definition of replicative fitness for an infectious disease.* Long-term persistence of a parasite population depends on successful transmission to new hosts. Thus speed of replication within the tissues of an individual host does not necessarily translate into high overall fitness. Similarly, rapidity of spread within a host population (the growth rate of an epidemic) does not necessarily correlate with long-term fitness. A more useful measure is the basic reproduction number, R_0, defined to be the typical number of secondary infections generated by a single primary infection in an entirely susceptible population (Anderson and May, 1991). However, as we will see below, even this is not entirely adequate when considering a complex multiple-strain pathogen population (Dieckmann, 2002).
2 *The rate at which new antigenic variants are generated or expressed by a pathogen as it replicates.* This determines the manner in which the fitness benefit of variation is expressed: if immunologically distinct variants are produced sufficiently rapidly to make it likely that they will emerge prior to elimination of the initial pathogen strain by host immunity (e.g. *Neisseria*

gonorrhoeae), then the likely consequence will be a lengthening of the duration of infection as each new strain requires a new host immune response to be raised. This increases the probability of successful transmission, and also makes it likely that the transmitted strain will be immunologically distinct from the strain with which the host was initially infected. If variants are generated on a slower timescale (e.g. influenza), new strains will only be generated in a minority of infected hosts, so the typical duration of infection will not be significantly greater than for a comparable antigenically static pathogen. However, the few new variants that do arise will have an advantage in terms of the herd-immunity they experience on transmission. Compared with existing strains to which a significant proportion of the host population is immune, the new strain will experience a (more) immunologically naive population and have a (frequency-dependent) transmission advantage, at least in the short term.

Determinants of pathogen fitness

To explore these concepts in more depth, it is helpful to frame them in the context of simple epidemic theory. Epidemic models divide the host or cell population according to states of disease and susceptibility. The rate at which susceptibles (S), whether cells or individual hosts, become infected is proportional to the number of infecteds (I) in the population. Consequently, infection is a non-linear density-dependent process, which in turn generates the complex dynamics exhibited by many disease systems. Individuals who recover from infection and gain permanent immunity and those who are killed by the infection move to a recovered/removed state (R), whilst if recovery is not accompanied by immunity from reinfection, individuals return to the susceptible state (S).

The transmission dynamics of a simple antigenically static pathogen which generates permanent host immunity are therefore described by the SIR model (Kermack and McKendrick, 1927; Bailey, 1975), where the nomenclature describes the set of disease states moved through in the course of a single infection. At the opposite extreme lie diseases that either cause lifelong infection (e.g. hepatitis B), described by the SI model, or cause extended infection, but where recovery is not associated with protective immunity to future infection, described by the SIS model.

For our purposes we will concentrate on comparing the behaviour of the SIR model, representing an antigenically static pathogen, with the SIS model, representing the extreme of an antigenically variable pathogen where the rate of variation within a single infection and the population as a whole is sufficiently great that clearance of one infection does not provide immunity against future infections. Model definitions are provided in the Appendix. The SI model can be considered as a special case of both the SIS and the SIR models, where the duration of infection is taken to be the rest of the host lifespan. Note that we are also ignoring certain key additional factors, in particular the virulence of the pathogen and its impact on host mortality. The complex relationship between pathogen virulence and fitness (Anderson and May, 1982; Dieckmann, 2002) is beyond the scope of this chapter.

Figure 18.1A and B shows the typical time-course of simple SIS and SIR epidemics. For the same epidemiological fitness (measured by R_0, the basic reproduction number), an SIS pathogen (e.g. *N. gonorrhoeae*) is able to maintain a much higher endemic prevalence than an infection generating permanent immunity, for which endemic infection incidence is limited by susceptible renewal (via host births). Figure 18.1C demonstrates how this difference translates into competitive success, showing the outcome of competition between two strains of a disease, both of which induce the same immune response, but only one of which is affected by this response. As expected, the strain not experiencing immunity rapidly eliminates the SIR strain from the population.

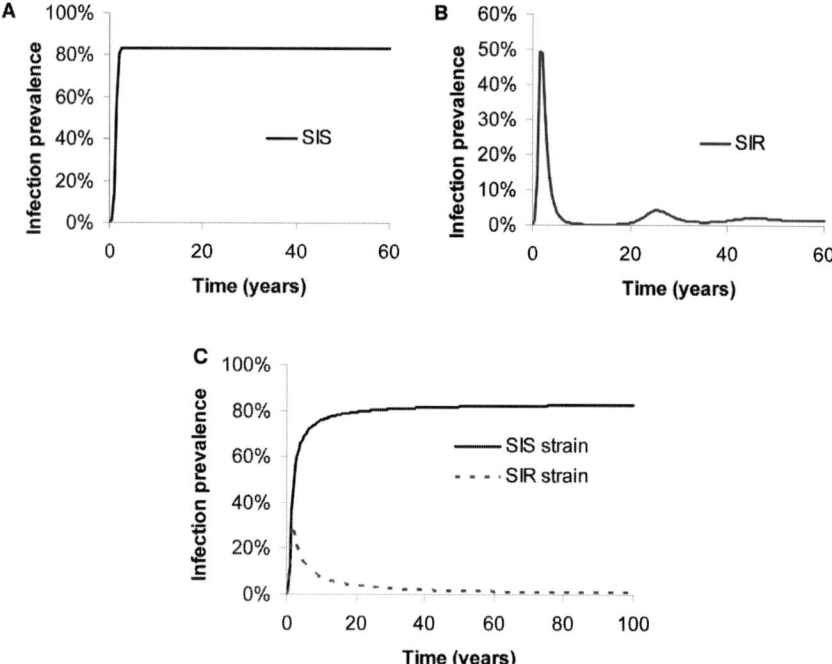

Figure 18.1 Dynamics of SIR and SIS simple epidemic models. (A) Prevalence through time of SIS infection following initial introduction into naive host population. (B) As (A) but for SIR infection. Note the classic epidemic peak, followed by a deep trough caused by susceptible burnout, then gradual equilibration to the endemic prevalence level. (C) Outcome of competition between SIR and SIS 'strains', where both strains induce immunity that is only experienced by the SIR strain. Both strains have the same transmission potential (R_0=4) but, unsurprisingly, in direct competition an SIS strain will outcompete the SIR one, driving it to extinction.

Endemic prevalence determines the ability of diseases to persist in finite host populations; the low infection prevalence of SIR infections (such as measles) makes such diseases much more likely to go temporarily extinct (Bolker and Grenfell, 1995; Grenfell *et al.*, 1995) in smaller host populations (e.g. 1 000 000 or less). Such diseases only persist at all by being amongst the most transmissible of pathogens (R_0>15). By comparison, infections exhibiting sufficient antigenic variation (or lack of immunogencity) to effectively escape herd immunity can persist in very small populations at much lower transmissibility (R_0=2–3) – the sexually transmitted diseases being good examples.

Of course, in reality many pathogens induce an immune response that is temporary or partial – due to antigenic changes in the pathogen over time, waning of immunity, or host variability. Figure 18.2 illustrates the relationship between equilibrium prevalence of infection, transmissibility and degree of immunity for a simple model able to span SIR and SIS behaviour. The non-linearity of the surface shown demonstrates that even partial immunity can be highly effective at reducing infection prevalence at a population level; conversely, the fitness/persistence advantage to a pathogen able to escape 80% of the effect of entirely protective host immunity is disproportionately greater than that for a pathogen able to escape, say, 50%. Indeed, the steepness of the slope

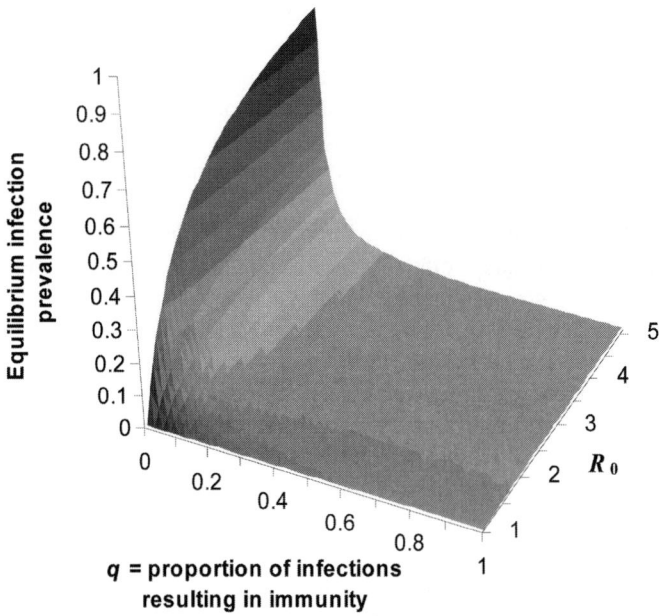

Figure 18.2 Relationship between endemic infection prevalence, p_I, the transmission potential of a pathogen (measured by R_0), and the proportion of infections resulting in protective immunity blocking future infection, q. The SIS and SIR models correspond to the $q = 0$ and $q = 1$ lines respectively. A 1-year duration of infection, D, and 50-year host lifespan, L, is assumed.

of the R_0 surface in Figure 18.2 along the q axis for low values of q is a measure of the intense selective pressure imposed by host immunity.

These results demonstrate that the basic reproduction number is not the sole determinant of long-term pathogen fitness at the level of the host population; whilst R_0 determines the ability of a disease to invade a previously unexposed population, the immunogenic characteristics of the organism also play a key role, such that pathogens can trade off transmissibility against lack of immunogenicity. The SIS and SIR models represent extreme limits of infectious disease behaviour – more generally, organisms can generate partial or waning host immunity, or achieve the same goal of immune escape through antigenic diversity. Evaluating overall fitness in such contexts – in the sense of identifying the fittest strain in a diverse population – is far from trivial (Dieckmann, 2002). As we will see later in the chapter, intense frequency-dependent selection induced by inter-strain competition (mediated by cross-immunity) is often the dominant determinant of fitness at a point in time, making the outcome of such competition often unpredictable.

Within-host infection dynamics and the fitness benefit of variation

From the perspective of transmission success, it is clearly advantageous for a pathogen to generate novel antigenic variants as rapidly as possible. What, therefore, are the factors constraining evolution from driving pathogens to ever-faster rates of diversification? Constraints imposed by the receptor-binding functions of most protein epitopes (in particular) obviously play a role – and will differ depending on the pathogen (virus, bacterium or helminth) and its cellular targets. Secondly, for clonal pathogens where mutation is the principal variant-generating mechanism, the increasing fitness cost of deleterious mutations will trade off against the benefits endowed by rapid generation of antigenic variants. However, the very large effective population sizes of viral and bacterial populations call into question the extent to which this represents a critical constraint. Furthermore, many more sophisticated pathogens (e.g. *N. gonorrhoeae*, *Trypanosoma cruzi*) have developed mechanisms (e.g. phase variation) to generate antigenic variation without high mutation rates, several of which are discussed extensively in other chapters of this volume.

In this section we therefore consider the less explored influence of within-host constraints on pathogen replication imposed by density- or resource-dependence and their interaction with strain-specific immune responses. We use a novel mathematical model (see Appendix) of the within-host dynamics of a generic pathogen and the immune response induced by that pathogen. Similar models have been used in the past to explore the possible role of immune escape mutants in HIV pathogenesis (Nowak and May, 1991; Nowak *et al.*, 1991, 1995), more general models of viral and immune-system dynamics (e.g. Wodarz and Nowak, 2000), and antigenic switching matrices in trypanosomes (Frank, 1999). The work presented here differs in using a highly general model of resource depletion, and specifically examining the relationship between mutation/variation rate and pathogen fitness. We characterize fitness in terms of the total amount of pathogen replication prior to clearance – the area under the parasitaemia curve. Under the assumption (discussed further below) that transmissibility is proportional to parasitaemia (i.e. the population size of the pathogen within the host), this measure corresponds to the total net transmission potential of the pathogen during a single infection.

The model describes growth of an initial infecting dose of a pathogen, development of the immune response against it, and the effect of that immune response in inhibiting pathogen replication. Pathogen growth is assumed to deplete some (renewable) resource (e.g. target cells or nutrients), and stochastic mutation of the pathogen is modelled as occurring at a fixed probability per pathogen particle (virion/bacterium) per replication cycle. New variants are assumed to require the same resource as the initial strain for replication, but to be immunologically distinct, requiring additional strain-specific immune responses to develop to inhibit their growth.

Figure 18.3 shows example output from the model demonstrating its ability to capture the basic dynamics of pathogen replication and immune clearance for a single strain. Note the model of immunity used is deliberately simple, representing the strain-specific immune response in terms of a simple fast-replicating and slow-decaying cell class that directly clears the pathogen; using a more sophisticated two-stage model (e.g. Ferguson *et al.*, 1999; Frank, 1999)

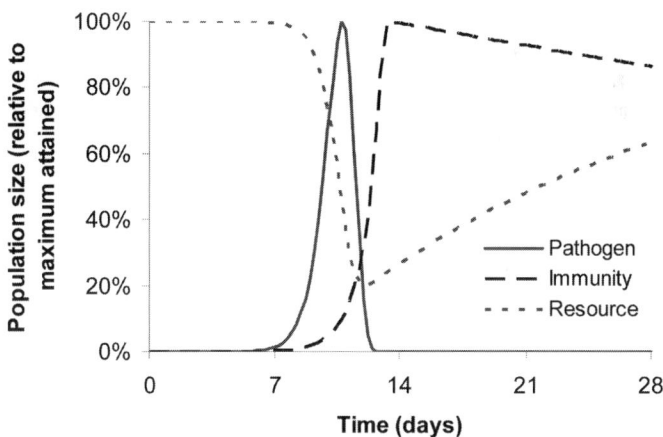

Figure 18.3 Typical response of simple within-host pathogen replication/immune response model to a single strain of a new infection. The pathogen is introduced at low dose at time 0, and begins to consume resource whilst replicating exponentially. In response to the pathogen, a strain-specific immune response also proliferates, but only begins to have a significant effect on pathogen growth after several days. Note that whilst resource is consumed rapidly, immunity (not resource depletion) plays the dominant role in limiting pathogen growth here.

of active and memory cellular responses does not qualitatively change the results.

Figure 18.4 shows an example of the dynamics of the same system once antigenic variation is introduced, here modelled as random mutation generating immunogenically novel strains. The dynamics are characterized by multiple waves of parasitaemia. Thus the model presented here is arguably most suited to describing pathogens (typically bacteria and protozoa) that exhibit antigenic variation through phenotypic switching of antigenic gene expression. Such pathogens include *Giardia lamblia* (Nash, 1997; Muller and Gottstein, 1998), *Borrelia burgdorferi* (Zhang *et al.*, 1998), *Babesia bovis* (O'Connor *et al.*, 1997), *N. gonorrhoeae* (Serkin and Seifert, 1998), *Neisseria meningitidis*, *Escherichia*

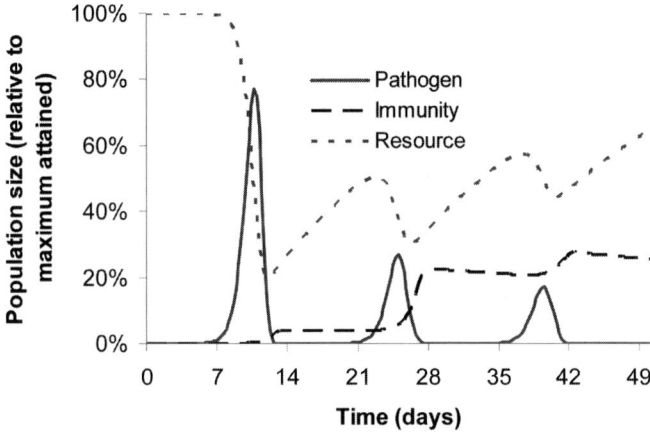

Figure 18.4 Example dynamics of simple within-host model when random mutation generating antigenic variants is included ($\delta = 2\text{e-}8$, see Appendix). The second peak in pathogen population size is composed of multiple pathogen strains, which are generated at the time of maximum replication of the original pathogen strain. The immune response shown also represents the sum across all strain-specific responses.

coli (McKenzie and Rosenberg, 2001), *Haemophilus influenzae* (Weiser *et al.*, 1989), *Campylobacter coli* (Park *et al.*, 2000), *Plasmodium falciparum* (Vanhamme *et al.*, 2001), *T. cruzi* (Tarleton and Kissinger, 2001) and *Trypanosoma brucei* (Vanhamme *et al.*, 2001). Trypanosomes escape from host immunity by (apparently) randomly switching expression of the variant surface glycoprotein (VSG) from a repertoire of an estimated one thousand distinct genes (Borst, 1991); the resulting sequential dominance of antigenic variant has been the topic of much previous debate and some modelling (Barry and Turner, 1991; Frank, 1999; Vanhamme *et al.*, 2001).

The extent to which sequential waves of parasitaemia are observed in bacterial or viral infections is difficult to ascertain; data do not exist to examine whether organisms such as *N. gonorrhoeae* exhibit similar dynamics. We see no reason why the model should not also be viewed as a first-order approximation to the dynamics of pathogens with short or intermediate durations of infection that generate antigenic variation directly through mutation and switching; however, in such cases the relationship between the (antigenic) variant-generation rate and underlying error rate in transcription is complex (see Appendix).

That said, the model is certainly too simple to describe the pathogenesis of chronic retroviral infections (e.g. HIV and hepatitis C) accurately. In such infections, whilst it is known from sequence and tetramer studies (McMichael *et al.*, 1995, 1996; Borrow *et al.*, 1997; Goulder *et al.*, 1997; Price *et al.*, 1997) that escape mutants do arise, it is also clear that single antigenic variants are able to persist as the dominant strain in the viral quasi-species for extended periods (months–years). To describe such dynamics, a more sophisticated model of the interaction between viral and immune-system dynamics is required, which permits the equilibrium persistence of infection in the host. However, preliminary analysis of more complex models indicates that the qualitative conclusions of the results presented below remain unchanged.

Having satisfied ourselves that the model is a reasonable representation of within-host infection dynamics, we can now examine how the simulated dynamics change as a function of the mutation (or variant generation) rate. Figure 18.5A shows how the average time-course of parasitaemia varies with mutation rate, and Figure 18.5B the corresponding time-course of resource depletion. The striking pattern highlighted by these figures is the relatively narrow band of mutation rates that permit significantly extended durations of infection. It is only within this band that parasitaemia is maximized (Figure 18.6): lower mutation rates do not generate new variants sufficiently often to make it likely that a second wave of parasitaemia will be generated, whilst higher mutation rates cause multiple variants to be generated too rapidly, resulting in multiple strains competing for limited resources simultaneously – lowering the net replication rate and exhausting the antigenic repertoire more rapidly.

It is worth considering the implication of these results specifically in the context of trypanosome infections. These results demonstrate that multiple waves of parasitaemia emerge naturally as an evolutionary optimum of an entirely random switching process once the stochastic nature of the transition process is fully accounted for; in contrast to past work, we do not find non-randomness in switching rates to be essential to explain observed patterns (though it may indeed further prolong the duration of infection) (Frank, 1999).

Whilst the results presented above give new insight into the relationship between antigenic variation and transmissibility and the limits imposed on the rate at which pathogens can diversify, they do not explain why certain types of infectious diseases show very little variation – particularly highly transmissible respiratory agents such as measles, rubella and pertussis. More specifically, why do relatively few highly variable pathogens adopt this niche? In this regard even diseases such as influenza can be viewed as relatively non-varying, given

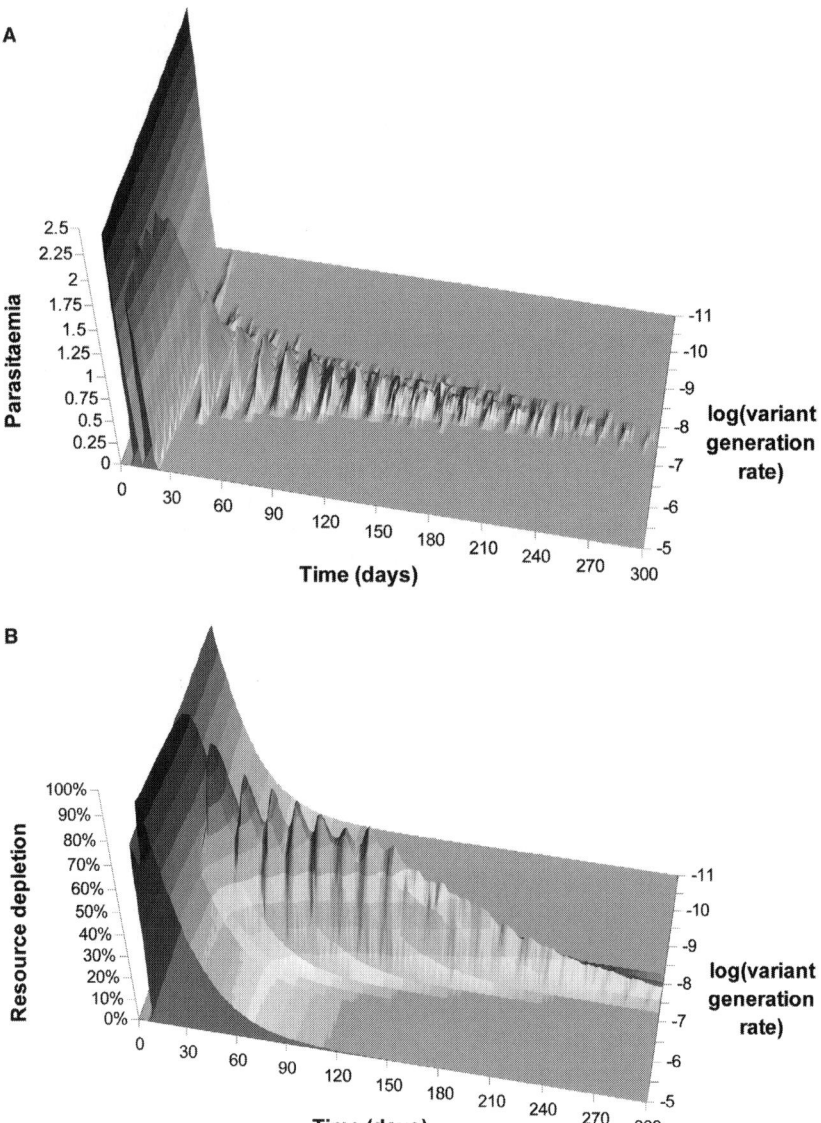

Figure 18.5 Average (over 50 simulations) time-course of (A) infection and (B) resource depletion, for a pathogen generating permanent strain-specific immunity, as a function of the mean rate at which new antigenic variants are generated. The graph should be interpreted as a series of time-series plots running from left to right, which have been plotted adjacent to each other in surface form to illustrate the dependence of model dynamics on variant generation rate. It can be seen that for a limited range of generation rates, significantly extended infections are possible (which maximize long-term usage of host resources), with successive waves of parasitaemia, which on average decrease in amplitude through time. The variant generation rate (parameter δ, see Appendix) is standardized for a pathogen with a peak population size in a single-strain infection of ~3e7. The value of the generation rate giving maximum parasitaemia varies as a function of the assumed within-host pathogen population size.

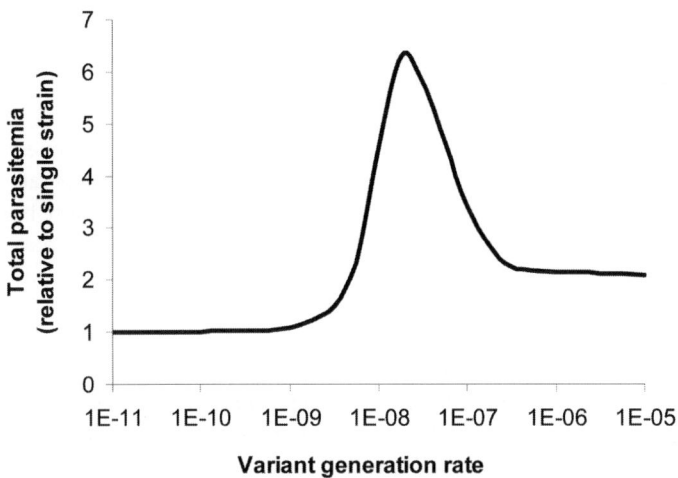

Figure 18.6 Predicted relationship between total cumulative parasitaemia during an infection and the rate at which new antigenic variants are generated. Note the very significant maximum in parasitaemia seen at intermediate values of the variant generation rate. The curve was calculated by taking the model output shown in Figure 18.5 and calculating the area under the parasitaemia curve for varying values of the variant generation rate.

antigenic change occurs on the timescale of years and after the virus has been transmitted through millions of hosts.

Addressing this issue requires consideration of the other determinants of transmission success as well as the rate at which a pathogen generates new variants within an infected host. Replication rate, the efficiency of the immune response, cross-immunity between strains and the nature of the transmission process all play key roles. The interaction between all these factors modifies the optimal mutation rate in a highly non-trivial manner, giving different outcomes depending on the ecological niche to which a particular pathogen is adapted. There is also a complex relationship between the rate at which new variants are generated within an infected host and the probability that the infections that a host generates in the wider population will be of a different strain from the original strain with which it was infected. Full exploration of these issues is the topic of ongoing work, but requires the model detailed above to be extended to explicitly consider transmission in a population of hosts, and is thus beyond the scope of this chapter.

INTER-STRAIN COMPETITION AND EPIDEMIOLOGICAL CONSTRAINTS ON DIVERSITY

Having discussed, albeit selectively, some of the constraints acting to limit the diversification of pathogens within an infected host, we now shift the focus of our discussion to the level of an entire host population, and explore how the transmission dynamics of diseases in host populations affects pathogen population structure. Four factors are key:

1 *Rate of generation of new variants* – new strains are generated through mutation or recombination. The rate at which such newly generated strains emerge in the population (i.e. infect at least one additional host besides the originating individual), and their typical antigenic distance from existing strains are critical in determining either the rate of antigenic diversification at the population level or the equilibrium level of diversity observed. The recombination rate depends upon the frequency of co-infected hosts (or vectors), as discussed below.

2 *Cross-immunity* – the antigenic relationship between strains in a population is characterized by the degree of cross-immunity induced against one strain as a result of previous exposure to another strain. It can be helpful to conceptualize strain relationships in terms of a strain 'space' (e.g. Smith *et al.*, 1997, 1999), in which strains are represented as points, and the distance between any two points represents the degree of antigenic overlap between the corresponding strains. Whilst the geometry of strain space has not been determined for any real pathogen, a number of theoretical frameworks have been proposed (Figure 18.7). However, any conceptualization of strain relationships in terms of a strain space derived from pairwise differences does not provide a complete representation of the competitive relationships between strains induced by cross-immunity; in reality the response of a host to exposure with any one strain depends upon that host's cumulative history of exposure to heterologous strains. This topic has only recently been explicitly examined in theoretical studies (Gomes *et al.*, 2002). Cross-immunity induces frequency-dependent selection on pathogen populations, since if a host population has had little exposure to a particular strain or to strains that are antigenically similar to it, then that strain will have a temporary transmission fitness advantage over strains with higher prevalence. The importance of such competition is supported by a wide range of empirical data (e.g. for malaria) (Hargreaves *et al.*, 1975; Taylor *et al.*, 1997; Gilbert *et al.*, 1998). The complex strain dynamics such competition can generate are discussed below.

3 *Coinfection* – this has been rather less frequently examined in theoretical studies (but see Adler and Losada, 2002; Gandon and Michalakis, 2002; Nowak and Sigmund, 2002), due to inherent technical challenges, but may be equally important in determining pathogen population structure and levels of diversity. For the great majority of pathogens, resource constraints (see the first section of this chapter) and non-specific immunity can be expected to reduce the susceptibility of an infected host to simultaneous infection with a different strain and/or to lower the infectiousness of the host (relative to a host with only one infection) with respect to one or both strains. For

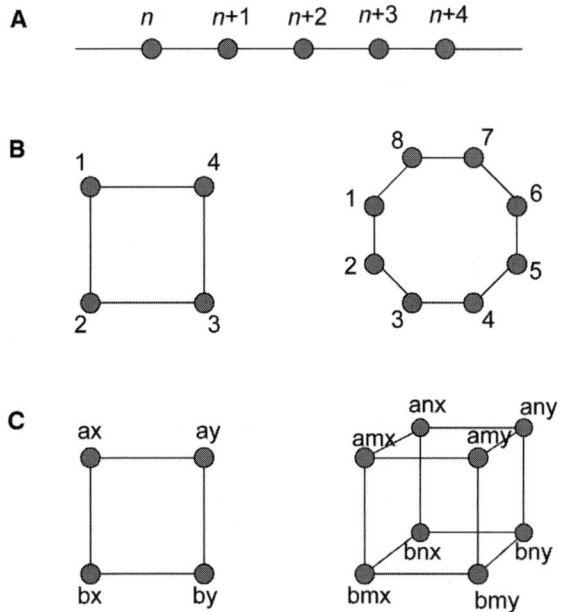

Figure 18.7 Hypothesized 'strain spaces' – networks of relationships between strains of a single pathogen. Nodes represent strains, and links couple strains which compete via cross-immunity. (A) The simplest model – a large number of strains lying on a line with only nearest-neighbour strains interacting through cross-immunity – has been used as a phenomenological model of influenza evolution (Andreasen *et al.*, 1996, 1997); (B) 'ring' models (Lin *et al.*, 1999; Gomes and Medley, 2002) of antigenic interactions (two models are shown, with four and eight strains respectively) permit long-term stable periodic cycles of strain prevalence to develop, and have been used to examine interactions between all pairs of strains on the ring, not just nearest neighbours, where the distance (in links) between two strains determines the degree of cross-immunity, though sometimes in a non-linear fashion; (C) 'genetic' models (Gupta *et al.*, 1996, 1997, 1998; Ferguson and Andreasen, 2002) assume immunity is determined by N antigenic loci, each with M alleles, where prior infection gives partial cross-protection against infection with any heterologous strain sharing one or more alleles with the earlier strain. In total, such systems have M^N strains, with any one strain having at least one allele with $M^N - (M-1)^N - 1$ others. The 2-locus 2-allele and 3-locus 2-allele systems are shown (the only ones easily represented as polyhedra). Note the former is identical to the four-strain 'ring' model. All the strain-space models shown share the characteristic of being symmetric with respect to strain interactions; no one strain has a 'privileged' position.

highly diverse pathogens where little cross-immunity is seen between strains (perhaps as a result of frequency-dependent selection), this form of competition would be predicted to play the dominant role in limiting overall population diversity. Within-host inter-strain competition is also critical to determining the frequency of recombination events and thus the rate of generation of novel variants.

4 *Strain extinction* – strains can be driven to extinction through either cross-immunity or coinfection-mediated competitive exclusion, or by random chance as a result of demographic stochasticity. In finite host populations both

effects are likely to play an important role, and indeed to interact – since frequency-dependent selection can temporarily drive strain frequencies to low enough levels that random extinction becomes much more probable. Other factors (e.g. seasonal fluctuations in transmission rates) can also interact with demographic stochasticity to generate genetic bottlenecking in some cases. Additionally, stochasticity has the effect of reducing the establishment probabilities of newly generated strains compared with what would be predicted from considerations of relative fitness alone. However, despite its obvious importance, few theoretical studies have considered the impact of stochastic effects on pathogen diversity within the framework of an epidemiologically realistic transmission model, in large part due to the analytical and computational challenges involved.

An integrated and generic analysis of all the factors above remains beyond the scope of both experimental and theoretical techniques; from the latter perspective the key problem is one of combinatorial complexity – when modelling N strains co-circulating in a population, except in special cases (discussed below), it is necessary to track of the order of $N!$ unique immune histories of exposure to all possible combinations of strains (Andreasen et al., 1997). Within a traditional compartmental model, formalism, even numerical exploration of such systems, therefore becomes impractical for more than about eight strains. Microsimulation methods can be used to examine more complex systems (Galvani et al., 2001), but are very computationally intensive, with robust identification of the key determinants of system behaviour being generally more difficult.

For these reasons, most work to date has examined highly simplified special cases, concentrating on exploring the action of cross-immunity-mediated inter-strain competition in structuring pathogen populations with a defined number of pre-existing strains. The aim of these analyses has been to explain the wide variety of patterns of strain structure seen in different pathogens. Antigenically variable pathogens exhibit an array of different patterns of temporal change in population structure, and phylogenetic analysis can help to characterize the temporal change in the antigenic composition of pathogen lineages (Bush et al., 1999a; Holmes and Burch, 2000). Four basic patterns are apparent, as crudely typified by dengue, *N. meningitidis* subtypes, HIV and influenza A respectively: static strain structure, fluctuating epistasis, increasing diversity and sequential strain replacement ('drift').

Aspects of these patterns have been successfully captured by a range of simple models (Andreasen et al., 1997; Gupta et al., 1998; White et al., 1998; Lin et al., 1999; Ferguson and Andreasen, 2002; Gomes and Medley, 2002; Gomes et al., 2002). Depending on the strength of cross-immunity between antigenically similar strains, models with limited (<10) strains share similar properties, with the dynamics generated being relatively insensitive to assumptions made about the structure of antigenic space and action of immunity (Ferguson and Andreasen, 2002; Gomes and Medley, 2002; Gomes et al., 2002):

1 At low levels of cross-immunity, all strains persist since inter-strain competition is insufficiently strong to cause exclusion.
2 At intermediate levels of cross-immunity, cyclical dynamics (fluctuating epistasis) are seen, with sequential replacement of one subset of strains (with minimal antigenic overlap) by another set. In this regime cross-immunity is sufficiently strong when all strains have near equal prevalence to begin the process of competitive exclusion of strains with antigenic overlap. However, as the subset of strains being excluded is driven to lower frequencies, the competitive advantage offered by negative frequency-dependent selection (i.e. the increasing numbers of hosts in the population without prior exposure to the low-frequency strains) slows and eventually reverses the fitness difference between high- and low-frequency subsets of strains, eventually leading to oscillations. The wide fluctuations in strain prevalence observed in serotypes of group A streptococcus (Anthony et al., 1976) is suggestive of such dynamics. In addition, different serogroups of *N. meningitidis* exhibit variation in both the elicitation of cross-immunity and their patterns of antigenic population structure (Caugant et al., 1990), which is consistent with this type of dynamical behaviour.
3 At high levels of cross-immunity, the period of the cycles described above increases dramatically, eventually becoming infinite, at which point the system self-organizes into a set of variants with non- or minimally overlapping antigenic repertoires (Gupta et al., 1996, 1998). This represents complete competitive exclusion, with frequency-dependent variation in fitness being unable to prevent the subset of excluded strains from being driven extinct. However, it should be noted that models also indicate that the timescale of competitive exclusion may be in the order of decades, making transient effects important (Porco and Blower, 1998).

Figure 18.8 shows results from such a model (Gupta et al., 1998; Ferguson and Andreasen, 2002) (see Appendix), illustrating this variation in dynamical behaviour. Figure 18.8A depicts the non-linear relationship between mean pathogen incidence (a proxy for fitness) and the strength of cross-immunity, with the self-organization of strains into non-overlapping subsets of oscillating frequency acting to minimize the effect of cross-immunity-induced competition. Figures 18.8B and C show how as cross-immunity becomes more intense, regular fixed-period cycles in infection prevalence are replaced by chaotic cycles of longer period and random amplitude, with some strains falling to very low prevalences for many years.

However, simple models using a 'genetic' strain space (see Figure 18.7) do not capture the pattern of antigenic change exhibited by influenza A well, even when extended to describe much larger numbers of strains. Inter-pandemic influenza A virus evolution is characterized by antigenic 'drift' (though this term is somewhat misleading, given the positive selection on antigenic loci) (Bush et al., 1999b), which results in a phylogenetic tree which is unusually linear compared with other viruses and even to the other influenza subtypes B and C (Bush et al.,

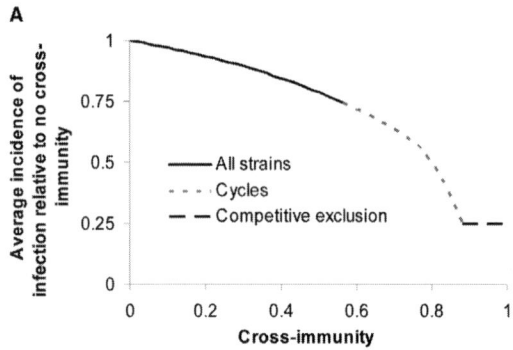
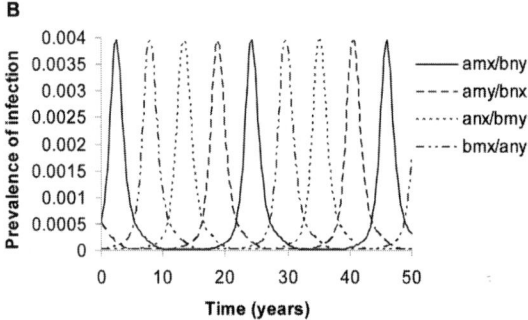
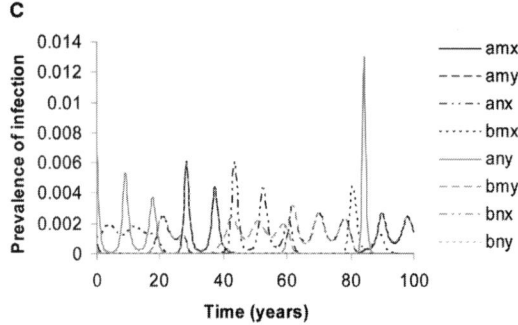

Figure 18.8 Dynamics of the 3-locus, 2-allele multi-strain epidemic model (see Appendix for model and parameter definitions) without mutation. (A) Average long-term total infection incidence of all strains as a function of the strength of cross-immunity (0 = none, 1 = complete) and represented as a fraction of the incidence in the zero cross-immunity case. The line pattern and colour is used to distinguish the behaviour of the model as a function of cross-immunity: for values <0.46, all strains coexist at identical equilibrium prevalence, from 0.46 to 0.88 cyclical strain dynamics are seen, and for >0.88, the systems self-organizes, with competitive exclusion of all but two antigenically non-overlapping ('discordant') strains. (B) Regular cycles in strain-specific infection incidence seen for cross-immunity of 0.6. (C) Chaotic fluctuations in strain incidence seen for cross-immunity of 0.75. Note that whilst discordant strains (e.g. amx and bny) have identical incidence through time in (B), this symmetry is broken in (C). The dynamics of these systems is discussed in more detail elsewhere (Gupta *et al.*, 1998; Ferguson and Andreasen, 2002).

1999a). This pattern reflects continual generation of new strains through mutation, and extinction of others through the combined effect of inter-strain competition and demographic stochasticity.

Deterministic models assuming a linear one-dimensional strain space where nearest-neighbour strains interact under high levels of cross-immunity (Andreasen et al., 1996, 1997) can reproduce this pattern to some extent, but the description they provide is relatively uninformative. This is because by construction such models automatically generate travelling waves of fixed diversity in strain space. In reality the dimensionality of strain space for influenza A is likely to be considerably greater than one, given the multiple antigenically active codons under selection (Bush et al., 1999a). Thus the other factors (mutation, stochastic extinction and coinfection) described above are likely to be important in explaining the limited observed diversity of influenza. Mathematical models need both to incorporate this complexity and to reproduce the observed pattern of antigenic variation to be truly explanatory (Ferguson et al., 2003).

The motivation to find such a description – despite the technical challenges involved – is far from purely academic. The continual emergence of new strains necessitates near-annual updating of influenza vaccine stocks, based on subjective prediction of the antigenic composition of the next major epidemic strain months in advance (WHO 00: 330; WHO 00: 61; WHO 00: 281). Robust quantitative techniques for improving prediction of influenza evolution (Bush et al., 1999a) could therefore have a major impact on the effectiveness of future vaccination programmes.

Lastly, it should be noted that cross-immunity need not always be protective. Co-operative interactions can also occur, the best known being antibody-dependent enhancement (ADE) – where host exposure to one strain facilitates the replication of another. This has been reported for a number of flaviviruses, most notably dengue (Holmes and Burch, 2000), and also in HBV and the bacterium *Dichelobacter nodosus* (Hunt et al., 1995). ADE is a key factor in severe disease pathogenesis (Halstead, 1988), and may underlie the observation that dengue haemorrhagic fever is more likely in multiple-strain infections (PAHO, 1994). Ecologically, ADE is likely to enhance strain coexistence and can precipitate complex dynamics, including chaotic or cyclic behaviour (Ferguson et al., 1999). Interestingly, the rate of antigenic diversification of dengue is slower than that of other RNA viruses that employ the same RNA polymerase (Zanotto et al., 1996), perhaps reflecting the effect of co-operative inter-strain interactions in reducing immune-mediated selection for diversification.

CONTROL OF ANTIGENICALLY VARIABLE PATHOGENS

Given the complex non-linear patterns generated by frequency-dependent selection between competing strains just discussed, it is clear that mathematical models are potentially valuable tools for assessing the impact of interventions targeting antigenically variable pathogens. The insights offered are at the level of both the host community and the interaction of the pathogen with the immune system of a single infected host. Here we briefly review some of the recent work in this area, concentrating on the use of models to predict the epidemiological outcome of vaccines targeting antigenically diverse bacterial populations, and in evaluating the potential benefits of immune therapies in slowing HIV progression within the treated patient.

The success of a vaccine depends on how pathogen populations change under the selective pressures imposed by its use. This inter-dependence poses challenges in predicting the long-term consequences of an intervention targeting an antigenically heterogeneous pathogen. Furthermore, at a more basic level antigenic variation clearly poses an obstacle to the development of effective vaccines, since a vaccine must be protective against multiple variants.

These challenges have been well appreciated for many pathogens, including HIV (Bloom, 1996; Heeney and Hahn, 2000; Ruprecht *et al.*, 2000), influenza (Treanor, 1998) and malaria (Howard and Pasloske, 1993). Understanding the causes and consequences of antigenic diversity is the first step in predicting the response of pathogen populations to drugs and vaccines. This understanding is made all the more essential given that misguided intervention might increase the morbidity in the long run.

The impact of immunization on disease incidence depends on the interaction between the antigenic composition of the vaccine and the population structure of the pathogen in question. Intervention can substantially alter competitive interactions between strains (McLean, 1995; Lipsitch, 1997; Porco and Blower, 1998), such that a vaccine or antiviral drug targeting only a subset of circulating strains could ultimately increase the burden of disease (Gupta *et al.*, 1997; Lipsitch, 1999) by enhancing the prevalence of other strains that previously had been suppressed by competition with the targeted strain.

Using the simple multi-strain epidemic model introduced above, Figure 18.9 illustrates these potentially adverse consequences by showing the effect of a vaccine targeting the dominant strains circulating in a population at a particular time. The result of the vaccination programme is seen to rely critically on whether the vaccine induces the same level of cross-protection against heterologous strains as infection itself; if not, the effect of the programme can be to enable the strains that had previously been suppressed by cross-immunity-mediated competition to emerge at higher frequencies in the host population – with potentially adverse consequences if the newly dominant variants are more virulent than those displaced. This new analysis complements past work examining vaccination which targets a subset of strains currently dominant in a population (Gupta *et al.*, 1997).

Clinical trials (Obaro *et al.*, 1996; Dagan *et al.*, 1998; Mbelle *et al.*, 1999) and experimental findings (Lipsitch *et al.*, 2000) demonstrating replacements of different antigenic variants of *Streptococcus pneumoniae* confirm the potential

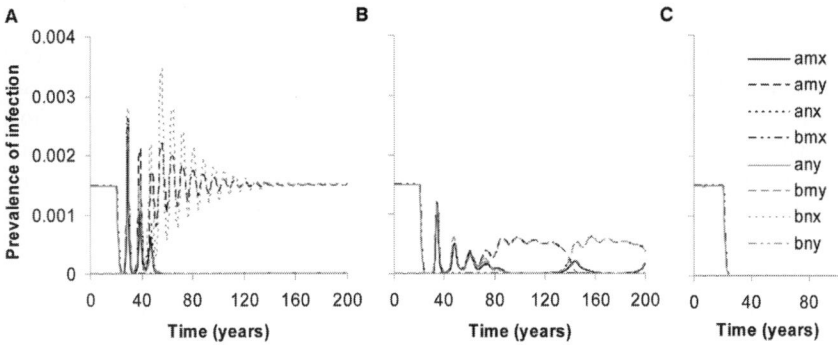

Figure 18.9 The impact of vaccination on the transmission dynamics of the 3-locus, 2-allele multi-strain model. Cross-immunity is assumed to take a value of 0.9 between strains sharing alleles, leading at equilibrium (see Figure 18.8) to competitive exclusion of all but two strains (bmx and any in this example). To simulate regeneration of strains following extinction, mutation is included in the model at a probability of 10^{-4} per infection. Multivalent vaccination against these two strains alone is introduced at time 20, with the equilibrium mean age at vaccination being 1 year. The three graphs explore the results of varying assumptions about the relative degree of cross-immunity induced against heterologous strains by vaccination relative to that caused by natural infection, f: (A) $f = 0$ – when no cross-immunity is induced another discordant set of two strains (which was previously competitively excluded by the previously dominant set) emerges, and the system eventually returns to the incidence levels seen prior to the introduction of vaccination; (B) $f = 0.25$ – again previously suppressed strains emerge, but at a lower incidence than previously, and with cycles in incidence between the three remaining discordant sets of two strains being seen. (C) $f = 0.5$ – at this level, vaccine-induced cross-protection is sufficient to cause continued suppression of the six low-prevalence strains, leading to eventual elimination of all strains.

that such theoretical predictions may be a cause for real concern. As polymorphic pathogens increasingly become the focus of vaccination research there is an increased need to develop corresponding theoretical tools for accurately predicting the likely consequences of widespread use of novel polyvalent vaccines – such as that being developed for *S. pneumoniae*, which only affords protection against 10% of the 90 identified pneumococcal antigenic variants (Mbelle *et al.*, 1999). In this context it is also worth noting that as well as giving insight into the potential for adverse outcomes, models can also aid our understanding of how the impact of interventions of pathogen population structure can be monitored (Lipsitch, 1997, 1999).

While induction of protective immunity in uninfected individuals is the principal goal of much HIV vaccine research (Boyer *et al.*, 2000; Dorrell *et al.*, 2000; Hanke and McMichael, 2000; McElrath *et al.*, 2000; Nitayaphan *et al.*, 2000), vaccine candidates also represent another potentially important therapeutic tool for viral control in infected patients. Unfortunately, it is currently unlikely that vaccines will be completely effective in either context (Girard *et al.*, 1999; Ortiz *et al.*, 1999; Dale and Kent, 2000; Lifson *et al.*, 2000; Rosenberg *et al.*, 2000), with eradication of virus in the infected patient looking an increasingly unrealistic goal. Thus the primary objective of current therapeutic vaccination research is long-term suppression of viral replication, an aim shared with antiviral therapy. However, since antimicrobials and immunization act via different pathways (epidemiologically, one reduces infectiousness, and the other susceptibility), basic epidemiological theory predicts that dual therapies are likely to act synergistically both within the host and at the level of the host population (Anderson and May, 1991). In addition, combining drug and vaccination therapies will reduce pathogen replication in the host – thus reducing the net mutation rate – as well as imposing a greater mutational barrier to the evolution of drug resistance or vaccine escape mutants.

A number of theoretical studies have provided quantitative predictions of dual treatment strategies involving both immunotherapy and antiviral use (Bonhoeffer et al., 1997; Fraser et al., 2000). Recent experimental work has also shown that here such dual strategies may offer the potential for autoimmunization using intermittent therapy to manipulate viral load during primary HIV infection, so as to maintain a broad CTL response (Rosenberg et al., 2000). Key to the success of structured treatment interruption is predicting the balance between achieving sufficient antigen presentation to elicit immunity and minimising HIV replication-associated impairment of the CTL response (Nowak and Bangham, 1996). Models (Wodarz et al., 1999; Wodarz and Nowak, 2000) have had a key role in exploring different hypotheses for the immune-dynamics mechanism that might explain the apparent success (Rosenberg et al., 2000) of such strategies.

CONCLUSIONS

This chapter has shown how the techniques of theoretical population biology can give insight into the adaptive advantage of antigenic variation and the constraints acting to limit diversity in real host–pathogen systems. The number of different factors that determine both the short-term (epidemiological) and long-term (evolutionary) dynamics of such systems – and their complex non-linear interactions – poses many challenges to any analysis attempting to provide a unified description of the behaviour of such systems. For this reason, most of the theoretical studies performed thus far have focused on individual aspects of system behaviour (e.g. within-host dynamics, cross-immunity-induced competition), usually using highly simplified representations of key biological processes and heterogeneity.

Inclusion of greater biological realism (e.g. relaxing the assumption that all strains are identical, incorporation of stochasticity) will become increasingly important to match model predictions more directly to newly available data, and incorporate advances in our understanding of the molecular mechanisms of variation. Again, however, the technical obstacles are formidable; assuming symmetry between strains, un-constrained coinfection and simple models of immune memory formation (Gomes and Medley, 2002) appear essential to compact model formulation within a compartmental framework (Ferguson and Andreasen, 2002). Microsimulation models are clearly one approach to the incorporation of greater complexity, but are computationally demanding and risk being analytically opaque. More powerful analytical tools are therefore urgently required – perhaps akin to a statistical mechanics of multiple-strain systems – to give insight into both equilibrium behaviour and the response of such systems to exogenous perturbation.

Much also remains to be understood about the evolutionary dynamics and stability of antigenic variation at a population level. Most theoretical studies of multiple-strain systems have explored the consequences of particular assumed forms (and rates) of antigenic variation on pathogen population structure and dynamics, with rather less emphasis on how such systems arise evolutionarily. We also still have little insight into why antigenic variation is not universal; it is inadequate to cite functional constraints of particular pathogen species – the question is how some relatively invariant species (e.g. many of the childhood diseases) came to occupy their niches without being displaced by more variable competitors.

We feel two factors may be key to addressing these issues. The first is the effect of constraints imposed on pathogens within an infected host on rates of antigenic change, transmissibility and duration of infection. Whilst the modelling of within-host dynamics has given much insight into disease pathogenesis (Frank, 1999; Nowak and May, 2000), less attention has been paid to evaluating the impact of the interaction between host immunity and pathogen replication on disease transmission and between-host population dynamics, particularly in the context of antigenic variation. In this regard, the results presented in the first section of this chapter are a first step towards such a description.

Secondly, explaining observed levels of pathogen diversity arguably requires an understanding not only of pathogen evolution in isolation, but of the coevolution of pathogen and host populations. Mathematical models have shown that heterogeneity in the ability of hosts to respond to different parasite antigens may be critical to pathogen population dynamics (Gupta and Hill, 1995; Gupta and Galvani, 1999) with the coexistence of antigenic variants depending on the precise composition of the immune responses in the host population. Host heterogeneity weakens competition between strains by providing separate niches for different types of parasites, but the smaller size of these niches subjects their inhabitants to larger ecological fluctuations. Such heterogeneity is the hallmark of the vertebrate immune system and is a reflection of the coevolutionary arms race between hosts and pathogens.

Nowhere is this better observed than in the variability of the major histocompatibility complex (MHC) – in humans known as the human leucocyte antigen (HLA) region – the mechanism underlying antigen recognition in vertebrates. The influence of MHC/HLA genotype and restriction on disease susceptibility and progression has been demonstrated for a wide variety of human pathogens, examples being influenza A (Gianfrani *et al.*, 2000; Nakajima *et al.*, 2000; Voeten *et al.*, 2001) and HIV (Scorza Smeraldi *et al.*, 1986; Steel *et al.*, 1988; Just, 1995; Kroner *et al.*, 1995; Snowden *et al.*, 1996; Carrington *et al.*, 1999). HLA is the most polymorphic region of the human genome (Hendrick *et al.*, 1991) and the overwhelming predominance of non-synonymous to synonymous mutations indicates intense selection (Parham *et al.*, 1989; Jeffery and Bangham, 2000). The mechanism driving this selection has been the subject of much theoretical analysis, with both heterozygous advantage in antigen recognition (Doherty and Zinkernagel, 1975; Hughes and Nei, 1989; Hughes *et al.*, 1994) and frequency dependence (i.e. rare HLA genotypes responding better against pathogens adapted to common HLA genotypes) being favoured hypotheses (Takahata and Nei, 1990; Slade and McCallum, 1992; Wills and Green, 1995). Evidence for both mechanisms has also been observed experimentally in a range of host–pathogen systems (Hill *et al.*, 1991; Gilbert *et al.*, 1998; Carrington *et al.*, 1999; Tang *et al.*, 1999), in agreement with more recent model results (Beltman *et al.*, 2002) and a range of more qualitative arguments (Parham *et al.*, 1989; Wills, 1991).

However, the interactions between the immune system and a wide variety of pathogens are so complex that most coevolutionary models to date have been highly simplified caricatures of reality. Again, much remains to be done to improve model realism, but we emphasize that such realism is not necessarily an 'optional extra'. Past analyses (Hamilton, 1980) of the role of host–pathogen coevolution and red-queen dynamics in stabilizing sex as an evolutionary strategy of hosts were later challenged by models with greater epidemiological detail (May and Anderson, 1983). A converse example is provided by studies exploring a related phenomenon: the role adaptive immunity might have in stabilizing sex in helminthic parasites, where recent work using microsimulation approaches (Galvani *et al.*, 2001) has shown, in contrast to earlier deterministic analyses (Lythgoe, 2000), that the benefit gained through enhanced maintenance of diversity can more than outweigh the twofold fitness cost of sex. Therefore, while simple models will always be important for generating hypotheses and providing unencumbered insight, analytical tractability should not be the sole goal of future theoretical research. Equally important is the development of frameworks capable of being rigorously tested against the rapidly growing base of experimental data.

ACKNOWLEDGEMENTS

N. M. F. thanks the Royal Society, Howard Hughes Medical Institute and Medical Research Council for research support. A. G. also thanks the Domus Merton Senior Scholarship for grant funding.

APPENDIX: MODEL DETAILS

The three models presented below address very different issues in the population biology of antigenically variable pathogens, so no attempt at defining a unique set of symbols and state variables across all the models has been attempted. Thus where parameters in two models share the same symbol, no equivalence is implied.

1. SIR and SIS models

These can be specified with a single set of ordinary differential equations describing an infectious disease that generates permanent immunity in a proportion q of infections:

$$\frac{dS}{dt} = \mu N + (1 - q)\nu I - \beta SI/N - \mu S$$

$$\frac{dI}{dt} = \beta SI/N - \nu I - \mu I$$

$$\frac{dR}{dt} = q \nu I - \mu R$$

The model has three other parameters, μ, host mortality (=1/L, where L = lifespan), ν, the recovery rate (= 1/D, where D = duration of infection), and β, transmissibility. S, I and R represent the number of susceptible, infected and recovered (immune) individuals in the population respectively. The basic reproduction number (the expected number of infections caused by a single infection in an otherwise susceptible population) is given by $R_0 = \beta/(\nu+\mu)$.

2. Multiple-strain disease pathogenesis model

We consider a population of N viral strains, of which only one exists in a host at the start of infection. Immunity acts on an entirely strain-specific basis: in response to the presence of any one strain, v_i ($i = 1...N$) the small pre-existing population of immune cells (i.e. T and/or B cells) specific to that strain, x_i, proliferate rapidly and suppress the pathogen population directly through pathogen/infected-cell killing. All strains are assumed to require the same resource (target cells, space or nutrients), c, for replication, with the resource itself being regulated according to simple birth–death dynamics. The dynamics of the model are governed by the following set of differential equations:

$$\frac{dv_i}{dt} = r\left[(1 - \delta)v_i + \delta\sum_j \frac{v_j}{N}\right] - (\psi + \sigma x_i)v_i \qquad r = \rho\,\frac{c}{c + C_{50}}$$

$$\frac{dx_i}{dt} = \gamma(X_0 - x_i) + \varepsilon_i x_i \qquad \varepsilon_i = \mu \frac{v_i}{v_i + V_{50}}$$

$$\frac{dc}{dt} = \kappa(C_0 - c) - \alpha r \sum_j v_j$$

Here ρ (= 2/day) is the maximum replication rate of the pathogen; C_{50} (= 0.1) is the level of resource at which pathogen replication is reduced by half; δ (varied between 0 and 10^{-5}) is the probability replication results in antigenic change or mutation; ψ (= 0.25/day) is the intrinsic death rate of the pathogen; σ (= 5 × 10^{-6}/day) specifies the efficiency with which the immune response clears the pathogen; γ (= 0.01/day) is the decay rate of a strain-specific immune response; X_0 (= 10) is the initial population size of a strain-specific response prior to infection; μ (= 1/day) is the maximum replication rate of a strain-specific subpopulation of immune cells; V_{50} (= 10) is the level of parasitaemia which triggers immune cell proliferation at half its maximum rate; κ (= 0.05/day) is the turnover rate of resource; C_0 (= 1) is the equilibrium level of resource in the absence of infection; and α (= 10^{-8}) is the amount of resource used by a pathogen particle in replication. The parameter values used to generate the results shown in the main text are given in parentheses above, and generate peak single-strain pathogen population sizes of $\sim 3 \times 10^7$, typical of some parasite infections and localized bacterial infections.

For simplicity mutation is shown as a deterministic process in these equations, but in reality it was modelled as a random process. The parameter δ is referred to as the variant-generating rate or mutation rate in the text, but it should be noted that even for viruses it only has a loose connection with the error rate in DNA or RNA transcription. This is because quoted error rates represent the per base pair probability of mutation, but for most viruses only of the order of 1 in 1000 of base pairs code for antigenically dominant epitopes. Thus an error rate of 10^{-5} corresponds to δ having a value of around 10^{-8}. Furthermore, in many cases it would be unlikely that a single amino acid change would give rise to an entirely immunologically novel variant. If multiple changes were required to fully escape existing immune response, this would further reduce δ relative to the intrinsic error rate. Lastly, since the total mutation rate of a population is proportional to the product of the mutation rate and the population size, for pathogens (e.g. many viral and bacterial infections) with much larger population sizes than the 3×10^7 modelled here, the level δ required to maximize total parasitaemia is correspondingly reduced.

In contrast to viruses with their small genomes, a number of pathogens, particularly bacteria and protozoa, generate antigenic variation through phenotypic switching of antigenic gene expression. Such pathogens include *G. lamblia* (Nash, 1997; Muller and Gottstein, 1998); Lyme disease *B. burgdorferi* (Zhang et al., 1998); *B. bovis* (O'Connor et al., 1997); *N. gonorrhoeae* (Serkin and

Seifert, 1998); *N. meningitidis* (Feavers, 1996); *E. coli* (McKenzie and Rosenberg, 2001); *H. influenzae* (Weiser *et al.*, 1989); *C. coli* (Park *et al.*, 2000); *P. falciparum* (Vanhamme *et al.*, 2001); *T. cruzi* (Tarleton and Kissinger, 2001); and *T. brucei* (Vanhamme *et al.*, 2001). The phenomenon of antigenic switching has been most extensively modelled in trypanosomes which escape from host immunity by switching expression of the variant surface glycoprotein (VSG) from an estimated repertoire of one thousand distinct genes (Borst, 1991; Frank, 1999; Vanhamme *et al.*, 2001). Phenotypic switching between antigens appears random and sequentially generates a diverse set of antigenic variants (Turner and Barry, 1989; Barry and Turner, 1991). An elegant recent study showed that in a deterministic framework, slight diversity in the rate at which each variant switches to another variant could account for this pattern and thus the switch rates might not be purely random (Frank, 1999). The results we present above, where variant emergence is modelled as a stochastic process, show that such deviations from equal switch rates are not strictly necessary to reproduce observed patterns, though such preferential switching may further optimize the total duration of infection.

3. Multiple-strain transmission model with cross-immunity

We consider the genetic model (Gupta *et al.*, 1998; Ferguson and Andreasen, 2002) of strain space illustrated in Figure 18.7, focusing on a system with three antigenic loci, each with two alleles. In all, the system therefore has eight genotypes: 1 = amx, 2 = amy, 3 = anx, 4 = bmx, 5 = any, 6 = bmy, 7 = anx, 8 = bny, where the letters denote the allele at each locus. Cross-immunity is exhibited through reduced infectivity of a host infected with a strain i ($= 1...8$), where that host has previously been infected with any strain sharing one or more alleles with i. Infection is assumed to give permanent immunity against reinfection with the same strain. All strains are assumed to have equal transmissibility ($R_0 = 4$), and the mean duration of infection is assumed to be 0.1 years. The host lifespan is modelled as 50 years.

Denoting z_i as the proportion of the population previously exposed to strain i, w_i as the proportion exposed to i or any strain sharing alleles with i, and y_i as the proportion infected and infectious with strain i, the dynamics of this system are determined by the following equations:

$$\frac{dz_i}{dt} = (1 - z_i)(\lambda_i + v_i) - \mu z_i \qquad i = 1, \ldots, 8$$

$$\frac{dw_i}{dt} = (1 - w_i)\left((1-f)v_i + \sum_j (\lambda_j + fv_j) - (\lambda_{9-i} + fv_{9-i})\right) - \mu w_i$$

$$\frac{dy_i}{dt} = [(1-z_i) + (1-\gamma)(w_i - z_i)]\lambda_i - (\sigma + \mu)y_i$$

$$\lambda_i = \beta\left[(1-\delta)y_i + \frac{\delta}{3}\left(\sum_j y_j - y_i - y_{9-i}\right)\right]$$

where β (= 40/year) is the transmission coefficient, μ (= 0.02/year) is the host death rate, σ (= 10/year) is the rate of recovery from infection, and γ is the level of cross-immunity (0 = none, 1 = complete). The model above also includes mutation, which is modelled as a probability δ of the allele at one locus changing to the other variant, and vaccination, which occurs at rate v_i for strain i. Vaccination is assumed to induce permanent immunity against the strain vaccinated against, and a proportion f of the cross-protective immunity against heterologous strains that natural infection induces.

It should be noted that this model assumes no inhibition of coinfection; i.e. the compartments of the model overlap, and an individual can therefore be simultaneously infected with multiple strains. Generalizing the model to relax this assumption dramatically increases its complexity (Ferguson and Andreasen, 2002).

REFERENCES

Adler, F. R. and Losada, J. M. (2002) In *Virulence Management: The Adaptive Dynamics of Pathogen–Host Interactions* (eds U. Dieckmann, H. Metz, M. Sabelis and K. Sigmund), Cambridge University Press, Cambridge, pp. 138–149.

Anderson, R. M. and May, R. M. (1982) *Parasitology* 86, 411–426.

Anderson, R. M. and May, R. M. (1991) *Infectious Diseases of Humans: Dynamics and Control*. Oxford University Press, Oxford.

Andreasen, V., Levin, S. A. and Lin, J. (1996) *Z. Angew. Math. Mech.* 52, 421–424.

Andreasen, V., Lin, J. and Levin, S. A. (1997) *J. Math. Biol.* 35, 825–842.

Anthony, B., Kaplan, E., Wannamaker, L. and Chapman, S. (1976) *Am. J. Epidemiol.* 104, 652–666.

Bailey, N. T. J. (1975) *The Mathematical Theory of Infectious Disease and its Applications*, C. Griffin and Co. Ltd, London.

Barry, J. D. and Turner, C. M. R. (1991) *Parasitol. Today* 7, 207–211.

Beltman, J. B., Borghans, J. A. M. and de Boer, R. J. (2002) In *Virulence Management: The Adaptive Dynamics of Pathogen–Host Interactions* (eds U. Dieckmann, H. Metz, M. Sabelis and K. Sigmund), Cambridge University Press, Cambridge, pp. 210–221.

Bloom, B. R. (1996) *Science* 272, 1888–1890.

Bolker, B. M. and Grenfell, B. T. (1995) *Proc. R. Soc. Lond. B* 348, 308–320.

Bonhoeffer, S., Coffin, J. M. and Nowak, M. A. (1997) *J. Virol.* 71, 3275–3278.

Borrow, P., Lewicki, H., Wei, X. P., Horwitz, M. S., Peffer, N., Myers, H., Nelson, J. A., Gairin, J. E., Hahn, B. H., Oldstone, M. B. A. and Shaw, G. M. (1997) *Nat. Med.* 3, 205–211.

Borst, P. (1991) *Immunol. Today* 12, A29–A33.
Boyer, J. D., Cohen, A. D., Vogt, S., Schumann, K., Nath, B., Ahn, L., Lacy, K., Bagarazzi, M. L., Higgins, T. J., Baine, Y., Ciccarelli, R. B., Ginsberg, R. S., MacGregor, R. R. and Weiner, D. B. (2000) *J. Infect. Dis.* 181, 476–483.
Bush, R., Bender, C., Subbarao, K., Cox, N. and Fitch, W. (1999a) *Science* 286, 1921–1925.
Bush, R. M., Fitch, W. M., Bender, C. A. and Cox, N. J. (1999b) *Mol. Biol. Evol.* 16, 1457–1465.
Carrington, M., Nelson, G. W., Martin, M. P., Kissner, T., Vlahov, D., Goedert, J. J., Kaslow, R., Buchbinder, S., Hoots, K. and O'Brien, S. J. (1999) *Science* 283, 1748–1752.
Caugant, D. A., Bol, P., Hoiby, E. A., Zanen, H. C. and Froholm, L. O. (1990) *J. Infect. Dis.* 162, 867–874.
Dagan, R., Givon, N., Yagupsky, P., Porat, N., Janco, J., Chang, I. *et al.* (1998) *Proceedings 38th ICAAC*, San Diego, CA.
Dale, C. J. and Kent, S. J. (2000) *Exp. Opin. Ther. Patients* 10, 1179–1188.
Dieckmann, U. (2002) In *Virulence Management: The Adaptive Dynamics of Pathogen–Host Interactions* (eds U. Dieckmann, H. Metz, M. Sabelis and K. Sigmund), Cambridge University Press, Cambridge, pp. 39–59.
Doherty, P. C. and Zinkernagel, R. M. (1975) *Lancet* 305, 1406–1409.
Dorrell, L., O'Callaghan, C. A., Britton, W., Hambleton, S., McMichael, A., Smith, G. L., Rowland-Jones, S. and Blanchard, T. J. (2000) *Vaccine* 19, 327–336.
Feavers, I. M., Fox, A. J., Grey, S., Jones, D. M. and Maiden, M. C. (1996) *Clin. Diag. Lab. Immunol.* 3, 444–450.
Ferguson, N. M. and Andreasen, V. (2002) In *Mathematical Approaches for Emerging and Re-emerging Infectious Diseases: Models, Methods and Theory*, Vol. 125 (eds S. M. Blower and C. Castillo-Chavez), Springer, New York, pp. 157–169.
Ferguson, N., Anderson, R. and Gupta, S. (1999) *Proc. Natl Acad. Sci. USA* 96, 790–794.
Ferguson, N. M., Galvani, A. P. and Bush, R. M. (2003) *Nature* 422, 428–433.
Frank, S. A. (1999) *Proc. R. Soc. Lond. B* 266, 1397–1401.
Frank, S. A. (2002) *Immunology and Evolution of Infectious Disease,* Princeton University Press, Princeton, NJ
Fraser, C., Ferguson, N. M., Ghani, A. C., Goudsmit, J., Lange, J., Anderson, R. M. and de Wolf, F. (2000) *AIDS* 14, 659–669.
Galvani, A. P., Coleman, R. M. and Ferguson, N. M. (2001) *Ann. Zool. Fennici* 38, 305–314.
Gandon, S. and Michalakis, Y. (2002) In *Virulence Management: The Adaptive Dynamics of Pathogen–Host Interactions* (eds U. Dieckmann, H. Metz, M. Sabelis and K. Sigmund), Cambridge University Press, Cambridge, pp. 150–164.
Gianfrani, C., Oseroff, C., Sidney, J., Chestnut, R. W. and Sette, A. (2000) *Hum. Immunol.* 61, 438–452.
Gilbert, S. C., Plebanski, M., Gupta, S., Morris, J., Cox, M., Aidoo, M., Kwiatkowski, D., Greenwood, B. M., Whittle, H. C. and Hill, A. V. S. (1998) *Science* 279, 1173–1177.
Girard, M., Habel, A. and Chanel, C. (1999) *CR Acad. Sci. Ser. III* 322, 959–966.
Gomes, M. G. M. and Medley, G. F. (2002) In *Mathematical Approaches for Emerging and Re-emerging Infectious Diseases: Models, Methods and Theory*, Vol. 125 (eds S. M. Blower and C. Castillo-Chavez), Springer, New York, pp. 171–191.
Gomes, M. G. M., Medley, G. F. and Nokes, D. J. (2002) *Proc. R. Soc. Lond. B* 269, 227–233.

Goulder, P. J. R., Phillips, R. E., Colbert, R. A., McAdam, S., Ogg, G., Nowak, M. A., Giagrande, P., Luzzi, G., Morgan, B., Edwards, A., McMichael, A. J. and Rowland-Jones, S. (1997) *Nat. Med.* 3, 212–217.

Grenfell, B. T., Bolker, B. M. and Kleczkowski, A. (1995) *Proc. R. Soc. Lond. B* 259, 97–103.

Gupta, S. and Galvani, A. P. (1999) *Phil. Trans. R. Soc.* 354, 711–719.

Gupta, S. and Hill, A. S. (1995) *Proc. R. Soc. Lond. B* 261, 271–277.

Gupta, S., Ferguson, N. and Anderson, R. (1998) *Science* 240, 912–915.

Gupta, S., Ferguson, N. M. and Anderson, R. M. (1997) *Proc. R. Soc. Lond. B* 264, 1435–1443.

Gupta, S., Maiden, M. C. J., Feavers, I. M., Nee, S., May, R. M. and Anderson, R. M. (1996) *New Med.* 2, 437–442.

Halstead, S. B. (1988) *Science* 239, 476–481.

Hamilton, W. D. (1980) *OIKOS* 35, 282–290.

Hanke, T. and McMichael, A. J. (2000) *Nature Med.* 6, 951–955.

Hargreaves, B. J., Yoeli, M., Nussenzweig, R. S., Walliker, D. and Carter, R. (1975) *Ann. Trop. Med. Parasitol.* 69, 289–299.

Heeney, J. L. and Hahn, B. H. (2000) *AIDS* 14, S125–S127.

Hendrick, P. W., Whittam, T. S. and Parham, P. (1991) *Proc. Natl Acad. Sci. USA* 88, 5897–5901.

Hill, A. S., Allsopp, C., Kwiatkowski, D., Anstey, N. M., Twumasi, P., Rowe, P., Bennett, S., Brewster, D., McMicheal, A. J. and Greenwood, B. M. (1991) *Nature* 352, 595–600.

Holmes, E. C. and Burch, S. S. (2000) *Trends Microbiol.* 8, 74–77.

Howard, R. J. and Pasloske, B. L. (1993) *Parasitol. Today* 9, 369–372.

Hughes, A. L. and Nei, M. (1989) *Genetics* 132, 863–864.

Hughes, A. L., Hughes, M. K., Howell, C. Y. and Nei, M. (1994) *Phil. Trans. R. Soc. Lond. B* 346, 359–367.

Hunt, J. D., Jackson, D. C., Wood, P. R., Stewart, D. J. and Brown, L. E. (1995) *Vaccine* 13, 1649–1657.

Jeffery, K. J. M. and Bangham, C. R. M. (2000) *Microb. Infect.* 2, 1335–1341.

Just, J. J. (1995) *Hum. Immunol.* 44, 156–169.

Kermack, W. O. and McKendrick, A. G. (1927) *Proc. R. Soc. Lond. A* 115, 700–721.

Koenig, S., Conley, A. J., Brewah, Y. A., Jones, G. M., Leath, S., Boots, L. J., Davey, V., Pantaleo, G., Demarest, J. F., Carter, C. *et al.* (1995) *Nat. Med.* 1, 330–336.

Kroner, B. L., Goedert, J. J., Blattner, W. A., Wilson, S. E., Carrington, M. N. and Mann, D. L. (1995) *AIDS* 9, 275–280.

Lifson, J. D., Rossio, J. L., Arnaout, R., Li, L., Parks, T. L., Schneider, D. M., Kiser, R. F., Coalter, V. J., Walsh, G., Imming, R., Fischer, B., Flynn, B. M., Bischofberger, M., Piatak, M. J., Hirsch, V. M., Nowak, M. A. and Wodarz, D. (2000) *J. Virol.* 74, 2584–2593.

Lin, J., Andreasen, V. and Levin, S. A. (1999) *Math. Biosci.* 162, 33–51.

Lipsitch, M. (1997) *Proc. Natl Acad. Sci. USA* 94, 6571–6576.

Lipsitch, M. (1999) *Emerg. Infect. Dis.* 5, 336–345.

Lipsitch, M., Dykes, J. K., Johnson, S. E., Ades, E. W., King, J., Briles, D. E. and Carloe, G. M. (2000) *Vaccine* 18, 2895–2901.

Lythgoe, K. A. (2000) *Evolution* 54, 1142–1156.

May, R. M. and Anderson, R. M. (1983) *Proc. R. Soc. Lond. B* 219, 281–313.

Mbelle, N., Huebner, R. E., Wasas, A. D., Kimura, A., Chang, I. and Klugman, K. P. (1999) *J. Infect. Dis.* 180, 1171–1176.
McElrath, M. J., Corey, L., Montefiori, D., Wolff, M., Schwartz, D., Keefer, M., Belshe, R., Graham, B. S., Matthews, T., Wright, P., Gorse, G., Dolin, R., Berman, P., Francis, D., Duliege, A. M., Bolognesi, D., Stablein, D., Ketter, N. and Fast, P. (2000) *AIDS Res. Hum. Retroviruses* 16, 907–919.
McKenzie, G. J. and Rosenberg, S. M. (2001) *Curr. Opin. Microbiol.* 4, 586–594.
McLean, A. R. (1995) *Proc. R. Soc. Lond. B* 261, 389–393.
McMichael, A., Goulder, P., Rowland-Jones, S., Nowak, M. and Phillips, R. (1996) *Immunology* 89 (Suppl.), 111.
McMichael, A., Rowland-Jones, S. and Klenerman, P. (1995) *J. Cell. Biochem.* 21A (Suppl.) 60.
Muller, N. and Gottstein, B. (1998) *Int. J. Parasitol.* 28, 1829–1839.
Nakajima, S., Nobusawa, E. and Nakajima, K. (2000) *Virology* 274, 220–231.
Nash, T. E. (1997) *Phil. Trans. R. Soc. Lond. B* 352, 1369–1375.
Nitayaphan, S., Khamboonruang, C., Sirisophana, N., Morgan, P., Chiu, J., Duliege, A. M., Chuenchitra, C., Supapongse, T., Rungruengthanakit, K., deSouza, M., Mascola, J. R., Boggio, K., Ratto–Kim, S., Markowitz, L. E., Birx, D., Suriyanon, V., McNeil, J. G., Brown, A. E. and Michael, R. A. (2000) *Vaccine* 18, 1448–1455.
Nowak, M. A. and Bangham, C. R. M. (1996) *Science* 272, 74–79.
Nowak, M. and May, R. M. (1991) *Math. Biosci.* 106, 1–21.
Nowak, M. A. and May, R. M. (2000) *Virus Dynamics: Mathematical Principles of Immunology and Virology,* Oxford University Press, Oxford.
Nowak, M. A. and Sigmund, K. (2002) In *Virulence Management: The Adaptive Dynamics of Pathogen–Host Interactions* (eds U. Dieckmann, H. Metz, M. Sabelis and K. Sigmund), Cambridge University Press, Cambridge, pp. 124–137.
Nowak, M. A., Anderson, R. M., Mclean, A. R., Wolfs, T. F., Goudsmit, J. and May, R. M. (1991) *Science* 254, 963–969.
Nowak, M. A., May, R. M., Phillips, R. E., Rowland-Jones, S., Lalloo, D. G., McAdam, S., Klenerman, P., Koppe, B., Sigmund, K., Bangham, C. R. M. and McMichael, A. J. (1995) *Nature* 375, 606–611.
Obaro, S. K., Adegbola, R. A., Banya, W. A. S. and Greenwood, B. M. (1996) *Lancet* 348, 271–272.
O'Connor, R. M., Lane, T. J., Stroup, S. E. and Allred, D. R. (1997) *Mol. Biochem. Parasitol.* 89, 259–270.
Ortiz, G. M., Nixon, D. F., Trkola, A., Binley, J., Jin, X., Bonhoeffer, S., Kuebler, P. J., Donahoe, S. M., Demoitie, M. A., Kakimoto, W. M., Ketas, T., Clas, B., Heymann, J. J., Zhang, L., Cao, Y., Hurley, A., Moore, J. P., Ho, D. D. and Markowitz, M. (1999) *J. Clin. Invest.* 104, R13–R18.
PAHO (1994) *Dengue and Dengue Hemorrhagic Fever in the Americas: Guidelines for Prevention and Control,* Pan American Health Organization, Washington, DC.
Parham, P., Lawlor, D. A., Lomen, C. E. and Ennis, P. D. (1989) *J. Immunol.* 142, 3937–3950.
Park, S. F., Purdy, D. and Leach, S. (2000) *J. Bacteriol.* 182, 207–210.
Phillips, R. E., Rowland–Jones, S., Nixon, D. F., Gotch, F. M., Edwards, J. P., Ogunlesi, A. O., Elvin, J. G., Rothbard, J. A., Bangham, C. R. M., Rizza, C. R. and McMichael, A. J. (1991) *Nature* 354, 453–459.
Porco, T. C. and Blower, S. M. (1998) *INTERFACES* 28, 167–190.

Price, D. A., Goulder, P. J. R., Klenerman, P., Sewell, A. K., Easterbrook, P. J., Troop, M., Bangham, C. R. M. and Phillips, R. E. (1997) *Proc. Natl Acad. Sci. USA* 94, 1890–1895.

Rosenberg, E. A., Altfeld, M., Poon, S. H., Phillips, M. N., Wilkes, B. M., Eldridge, R. L., Robbins, G. K., D'Aquila, R. T., Goulder, P. J. and Walker, B. D. (2000) *Nature* 407, 523–526.

Ruprecht, R. M., Hofmann-Lehmann, R., Rasmussen, R. A., Vlasak, J. and Xu, W. (2000) *J. Hum. Virol.* 3, 88–93.

Scorza Smeraldi, R., Fabio, G., Lazzarin, A., Eisera, N. B., Moroni, M. and Zanussi, C. (1986) *Lancet* 22, 1187–1189.

Serkin, C. D. and Seifert, H. S. (1998) *J. Bacteriol.* 180, 1955–1958.

Slade, R. W. and McCallum, H. I. (1992) *Genetics* 132, 861–864.

Smith, D. J., Forrest, S., Ackley, D. H. and Perelson, A. S. (1999) *Proc. Natl Acad. Sci. USA* 96, 14001–14006.

Smith, D. J., Forrest, S., Hightower, R. R. and Perelson, A. S. (1997) *J. Theoret. Biol.* 189, 141–150.

Snowden, N., Pepper, L., HKhoo, S., Hajeer, A., Worthington, J., Mandal, B. K. and Ollier, W. (1996) *Hum. Immunol.* 47, 119.

Steel, C. M., Ludlam, C. A., Beatson, D., Peutherer, J. F., Cuthbert, R. J., Simmonds, P., Morrison, H. and Jones, M. (1988) *Lancet* 28, 1185–1188.

Takahata, N. and Nei, M. (1990) *Genetics* 124, 967–978.

Tang, J., Costello, C., Keet, I. P., Rivers, C., Leblanc, S., Karita, E., Allen, S. and Kaslow, R. A. (1999) *AIDS Res. Hum. Retroviruses* 15, 317–324.

Tarleton, R. L. and Kissinger, J. (2001) *Curr. Opin. Microbiol.* 13, 395–402.

Taylor, L. H., Walliker, D. and Read, A. F. (1997) *Proc. R. Soc. Lond. B* 264, 927–935.

Treanor, J. J. (1998) *Infect. Med.* 15, 487–492.

Turner, C. M. R. and Barry, J. D. (1989) *Parasitology* 99, 67–75.

Vanhamme, L., Pays, E., McCulloch, R. and Barry, J. D. (2001) *Trends Parasitol.* 17, 338–343.

Voeten, J. T. M., Rimmelzwaan, G. F., Nieuwkoop, N. J., Fouchier, R. A. M. and Osterhaus, A. D. M. E. (2001) *Clin. Exp. Immunol.* 125, 423–431.

Weiser, J. N., Love, J. M. and Moxon, E. R. (1989) *Cell* 59, 657–665.

White, L. J., Cox, M. J. and Medley, G. F. (1998) *IMA J. Math. Appl. Med. Biol.* 15, 211–233.

Wills, C. (1991) *Immunol. Rev.* 124, 165–220.

Wills, C. and Green, D. R. (1995) *Immunol. Rev.* 143, 263–292.

Wodarz, D. and Nowak, M. A. (2000) *Eur. J. Immunol.* 30, 2704–2712.

Wodarz, D., Lloyd, A. L., Jansen, V. A. A. and Nowak, M. A. (1999) *J. Theor. Biol.* 96, 101–113.

Zanotto, P. M., Gould, E. A., Gao, G. F., Harvey, P. H. and Holmes, E. C. (1996) *Proc. Natl Acad. Sci. USA* 93, 548–553.

Zhang, J. R., Hardham, J. M., Barbour, A. G. and Norris, S. J. (1998) *Cell* 89, 275–285.

INDEX

Active chromatin state, 11
ADE2 gene, 186, 187
α2-*fucT* gene, phase variation, 130–1
α3-*fucT* gene, phase variation, 128–30
    mechanisms of, 129
Anaplasma caudatum, 243
Anaplasma centrale, 243, 250
Anaplasma marginale, 243–72, 285
    antigenic variation in, 247
    disease pathogenesis, 245–7
    inclusion appendage, 247
    major surface protein, 248–9
    MSP 2 protein, 250
        antigenically variant, expression of, 250–3
        expression in cyclic transmission of, 255–6
        molecular mechanisms of antigenic variation, 253–5
    MSP 3 protein, 250
        antigenic variation, 256–7
    MSP 4 protein, 257–8
    MSP 5 protein, 257–8
    outer membrane proteins, 247–8
    phylogenetic re-classification, 244–5
Anaplasma ovis, 243, 250

Antibodies
    influenza, 63–4
    Plasmodium spp., 294–5
Antibody-dependent enhancement, 419
Antigenic diversity, 312
Antigenic drift
    biological factors driving, 76–9
    influenza, 60, 67, 71, 76–9
    rotavirus, 93
Antigenic shift, 60, 65
Asian influenza, 71
Aspergillus nidulans, 177
Avian influenza, 69, 75

Babesia bigemina, 273
Babesia bovis, 273–90
    antigenic variation and pathology, 286–7, 410
    mechanisms of antigenic variation in, 281–5
    phenotype of antigenic variation in, 274–81
    variant antigens in, 277–8
    ves multigene family, 278–9
    *ves1*α gene and VESA1a polypeptide, 279–81

Babesia rodhaini, 274
Bacterial adhesion, and lipopolysaccharide phase variation, 135–8
β*3-galT* gene, 132
BIR protein, 310
Bloodstream expression sites, 229, 230
Bordetella pertussis, 8
Borrelia spp., 3, 7, 319–56
  antigenic variation, 410
  cellular organization, 324
  immunity, 331–2
  mechanisms of antigenic variation, 340–9
  morphology and physiology, 323–4
  phylogeny and genetics, 325–7
  variable antigens, 332–7
    evolution of, 337–40
Borrelia afzelii, 326
Borrelia anserina, 326
Borrelia bissettii, 326
Borrelia burgdorferi, 322, 326, 330
  horizontal transfer, 338
Borrelia crocidurae, 326, 332
Borrelia duttonii, 326, 329
Borrelia garinii, 326
Borrelia hermsii, 165, 186, 232, 321, 326, 332, 334
  antigenic variation, 344
Borrelia lonestari, 326
Borrelia miyamotoi, 326
Borrelia persica, 326
Borrelia recurrentis, 321, 332
Borrelia theileri, 326
Borrelia turdi, 326
Borrelia turicatae, 326, 328, 329, 332, 333
  antigenic variation, 344, 346
Borrelia venezuelensis, 326
Bovine leukaemia virus, point mutation rates, 20

C-reactive protein, 113
C-tracts, 123
Caliciviruses, 33–51
  antigenic diversity, 40
  see also Feline calicivirus

Campylobacter spp., 6
Campylobacter coli, 410
Campylobacter jejuni, 115, 122
Candida albicans, 165–201
  gene regulation during switching, 179–85
    functional analyses of phase-specific gene promoters, 181
    promoters of opaque phase-specific genes, 183–5
    promoters of white phase-specific genes, 181–3
  phase-specific trans-acting factors, 185–6
  switching in
    differential gene expression, 176–9
    discovery of, 166–8
    gene regulation, 179–85
    gross chromosomal rearrangements, 187
    heat-induced mass conversion, 189–91
    possible mechanisms, 186–7
    role of deacetylases, 187–9
    significance in pathogenesis, 191–3
  white-opaque transition, 168
    antigenic changes in, 170–5
    effect on budding yeast cell morphology, 168–70
Candida glabrata, 166
  switching in, 193–7
Capsule phase variation, 103
Cassette mechanism, 6, 149–51
CD4 cells, 25, 27
  response to influenza, 64–5
CD8 cells, 25
  response to influenza, 64–5
CDR3 gene, 177, 179
CDR4 gene, 177, 179
Ciliates *see Ichthyophthirius* spp.; *Paramecium* spp.
CIR protein, 310
Circumsporozoite protein, 312
CKRD, 280
clag, 305–6
Coinfection, 414–15
Contingency genes, 13

Control of antigenically variable pathogens, 419–22
Copy choice, 20
Cross-immunity, 414, 417
Cryptococcus neoformans, 166
Cytoadhesion, 287
Cytotoxic T lymphocytes, 25

Deacetylases, and suppression of switching, 187–9
Descent with modification, 2
Dichelobacter nodosus, 419
Dictylocaulus viviparus, Lewis antigen expression, 126
Didinium spp., 398
Diversification, constraints on, 413–19
DNA recombination, and antigenic variation, 231–6
DNA transformation, *Neisseria* spp., 143–5
Drift, 416
Duffy binding ligand, 298
Duplicative transposition *see* Gene conversion

EFG1 gene, 177, 181, 183
Ehrlichia canis, 262, 265
Ehrlichia chaffeensis, 262, 265
Ehrlichia ruminantium, 243–72
   antigenic variation in, 259–60
   disease pathogenesis, 245–7
   MAP proteins 1, 260
      homology, 264–6
      and newly cloned *E. rumanantium*, 266–7
   molecular characterization of *map 1* gene, 260–4
   outer membrane proteins, 259
   phylogenetic re-classification, 244–5
Env protein, 22, 26
Epidemic influenza, 65–8
Erythema migrans, 322
Escherichia coli, 147, 186
   antigenic variation, 410
European brown hare syndrome virus, 33, 40

Expression site associated genes, 229, 230
Expression site body, 11, 231

Feline calicivirus
   antigenic structure of capsid, 41–2
   antigenic variation, 38–9
   carrier state, 42–4
   clinical disease, 36–8
   epidemiology, 43
   genetic variation, 39–40
   genomic and antigenic structure, 35
   phylogenetic tree, 39
   vaccination, 46–7
   virus evolution, 42–7
      individual, 42–4
      population, 44–6
Feline immunodeficiency virus, 44
Fimbriae, 105
Fixation rate, 21
FlaB protein, 332
Fluctuating epistasis, 417
Fungi *see Candida albicans*; *Pneumocystis carinii*

Gag proteins, 18, 22
galE gene, 116
Gene conversion, 6, 233
Gene silencing, 11
Genetic variation
   functional significance, 155–9
   illegitimate route, 152–4
   versus gene regulation, 145–6
   via homologous recombination, 146–7
Giardia spp., 13
Giardia lamblia, 357–74, 390
   antigenic variation, 358–65, 410
      biological role *in vivo*, 365–9
      molecular mechanisms of, 369–71
   location of expressed genes, 10
   mechanism to change gene expressed, 10
   number of genes, 10
   surface antigen, 10
GlcNAcT gene, 131–2

GPI-anchored proteins, 395–6
GPM gene, 181

Haemagglutination inhibition test, 52
Haemagglutinin, influenza, 52
  antigenic mapping of, 56–9
  biological properties of antibodies to, 63–4
  genetic variation of, 60–1
  structure, 57
  X-ray crystallography of, 59–60
Haemophilus spp., 13
Haemophilus influenzae, 9, 102–21
  capsule phase variation, 103
  non-typeable strains, 102
  phase variation
    by modulation of transcription, 105–6
    by simple nucleotide repeats, 103–5
    by tetranucleotide repeats, 106–17
    factors influencing, 117–19
  see also Influenza
HDA1 gene, 188
Helicobacter spp., 13
Helicobacter pylori, 9
  bacterial adhesion, 135–8
  Lewis antigens, 126–33
    mimicry and immune evasion, 134–5
  phase variation in lipopolysaccharide, 122–41
    biological role of, 133
    expression of blood group antigens, 124–6
hgpA gene, 108
hhuA gene, 108
HIV see Human immunodeficiency virus
HIV-1, 17
  point mutation rates, 20
  recombination rate, 20
HIV-2, 17
hmw2A gene, 106
Homologous recombination, 146–7
Hong Kong influenza pandemic, 73–6
HOS1 gene, 188
HOS2 gene, 188

HOS3 gene, 177, 181
Human immunodeficiency virus, 3, 16–32
  fixation, 18–21
  genetic variation as noise, 29
  genomic organization, 18
  HIV lentiviruses, 16–18
  HIV-1, 17, 20
  HIV-2, 17
  mutation, 18–21
  persistence of, 28
  phylogenic tree, 19
  point mutation rate, 24
  recombination, 18–21, 25–8
  replication, 22–4

i-antigens
  Ichthyophthirius
    antigenic variation, 397–9
    clustering and parasite behaviour, 396–7
    immune evasion, 394–5
    structural features, 390–2
    as targets of humoral immunity, 388–9
    transmembrane signalling, 395–6
    variation in natural parasite populations, 392–4
  infection and immunity, 387–90
  Paramecium, 376, 377
    isolation and analysis of mutants, 382
    regulation of expression, 381–5
    role in regulation of mutual exclusion, 383–5
    stable genes without introns, 381
    structure and variation, 378–80
    transcriptional control of antigen expression, 382–3
IAG gene family, 392–4
Ichthyophthirius spp., 375, 385
  i-antigens
    antigenic variation, 397–9
    clustering and parasite behaviour, 396–7
    immune evasion, 394–5
    structural features, 390–2

as targets of humoral immunity, 388–9
transmembrane signalling, 395–6
variation in natural parasite populations, 392–4
see also Paramecium spp.
Ichthyophthirius multifiliis, 360, 387
humoral immunity, 388–9
Immune evasion
H. pylori, 134–5
Ichthyophthirius spp., 394–5
Immunity
Borrelia spp., 331–2
cross-immunity, 414, 417
i-antigens in, 387–90
Ichthyophthirius spp., 388–9
Paramecium spp., 387–99
rotavirus, 92–8
In situ activation, 8–10
Inclusion appendage, 247
Infection
i-antigens in, 387–90
Ichthyophthirius spp., 387–99
Plasmodium spp., 293
role of tetranucleotide repeats, 117
within-host dynamics, 408–13
Influenza, 21–2, 52–83
avian, 69, 75
biological factors driving antigenic drift, 76–9
clinical impact of, 55–6
H1N1 type, reappearance of, 73–4
haemagglutinin, 52
antigenic mapping, 56–9
biological properties of antibodies to, 63–4
genetic variation, 60–1
structure, 57
X-ray crystallography of, 59–60
mortality due to, 56, 70
neuraminidase, 61–2
nucleotide sequence analysis, 74–5
pandemics and epidemics, 65–8
1918/19 pandemic, 68–71
1957 Asian pandemic, 71
Hong Kong pandemic, 73–4, 75–6
theories of origin, 68
vanishing pandemic virus, 72

significance of T cell response to, 64–5
type A, 55–6
type B, 55–6
vaccines, 79–80
recommended changes in composition, 58
see also Haemophilus influenzae
Integrase, 22
Integrins, 329–30
Inter-strain competition, 413–19
Iron-binding proteins, 108–9
Ixodes pacificus, 321
Ixodes persulcatus, 32
Ixodes ricinus, 321
Ixodes scapularis, 321

KAHRP1 protein, 303
Knob-associated histidine rich protein, 299

Lagoviruses, 33, 34
Lembadion bullinum, 360
Lentiviruses, 16–18
Lewis antigens, 124–6
mimicry, 134–5
phase variation in, 126–33
alpha2-fucT, 130–1
alpha3-fucT, 128–30
beta3-galT, 132
double switch, 133
G1cNAcT gene, 131–2
sialyl-Le^x, 132–3
lgtC gene, 110, 113–15
lic1 gene, 110, 111–13
lic2A gene, 113–15
lic3A gene, 115–16
Lipid rafts, 396
Lipo-oligosaccharides, 109
Lipopolysaccharide phase variation
H. influenzae, 109–11
H. pylori, 122–41
and bacterial adhesion, 135–8
biological role of, 133
expression of blood group antigens, 124–6

Lipopolysaccharide phase variation – *continued*
  H. pylori – *continued*
    Lewis antigen mimicry and immune evasion, 134–5
    phase variation in Lewis antigens, 126–33
Looping-out model, 151–2
Lyme borreliosis, 319–56
  antigenic variation, 348–9
  clinical manifestations and pathology, 322–3
  ecology and epidemiology, 320–2
  pathogenesis, 330–1
  see also Borrelia spp.

MADS box consensus binding site, 184–5
Major antigenic proteins *see* MAP proteins
Major histocompatibility complex, 423
Major surface proteins, 246
  Anaplasma marginale, 248–9
Malaria *see Plasmodium* spp.
map gene family, 260–4, 266
MAP proteins, 260, 266–7
  shared homology, 264–6
Matrix protein, 52
MCM1 gene, 181
Measles virus, point mutation rates, 20
Metacyclic expression sites, 229, 230
Metacyclic VAT, 227
Metacyclic *VSG*, 227
  transcriptional control of, 237–8
Microsatellites, 103
Mini-cassette exchange model, 149–51
Minichromosomes, 229, 232
mod gene, 117–18
Mono-allelic expression, 12
Moraxella catarrhalis, 115
MSG gene family, 202–23
  and antigenic variation, 204–8, 216–19
  co-evolution of protease family with, 214
  expression of, 208–12, 213
  isolation of, 205

  open reading frames, 207
  translation and transport, 212–14
  upstream conserved sequence (UCS), 209–14
  and MSG isoform expression, 215–16
MSG-related gene family, 214–15
msp 2 gene family
  conserved sequence blocks, 252
  genomic structure and orientation, 257
  structure of, 254
msp 3 gene family, 256–7
  conserved sequence blocks, 252
  genomic structure and orientation, 257
MSP 1 protein, 248–9, 312–13
MSP 2 protein, 248
  antigenic variation, 250
    molecular mechanisms of, 253–5
  expression by intra-erythrocytic *A. marginale*, 250–3
  expression in cyclic transmission of *A. marginale*, 255–6
MSP 3 protein, 248
  and antigenic variation, 250, 256–7
MSP 4 protein, 248, 257–8
MSP 5 protein, 248, 257–8
MSP 7 protein, 248
Multiple strain disease pathogenesis model, 425–7
Multiple strain transmission model with cross-immunity, 427–8
Mutual exclusion, 383–5
Mycoplasma hyorhinis, 8

Nef protein, 22
Neisseria spp., 122, 142–64
  functional significance of genetic variation, 155–9
    cell tropisms conferred by Opa proteins, 156–8
    neisserial pili, 155–6
    optional resistance to extracellular environment, 158–9
  genetic diversity and evolutionary adaptability, 143–5

genetic variation via illegitimate
route, 152–4
in situ activation, 8
molecular models for *pilE* variation,
149–52
looping-out model, 151–2
mini-cassette exchange model,
149–51
natural competence for
transformation, 143–5
opa gene family, 8, 154
rapid micro-environmental adaptation,
145–9
genetic requirements for
homologous recombination,
147–9
genetic variation versus gene
regulation, 145–6
genetic variation via homologous
recombination, 146–7
Neisseria gonorrhoeae, 126, 142
antigenic variation, 410
pathogen fitness, 406, 408
Neisseria meningitidis, 142
antigenic variation, 410
nucleotide repeats, 107
Neuraminidase, influenza, 61–2
NIK1 gene, 178
Noroviruses, 33, 34
NOS3, 181
Nucleoprotein, 52

OP4 gene, 176, 178
opa gene family, 8, 154
Opa proteins, 156–8
Opaque-phase specific gene promoters,
182, 183–5
Open reading frames
calicivirus, 35
lic1 locus, 111–12
MSG gene family, 207
msp 2 locus, 253
rotavirus, 95
Original antigenic sin, 63
Ornithodoros hermsi, 321
ospA gene, 334, 335, 340
OspA protein, 334

ospB gene, 334
OspB protein, 334
ospC gene, 335, 338, 340
OspC protein, 334, 335
Outer membrane proteins, 246
Anaplasma marginale, 247–8
Ehrlichia ruminantium, 259

Pahase variation, 1, 3–4
Pandemic influenza, 65–8
1918/19 pandemic, 68–71
age-related mortality in, 70
nucleotide sequencing, 74–5
1957 Asian pandemic, 71
Hong Kong pandemic, 73–4
as predictor of new pandemic,
75–6
theories of origin, 68
vanishing pandemic virus, 72
pap system, 10
Paralogous genes, 1
Paramecium spp., 375–402
antigen structure and variation,
378–80
genetics and molecular biology,
376–7
i-antigens, 376, 377
isolation and analysis of mutants,
382
regulation of expression, 381–5
role in regulation of mutual
exclusion, 383–5
stable genes without introns, 381
structure and variation, 378–80
transcriptional control of antigen
expression, 382–3
variable surface antigens, 376–85
see also Ichthyophthirius spp.
Paramecium primaurelia, 378, 381
Paramecium tetraurelia, 376, 378, 381
Pathogen fitness, 404–13
benefit of antigenic variation, 408–13
determinants of, 405–8
Pediculus humanus, 321
PfEMP1 protein, 295, 296–300
role in pathology, 295–6
PFK gene, 181

Phase variation
  C-tracts in, 123
  in fungi, 197–8
  Haemophilus influenzae
    factors influencing, 117–19
    modulation of transcription, 105–6
    simple nucleotide repeats, 103–5
    tetranucleotide repeats, 106–17
  Helicobacter pylori
      lipopolysaccharide, 122–44
    and bacterial adhesion, 135–8
    biological role of, 133
    expression of blood group antigens, 124–6
    Lewis antigen mimicry and immune evasion, 134–5
    Lewis antigens, 126–33
  as survival strategy, 12–13
Phase-specific trans-acting factors, 185–6
Phosphorylcholine, 111–13
PilC1 protein, 155–6
PilC2 protein, 155–6
pilE gene, 146–9
  molecular models for variation, 149–52
    looping-out model, 151–2
    mini-cassette exchange model, 149–51
PilE protein, 155, 156
Pili
  H. influenzae, 105
  Neisseria spp., 155–6
Plasmodium spp.
  antigenic diversity, 312–13
  antigenic variation of infected red blood cells, 292–310
    role of antibody in protection, 294–5
    switching of variant antigens and role of spleen, 294
  chronic infection, 293
  phenotypic/antigenic variation of merozoites, 311–12
Plasmodium berghei, 309
  BIR protein, 310
  STEVOR protein, 310
Plasmodium chabaudi, 294, 309
  CIR protein, 310

Plasmodium falciparum, 295–306
  antigenic variation, 410
  clag, 305–6
  host red cell surface modifications, 296
  KAHRP1 protein, 303
  location of expressed genes, 10
  mechanism to change gene expressed, 10
  number of genes, 10
  pathology and antigenic variation in, 295–6
  PfEMP1 protein, 295, 296–300, 310
  rif/RIFIN and *stevor*, 304–5
  RIFINS, 298, 304–5, 310
  surface antigen, 10
  switching/transcriptional control of *var* genes, 300–3
  var multigene family, 296–300
Plasmodium fragile, 294
Plasmodium knowlesi, 294, 306–8
  SICA proteins, 286, 310
  SICAvar gene family, 307–8
Plasmodium vivax, 306
  Duffy binding ligand, 298
  VIR protein, 310
Plasmodium yoelii, 309
  PY235 protein, 310
  YIR protein, 310
Platelet adherence factor receptor, 113
Pneumocystis spp., 6, 7, 13
Pneumocystis carinii, 202–23
  and laboratory rats, 203–4
  location of expressed genes, 10
  mechanism to change gene expressed, 10
  number of genes, 10
  potential for antigenic variation, 204–8
  surface antigen, 10
  see also MSG gene family
Pneumocystis jiroveci, 203
Point mutation rate, 20, 23
  HIV, 24
  rotavirus, 94–5
pol genes, 19, 118
Pol protein, 22
Poliovirus, point mutation rates, 20

Polymerase proteins, 52
Privileged location model, 12
Programmed variation *see* Phase variation
PRT1 gene family, 214
Py235 gene family, 310
PY235 protein, 310

Quasi-species, 2

Rabbit haemorrhagic disease virus, 33, 40
Random variation, 1, 2–3
rec genes, 147–9
Recombination
  HIV, 18–21, 25–8
  Neisseria spp., 147–9
  rates of, 20
Recombinational mechanisms, 4–8
Relapsing fever, 319–56
  antigenic variation, 342–8
  clinical manifestations and pathology, 322–3
  ecology and epidemiology, 320–2
  pathogenesis, 327–30
  see also Borrelia spp.
Renilla reniformus, 181, 183
Resting memory T cells, 28
Restriction enzyme mediated integration, 196
Retroviral reverse transcriptase, 18
Retroviruses, 3, 16–17
  point mutation rates, 20
Rev protein, 22
Reverse transcription, 16
rif gene family, 304–5, 310
RIFIN protein, 298, 304–5, 310
RNA viruses, point mutation rates, 20
Rotavirus, 84–101
  antigens and epidemiological markers, 86–91
  classification and structure, 84–5
  electropherotypes, 90
  genogroups, 90–1
  genome and gene products, 86
  group, 86–7

  pathogenesis and immunity, 92–8
  consequences of variability, 98
  genetic and antigenic change, 93–4
  genome rearrangements, 95
  genome reassortment, 96
  interspecies transmission, 96–7
  sequential point mutations, 94–5
  reservoirs, 87
  serotype, 87–9
    G-types, 87–8, 97
    P-types, 88–9
  structure, 86
  subgroup, 87
RPD3 gene, 188

Saccharomyces cerevisiae, 147, 166, 177, 184, 186
Salmonella typhimurium, 147, 165, 186
SAP1 gene, 176, 178, 181
SAP2 gene, 181
SAP3 gene, 176
Sapoviruses, 33, 34
Schistosoma mansoni, Lewis antigen expression, 126
Schizont infected cell agglutination (SICA) test, 292, 307
Schizosaccharomyces pombe, 166, 186
Sequence variation, 93–4
Sequential strain replacement, 416
Serotypes, 328
  rotavirus, 87–9, 97
Serum resistance, 115–16
Shufflons, 4, 5
Sialyl transferase, 115–16
Sialyl-Le^x, phase variation in, 132–3
SICA proteins, 286, 310
SICA test *see* schizont infected cell agglutination test
SICAvar gene family, 307–9, 310
Silenced chromatin state, 11
Silent expression sites, 7
Silent genes, 229, 232
Simian immunodeficiency virus, 18
  phylogenic tree, 19
Simple nucleotide repeats, 103–5
SIR model, 425
SIR2 gene, 186

SIS model, 425
Sleeping sickness *see Trypanosoma* spp.
Small round structured viruses, 34
Somatic mutation, 3
Spanish influenza, 68–71
Spirochetaemia, 322
Spleen necrosis virus
  point mutation rates, 20
  recombination rate, 20
stevor gene family, 304–5, 310
STEVOR protein, 310
Strain extinction, 415–16
Strain-spaces, 415
Swine influenza virus, 66, 67
Switching in *C. albicans*, 166–8
  deacetylases in suppression of, 187–9
  differential gene expression in, 176–9
    CDR3 and *CDR4*, 177
    EFG1, 177
    HOS3, 177
    OP4, 176
    SAP1 and *SAP3*, 176
    WH11, 177
  gene regulation during, 179–85
    functional analyses of phase-specific gene promoters, 181
    promoters of opaque phase-specific genes, 183–5
    promoters of white phase-specific genes, 181–3
  model for, 189–91
  possible mechanisms, 186–7
  role in pathogenesis, 191–3
Switching in *C. glabrata*, 193–7

T cells
  CD4, 25, 27, 64–5
  CD8, 25, 64–5
  response in influenza, 64–5
  resting memory, 28
TAR loop, 22
Tat protein, 22
Telomeres, 11
Telomeric silencing, 11
Tetrahymena spp., 360, 375, 387, 398
Tetrahymena thermophila, 390
Tetranucleotide repeats, 106–17

iron-binding proteins, 108–9
lic1 locus and phosphorylcholine, 111–13
lic2A and *lgtC* genes, 113–15
lic3A, 115–16
lipopolysaccharide biosynthesis, 109–11
phase variation and infection, 117
Texas redwater fever, 273
Transcription modulation, 105–6
Transmembrane signalling, 395–6
Treponema pallidum, 325, 326
Trichostatin A, 188
Trypanosoma spp., 224–42
  DNA recombination and antigenic variation, 231–6
  Rad51 mutant lab-adapted, 236
  secondary phenotypes associated with antigenic variation, 238
  transcription of *VSG*, 231
  transcriptional control of metacyclic *VSG*, 237–8
  variable antigen type, 226
  variant surface glycoprotein, 225–6
  VSG genes and genome, 228–30
Trypanosoma brucei, 3, 7, 165, 186, 224
  antigenic variation, 227
  life cycle, 225
  location of expressed genes, 10
  mechanism to change gene expressed, 10
  number of genes, 10
  surface antigen, 10
Trypanosoma congolense, 225
  minichromosomes, 232
Trypanosoma cruzi
  antigenic variation, 410
  pathogen fitness, 408
Trypanosoma gambiense, 225
Trypanosoma vivax, 225, 226

Vaccines, 420–1
  feline calicivirus, 46–7
  influenza, 79–80
    recommended changes in composition, 58

var multigene family, 296–300, 310
  switching and transcriptional control, 300–3
Variable antigens
  Borrelia spp., 332
  Trypanosoma spp., 226–8
Variable large proteins, 332
Variable small proteins, 332
Variant surface glycoprotein
  Giardia lamblia, 359–65
  Trypanosoma spp., 225–6
VAT *see* Variable antigen type
ves multigene family, 278
*ves1*α gene, 279–81
VESA1 polypeptides, 277, 279–81
Vesicular stomatitis virus, point mutation rates, 20
Vesiviruses, 33, 34
vir gene family, 306, 310
  homologues of, 309
VIR protein, 310
Viraemia, 25
Viral burst size, 24
Virological Richter scale, 17
Viruses, 3
  bovine leukaemia virus, 20
  caliciviruses, 33–51
  European brown hare syndrome virus, 33
  feline immunodeficiency virus, 44
  HIV, 3, 16–32
  influenza *see* Influenza
  lagoviruses, 33, 34
  lentiviruses, 16–18
  measles virus, 20
  Noroviruses, 33, 34
  point mutation rates, 20
  poliovirus, 20
  rabbit haemorrhagic disease virus, 33
  recombination rates, 20
  retroviruses, 3, 16–17, 20
  RNA viruses, 20
  rotavirus, 84–101
  Sapoviruses, 33, 34
  simian immunodeficiency virus, 18, 19
  small round structured, 34

spleen necrosis virus, 20
swine influenza virus, 66, 67
vesicular stomatitis virus, 20
vesiviruses, 33, 34
vlp gene family, 7, 340
Vlp protein, 332, 336
vls gene family, 335
Vls protein, 336
VSG *see* Variant surface glycoprotein
VSG gene family, 227, 302
  bloodstream expression sites, 229, 230
  expression site associated genes, 229, 230
  expression site body, 231
  and genome, 228–30
  metacyclic expression sites, 229, 230
  minichromosomes, 229
  secondary phenotypes, 238
  silent genes, 229, 232
  telomeric, 234
  transcription of, 231
vsp gene family, 7
  Borrelia spp., 342–8
  Giardia lamblia, 363, 392
Vsp protein, 332–3, 335
vtp gene family, 340, 342–8

WH11 gene, 177, 181–3
White–opaque transition, 168
  antigenic changes in, 170–5
  effect on budding yeast cell morphology, 168–70
White-phase specific gene promoters, 181–3
White-spot *see* *Ichthyophthirius multifiliis*

X-ray crystallography
  influenza haemagglutinin, 59–60
  influenza neuraminidase, 61–2

Yeasts *see* *Candida albicans*
YIR protein, 310